ENHANCED WIRELESS NETWORKING CERTIFICATION

Editor in Chief: Stephen Helba
Assistant Vice President and Publisher: Charles E. Stewart, Jr.
Production Editor: Alexandrina Benedicto Wolf
Design Coordinator: Diane Ernsberger
Cover Designer: Michael R. Hall
Cover Art: Michael R. Hall
Illustrations: Michael R. Hall and Cathy J. Boulay
Production Manager: Matthew Ottenweller

This book was set in Times New Roman and Arial by Cathy J. Boulay, Marcraft International, Inc. It was printed and bound by R.R. Donnelley & Sons Company. The cover was printed by Phoenix Color Corp.

Pearson Education Ltd., *London*
Pearson Education Australia Pty. Limited, *Sydney*
Pearson Education Singapore, Pte. Ltd.
Pearson Education North Asia Ltd., *Hong Kong*
Pearson Education Canada, Ltd., *Toronto*
Pearson Educacion de Mexico, S.A. de C.V.
Pearson Education–Japan, *Tokyo*
Pearson Education Malaysia, Pte. Ltd.
Pearson Education, *Upper Saddle River, New Jersey*

Written by Max Main

10 9 8 7 6 5 4 3 2 1

ISBN 0-13-093015-6

Trademark Acknowledgments

Preface

PURPOSE

Wireless communications is one of the rapidly emerging technologies of the 21st century. The first phase of the wireless explosion has already arrived with worldwide demand for cellular mobile phone service exceeding all projections forecast earlier in the 1990s. The second generation of wireless cellular service introduced digital voice and data service in the late 1990s.

The convergence of global third generation (3G) wireless cellular network standards with more powerful handheld computing platforms is opening the path for high-speed broadband mobile access to the Internet and m-commerce worldwide. Future 3G wireless personal communications services will eventually surpass wired Internet access media in cost effectiveness and performance.

Wireless technology is also adding a new dimension to network computing. Wireless local area networks are changing the way corporations upgrade their information technology. The network computer workstation has now gone mobile.

Wireless is the new core technology leading the advancement of public safety, security, and the safeguarding of human life. Advanced satellite and maritime wireless communications systems now provide worldwide instant weather alerts, search and rescue support, navigation aids, and radiotelephone service for the marine industry. Federal communications policy and international treaty agreements require FCC licensed wireless technicians to install, operate, and maintain all certified marine communications equipment on board commercial passenger and cargo vessels. Satellites now assist in the search for downed aircraft and the rescue of survivors. Emergency location and positioning enhancements for cellular networks to assist E911 emergency call centers in locating a mobile caller are now mandated by the Federal Communications Commission. The U.S. Department of Defense Global Positioning System (GPS) now supports a multitude of civilian oriented services. Vehicle navigation, anti-theft security, emergency road services, and m-commerce are only a few of the wireless services now available for motorists and world travelers.

The future demand for technically skilled technicians in the emerging wireless generation is potentially as explosive as the industry itself. The cellular base station equipment installer, wireless network administrator, mobile computer network technician, microwave radio equipment installer and repairman, satellite ground station installer, maritime equipment technician, wireless sales technical support, and fixed wireless installation technician are only a few of the many positions in the future job market that will emerge for wireless technicians.

The broad scope of the wireless technical skills that will be needed in the future was the focus and inspiration for this *Enhanced Wireless Networking Certification* study guide. Several textbooks and courses have addressed popular topics, such as wireless local area networking standards and design theory, while omitting the basics of digital modulation, lab exercises, radio propagation theory, and other essential prerequisites needed by a wireless technician. This study guide provides the history of wireless and how we arrived at the modern art of personal communications by radio. The basics of wireless standardization and radio propagation will prepare the student for the later advanced topics in the course such as cellular radio, fixed wireless networks, wideband wireless networks, satellite systems, and advanced wireless concepts.

Wireless Certification

The Marcraft *Enhanced Wireless Networking Certification* training guide provides students with the information and skills necessary to pass the Wireless Network Installer Certification (WNIC) Exam. The WNIC is a comprehensive exam covering the major wireless technologies in the industry. It is recognized nationwide as the premier program for certifying wireless technicians. It has become the hiring criterion used by major wireless telecommunications companies. It is also widely recognized by those working in the industry as a step to improving their skills and keeping abreast of the latest wireless standards and technology.

Marcraft International has selected the WNIC exam because it establishes a high standard for excellence consistent with the quality of courses offered by Marcraft. This course will not only prepare students to successfully challenge the WNIC examination, but also provide an extensive study program for upgrading skills for those now working in a wireless or wireless related sector of the industry.

The Wireless Network Installer Certification exam is administered by Evolving Technologies Association International Incorporated (ETA International^{TM}), an organization that establishes certification examination requirements for wireless communications and a number of other telecommunications areas. ETA International issues the certification to candidates who successfully pass this examination, which is designed to certify wireless communications technicians in the theory, operation, installation, and testing of wireless systems. For more information on registering with ETAI for the Wireless Network Installer Certification exam, visit *http://www.evolvingcerts.org*.

This text traces the wireless industry from its infancy to advanced third generation wireless systems and future technologies with detailed explanations of how they are designed, how they are used, and how they evolved from earlier wireless networks. Detailed graphics and text explain the sequence of events for cellular call initiation, registration, and authentication. The radio interface access protocols used in cellular systems and wireless computer networks are also covered.

Wireless local area network standards, protocols, and topologies are explained with the help of graphics and tables. An integrated lab provides hands-on training for installing, testing, and troubleshooting a wireless local area network and interfacing wireless components with a wired Ethernet local area network. The text also covers the fundamentals of radio propagation to aid the student in understanding how wireless system problems and anomalies occur, how to avoid them, and how to analyze and troubleshoot wireless transmission problems.

The FCC allocation of the radio spectrum for the various wireless services is covered with common naming conventions and acronyms explained. Other lab exercises are included to provide hands-on training for setting up and using a wireless messaging terminal, a GPS handheld receiver, a network interface card, and a modem. A selection matrix guides the student through additional labs for selecting wireless phones, cellular service plans, and GPS handheld navigation equipment.

The *Enhanced Wireless Networking Certification* training guide with integrated lab procedures and accompanying test preparation materials are intended for anyone interested in pursuing the Wireless Network Installer Certification. While the textbook contains all of the pedagogical support materials required for use in a classroom environment, it can also be used by experienced technicians who wish to prepare for the certification exam in a self-study mode.

Key Features

The pedagogical features of this book were carefully developed to provide readers with key content information, as well as review and testing opportunities. Each chapter begins with a list of learning objectives that establishes a foundation and systematic preview of the chapter and concludes with a summary of the points and objectives to be accomplished. Key terms are presented in bold throughout the text.

Appendices

Appendix A is a comprehensive glossary of words and meanings to provide quick, easy access to definitions of key terms that appear in each chapter. Appendix B provides a list of common acronyms used in the wireless industry.

Evaluation and Test Material

An abundance of test materials is available with this course. At the end of each chapter, there is a question section with open-ended review questions, which are designed to test critical thinking. Additional test material in the form of multiple-choice questions can be found on the interactive CD-ROM.

Lab Exercises

Applying the concepts of the chapter to hands-on exercises is crucial in preparing for a successful career as a wireless installer. Some chapters include integrated lab sections that call upon the students to perform exercises that reinforce the chapter content via hands-on exploration. Questions appear at the end of each lab enabling students to assess their understanding of the lab.

Interactive CD-ROM

The *Enhanced Wireless Networking Certification* training guide is accompanied by a comprehensive, electronic practice test bank on a CD-ROM, which is sealed on the back cover of the book. This CD testing material was developed to simulate the Wireless Certification Exam process and to allow students to complete mock tests, determine their weak points, and study more strategically.

During the question review, the correct answer is presented on the screen, along with the reference heading where the material can be found in the text. A single mouse click will take you directly to the corresponding section of the embedded electronic textbook.

ORGANIZATION

Chapter 1 – *Introduction to Wireless Communications* presents a chronological history of wireless communications and the emergence of government regulatory policy for the use of the radio spectrum. The chapter traces early key wireless inventions and events, which later motivated government agencies, industry groups, and the United Nations to establish standards and international treaties related to wireless communications. The maritime industry and related technologies that have arisen out of the need for safety at sea, certification of radio equipment, and licensing of maritime radio operators are also covered in this chapter.

Chapter 2 – *Wired Network Architectures* is a basic review of the fundamentals of wired computer networks, network services, and design concepts. This chapter covers the basic networking prerequisites for later chapters covering wired and wireless network design concepts and architectures.

The 7-layer Open Systems Interconnect (OSI) reference model is covered as well as the related features of the DOD reference model. The chapter introduces the features and limitations of wired systems and highlights the advantages that are available with wireless mobile station extensions to existing fixed wired networks. Readers who have extensive working knowledge of local area networks and standards may elect to bypass this chapter.

Chapter 3 – *Wired Network Protocols and Standards* builds on the fundamentals of wired network architectures given in Chapter 2. The chapter begins with a description of network media for both wired and wireless applications. The concept of hybrid media networks is explained.

The specifics of the evolution of Ethernet local area network standards are covered in this chapter. A basic understanding of Ethernet architectures and access protocols is essential for the study of advanced concepts of wireless local area networking discussed later in Chapter 6. The basics of wireless local area networking are introduced with a list of the advantages offered over wired local area networks.

Wired networks provide a means for sharing information among other users in a local area network as well as accessing the Internet over the public switched telephone network. A lab is included in this chapter, which provides a step-by-step procedure for installing a modem and an Ethernet network interface card. This lab demonstrates how information can be shared between network computer workstations as well as how to connect a computer workstation to the Internet.

Chapter 4 – *Radio Communications Fundamentals* begins with a description of the radio frequency band designations and terminology used in technical literature, industry standards, and FCC rules and regulations. Propagation characteristics of three major categories of the electromagnetic spectrum are covered.

The chapter continues with a description of how radio propagation for wireless commercial applications is affected by the atmosphere, distance, terrestrial features, and weather. Sample problems are included to emphasize and illustrate key points.

A model of basic radio system architecture is included as well as a detailed explanation of the types of digital modulation and voice processing techniques employed in modern wireless communications systems. A lab in this chapter models the requirements, procedures, and steps necessary for planning the installation of a basic point-to-point microwave transmission system.

Chapter 5 – *Cellular Radio* introduces a detailed study and analysis of one of the fastest growing areas of wireless technology: The chapter provides the background on how early pioneering mobile radio systems, such the Mobile Telephone Service, provided the groundwork for subsequent cellular radio service.

The concept of cellular reuse of radio frequencies is explained with examples of 7-cell reuse plans. The architectural components of a typical cellular telephone system are included. Cell sizing and sectoring concepts are presented in this chapter with analysis of how cellular subscriber density can be increased in high-capacity urban areas.

Wireless access protocols are covered with definitions of numerous acronyms and standards used in the industry. The basics of Frequency Division Multiple Access (FDMA), Time Division Multiple Access (TDMA), and Code Division Multiple Access (CDMA), which are used as the air interface protocols for cellular radio systems, are presented in easy-to-understand language and graphical examples. The spectrum plan and overview of the Personal Communications Service (PCS) is presented, including the narrow band and unlicensed services.

The chapter includes a lab on selecting a cell phone and service plan. Cross-reference matrices are included for comparing service plans and cell phone features. Steps are provided for comparing features and completing a cost-benefit analysis of cellular products.

Chapter 6 – *Wireless Networks* begins with a detailed description of the intersystem network standard ANSI-41 and the architecture used in North American cellular networks. Features of the ANSI-41 cellular network standard are discussed with emphasis on how identification, security, and authentication of wireless mobile stations are accomplished in a cellular system. The network elements and various databases used in mobile management and subscriber inter-system roaming are also covered in this chapter.

Mobile IP is another example of the integration of fixed enterprise local area networks using the TCP/IP protocol suite and mobile cellular networks using enhancements to the ANSI-41 and IP protocol standard. Examples in this chapter show how mobile IP works using enhanced IP tunneling and encapsulation protocols. Also included is an example of both portable computing and mobile computing using the enhanced features of mobile IP and mobile IP–capable routers deployed within a cellular network.

Wireless local area networks are another rapidly growing sector of the wireless industry. The IEEE 802.11 series of standards for wireless local area networks (wireless LANs) have provided an incentive for more vendors to enter the market with wireless LAN products. This chapter provides a brief tutorial on the IEEE 802.11 wireless LAN standard with examples of how wireless LANs can add mobility and versatility to existing wired LANs.

Interactive paging and messaging networks are included with a description of a mobitex-based network operated in the U.S. by BellSouth Wireless Data.

This chapter concludes with a series of five labs that provide hands-on experience with installing, testing, and using wireless network equipment. The first two labs include installing a wireless LAN network interface card in a portable computer, then proceeding with the installation of a wireless access point in a wired LAN. In the third lab a wireless LAN tester is used to optimize the location and performance of the lab-installed wireless LAN components.

In the final two wireless labs in this chapter, a personal digital assistant (PDA) handheld terminal with an integrated wireless data network capability is set up and software is installed with a computer. This software is included for synchronizing the desktop computer databases with the handheld terminal applications and databases. Wireless Internet access service is initiated with the data network as an option to demonstrate state of the art two-way messaging, Internet access, and e-mail services using a data-only cellular network and a handheld data terminal.

Chapter 7 – *Wireless Broadband Networks* discusses another type of fixed-service wireless communications system. Both wired and wireless broadband services are covered in this chapter with comparisons between each technology.

The topics in this chapter address various segments of the broadband service, which not only carries the largest amount of telecommunications traffic in the wireless industry, but also uses more of the electromagnetic spectrum than any other service.

This chapter concentrates on the two primary services in the broadband technology areas. The two types of service are the Multichannel Multipoint Distribution Service (MMDS) and the Local Multipoint Distribution Service (LMDS), which are the dominant fixed wireless broadband access technologies.

Additional information on emerging broadband technologies is also included. The features and benefits of the advanced technology 39 GHz band, free-space optical systems, and wireless local loop systems are discussed.

Chapter 8 – *Satellite Communications Systems* begins with an analysis and definition of the three main elements of a satellite communications system. The chapter describes how these three elements interact to provide long distance wireless communications service.

Various types of satellite orbits are described with a comprehensive analysis of how each is used for a specific application. The frequency bands used in the various services are described with an introduction on how the international community manages the allocation of orbital slots and frequency bands to avoid intersystem interference.

The chapter includes a detailed explanation of the subsystems and functions of a satellite with references to typical on-board control systems and bus architectures. Earth station configurations are explained with example problems in calculating antenna gain and beamwidth for various types of earth stations. The calculations are added to provide necessary background information for technicians involved in installing and orienting earth station antenna systems.

Position location services using satellite constellations, such as the Global Positioning System (GPS), are also covered. The chapter includes two labs on comparing features and costs of GPS receivers and on the setup and operation of a typical GPS receiver.

Chapter 9 – *Advanced Wireless Systems* focuses on the advanced types of wireless communications systems that will revolutionize the way we communicate not only with personal cellular-type phones but within business organizations as well. This chapter provides an in-depth study, explanation and technical review of the international effort to establish a set of global standards for the future called International Mobile Telecommunications (IMT) 2000. This effort has been labeled as third generation, or 3G, for wireless communications. Included in the chapter are tutorials on the third generation Wideband CDMA (W-CDMA) and ANSI-2000 CDMA wireless standards.

The chapter also provides a summary of how the various world regulatory and standards bodies are working together to establish a frequency band plan that will allow international roaming by wireless 3G-equipped subscribers. You will discover the rationale behind the universal coordination of a global spectrum plan for personal third generation wireless telecommunications.

Three revolutionary new wireless technologies are introduced in this concluding chapter that have the potential for dramatic technical advances in high-speed wireless networks for both business and personal subscribers. The basic principles behind these emerging technologies are discussed in easy-to-understand terms.

Teacher Support

A full-featured instructor's guide is available for the course. Answers for all of the end-of-chapter review questions are included along with a paragraph reference in the chapter where that item is covered. Sample schedules are included as guidelines for possible course implementations. Answers to all lab questions are provided so that there is an indication of what the expected outcomes are.

AUTHOR

Max Main has several years of experience in the computer and wireless industry. He has held senior management positions with Computer Sciences Corporation and Linkabit Corporation. He also served as Vice President of Marketing for California Microwave Defense Systems. He is the author, developer, and National Chairman of the Certified Network Systems Technician Program originally sponsored by the Texas Engineering Extension Service, a member of the Texas A&M University System.

He is a member of the faculty of Telecommunications Research Associates, where he specializes in teaching courses in wireless technology.

ACKNOWLEDGMENTS

First to my wife, Mary J. Main, for her patience, understanding, and endurance with the numerous late hours during the preparation of the draft of this book.

I would also like to thank Charles J. Brooks, President of Marcraft International, for his encouragement and support for me to take leave from teaching to write a book about wireless networking technology.

Thanks to Paul Gannon of SBC Communications who coaxed me from temporary retirement to begin a second career teaching computer courses and developing a computer network certification program.

To Ray Stroud who kept me busy teaching networking technology courses and training instructors for several years.

I also want to thank Cathy J. Boulay and Michael R. Hall, at Marcraft International Corporation, who provided the layout, graphics, editing, and valuable consultation on the style and format for this book. Also, special thanks to Caleb Sarka, who provided text, materials, valuable technical support, and advice for all of the wireless labs.

Table of Contents

CHAPTER 1 INTRODUCTION TO WIRELESS COMMUNICATIONS

CHAPTER 2   WIRED NETWORK ARCHITECTURES

CHAPTER 3   WIRED NETWORK PROTOCOLS AND STANDARDS

CHAPTER 4   RADIO COMMUNICATIONS FUNDAMENTALS

CHAPTER 5 CELLULAR RADIO

CHAPTER 6   WIRELESS NETWORKS

CHAPTER 7   WIRELESS BROADBAND NETWORKS

CHAPTER 8   SATELLITE COMMUNICATIONS SYSTEMS

CHAPTER 9 ADVANCED WIRELESS SYSTEMS

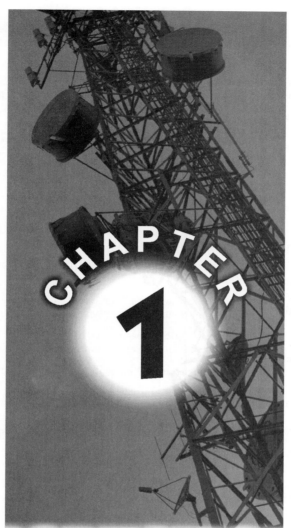

CHAPTER

1

INTRODUCTION TO WIRELESS COMMUNICATIONS

LEARNING OBJECTIVES

LEARNING
OBJECTIVES

In this chapter, you will gain an understanding of the basic discoveries that founded the wireless industry we know today. Upon completion of this chapter, you should be able to perform the following tasks:

1. Explain the theory of Amplitude Modulation (AM) and Frequency Modulation (FM).

2. Explain the features of the early basic radio circuits used to receive and demodulate radio signals.

3. Describe the advantages of FM over AM radio technology.

4. Describe the names and titles of important standards associated with the main wireless technologies in the world today.

5. Identify the charters and purpose of the major wireless standards bodies of the world.

6. Describe the international and domestic organizations responsible for maintaining treaties and safety standards for the maritime industry.

7. Describe the different elements of the GMDSS and how they interact to provide search and rescue operations on a global basis.

8. Describe the technical licensing requirements specified by the Federal Communications Commission (FCC) for all types of radio, telephone, and GMDSS operator requirements.

9. Describe the requirements and operating procedures for various search and rescue subsystems and how they use the radio spectrum.

Introduction to Wireless Communications

INTRODUCTION

This introductory section reviews the history of the discovery and development of wireless communications. The term "wireless" has come full circle from the beginning of the early 20th century when the term was widely used to describe a new invention for sending telegraph messages over long distances without wires. The term became the standard name for shipboard telegraphy. The wireless era evolved into the radio broadcasting era, when commercial broadcasting was the dominant industry. Wireless became obsolete in the new radio age. And 100 years later, the explosive growth of and industry based on radio technology has become the new generation of "wireless," which promises to again provide mobility and freedom from a wired telecommunications infrastructure. In this chapter, you will be introduced to the pioneers of the first **wireless communications** era. This brief history traces the development of wireless communications by Guglielmo Marconi, Lee DeForest, Edwin Howard Armstrong, and others. Also, the development of early basic wireless telegraphy and radio circuits is described.

The early moves toward regulation of the wireless communications industry are explained. The major legislative events that established the **Federal Communications Commission (FCC)** are described. The two principal organizations that have developed and promoted wireless communications standards are covered. Since passage of the Telecommunications Act of 1934, several major changes have been made in the regulatory environment to protect privacy, increase competition, and divide the assets of the AT&T monopoly into separate operating companies. These events all had a role in shaping the future of the wireless industry and are explained in more detail in the following sections.

The wireless industry and how it is managed from a technology point of view is discussed. As the industry has grown, the need for global standards has become paramount. Each country is promoting standards for adoption before the **International Telecommunications Union (ITU)** and other standards review groups. The ultimate goal is the adoption of standards to permit wireless subscribers to "roam" seamlessly between nations and service providers on a global scale. We will look at the charters and goals of the principal standards bodies that you will need to know about and become familiar with in the future.

One of the most interesting, as well as life-saving elements of wireless technology is discussed. You will discover how maritime wireless systems work aboard ships using standard short wave radio, advanced digital communications modes, and satellites that monitor all of the ocean areas for ships or aircraft in distress. Included also are the FCC operator and equipment maintainer licensing requirements as applicable for maritime telecommunications equipment. This chapter provides important information to technicians who are preparing for the FCC commercial radiotelephone operator examination and associated maritime endorsements.

wireless communications

Federal Communications Commission (FCC)

International Telecommunications Union (ITU)

The Development of Wireless Telegraphy

The wireless industry we know today began at the dawn of the twentieth century. The telegraph and telephone had established a means for sending information over copper wires and spanned great distances; however, they were tethered to locations that could accommodate wires.

Marconi and the Wireless Telegraph

In 1895, a brilliant young Italian engineer, Guglielmo Marconi (1874–1937), shown in Figure 1-1, began laboratory experiments at his father's estate in Pontecchio, Italy, where he succeeded in sending wireless signals over a distance of one and one-half miles.

wireless telegraphy

spark gap transmitter

radiated electromagnetic wave

Marconi's transmitting device was adapted from early experiments conducted by the German scientist, Heinrich Hertz (1857–1894), who had demonstrated the existence of electromagnetic waves. This event established Marconi as the inventor of the first practical system of **wireless telegraphy**. Marconi's early **spark gap transmitter**, depicted in Figure 1-2, generated a spark by opening and closing the primary windings of a transformer with a telegraph key. The collapsing magnetic field created a high voltage in the transformer secondary winding, which generated a spark across the gap. This created a **radiated electromagnetic wave**.

Figure 1-1: Guglielmo Marconi

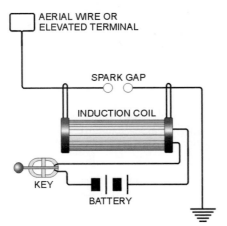

Figure 1-2: Spark Gap Transmitter Schematic

In 1896 Marconi took his equipment to England and later that year was granted the world's first patent for a system of wireless telegraphy. He demonstrated his system successfully in London, on Salisbury Plain and across the Bristol Channel, and in 1897 formed The Wireless Telegraph & Signal Company Limited (renamed Marconi's Wireless Telegraph Company Limited in 1900). In the same year he gave a demonstration to the Italian government in which wireless signals were sent over a distance of twelve miles. In 1899 he established wireless communication between France and England across the English Channel.

Marconi arrived in New York in the fall of 1899 to demonstrate his new invention, the "wireless" telegraph, to scientists in the U.S. His invention promised to forever remove the tether from the telegraph and transmit information through the atmosphere using **electromagnetic waves**. Marconi showed American scientists how the wireless telegraph transmitter used a spark gap instrument that could be turned on and off with a telegraph key. The spark created a wideband electromagnetic radiation whose frequency could be controlled in a crude manner with a tuned circuit consisting of a coil and spark gap. The instrument also included a telegraph key, an antenna, and a power supply. Figure 1-3 shows how a fixed condenser was incorporated with the large inductor coil to form a tuned circuit. These components established the basis for a primitive wireless transmitter.

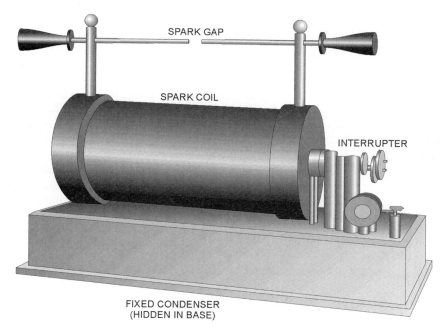

Figure 1-3: Spark Gap Transmitter with Large Coil

In 1900 Marconi took out patent No. 7777 for "tuned or syntonic telegraphy." On a historic day in 1901, determined to prove that wireless waves were not affected by the curvature of the Earth, he used his wireless telegraph system for transmitting the first wireless signals across the Atlantic between Poldhu, Cornwall, and St. John's, Newfoundland, a distance of over 2,200 miles (the U.S. government would not grant permission at that time to locate a transmitter site in the United States). In 1905 he patented his horizontal directional aerial and in 1912 a "timed spark" system for generating continuous electromagnetic waves.

The "wireless age" had finally arrived!

The early use of the term "wireless" referred exclusively to the evolution of the "wired" telegraph. The use of the term "radio" would not appear until after the concept of commercial broadcasting was established by Reginald Fessenden.

The goal of Marconi's invention was to transmit messages between two fixed points, thus replacing the telegraph ("wireless" telegraphy). He did not envision a commercial market for broadcasting voices or music to a wide audience. This concept would be developed later by other wireless entrepreneurs in the U.S.

The Vacuum Tube Advances the Wireless Age

The wireless age gained new ground with the invention and development of the **vacuum tube**.

Figure 1-4: Sir John Fleming

In England, around 1901, inventor and engineer Sir John Fleming, depicted in Figure 1-4, had been working in a laboratory with devices he hoped would improve the detection of electromagnetic waves. He had observed a phenomenon known as the Edison effect, wherein the inside of an early light bulb developed a dark coating on the inside of the glass. This was caused by the current flowing in a single direction. Fleming decided to experiment with a similar glass bulb outfitted with two electrodes. One of the electrodes was heated sufficiently to cause it to emit electrons from a negatively charged cathode to a positively charged anode. This provided rectification of incoming waves of electromagnetic energy, thereby converting the oscillating waves into useable direct current. This was an important step in developing a method for detecting very weak signals from early spark-type wireless transmitters. John Fleming subsequently patented his device (803,684 in November 1905), which he called the Fleming valve. He would later make numerous contributions to the development of wireless telegraphy.

Voice Communications and Continuous Waves

Spark gap transmitters were able to generate the dots and dashes used to send telegraph information over the air; however, they were not suitable for carrying the constant variations of the human voice over electromagnetic waves. The next evolution of wireless research would seek methods for transmitting audio sounds using continuous waves.

Reginald Fessenden, a Canadian engineer shown in Figure 1-5, had witnessed a demonstration of the telephone by Alexander Graham Bell when he was a young man in Brantford, Ontario. Fessenden continued Bell's work and maintained a goal to transmit sounds of the human voice over electromagnetic waves. He carefully studied the alternating current generator that had been developed by Nikola Tesla to provide power for lighting. These devices were rotary machines that developed a current that reversed its direction of flow at periodic intervals (alternating current or AC).

Figure 1-5: Reginald Fessenden

Fessenden reasoned that if it were possible to increase the speed of the alternator to a higher number of revolutions per second, it could be made to generate waves of 100,000 cycles or higher, which would be continuous rather than the pulsed oscillations of the spark gap transmitter. Waves with this much higher periodic rate could serve as electromagnetic "carrier waves" and be the basis for handling audio sounds that needed a constant train of high frequency waves. This of course seemed reasonable in theory, but building it was an engineering challenge. Fessenden decided to work with the General Electric Company and Charles Steinmetz to develop the Fessenden alternator. This device would rotate at the amazing speed of 20,000 revolutions per minute and would generate 100,000 cycles per second of electricity. To accomplish the design of the alternator, Steinmetz would use the talents of Ernst Alexanderson, shown in Figure 1-6, a new engineer from Sweden who had recently joined General Electric. Alexanderson was successful in producing the machine that would allow Fessenden to realize his idea of transmitting voice sounds through the air. The development of the Alexanderson alternator was put to the test in 1906. The first model was

Figure 1-6: Ernst Alexanderson

installed at Fessenden's station in Massachusetts. He was soon broadcasting music as well as voice sounds. Ship wireless operators and radio amateurs were startled to hear these new sounds arriving at various stations in the vicinity. Fessenden had established a new use of the airwaves called broadcasting (sending information to a large audience rather than a point-to-point telegraph message). To define this new concept of use of the electromagnetic spectrum, the term wireless was replaced by the term **radio** (most likely derived from the idea of radiation on a continuous basis). It also served as a more suitable marketing term for commercializing the concept of wireless communication.

The Evolution from Wireless Telegraphy to Radio Telephony

Figure 1-7: Edwin Howard Armstrong

Fessenden's machine provided the transition from using electromagnetic waves for telegraphy to transmitting audio. Since voice information could now be used on continuous carrier waves, it was logical that the term **radio telephony** would be used to describe this new media. Although the alternator was a major breakthrough in radio technology, it was also very large and expensive to install and maintain. The solution to this problem would come from a laboratory at Columbia University and a bright undergraduate student by the name of Edwin Howard Armstrong, shown in Figure 1-7.

regenerative
feedback

TEST TIP

Be familiar with the theory of regenerative feedback.

Later, another inventor and young entrepreneur in the U.S., Lee DeForest, was working to improve a wireless telegraph receiver. He modified the concept of the Fleming valve by creating what he later named the Audion. The Audion was a vacuum tube that contained some gas, a filament, an anode, and a grid element inserted between the filament and the plate. DeForest believed he could use the Audion to improve wireless reception. Unfortunately, he did not recognize the potential for his work. As late as 1912, no one had done much with the Audion tube. DeForest himself was uncertain on the theory of how it really worked. Edwin Howard Armstrong, while still an undergraduate in electrical engineering at Columbia University, discovered that the gain of a triode amplifier tube could be enormously increased by feeding some of the amplifier output back into the input. This would later be called **regenerative feedback**. Given enough regeneration or feedback, the amplifier became a more stable and powerful oscillator, perfect for driving additional amplifier stages in large radio transmitters. Given a little less feedback, the amplifier became a more sensitive radio receiver, eventually able to be heard with a speaker instead of mere headphones. Other people had discovered feedback at about the same time, but Armstrong characterized it and understood it in a way that made it practical for real use.

DeForest heard of Armstrong's work, and immediately directed his own research into regenerative techniques. His three-element vacuum tube was to become the key device for producing continuous wave oscillations. But when DeForest applied for a patent for his Audion tube in 1915, he found that a similar tube had already been patented by Armstrong. DeForest and Armstrong would wage a long and bitter fight in the courts over who actually invented the regenerative principle with DeForest eventually winning. Most engineering experts still credit Armstrong's work as the important pioneering effort that led to advanced wireless transmission and receiving equipment design.

Frequency Modulation
(FM)

Armstrong later developed the concept of **Frequency Modulation (FM)** that is still used today in the first generation of cellular phones, mobile radio equipment, military radios, and commercial radio broadcasting.

Development of the Superheterodyne Circuit

Radio technology advanced rapidly in the period from 1915 to 1930, and many new applications would evolve from Armstrong's inventions. Radio commercial broadcasting dominated this period. Radio was also recognized for its potential to provide communications between ships and shore stations, aviation and radio navigation aids, as well as communication for marine safety and rescue operations. After the Titanic sank in 1912, radio procedures for wireless operators at sea saw dramatic changes for distress and radio warning procedures. If such procedures had been in place in 1912, many lives could have been saved since a vessel was close enough at the time to have made rescue a possibility.

None of these early applications would have been possible without another important invention by Armstrong.

superheterodyne

During his enlistment in the Army during World War I, Armstrong came up with the **superheterodyne**, a subtle but elegant technique for improving reception and making the radio offer features called improved selectivity and sensitivity.

It's difficult to build an amplifier that will work at the high frequencies that radio uses. It is also difficult to build a tuning filter that can select a narrow band of frequencies and yet be adjusted across a wide range of frequencies. The filter must tune in one station and reject all others, but then change to tune in other stations. It's easy, however, to build a tunable oscillator. If the oscillator signal is added to an incoming radio signal, the two signals will mix and produce two new frequencies. One of the new frequencies is the difference between the two and the other frequency is the sum of the two. A fixed filter, therefore, can be built to narrowly select the difference frequency, and pass it on to a low-frequency amplifier. As the oscillator frequency is varied, different radio frequencies will be moved down to the lower frequency. The net result is a variable oscillator, and a fixed narrow filter can do the work of a variable narrow filter. In order to appreciate the advantages of the superheterodyne, it is necessary to first look into radio modulation theory and further examine the types of circuits used to recover the transmitted audio information.

Amplitude Modulation

Amplitude Modulation (AM)

carrier wave

Amplitude Modulation (AM) is a type of modulation used to transport information on a continuous wave radio signal called the carrier. This is the method used to impose an audio sound wave onto a high-frequency radio **carrier wave**.

The carrier wave is used because its high frequency gives it the ability to travel at the speed of light over long distances. The carrier wave by itself contains no information. Only when combined with the audio information to be transmitted in a circuit called a modulator will any information be conveyed to the receiver. The receiver must be able to recover the transmitted signal from the carrier wave. The AM transmitter at the transmitting station generates the radio frequency carrier wave and passes it to the modulator, where it is combined with the audio information to be transmitted. A typical model for an AM broadcast transmitter is illustrated in Figure 1-8.

The modulated signal is then amplified to the required power level and coupled to the transmitter antenna. (The modulator and amplifier may be combined into a single stage in simple AM transmitters.)

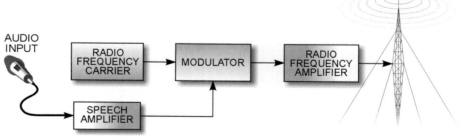

Figure 1-8: Block Diagram of an AM Radio Transmitter

Figure 1-9 represents the audio frequency wave and the high frequency carrier wave. In the process of modulation, the amplitude (height) of the carrier wave is modified so that it's peaks follow the shape of the sound wave. The radio receiver is designed to pick up this distant and weak signal, separate it from all of the other radio signals, amplify it greatly, and recover the sound wave back from the carrier wave. This audio is then amplified further, and reproduced by a speaker or headsets.

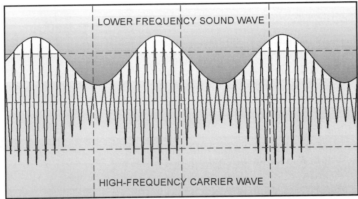

Figure 1-9: Amplitude Modulation

In this type of radio system, each station is assigned a radio channel using a specific radio carrier frequency. Each station is required to maintain the exact same transmit frequency while transmitting in order to avoid interference with stations operating on different assigned frequencies. This type of radio access is called **Frequency Division Multiple Access** or **FDMA**. It is one of several access methods that allow several transmitters to "access" the radio frequency spectrum in accordance with an organized set of rules or "protocols" while still allowing each transmitter equal and interference-free utilization. Other types of radio access protocols will be covered in later sections.

The Tuned Radio Frequency Radio Receiver

The **Tuned Radio Frequency (TRF) receiver** utilized a series of tuned amplifier stages. This was a major improvement over the wideband receiving systems that were used by the spark-gap wireless telegraph systems.

Early on, broadcast AM stations could be heard, and they sounded clearer because the off frequency noise was blocked better. This great performance was possible because of a fundamental improvement in the method used to amplify the received signal. In a TRF system, the radio signal is picked up and amplified many times at its own specific radio channel frequency. The radio frequency that is tuned in is selected from the others by designing a series of amplifier stages, shown in Figure 1-10, to prefer the selected frequency to the rest. This is done by adding **resonant (frequency selective) circuits** to each amplifier. This makes the total amplification much stronger for the one desired frequency and provides a degree of selectivity over stations transmitting on nearby frequencies. When the signal is strong enough, the audio wave is "detected" from the carrier, and sent to the audio amplifier and to the speaker.

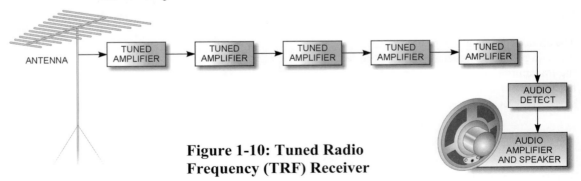

**Figure 1-10: Tuned Radio
Frequency (TRF) Receiver**

Limitations of the TRF Receiver

A problem with TRF was that it was very difficult to design a radio frequency amplifier that was sensitive across the entire radio broadcast band (500 kHz to 1600 kHz).

The simple three-element vacuum tubes that were available at the time had reasonable amplification at the lower frequencies, but less amplification at the higher radio frequencies. This was due to frequency dependent limitations internal to the tube that were difficult to overcome. If attempts were made to increase this gain, severe oscillation or squealing would result at the high end of the frequency band, which also turned the radio receiver into an uncontrolled transmitter, duplicating the experiment conducted by Edwin Armstrong. When the gain was reduced enough to prevent oscillation at the high frequencies, the sensitivity (overall amplification) was pretty low across the rest of the broadcast band. Usually three or four of these relatively low gain amplifier stages were required to step the signal up for acceptable sensitivity.

The Superheterodyne Circuit

Armstrong studied ways to overcome the problems associated with the TRF receiver. He reasoned that reducing the received carrier frequency to a lower frequency and then amplifying the signal would solve the problem.

superheterodyne circuit

local oscillator frequency

The **superheterodyne circuit**, shown in Figure 1-11, mixes the incoming received frequency with a locally generated carrier. The **local oscillator frequency** was designed to be tunable so that the output of the mixed frequencies would be constant over the entire band of frequencies.

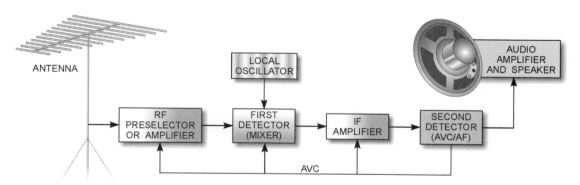

Figure 1-11: The Armstrong Superheterodyne Circuit

This way, the amplifier could have a much higher gain using fewer tubes, and without the oscillation problem of the TRF receiver. It was also desirable that this reduced carrier frequency somehow be kept constant for all received radio stations. The superheterodyne receiver accomplished this goal. In the superheterodyne circuit, the received or incoming signal is picked up by the antenna and coupled to the preselector, which may be a stage of radio frequency amplification or a filter circuit. The local oscillator circuit generates the slightly higher frequency local signal that will be combined with the incoming radio signal in the mixer stage. The frequency of this local oscillator maintains a fixed distance above the radio frequency because of the way the tuning capacitor is built. It has two capacitor sections that are mechanically connected to the same shaft and move together. One section controls the preselector, the other the oscillator frequency. The key design feature of the superheterodyne is the tunable local oscillator. The frequency of this oscillator signal is designed to change as you tune the radio to different stations. The local oscillator frequency is always maintained at a small but fixed difference higher than the radio carrier frequency that you are tuning to receive.

─ TEST TIP ─────────────────────

Know how the local oscillator produces a constant intermediate frequency (IF).

The lower frequency carrier is amplified with a set of fixed-frequency amplifiers (the "intermediate frequency" amplifiers). The signal is then passed through a detector and a low-pass filter. The recovered audio information is then amplified again to provide sufficient power for a loudspeaker. In order to maintain a sufficient amount of amplification across different levels of received signals, an automatic feedback circuit called an **Automatic Volume Control** (**AVC**) is included. The amplification is therefore made to vary between weak stations and strong stations. This keeps a near constant volume at the output of the audio amplifier.

Automatic Volume Control (AVC)

In Table 1-1, we see that the local oscillator tracks the incoming desired radio carrier frequency when the radio is tuned. The oscillator tuning rate produces a constant difference between the received carrier, so the local oscillator remains constant as well as the frequency of the difference signal, no matter what station the radio is tuned to. The first detector/mixer produces not only the desired low frequency product (the difference frequency) but also the undesired sum frequency. The sum frequency is eliminated since the IF amplifiers are designed to pass only the difference frequency of 455 kHz. The audio signal that was modulated onto the original broadcast carrier is also present in this new carrier frequency. Since it is a constant lower frequency, it is much easier to amplify.

Table 1-1: Local Oscillator Tuning

When the radio is tuned to 700 kHz:	When the radio is tuned to 1000 kHz:
The inbound received signal — 700 kHz	The inbound received signal — 1000 kHz
The local oscillator frequency — 1155 kHz	The local oscillator frequency — 1455 kHz
The sum of the two frequencies — 1855 kHz	The sum of the two frequencies — 2455 kHz
The difference of the two frequencies — 455 kHz	The difference of the two frequencies — 455 kHz

The Development of Frequency Modulation

TEST TIP

Know the advantages of frequency modulation.

Armstrong was interested in solving the problem of static associated with radio reception. He worked for many years on new theories and maintained his determination to eliminate static from radio signals. He was able to show that a dramatically different type of modulation, **Frequency Modulation (FM)**, could be used not only to eliminate the noise, but also to improve the quality of the received audio information.

Frequency
Modulation (FM)

Armstrong completed his tests in 1934 from a transmitter located atop the Empire State Building in New York. This important invention would also be employed in military communications in WW II. FM changed the frequency in accordance with the audio signal instead of the amplitude of the carrier wave as had been the case with AM radio. The received signal could therefore be processed using a limiter circuit, which aided in eliminating the static. A typical FM receiver block diagram is shown in Figure 1-12. Some components of the FM radio transmitter and receiver are quite similar to the AM components. As shown in the block diagram, the FM receiver uses a conversion to a lower intermediate frequency, similar to AM; however, the detector in the FM receiver uses a circuit called a discriminator to remove the audio information from the carrier wave.

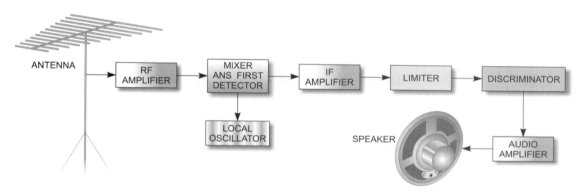

FM not only eliminated static, it also produced a better sound.

Figure 1-12: FM Receiver Block Diagram

In the period following Armstrong's first broadcast, FM radio became the preferred modulation for two-way mobile communication, and in the 1970s it would become the radio interface standard for the first generation of cellular mobile telephones.

U.S. Regulatory Background

The activities and experiments of the early wireless pioneers were described earlier. During this period, all types of wireless transmissions were unregulated. The potential for interference was of little concern since no specific use of the radio spectrum had been planned by any organization. Eventually, amateur radio enthusiasts and radio broadcasters used the airwaves to develop the new science of wireless in increasing numbers. With no coordination, interference soon became a problem, particularly among radio broadcasters. The Federal Government took its first step toward regulating the use of the radio spectrum with an act of legislation passed on August 12, 1912.

The Initial Steps Toward Regulation

Governments around the world became aware of the importance of some type of formal regulation of wireless communications after the loss of the Titanic in 1912.

Subsequent investigation of the Titanic disaster showed that another vessel equipped with a wireless telegraph system was nearby but did not have the receiving system turned on, and therefore failed to hear the distress signals transmitted by the Titanic's radio operator. It was only a matter of time that wireless communication aboard ships would become a matter for international concern. Formal government regulation of not only the radio spectrum but also the procedures for terrestrial use and international maritime safety would soon follow.

The U.S. Establishes a Regulatory Agency

The U.S. Government established a Radio Service Bureau under the management of the Department of Commerce. The 1912 Radio Wireless Act divided the spectrum of electromagnetic waves into four different parts. Wireless communications was then defined as the "radiotelephone" for voice communications and the "radiotelegraph" for Morse code communications. Nations had begun international conferences to regulate the use of radio spectrum with a convention in London in 1912. The U.S. followed the recommendations set forth by the London convention for the initial four-part spectrum partitioning. All of the radio frequencies that were considered useful were considered to be **ground wave** frequencies below 1500 kHz or with wavelengths less than 200 meters. Marconi had demonstrated the existence of ground waves earlier in contrast to popular belief that radio signals traveled in straight lines instead of following the curvature of the earth. Frequencies above 1500 kHz were assumed to have no practical use since they appeared to propagate in straight lines. These frequencies were allocated to the radio amateurs who would later demonstrate the true value of the upper frequency bands for worldwide **short wavelength** communications. Many years of research would follow leading to a better understanding of the true value and potential of the higher frequency bands.

ground wave

short wavelength

The Federal Communications Commission

During the early 1930's, the wireless communications industry was dominated by the growth of commercial broadcasting. Other uses of electromagnetic spectrum appeared as the technology expanded into dispatch for taxi, law enforcement, and other mobile radio systems. The Communications Act of 1934, enacted by the U.S. Congress established the Federal Communications Commission (FCC) often referred to as "The Commission" by the wireless industry.

Congress appropriates money to fund the Commission and its activities, although recently the FCC raised revenues through an auction process for the **Personal Communications Service (PCS)** frequency spectrum.

The Communications Act of 1934 is divided into titles and sections, which describe various powers and concerns of the Commission. Title I describes the administration, formation, and powers of the Federal Communications Commission. The 1934 Act called for a Commission consisting of seven members (reduced to five in 1983) appointed by the President and approved by Senate. The President designates one member to serve as chairman. The chairman sets the agenda for the agency and appoints bureau and department heads. Commissioners serve for a period of five years. The President cannot appoint more than three members of one political party to the Commission. Title I empowers the Commission to create divisions or bureaus responsible for various specific work assigned.

Title II concerns "common carrier" regulation. The Act limits FCC regulation to interstate and international common carriers, which are communication companies that provide facilities for transmission but do not originate messages. Telephone and microwave communications are examples of common carriers.

Title III of the Act deals with broadcast station requirements. Many decisions about broadcasting regulations were made prior to 1934 by the Federal Radio Commission, and most provisions of the Radio Act of 1927 were subsumed into Title III of the 1934 Communications Act. Sections 303–307 define many of the powers given to the Commission with respect to broadcasting. Other sections define limitations placed upon the Commission. For example, section 326 within Title III prevents the Commission from exercising censorship over broadcast stations. Provisions in the U.S. code link to the Communications Act; for example, 18 U.S.C. 464 bars individuals from uttering obscene or indecent language over a broadcast station. Section 315, the Equal Time Rule, requires broadcasters to afford equal opportunity to candidates seeking political office and formally includes provisions for rebuttal of controversial viewpoints under the controversial Fairness Doctrine.

Titles IV and V deal with judicial review and enforcement of the Act. Title VI describes miscellaneous provisions of the Act including amendments to the Act, and the emergency war powers of the President. Title VI also extends FCC power to regulate cable television.

The 1934 Act has been amended considerably since its passage. For example, the Communications Satellite Act of 1962 gave the FCC new responsibilities for satellite regulation. The passage of the Cable Act of 1992 required similar revisions to the 1934 Act, but the flexibility incorporated into the general provisions has allowed the agency to survive for over sixty years. Numerous technical changes in communications have taken place during the FCC's history including the introduction of television, satellite and microwave communications, cable television, and cellular telephone. Though the FCC's responsibilities have broadened to include oversight of these new technologies, it now shares regulatory power with other federal, executive, and judicial agencies.

The FCC has broad oversight over all broadcasting regulation. The FCC can license operators of various services and has recently used auctions to award licenses for Personal Communications Service (PCS) in the 2 GHz frequency band. The Commission enforces various requirements for wire and wireless communication through the promulgation of rules and regulations. Major issues can come before the entire Commission at monthly meetings; less important issues are circulated among commissioners for action. Individuals or parties of interest can challenge the legitimacy of the regulations without affecting the validity or constitutionality of the Act itself. The language of the Act is general enough to serve as a framework for the Commission to establish new rules and regulations related to a wide variety of technologies and services.

The Federal Communications Commission is divided into several branches and divisions. Its organization is shown in Figure 1-13. The FCC handles the management, licensing, and administration of the various communication technology areas.

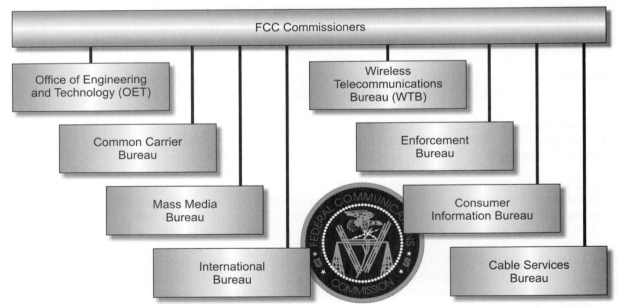

Figure 1-13: FCC Organization Chart

In February 1996, the Telecommunications Act of 1996 was signed into law, representing the first major overhaul of our nation's telecommunications policies in over 60 years. Since 1990, the FCC has taken on a large workload associated with the rapidly growing wireless industry.

The Wireless Telecommunications Bureau

The **Wireless Telecommunications Bureau (WTB)** is responsible for all FCC domestic wireless telecommunications programs and policies except those involving satellite communications or broadcasting. Wireless communications services include cellular telephones, paging, personal communications services (PCS), public safety, microwave, land mobile, and other commercial and private communications services. The WTB is also responsible for spectrum auctions for the PCS spectrum mentioned previously.

We now will examine the organizations that establish, maintain, and publish standards for the wireless industry.

Wireless
Telecommunications
Bureau (WTB)

The American National Standards Institute

The **American National Standards Institute (ANSI)** has served as administrator and coordinator of the United States private sector voluntary standardization system for more than 80 years.

American National
Standards Institute
(ANSI)

Founded in 1918 by five engineering societies and three government agencies, the Institute remains as a private, nonprofit membership organization supported by a diverse group of private and public sector organizations.

ANSI does not itself develop American National Standards (ANSs); rather it facilitates development by establishing consensus among qualified groups such as the Telecommunications Industry Association (TIA).

The Telecommunications Industry Association

The **Telecommunications Industry Association (TIA)** has developed and published the most recent set of standards associated with wireless digital cellular and PCS standards in the U.S. TIA is accredited by the American National Standards Institute (ANSI) to develop voluntary industry standards. The TIA's Standards and Technology Department is composed of five divisions, which sponsor more than 70 standards-setting formulating groups. The committees and subcommittees sponsored by the five divisions include:

Telecommunications
Industry Association
(TIA)

- Fiber Optics
- User Premises Equipment
- Wireless Communications

- Satellite Communications
- User Premises Equipment
- Network Equipment

TIA also participates in international standards-setting activities, such as the International Telecommunication Union (ITU), the Inter-American Telecommunication Commission (CITEL), and the International Electrotechnical Commission.

Regulatory Changes in the Telecommunications Industry

As shown in Figure 1-14, major changes occurred in the telecommunications regulatory arena from 1980 to 1996. These events have had a substantial impact on the industry as a whole and specifically on the wireless telecommunications sector. In this section, the three major legislative and judicial events that shaped the future wireless industry are explained.

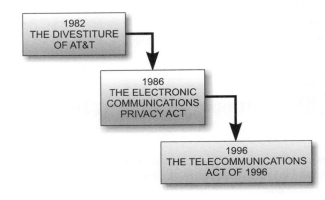

Figure 1-14: Regulatory Changes in the Telecommunications Industry

Event 1: The Divestiture of AT&T and the Bell System

In its early history, AT&T functioned as a legally sanctioned and regulated federal monopoly. There were many reasons why this was believed to be a good plan. The government and many other interested agencies and corporations thought that the telephone network would function most efficiently as a monopoly, providing seamless service throughout the U.S.

The United States government initially promoted this idea, informally at first, and then legislatively. But later on, as political philosophy evolved, federal administrations investigated the telephone monopoly in the light of general antitrust law and alleged company abuses. One notable result was an antitrust suit filed in 1949, which led in 1956 to a consent decree signed by AT&T and the Department of Justice, and filed in court, whereby AT&T agreed to restrict its activities to the regulated business of the national telephone system and government work. In other words, AT&T agreed to adhere to its core business.

In the late 1940s, new technologies appeared that provided alternatives to copper wires for long-distance telephone transmission. AT&T opened its first microwave relay system between the cities of New York and Chicago in 1950, and over the succeeding three decades added considerable microwave capacity to its nationwide long-distance network. In 1962, AT&T placed the first commercial communications satellite, Telstar I, in orbit, offering an additional alternative especially suited to international communications.

Technology changes began to occur at an exponential rate during the 1960s. With the advance of computing technology, data communications over existing voice channels became a reality. These new computer-based data technologies lowered the technological barriers to entry by would-be competitors to the Bell System. Slowly, over several decades, the Federal Communications Commission (FCC), the regulatory agency that oversees telecommunications in the United States, allowed some competition using these technologies at the edges of the network. By the mid-1970s, competition had advanced to general long-distance service. The changes in telecommunications during these years eventually led to an antitrust suit by the U.S. government against AT&T. The suit began in 1974 and was finally settled in January 1982 when AT&T agreed to divest itself of the wholly owned Bell operating companies that provided local exchange service. In return, the U.S. Department of Justice agreed to lift the constraints of the 1956 decree. Divestiture took place on January 1, 1984, and the Bell System was dead. In its place was a new AT&T and seven regional telephone holding companies.

This event, in an indirect way, has provided the framework for the growth of the wireless industry in the U.S. as well as internationally. We now have literally hundreds of wireless communications equipment suppliers and service providers.

The FCC has encouraged the evolution of many wireless radio interface standards for cellular and PCS. Having a free and open market for wireless technologies and multiple standards, however, has resulted in what many call the "Tower of Babble," which has made interoperability among wireless networks impractical in the second generation of wireless telecommunications. We will cover this issue of roaming and internetworking within wireless networks in a later chapter.

Event 2 : The Electronic Communications Privacy Act of 1986

The Electronic Communications Privacy Act (ECPA) of 1986 was adopted to address the legal privacy issues that were evolving with the growing use of computers and other innovations, such as wireless telephony, in electronic communications. The ECPA updated legislation passed in 1968 that had been designed to clarify what constitutes invasion of privacy when electronic surveillance is involved. The ECPA extended privacy protection outlined in the earlier legislation to apply to radio paging devices, electronic mail, cellular telephones, private communication carriers, and computer transmissions. This act has had a substantial impact on how cellular and PCS wireless networks have been designed to incorporate features that provide privacy and protection from fraud.

President Reagan signed the Electronic Communications Privacy Act into law on October 21, 1986. The ECPA was designed to expand Title III privacy protection to apply to radio paging devices, electronic mail, cellular telephones, private communication carriers, and computer transmissions. The Act also identified specific situations and types of transmissions that would not be protected, most notably an employer's monitoring of employees' electronic mail on the employer's system.

Cellular Telephones

Hall vs. U.S., a 1973 decision, held that mobile telephone conversations are protected under Title III when part of a communication is carried to or from a "landline" telephone. This decision failed to clarify whether protection applies to all cellular and cordless telephone conversations, however. The ECPA of 1986 amended the definition of protected "wire communication" to "include communications utilizing wires, cables, or other line connections within a switching office regardless of whether the communications are between two cellular phones or between a cellular telephone and a 'landline' telephone." However, in order to encourage the use of technological means of protection like scrambling and encryption, the Act reduced the criminal penalty for the interception of unencrypted, unscrambled cellular phone calls from a felony that could carry up to five years of imprisonment to a $500 fine.

NOTE

The Electronic Communications Privacy Act provides protection for the wire portion of cordless phone conversations, but specifically notes that "wire communication" protected under Title III "does not include the radio portion of a telephone that is transmitted between the cordless telephone handset and the base unit."

Radio Paging

radio paging

The ECPA clarified privacy protection relating to the use of **radio paging** devices. The Department of Justice defined voice and digital display pagers as "a continuation of an original wire communication" that should therefore be subject to Title III protection. The legislation also specifically identified the "tone-only" pager as a device whose use is not protected under Title III.

Customer Records

The ECPA of 1986 restricted government access to subscriber and customer records belonging to electronic service providers. Government agencies must first secure a search warrant, court order, or an authorized administrative or grand jury subpoena to access service provider records without first notifying a subscriber or customer.

Satellite Transmissions

The ECPA of 1986 identified the Cable Communications Policy Act of 1984 as the exclusive source of protection policy governing home reception of unencrypted cable satellite programming. The 1984 Act established a separate set of specialized policies to address cable satellite reception issues that related more to the conduct of commercial enterprise than to privacy issues. The ECPA also increased criminal penalties for malicious or intentional interference that impedes the delivery of satellite transmissions.

Event 3: The Telecommunications Act of 1996

On February 1, 1996, after more than a year of intensive negotiations and political wrangling, Congress passed the **Telecommunications Act of 1996** (the "Act"). The Act is the first comprehensive rewrite of the Communications Act of 1934 discussed earlier and dramatically changes the ground rules for competition and regulation in virtually all sectors of the communications industry, from local and long-distance telephone services, to cable television, broadcasting, and equipment manufacturing. President Clinton signed the Act on February 8, 1996, and its provisions became effective immediately.

The Telecommunications Act of 1996 is a major event in the development of the United States communications industry. For decades, communications policy — including ownership and service restrictions that maintained protected monopolies at both the state and federal levels — was set largely by the FCC and state public utility commissions ("PUCs"), and the federal courts' enforcement of the 1984 antitrust consent decree that dismantled the Bell System. Major strides were made, particularly in relaxing federal regulation and in ensuring fair competition in the long-distance telephone market, but the ambiguity inherent in enforcing a 62-year old statute led to legal uncertainty and conflicting interpretations. With the Act, Congress has reasserted primacy in setting U.S. communications policy, and has set a course that clearly adopts competition as the basic charter for all telecommunications markets.

The Act's provisions fall into five major areas:

- Telephone service

- Telecommunications equipment manufacturing

- Cable television

- Radio and television broadcasting

- The Internet and online computer services

In each of these areas, cross-market entry barriers have been eliminated, concentration and merger rules relaxed, and massive new implementation obligations placed on the FCC and state regulators. In some areas, specifically television violence and "indecent" online communications, Congress acted more to promote its current views of appropriate social and moral behavior than to unleash competitive market forces. Wireless industry customers benefitted substantially through the introduction of competition in the cellular/PCS market.

The Role of International Standards Organizations

There are numerous standards organizations and industry support groups that contribute to the development of standards for wireless communications.

Each plays an important role in shaping the future of the industry.

International Wireless Standards

The existence of different standards for similar radio-based technologies in different countries or regions can contribute to so-called "technical barriers to trade" in the wireless industries.

International
Organization for
Standardization (ISO)

This was one of the reasons for the formation of the European Telecommunications Standards Institute (ETSI) in Europe, which led to the development of Global Systems for Mobile Communications (GSM). Export-minded industries in Europe and Asia have long sensed the need to agree on world standards to help rationalize the international trading process. This was also the basis for the establishment of the **International Organization for Standardization (ISO)** as one of the first nonprofit organizations to promote the development of global telecommunications standards.

Advantages for International Standardization

Industry-wide standardization is a condition existing within a particular industrial sector when the large majority of products or services conform to the same standards. It results from consensus agreements reached between all players in that industrial sectors— suppliers, users, and often governments. They agree on specifications and criteria to be applied consistently in the choice and classification of materials, the manufacture of products, and the provision of services. The aim is to facilitate trade, exchange, and technology transfer through:

- Enhanced product quality and reliability at a reasonable price

- Improved health, safety and environmental protection, and reduction of waste

- Greater compatibility and interoperability of goods and services

- Simplification for improved usability

- Reduction in the number of models, and thus reduction in costs

- Increased distribution efficiency and ease of maintenance

Users have more confidence in products and services that conform to international standards. Assurance of conformity can be provided by manufacturers' declarations, or by audits carried out by independent bodies.

Progress Toward Global Wireless Standardization

The wireless industry has historically developed standards primarily on a regional basis.

Prior to the year 2000, very little work had been done towards the promotion of global mobile wireless standards. This is changing, as we shall see in subsequent sections on wireless technology evolution. Although much work needs to be done among standards bodies to achieve true global roaming, work is progressing towards a goal called third generation (3G) wireless standards for mobile subscriber systems. This effort is being led by the International Telecommunications Union (ITU) under the banner called **International Mobile Telecommunications (IMT) 2000**.

International Mobile
Telecommunications
(IMT) 2000

Wireless Subscriber Roaming

International standards and regulations provide the capability for mobile wireless subscribers to **roam** among different wireless service providers' service areas.

roam

Imagine the chaos that would result if your cellular phone only worked within a network of base station transceivers manufactured by the same company that built your wireless phone. Standards are now in place in the United States that enable mobile wireless voice and data subscribers to receive service when they are outside their primary home service area. A similar standard exists within Europe. One can imagine several situations in a non-standard world in which, for example, each auto manufacturer would require a special size tire for a specific product line that would be different from all other auto tire sizes; or perhaps how radio and television broadcasting could have evolved if each broadcasting station used a different standard for transmission. Of course standards are not new. They are a key element of the entire telecommunications industry, dating back to the infancy of the public telephone network.

Convergence of Wireless Industry Standards

The wireless industry has grown very rapidly. Setting standards for wireless communications systems has become an industry unto itself. As we shall see later, many standards have evolved for different sectors of the industry. Many experts believe we now have too many standards. Of course, the objective for standardization is just the opposite. Developers would like to see fewer standards converging in support of global internetworking. As you will learn in this section, global standardization of wireless networks and radio interfaces is essential. Without international standards, we would never have the potential for global wireless network connectivity.

Now we will review how various international and domestic standards bodies work together to support a future global wireless infrastructure. Major organizations that are shaping the future of wireless technology and standards include:

- American National Standards Institute (ANSI)

- Cellular Telecommunications & Internet Association (CTIA)

- European Telecommunications Industry Association (ETSI)

- Institute for Electrical and Electronic Engineers (IEEE)

- International Organization for Standardization (ISO)

- International Telecommunications Union (ITU)

- National Telecommunications and Information Administration (NTIA)

- Telecommunications Industry Association (TIA)

- World Radio Communications Conference (WRC)

American National Standards Institute

The **American National Standards Institute (ANSI)** was founded in 1918. ANSI is a private, nonprofit membership organization. The ANSI organization does not develop standards itself; rather it works with other standards groups such as the TIA to obtain agreements on standards across various wireless support organizations.

Standards groups obtain accreditation from ANSI. The Institute represents the interests of its nearly 1,000 company, organization, government agency, institutional, and international members through its office in New York, and its headquarter in Washington, D.C.

Currently, the approved cellular radio interface standards are labeled as ANSI standards although the standards were actually developed by the TIA, an ANSI-accredited organization.

Telecommunications Industry Association

Accredited by the American National Standards Institute (ANSI), **Telecommunications Industry Association (TIA)** is a major contributor of voluntary industry wireless standards.

These standards are the primary radio and network standards that govern the cellular industry in the U.S. The standards we will be concerned with in future chapters, related to cellular and PCS networks, were developed by TIA subcommittees. The TIA TR-45 General Purpose Standards include:

- **TR-45.1 Analog Cellular Radio Interface Standards** The TR45.1 subcommittee is responsible for the analog Advanced Mobile Phone System (AMPS) standard. This was the first generation of cellular radio interface standards.

- **TR-45-2 Wireless Network Standards** The TR-45.2 subcommittee is responsible for all network standards that support the air interfaces, both analog and digital (AMPS, TDMA, and CDMA), developed by other subcommittees.

- **TR-45.3 TDMA Digital Cellular/PCS Standards** The TR-45.3 subcommittee is responsible for all Time Division Multiple Access (TDMA) digital standards including ANSI-136 and ANSI-54.

- **TR-45.4 MSC-BS Interface Standards** The TR-45.4 subcommittee is responsible for the development of the Mobile Switching Center (MSC) to Cellular Base station (BS) standards and the Wireless Local Loop (WLL) standards.

- **TR-45.5 CDMA Digital Standards** The 45.5 subcommittee is responsible for the development of Code Division Multiple Access (CDMA) digital air interface standards including ANSI-95 and the family of other related CDMA standards.

- **TR-45.6 Cellular Digital Packet Data (CDPD) Standards** The 45.6 subcommittee is responsible for the development of all wireless data packet standards used in the U.S. cellular markets.

- **TR-45.7 Network Management** The TR-45.7 subcommittee was formed in 1998 to study network management standards for wireless networks.

> **NOTE**
>
> The interim standards (IS) designation has been replaced with ANSI-136 and ANSI-54 since they are approved standards and, therefore, are no longer considered interim standards.

The International Organization for Standardization

> The **International Organization for Standardization (ISO)** is a worldwide group of national standards bodies from 130 countries, one from each country.

International Organization for Standardization (ISO)

ISO is a nongovernmental organization established in 1947. The mission of ISO is to promote the development of standardization and related activities in the world with a view to facilitating the international exchange of goods and services.

Origin of the Name "ISO"

You probably have noticed a lack of consistency between the full official title of International Organization for Standardization, and the short form, ISO. Shouldn't the acronym be "IOS"? Yes, if it were an acronym— which it is not. ISO is often (incorrectly) referred to in various publications as an acronym for International Standards Organization.

In fact, "ISO" is a word, derived from the Greek word "isos," meaning "equal," which is the root of the prefix "iso-" that occurs in a host of terms, such as isometric (of equal measure or dimensions) and isonomy (equality of laws, or of people before the law).

From "equal" to "standard," the line of thinking that led to the choice of "ISO" as the name of the organization is easy to follow. In addition, the name ISO is used around the world to denote the organization, thus avoiding the plethora of acronyms resulting from the translation of "International Organization for Standardization" into the different national languages of members, e.g., IOS in English, OIN in French (from Organization Internationale de Normalisation), and so forth. Whatever the country, the short form of the Organization's name is always ISO.

ISO's work results in international agreements, which are published as International Standards. One of the popular standards developed by ISO is the Open Standards Interconnect (OSI) model, which will be explained in more detail in Chapter 2.

International Telecommunications Union

Wireless communications has become a global business. Each country is developing new wireless technologies at a very rapid pace.

> One of the most important organizations involved in the coordination of global standards is the **International Telecommunications Union (ITU)**.

It was founded in Paris in 1865 as the International Telegraph Union and took its present name in 1934 and became a specialized agency of the United Nations in 1947. Let's examine the organization and purpose of the ITU.

There are three sectors that make up the ITU structure:

- **The Radiocommunication Sector or the ITU-R** deals with all global radio standards bodies from each of the member nations including world and regional radiocommunication conferences, radiocommunication assemblies, and the Radio Regulations Board.

- **The Telecommunication Standardization Sector or the ITU-T** coordinates the efforts of member nations promoting world telecommunication standardization.

- **The Telecommunication Development Sector or ITU-D** including world and regional telecommunication development conferences.

The ITU is an intergovernmental organization, within which the public and private sectors cooperate for the development of telecommunications. The ITU adopts international regulations and treaties governing all terrestrial and space uses of the frequency spectrum as well as the use of the geostationary-satellite orbit, within which countries adopt their national legislation. It also develops standards to facilitate the interconnection of telecommunication systems on a worldwide scale regardless of the type of technology used.

Spearheading telecommunications development on a world scale, the ITU fosters the development of telecommunications in developing countries by establishing medium-term development policies and strategies in consultation with other partners in the sector and by providing specialized technical assistance in the areas of telecommunication policies, the choice and transfer of technologies, financing of investment projects and mobilization of resources, the installation and maintenance of networks, the management of human resources, and research and development.

The World Radio Communications Conference

The ITU radio regulations are established by decisions reached at **World Radio Communications Conferences (WRCs)**. WRC 92, WRC 95, WRC 97, and WRC 2000 were the most recent conferences where major spectrum allocations decisions were reached. The 1996 spectrum chart reflects those decisions reached at WRC 92 that have been implemented by the NTIA and the FCC for radio services operated in the United States. Work is being planned to update the 1996 chart.

National Telecommunications and Information Administration

The **National Telecommunications and Information Administration (NTIA)**, an agency of the U.S. Department of Commerce, is the executive branch's principal advisor on domestic and international telecommunications and information technology issues.

One of the branches of the NTIA that is involved in the management of radio frequencies is called the **Office of Spectrum Management (OSM)**.

Under the provisions of the United States Communications Act of 1934, as revised, authority for managing the use of the radio frequency spectrum within the United States is partitioned between the NTIA (specifically the OSM branch) and the Federal Communications Commission (FCC). Therefore, to establish which radio services will be allowed to operate in the United States in a given frequency band requires that radio frequency spectrum management policies be established by both the NTIA and the FCC.

Under the provisions of the International Telecommunication Union (ITU) treaty the U.S. is obligated to comply with the spectrum allocations specified in the ITU Radio Regulations' Article S 5 (International Table of Frequency Allocations). However, U.S. domestic spectrum uses may differ from the international allocations provided these domestic uses do not conflict with our neighbors' spectrum uses that do comply with international regulations.

Cellular Telecommunications & Internet Association

Founded in 1984, the **Cellular Telecommunications & Internet Association (CTIA)** is another international organization that represents all elements of wireless communication such as cellular, personal communication services, enhanced specialized mobile radio, and mobile satellite services serving the interests of service providers, manufacturers, and others.

The Cellular Telecommunications & Internet Association changed its name in October 2000. It was previously known as the Cellular Telecommunications Industry Association. The name change followed the CTIA's merger with the Wireless Data Forum (WDF).

CTIA has always been a major player within the wireless industry representing its members before the executive branch, in the Federal Communications Commission, and in Congress. CTIA's industry committees provide leadership in the area of taxation, roaming, safety, regulations, fraud prevention, and technology.

CTIA often serves as the press agent for the wireless industry as a whole. It distributes information to members, policymakers, the investment community, customers, and the news media on the latest policy and technical developments related to cellular mobile communications. The CTIA also operates an equipment testing and certification program to ensure high quality and reliability for consumers.

European Telecommunications Industry Association

European Telecommunications Industry Association (ETSI) is a non-profit organization whose mission is to produce the telecommunications standards that will be used throughout Europe.

Based in Sophia Antipolis, a high-tech research park in southern France, ETSI unites 789 members from 52 countries inside and outside Europe, and represents administrations, network operators, manufacturers, service providers, research bodies, and users. ETSI led the development and adoption of the single wireless standard for Europe called Global Systems for Mobile Communication (GSM). ETSI is also a leading organization for the development of 3G wireless standards evolving from GSM that will be addressed in later chapters.

Institute for Electrical and Electronic Engineers

The **Institute for Electrical and Electronic Engineers (IEEE)** is another nonprofit technical and professional association that publishes a large volume of information related to electrical engineering disciplines. The IEEE has more than 350,000 members in 150 countries.

The major activity of the IEEE related to the topics and scope of this book is the development of wireless and wired local area network (LAN) standards.

The growth of Ethernet as the dominant LAN standard can largely be credited to the "802 standards family" developed by the IEEE. The IEEE has also developed the 802.11 wireless LAN standard, which defines the radio technology for wireless computer networking. This will be covered in more detail in part II of this book.

The Importance of Maritime Telecommunications

Since the early 1980's, wireless telecommunications technology has advanced rapidly for land-based subscribers. Cellular telephones and wireless Personal Digital Assistants (PDAs) are widely available to everyone in the more populated centers of every nation in the world.

Ships at sea, however, do not have access to the same infrastructure persons enjoy on land. Mariners need not only to access international shore telephone and data public switched networks, but also to be able to communicate with other ships of any size or nationality, to receive and send urgent maritime safety information, and to send or receive distress alerts in an emergency to or from rescue coordination centers ashore and nearby ships anywhere in the world. They need to have reliable wireless communications when operating at sea far from coastal radio stations.

In order to provide coverage across oceans, special wireless technologies, standards, and procedures are needed. Unlike land-based cellular and PCS systems used in the United States, shipboard radio systems must be internationally interoperable. Above all, ship communications must be reliable because the lives of crew and passengers may be at stake during weather emergencies or other unforeseen emergencies when vessels are far from land. Bringing new telecommunications technology to mariners can be difficult, since to be interoperable, the technology must be affordable, acceptable, and available to most ships and maritime countries.

Many of the maritime communications systems described in this chapter will appear to be redundant. This was part of the master plan by the International Maritime Organization (IMO) and participating nations to the IMO treaty. This redundancy reflects added emphasis and importance for the protection of human life and assurance that search and rescue efforts will achieve a high rate of success. Maritime shipping resources should not have to depend upon one system, but rather have the benefit of several backup systems in order to achieve the maximum degree of reliability.

Unfortunately, a few nations have not yet fully complied with the international maritime treaties set forth by the IMO. There are several vessels on the high seas that do not meet minimum IMO requirements for wireless communications equipment onboard their home flag ships.

Maritime Wireless Communications Governing and Regulatory Agencies

The Federal Communications Commission

The United States **Federal Communications Commission (FCC)** regulates all use of radio onboard any recreational, commercial, state and local government, and foreign vessel in U.S. ports and waters.

Federal
Communications
Commission (FCC)

These regulations can be found in Title 47, Code of Federal Regulations, Part 80.

The National Telecommunications and Information Administration

National Telecommunications and Information Administration (NTIA)

The **National Telecommunications and Information Administration (NTIA)** regulates all use of radio onboard any federal government vessel, including military vessels, in U.S. ports and waters. NTIA rules do not apply outside the federal government. NTIA rules, applied in Title 47, Code of Federal Regulations (CFR), Part 300, are contained in the Manual of Regulations and Procedures for Federal Radio Frequency Management (NTIA Manual).

The U.S. Coast Guard

U.S. Coast Guard (USCG)

The **U.S. Coast Guard (USCG)** regulates the requirements and procedures for radio transmitting and receiving equipment on commercial fishing vessels and foreign vessels in U.S. waters.

Coast Guard radio watch keeping rules only apply to vessels participating in vessel traffic service (VTS) areas. Coast Guard rules affecting radio use are contained in Titles 33 and 46, Code of Federal Regulations (CFR). Morse wireless telegraphy, used by ships for distress and safety communications since the beginning of the century, was discontinued by the USCG in 1995 and ended worldwide on February 1, 1999. Many people owe their lives to this system.

The International Telecommunications Union

The International Telecommunications Union (ITU), an agency sponsored by the U.N., regulates all use of radio by any person or vessel outside U.S. waters. ITU rules affecting radio, which have treaty status in the U.S., are published in the ITU Radio Regulations.

The International Maritime Organization

The huge loss of lives caused by the sinking of the Titanic in 1912 triggered a concerted effort among nations of the world to adopt maritime wireless procedures for all ocean-going vessels.

International Maritime Organization (IMO)

Several countries proposed that a permanent international body should be established to promote maritime safety more effectively, but it was not until the establishment of the United Nations that these hopes were realized. In 1948 an international conference in Geneva, Switzerland adopted a convention formally establishing the **International Maritime Organization (IMO)**. It entered into force in 1958 and the new Organization met for the first time the following year.

Its first task was to adopt a new version of the International Convention for the Safety of Life at Sea (SOLAS), the most important of all treaties dating back to the 1920's dealing with maritime safety.

Most IMO radio regulations affect passenger ships and other types of ships of weight 300 tons and over. IMO rules affecting radio, which have treaty status in the U.S., are included in the Safety of Life at Sea (SOLAS) Convention. The current system adopted by these two United Nations Organizations is the Global Maritime Distress and Safety System, or GMDSS. GMDSS is an important development and is discussed later in this section.

Global Maritime Communications Standard

Since the invention of radio at the end of the 19th century, ships at sea have relied on Morse code. The need for ship and coast radio stations to have and use radiotelegraph equipment, and to listen to a common radio frequency for Morse encoded distress calls, was recognized after the sinking of the Titanic in the North Atlantic in 1912 with a great loss of life. If proper use had been made of the new "wireless" telegraph aboard ships at the time, a great number of lives could have been saved by nearby vessels. The U.S. Congress enacted legislation soon after the Titanic disaster, requiring U.S. ships to use Morse code radiotelegraph equipment (it was called the "the wireless") for transmitting and monitoring distress calls. The International Telecommunications Union (ITU) followed suit for ships of all nations at a later date. Morse encoded wireless telegraph distress calling has saved thousands of lives since its inception, but its use requires skilled radio operators spending many hours listening to the radio distress frequency. Its range on the medium frequency (MF) distress band (500 kHz) is limited, and the amount of traffic Morse signals can carry is also limited. As stated previously, the use of Morse code for maritime communications was finally discontinued worldwide February 1,1999. The basis for this change was the development of the GMDSS.

The Global Maritime Distress and Safety System

The **Global Maritime Distress and Safety System (GMDSS)** consists of several cooperative terrestrial and satellite radio communications systems.

The major difference between the GMDSS and previous maritime communications systems is that the radio communications equipment to be installed on a GMDSS ship is determined by the ship's area of operation, rather than by its size. Because the various radio systems used in the GMDSS have different limitations with regards to range and services provided, the system divides the world's oceans into four areas. The area traveled by a ship governs the type of equipment required aboard each vessel:

- **Area A1** lies within range of shore-based VHF coast stations (20 to 30 nautical miles).

- **Area A2** lies within range of shore-based MF coast stations excluding A1 areas (approximately 100–150 nautical miles).

- **Area A3** lies within the coverage area of Inmarsat communications satellites (excluding A1 and A2 areas — approximately latitude 70 degrees north to latitude 70 degrees south).

- **Area A4** comprises the remaining sea areas outside areas A1, A2, and A3 (the polar regions).

Ships covered by the GMDSS International Agreement:

- Cargo ships of 300 gross tons and over when traveling on international voyages or in the open sea.

- All passenger ships carrying more than twelve passengers when traveling on international voyages or in the open sea.

- The GMDSS regulations do not apply to vessels operating exclusively on the Great Lakes.

The GMDSS System Components

The core systems for the GMDSS consist of two satellite networks called the COSPAS-SARSAT system and the Inmarsat system. A terrestrial system called NAVTEX is used to distribute marine alert information on medium wave (518 kHz) radio transmitting stations. Another replacement for the older radiotelephone and radiotelegraph system is called the Digital Selective Calling (DSC) system.

Ships are also required to carry a Search and Rescue Radar Transponder (SART). We will now take a closer look at how each of the component systems of the GMDSS work together.

The COSPAS-SARSAT System

The SARSAT system was developed in a joint effort by the United States, Canada, and France. The COSPAS system was developed by the former Soviet Union and is now supported by Russia. These four nations banded together in 1979 to form COSPAS-SARSAT. In 1982, the first satellite was launched, and by 1984 the system was declared fully operational. Although COSPAS-SARSAT satellites were primarily designed to receive and relay information from the much improved 406 MHz beacons, they still had to make a provision for the thousands of 121.5 MHz beacons already in use aboard aircraft. The IMO has stated that the 121.5 MHz and also the 243 MHz beacons will be phased out over several years in favor of the more reliable 406 MHz beacons. The IMO has specifically stated that the COSPAS-SARSAT system will terminate processing of these two older beacons on February 1, 2009.

COSPAS-SARSAT

COSPAS-SARSAT is the acronym for two separate satellite systems used for search and rescue. COSPAS is an acronym for the Russian words "Cosmicheskaya Sistyema Poiska Avariynich Sudov," meaning Space System for the Search of Vessels in Distress. SARSAT stands for Search and Rescue Satellite-Aided Tracking.

Following the successful completion of the demonstration and evaluation phase started in September 1982, a second Memorandum of Understanding was signed on October 5, 1984 by the Centre National d'Etudes Spatiales (CNES) of France, the Department of National Defense (DND) of Canada, the Ministry of Merchant Marine (MORFLOT) of the former USSR, and the National Oceanic and Atmospheric Administration (NOAA) of the USA.

On July 1, 1988, the four countries providing the space segment signed the International COSPAS - SARSAT agreement, which ensures the continuity of the system and its availability to all nations on a nondiscriminatory basis. In January of 1992, Russia assumed responsibility for the obligations of the former Soviet Union. A number of countries that are not parties to the agreement have also associated themselves with the COSPAS-SARSAT program.

The four founding countries jointly developed a 406 MHz **Emergency Position Indicating Radio Beacon (EPIRB)** that is used by terrestrial units to transmit distress information and location to the satellites. These automatic-activating EPIRBs, now required on a certain class of ships, are designed to transmit vessel identification and an accurate location of the vessel from anywhere in the world to a rescue coordination center.

The COSPAS-SARSAT satellites are illustrated in Figure 1-15.

The heart of the of the satellite system is the **Low Earth Orbit (LEO)** satellites that circle the earth in near polar orbits. The SARSAT instruments are flown on the U.S. NOAA series of environmental satellites. The NOAA satellites orbit at an altitude of 528 miles at an inclination angle of 99 degrees. They complete one orbit every 100 minutes. Each satellite carries equipment capable of receiving emergency transmissions on 121.5 MHz, 243 MHz, and 406 MHz whenever the ground stations are within the view of the satellites. The COSPAS instrument is flown aboard the Russian NADEZHDA satellites orbiting the earth at an altitude of 620 miles at an inclination of 83 degrees. They complete one orbit every 105 minutes. The COSPAS instrument was built by the former Soviet Union. Other payloads onboard the NOAA satellites were provided by the Canadian and French governments. The satellites are traveling at a high rate of speed relative to the ground station and use the Doppler shift of the radio carrier wave to aid in determining the position of the distress signal.

Figure 1-15: The COSPAS-SARSAT Satellites

Complete coverage of the earth is achieved by the polar orbiting spacecraft; however, since the satellites are in low earth orbit, the instruments can only observe a small portion of the earth at any given time. It is therefore not possible to receive the radio beacon signal until the satellite "footprint" passes over the beacon. The satellite has a processor module that can store the 406 MHz beacon information because it may not be able to relay the information to a LEO Local User Terminal (LEOLUT) if it is not within the footprint of the satellite. If the ground LEOLUT station is in view at the time, the message is relayed immediately. If it is not in view, it is stored and downloaded later when the ground station is in view, thereby providing global coverage. The older beacons operating on 121.5 MHz do not have associated onboard processors. The COSPAS-SARSAT satellites relay the older distress signal to a ground LEOLUT when and if it is in view at the same time the signal is received. The coverage is therefore not continuous for the 121.5 ELTs.

Figure 1-16 depicts the COSPAS - SARSAT satellites placed in both Geostationary and Low Earth (polar) Orbits.

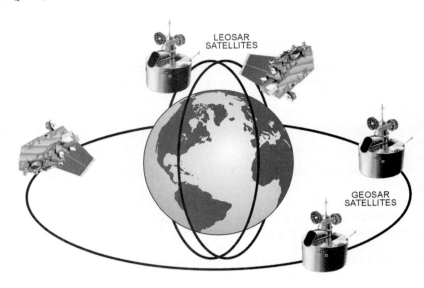

Figure 1-16: LEOSAR and GEOSAR Satellites

The COSPAS-SARSAT organization is also evaluating the use of 406 MHz receivers on board **geostationary (GEO)** satellites. The payload is carried aboard the U.S. Geostationary Operational Environmental Satellites (GOES) that also provide spectacular pictures of the weather often seen on evening television news broadcasts. The GOES satellites are at an altitude of 22,300 miles and are stationary with respect to the earth. They are therefore not able to use Doppler location processing since there is no relative motion between the satellite and the emergency beacon. Consequently, they are unable to provide the exact location of the beacon. On the positive side, the GEO satellites provide an immediate alert to search and rescue personnel on the ground. This allows them to start the initial verification process using the NOAA beacon registration database. This in turn allows the ground forces to begin a general estimation of the location of the aircraft or vessel in distress. Typically, the COSPAS-SARSAT satellites will fly over the same location on a LEO pass within minutes after the GEO alert, calculate the location using Doppler processing, and forward more precise location data to the search teams.

The COSPAS-SARSAT Satellite and Ground Support System is depicted in Figure 1-17.

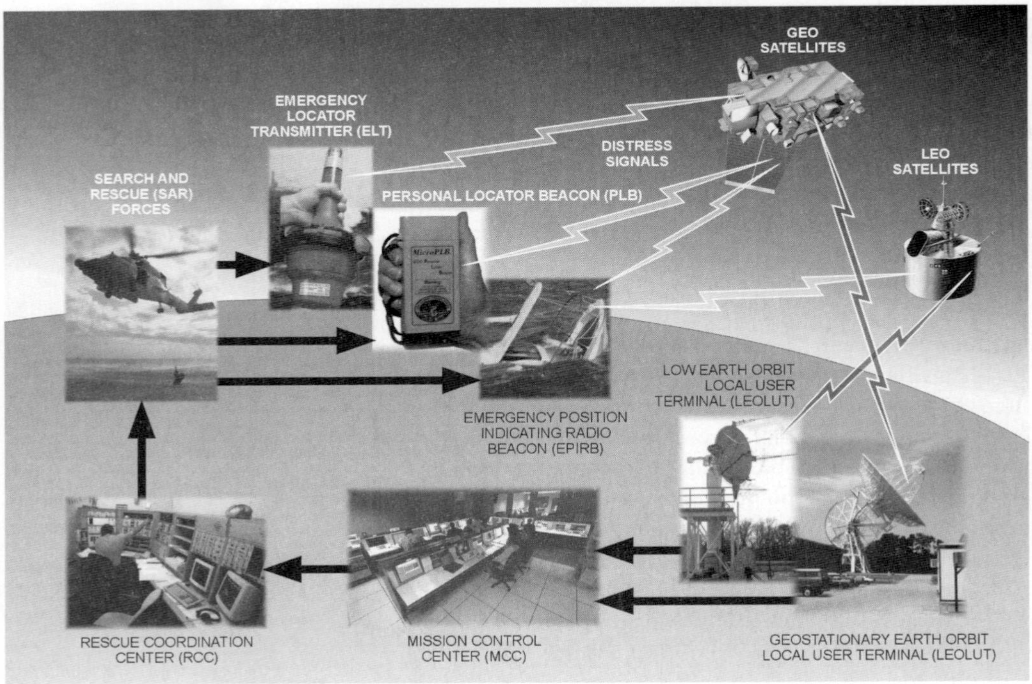

Figure 1-17: Ground Support System

We will now look closer at the tasks performed by the various components of the COSPAS-SARSAT system illustrated in Figure 1-15. The system is composed of:

- Three types of distress radio beacons. EPIRBs for maritime use, ELTs for aviation use, and **Personal Locator Beacons (PLBs)** for personal use. Each of these devices transmits signals automatically when the host is in a distress situation.

- Instruments onboard the COSPAS-SARSAT Geostationary (GEO) and Low Earth Orbit (LEO) satellites. They are designed to detect the signals transmitted by the radio beacons.

- Rescue Coordination Centers (RCCs), Mission Control Centers (MCCs), GEO Local User Terminals (GEOLUT), and LEO Local User Terminals (LEOLUT).

- Search and Rescue (SAR) forces

When an emergency beacon is activated, the signal is received by a satellite and relayed to the nearest available ground station. The ground station, called a **Local User Terminal (LUT)**, processes the signal and calculates the position from which it originated. This position is transmitted to a **Mission Control Center (MCC)** where it is joined with identification data and other information on that beacon. The Mission Control Center then transmits an alert message to the appropriate rescue coordination center based on the geographic location of the beacon. If the location of the beacon is in another country's service area, then the alert is transmitted to that country's MCC. All beacons transmit on 406 MHz.

The NAVTEX System

Navigational Telex (NAVTEX) is an international, automated system for instantly distributing maritime navigational warnings, weather forecasts and warnings, search and rescue notices, and similar information to ships.

NAVTEX uses transmitters located on the ground only. A small, low-cost digital receiver is used on ships to receive periodic transmissions. All NAVTEX transmissions are made on 518 kHz, which is just below the standard AM radio commercial broadcast band. No person needs to be present during a broadcast to receive the information. In the United States, NAVTEX is broadcast from Coast Guard facilities located at eight coastal stations plus Honolulu, HI; San Juan, PR; Kodiak, AK; and Guam. The Coast Guard began operating NAVTEX from Boston, MA in 1983.

The Inmarsat System

The Inmarsat satellite system was founded in 1979 as the **International Maritime Satellite Organization (Inmarsat)** to provide worldwide mobile communications capability for marine customers.

The agency changed its name in 1994 to the International Mobile Satellite Organization to reflect the growing group of land-based users. Inmarsat is now an important element of the GMDSS.

> INMARSAT was formerly an acronym for the International Maritime Satellite Organization. After the organization name was changed to the International Mobile Satellite Organization, the Inmarsat name remained although no longer as an acronym. It is therefore spelled in lowercase letters.

Inmarsat was originally formed as a treaty organization with several countries forming a partnership to finance the cost of launching satellites for the purpose of supporting maritime safety. In April 1999, Inmarsat was privatized and became the first of many international treaty organizations to be transformed into a commercial organization.

Inmarsat satellites are now in the fourth generation of spacecraft design. The satellites are placed in geostationary orbit. They remain stationary with respect to observers on the earth. They are located over the Atlantic, Pacific, and Indian Ocean areas.

Three types of **Inmarsat earth station terminals** are recognized by the GMDSS: the Inmarsat A, B, and C. The Inmarsat A and B (an updated version of the A) provide ship/shore, ship/ship, and shore/ship telephone, telex, and high-speed data services, including a distress priority telephone and telex service to and from rescue coordination centers. The Inmarsat C type terminal provides ship/shore, shore/ship, and ship/ship store-and-forward data and telex messaging and the capability for sending preformatted distress messages to a rescue coordination center. The Inmarsat C SafetyNET service is a satellite-based worldwide maritime safety information broadcast service. It provides high seas weather warnings, navigational warnings, ice reports and warnings generated by the Coast Guard conducted International Ice Patrol, and other similar information not provided by NAVTEX.

Inmarsat C equipment is relatively small and lightweight, and costs much less than an Inmarsat A or B earth station. Inmarsat A and B ship earth stations require relatively large gyro-stabilized antennas. The gyroscopes keep the antenna pointed at the Inmarsat satellite when the ship is rolling with the wave motion. The antenna size of the Inmarsat C is much smaller. Inmarsat also operates an EPIRB system, the Inmarsat L, which is similar to that operated by COSPAS-SARSAT.

The Digital Selective Calling System

Digital Selective Calling (**DSC**) is another terrestrial radio system component of the GMDSS. DSC uses radio frequencies in the Medium Frequency (MF), High Frequency (HF), and Very High Frequency (VHF) radio bands. Since different frequencies have different propagation and range attributes depending upon the time of day, the system is able to cover a wide range of available frequencies for maximum reliability. DSC equipment is installed aboard ships to send distress signals to shore stations.

DSC can be considered to be a replacement for the radiotelephone and radiotelegraph (Morse) alarm signals that would be used on older vessels. Rather than just indicate that the sending station is in distress, the DSC system allows a great deal more information to be transmitted, including:

- Priority of the call — DISTRESS, URGENCY, SAFETY, or ROUTINE

- Position of the ship in distress; and the nature of the distress

- Address

- Identification of the ship in distress

The ITU has allocated a DSC distress and safety channel in the MF, the HF, and the VHF marine radio bands. These are:

- 2187.5 kHz

- 4207.5 kHz

- 6312.0 kHz

- 8414.5 kHz

- 12,577.0 kHz

- 16,804.5 kHz

- VHF Marine Channel 70

> **NOTE**
>
> Voice transmissions are prohibited on the DSC channels.

GMDSS **DSC equipment** is normally comprised of a stand-alone control unit, with an alphanumeric display screen and a keyboard on which to compose messages, as shown in Figure 1-18.

DSC equipment

Figure 1-18: A Typical Digital Selective Calling Console

Search and Rescue Radar Transponder

A device required aboard ships to aid in the location of lifeboats or rafts is called **Search And Rescue (Radar) Transponders (SART)**, as shown in Figure 1-19. A SART is a self-contained, portable, and buoyant Radar Transponder (receiver and transmitter). SARTs operate in the 9 GHz marine radar band, and when interrogated by a searching ship's radar, respond with a signal that is displayed as a series of dots on a radar display. Although SARTs are primarily designed to be used in lifeboats or life rafts, they can be deployed onboard a ship, or even in the water. SARTs are powered by batteries, which provide up to 96 hours of operation.

Figure 1-19: Search and Rescue Radar Transponder

When activated, a SART responds to a searching radar interrogation by generating a swept frequency signal that is displayed on a radar display as a line of 12 dots extending outward from the SART's position along its line of bearing. The spacing between each dot is 0.6 nautical miles. As the searching vessel approaches the SART, the radar display will change to wide arcs. These may eventually change to complete circles as the SART becomes continually triggered by the searching ship's radar.

Technician Licensing for Wireless Marine Communications Systems

Commercial licensing by the FCC for technical personnel for operation, repair, and maintenance of commercial radio transmitting equipment has existed for many years.

Due to advancements in technology within all segments of the wireless telecommunications industry, FCC licensing policy has also changed accordingly. The development of more advanced wireless communications such as the GMDSS has created a demand for experienced and qualified technicians. In this section, the various commercial licensing and FCC examination requirements for GMDSS and other associated skills are explained.

FCC Requirements for Communications Personnel Aboard GMDSS Vessels

The FCC requires two licensed radio operators to be aboard all GMDSS certified ships, one of whom must be dedicated to communications during a distress situation.

The radio operators must be holders of a GMDSS Radio Operator's License. The GMDSS radio operator is an individual licensed to handle radio communications aboard ships in compliance with the GMDSS regulations, including basic equipment and antenna adjustments. The GMDSS radio operator need not be a radio officer.

The IMO requires all masters and mates to hold the GMDSS Radio Operator's License, attend a two-week training course, and demonstrate competency with operation of the GMDSS equipment. These requirements would also apply to any person employed specifically to act as a dedicated radio operator if the ship elected to carry such a position.

Identical to the international GMDSS regulations, the United States FCC regulations provide three methods to ensure that radio equipment is functionally capable of providing communications. The three methods approved for GMDSS ships are (two of the three methods are required for most ocean voyages):

- Shore-based maintenance

- At-sea maintenance

- Duplication of equipment

TEST TIP

Know the methods for ensuring the capability of GMDSS ships.

The GMDSS regulations require that GMDSS ships that choose at-sea maintenance carry a licensed GMDSS radio maintainer. In addition to those individuals holding a GMDSS Radio Maintainer's License, the First or Second Class Radiotelegraph Operator's Certificate and the General Radiotelephone Operator License are acceptable.

Types of FCC Commercial Licenses

Commercial licensing by the FCC for radio operators and radio system maintainers has traditionally been required for all terrestrial and shipboard radio and radar systems that fall within certain classes of operation. In recent years, the requirement for operation/maintenance of land-based commercial stations has been relaxed. Requirements still exist, however, for commercial operator testing and licensing for shipboard and aviation transmitting equipment. The FCC has also maintained the licensing and examination criteria for commercial radiotelegraph operators, primarily due to the Morse code proficiency requirement for ship radio operators. The 1990s saw major advancements in automatic digital radio transmission systems. This is in turn has led to the phasing out of shipboard wireless radiotelegraph systems. In February 1999, the requirement for ships to carry radiotelegraph radio systems ended worldwide, concluding almost a century of "brass pounders" on the high seas. The radiotelegraph operator exam remains for the time being, but may not be supported in the future.

The FCC Wireless Telecommunications Bureau maintains examinations and is the issuing authority for eight types of commercial radio operator licenses and two types of endorsements. Each of these is named below:

- Restricted Radiotelephone Operator Permit
- Marine Radio Operator Permit
- General Radiotelephone Operator License
- GMDSS Radio Maintainer's License
- Third Class Radiotelegraph Operator Certificate
- Second Class Radiotelegraph Operator Certificate
- First Class Radiotelegraph Operator Certificate
- GMDSS Radio Operator License
- Ship Radar Endorsement
- Six Month Service Endorsement

The FCC no longer issues the following commercial licenses or endorsements:

- First Class Radiotelephone Operator License
- Second Class Radiotelephone Operator License
- Third Class Radiotelephone Operator Permit
- Broadcast Endorsement
- Aircraft Radiotelegraph Endorsement (this endorsement will be dropped from any renewed radiotelegraph certificate)

The General Radiotelephone and Radiotelegraph Operator exam element requirements are shown in Table 1-2.

> **NOTE**
>
> The GMDSS licensing requirements are covered in written elements 7 and 9.

TYPE OF LICENSE	WRITTEN ELEMENTS							TELEGRAPHY ELEMENTS			
	1	3	5	6	7	8	9	1	2	3	4
First Class Radiotelegraph Operator Certificate	✓		✓	✓						✓	✓
Second Class Radiotelegraph Operator Certificate	✓		✓	✓				✓	✓		
Third Class Radiotelegraph Operator Certificate	✓		✓					✓	✓		
General Radiotelephone Operator License	✓	✓									
Marine Radio Operator License	✓										
GMDSS Radio Operator License	✓				✓						
GMDSS Radio Maintenance License	✓	✓					✓				
Ship Radar Endorsement						✓					

Table 1-2: Examination Elements for the Various Types of Commercial FCC Licenses

Requirements for Commercial Radiotelephone Operators

Radio Operations

The FCC policy states that a commercial radio operator license is required to operate ship radio stations subject to the following conditions:

- The vessel carries more than six passengers for hire.

- The radio operates on medium or high frequencies.

- The ship sails to foreign ports.

- The ship is larger than 300 gross tons and is required to carry a radio station for safety purposes.

TEST TIP

Know the FCC rules regarding the types of stations requiring a licensed radio operator.

You need a commercial radio operator license to operate the following coastal stations:

- Coast stations that operate on medium or high frequencies, or operate with more than 1,500 watts of peak envelope power

- Aircraft radio stations, except those that use only VHF frequencies on domestic flights

- International fixed public radiotelephone and radiotelegraph stations

- Coast and ship stations transmitting radiotelegraphy. As of February 1999, ships are no longer required to use radiotelegraphy as a mode of communication. This does not imply that onboard radiotelegraph transmitters may be operated past this period without a licensed operator in attendance.

You do NOT need a commercial radio operator license to operate the following:

- Coast stations operating on VHF frequencies with 250 watts or less of carrier power

- Ship stations operating only on VHF frequencies while sailing on domestic voyages

- Aircraft stations that operate only on VHF frequencies and do not make foreign flights

Radio Maintenance and Repair

A radio operator license is required to repair and maintain the following:

- All ship radio and radar stations

- All coast stations

- All hand-carried units used to communicate with ships and coast stations on marine frequencies

- All aircraft stations and aeronautical ground stations (including hand-carried portable units) used to communicate with aircraft

- International fixed public radiotelephone and radiotelegraph stations

A commercial radio operator license is NOT required to operate, repair, or maintain any of the following types of stations:

- Two-way land mobile radio equipment, such as that used by police and fire departments, taxicabs and truckers, businesses and industries, ambulances and rescue squads, and local, state, and federal government agencies

- Personal radio equipment used in the Citizens Band, Radio Control, and General Mobile radio services

- Auxiliary broadcast stations, such as remote pickup stations

- Domestic public fixed and mobile radio systems, such as mobile telephone systems, cellular systems, rural radio systems, point-to-point microwave systems, multipoint distribution systems, etc.

- Stations that operate in the Cable Television Relay Service

- Satellite earth stations of all types

The FCC policy on commercial licensing emphasizes two important points:

- Possession of a commercial radio operator license or permit does not authorize an individual to operate amateur radio stations. Only a person holding an amateur radio operator license may operate an amateur radio station.

- A commercial operator license does not constitute or imply FCC authorization to transmit radio signals. Before a technician may operate any radio station, it is necessary to make certain that the station is licensed as required by the FCC. Citizens Band and Radio Control radio stations do not require individual station licenses.

Key Points Review

In this chapter, the following key points were introduced concerning the history of the discovery and development of wireless communications. Review the following key points before moving into the Review Questions section to make sure you are comfortable with each point and to verify your knowledge of the information.

- The wireless industry we know today actually began at the dawn of the twentieth century. The telegraph and telephone had established a means for sending information over copper wires and spanned great distances; however, they were tethered to locations that could accommodate wires.

- In 1895, a brilliant, young Italian engineer, Guglielmo Marconi (1874-1937), began laboratory experiments at his father's estate in Pontecchio, Italy where he succeeded in sending wireless signals over a distance of one and one-half miles.

- The early use of the term "wireless" referred exclusively to the evolution of the "wired" telegraph. The use of the term "radio" would not appear until after the concept of commercial broadcasting was established by Reginald Fessenden.

- The wireless age would gain new ground with the invention and development of the vacuum tube in England, around 1901.

- Fessenden's alternator provided the transition from using electromagnetic waves for telegraphy to transmitting audio frequency information. Since voice information could now be used on continuous carrier waves, it was logical that the term "radio telephony" would be used to describe this new media. Although the alternator was a major breakthrough in radio technology, it was also very large and expensive to install and maintain. The solution to this problem would come from a laboratory at Columbia University and a bright undergraduate student by the name of Edwin Howard Armstrong.

- Armstrong later developed the concept of Frequency Modulation (FM) that is still used today in the first generation of cellular phones, mobile radio equipment, military radios, and commercial radio broadcasting.

- During his enlistment in the Army during World War I, Armstrong came up with the superheterodyne, a subtle but elegant technique for improving reception and making the radio offer features called selectivity and sensitivity.

- Amplitude Modulation (AM) is a type of modulation used to transport information on a continuous wave radio signal called the carrier. This is the method used to impose an audio sound wave onto a high-frequency radio "carrier wave."

- The Tuned Radio Frequency (TRF) receiver utilized a series of tuned amplifier stages. This was a major improvement over the wideband receiving systems that were used by the spark-gap wireless telegraph systems.

- A problem with TRF was that it was very difficult to design a radio frequency amplifier that was sensitive across the entire radio broadcast band (500 kHz to 1600 kHz).

- Armstrong studied ways to overcome the problems associated with the TRF receiver. He reasoned that reducing the received carrier frequency to a lower frequency and then amplifying the signal would solve the problem.

- Armstrong was interested in solving the problem of static associated with radio reception.

- Armstrong worked for many years on new theories and maintained his determination to eliminate static from radio signals. He was able to show that a dramatically different type of modulation, Frequency Modulation (FM), could be used not only to eliminate the noise, but also to improve the quality of the received audio information.

- After Armstrong's first broadcast FM radio would become the preferred modulation for 2-way mobile communication, and finally in the 1970s it would become the radio interface standard for the first generation of cellular mobile telephones.

- In February 1996, the Telecommunications Act of 1996 was signed into law, representing the first major overhaul of our nation's telecommunications policies in over 60 years.

- The American National Standards Institute (ANSI) has served in its capacity as administrator and coordinator of the United States private sector voluntary standardization system for more than 80 years.

- There are numerous standards organizations and industry support groups that contribute to the development of standards for wireless communications. Each plays an important role in shaping the future of the industry.

- The existence of different standards for similar radio-based technologies in different countries or regions can contribute to so-called "technical barriers to trade" in the wireless industries.

- The wireless industry has historically developed standards primarily on a regional basis.

- International standards and regulations provide the capability for mobile wireless subscribers to "roam" among different wireless service providers' service areas.

- The American National Standards Institute (ANSI) was founded in 1918. ANSI is a private, nonprofit membership organization. The ANSI organization does not develop standards itself; rather it works with other standards groups such as the TIA to obtain agreements on standards across various wireless support organizations.

- Accredited by the American National Standards Institute (ANSI), the Telecommunications Industry Association (TIA) is a major contributor of voluntary industry wireless standards.

- The International Organization for Standardization (ISO) is a worldwide group of national standards bodies from 130 countries, one from each country.

- One of the most important organizations involved in the coordination of global standards is the International Telecommunications Union (ITU).

- Since the early 1980's, wireless telecommunications technology has advanced rapidly for land-based subscribers. Cellular telephones and wireless Personal Digital Assistants (PDAs) are widely available to everyone in the more populated centers of every nation in the world.

- The United States Federal Communications Commission (FCC) regulates all use of radio onboard any recreational, commercial, state and local government, and foreign vessel in U.S. ports and waters.

- The U.S. Coast Guard (USCG) regulates the requirements and procedures for radio transmitting and receiving equipment on commercial fishing vessels and foreign vessels in U.S. waters.

- The huge loss of lives caused by the sinking of the Titanic in 1912 triggered a concerted effort among nations of the world to adopt maritime wireless procedures for all ocean-going vessels.

- The Global Maritime Distress and Safety System (GMDSS) consists of several cooperative terrestrial and satellite radio communications systems.

- COSPAS-SARSAT is the acronym for two separate satellite systems used for search and rescue. COSPAS is an acronym for the Russian words "Cosmicheskaya Sistyema Poiska Avariynich Sudov," meaning Space System for the Search of Vessels in Distress. SARSAT stands for Search and Rescue Satellite-Aided Tracking.

- Navigational Telex (NAVTEX) is an international, automated system for instantly distributing maritime navigational warnings, weather forecasts and warnings, search and rescue notices, and similar information to ships.

- The Inmarsat satellite system was founded in 1979 as the International Maritime Satellite Organization (Inmarsat) to provide worldwide mobile communications capability for marine customers.

- INMARSAT was formerly an acronym for the International Maritime Satellite Organization. After the organization name was changed to the International Mobile Satellite Organization, the Inmarsat name remained although no longer as an acronym. It is therefore spelled in lowercase letters.

- The FCC requires two licensed radio operators to be aboard all GMDSS certified ships, one of whom must be dedicated to communications during a distress situation.

The following questions test your knowledge of the material presented in this chapter:

1. What would you conclude was the most important discovery or invention relating to wireless technology from the information described in this chapter?

2. Describe three advantages of the superheterodyne receiver over the TRF receiver.

3. Describe the purpose of a discriminator circuit.

4. What type of modulation described in this chapter is still used in today's cellular telephone networks?

5. What institute within the FCC is responsible for administering cellular and PCS services?

6. What law provides protection against illegal monitoring of wireless communications?

7. What piece of legislation, among several new initiatives, provided for the opening up of the long distance market to competition in the U.S.?

8. List three advantages of the use of global wireless standards.

9. Briefly describe the term "roaming" as applied to mobile wireless subscribers.

10. What organization is also known by the short name of "ISO"?

11. What type of GMDSS equipment is used to enhance the radar display of a search and rescue surface vessel engaged in a search operation?

12. What GMDSS system has the primary mission for transmitting weather warning information to ships? What frequency is used for such transmissions?

13. What facility within the GMDSS system is responsible for directing SAR resources to the correct location?

14. What type of GMDSS equipment uses the Medium Frequency (MF), High Frequency (HF), and Very High Frequency (VHF) radio frequency bands for ship-to-shore communications?

15. What two international treaty organizations sanctioned by the United Nations are responsible for setting standards for the GMDSS?

CHAPTER 2

WIRED NETWORK ARCHITECTURES

LEARNING
OBJECTIVES

LEARNING OBJECTIVES

In this chapter, you will gain an understanding of the various basic configurations of wired computer data networks. Upon completion of this chapter, you should be able to perform the following tasks:

1. Describe the four main topologies used in computer networks.

2. Describe the difference between peer networks and server-based networks.

3. Develop a list of advantages of wireless networks compared to the characteristics of wired networks.

4. Discuss the features and basic concepts of the OSI 7-layer standard reference model.

5. Describe the differences between the DOD reference model and the OSI reference model, including the tasks performed by each of the layers.

Wired Network Architectures

INTRODUCTION

In this chapter, the basic structure and characteristics of wired networks are explained. The reasons for including a section on wired networks when the subject of this book is wireless communications are as follows:

- Wired networks provide the main transport for data and voice communications throughout the world. Wireless access to both public and private networks for mobile users is a new exploding technology.

- In order to understand how the two technologies are merging and working to become interoperable with each other, you first must become familiar with the basics of wired networks.

- In this chapter, you will become familiar with the types of public and private networks based upon their geographic definition as well as the types of services provided by each type of network architecture.

- The advantages of emerging wired local area networks (LANs) that will be explained

The 7-layer standard reference model, covered in this section, is a fundamental principle on which all effective communications is handled in the wired and wireless network world. It is often referred to as the primary reference model for all networks.

The Open Systems Interconnection (OSI) model was created in the early 1980s by the International Organization for Standardization. The OSI model is synonymous with the 7-layer reference model. ISO is often (incorrectly) referred to in various publications as an acronym for International Standards Organization.

Another reference model you will need to know about when dealing with network terminology is the Department of Defense (DOD) reference model. The DOD model has four, instead of seven layers. The DOD reference model is derived from the work of the early developers of the ARPANET, which later became the Internet we use today. The DOD reference model was designed to provide a layered model for the TCP/IP suite of protocols.

The top three layers of the OSI model are consolidated into a single Application layer in the DOD model. Also, the OSI Transport layer is called the Host-to-Host layer in the DOD model and the OSI Network Layer is called the Internet layer in the DOD model. The bottom two layers of the OSI model are merely referred to as the Network Access layer in the DOD model.

You will need to understand these differences in terminology when dealing with the DOD model and the TCP/IP protocol stack vs. the OSI reference model.

Definition of a Computer Network

A **computer network**, in its simplest form, can be defined as two or more computers connected together with the ability to share data or peripheral equipment.

Although computer networks are typically more complex than this basic definition suggests, they are all extensions of this fundamental concept. Computers and other components, such as networking software, printers, routers, bridges, switches, and repeaters, can be connected in a variety of configurations. In networking language, these configurations are referred to as topologies.

Computers may be connected using various types of media. The type of connection depends on the specific environment and the needs of the organization.

Networks can be connected using copper wire, fiber-optic cable, microwave radio, wireless network interface cards, or infrared light waves. The list of advantages for using wireless as the media for connecting network components is growing, and wireless should become a preferred technology as the costs continue to decline for wireless equipment. A more detailed description of wireless local area networking technologies will be given in Chapter 6. The purpose here is to examine some fundamental concepts of wired and wireless networks.

The real motivation for connecting computers together is the need to improve productivity and share data and computer peripheral devices among members of an organization. A stand-alone computer, like the one depicted in Figure 2-1, has no connection to any other computer; therefore, it is limited to running applications software for a single user. Stand-alone computers do not have the capability for sharing data or peripherals on a network. A stand-alone computer would rarely be of much value in any organization with more than a dozen employees. Stand-alone computers in a corporation would not be able to benefit from the large offering of new applications designed to share information and increase the productivity of the organization. We will examine how efficiencies and productivity enhancements can be realized using computer networks in the following sections of this chapter.

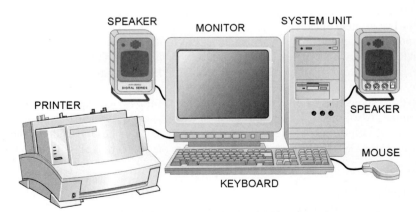

Figure 2-1: A Stand-Alone Computer

Types of Computer Networks

Computer networks are usually classified by the physical arrangement and distance they serve as well as the type of services they are designed to provide. Three types of networks classified according to physical size are Local Area Networks (LANs), Metropolitan Area Networks (MANs), and Wide Area Networks (WANs).

Networks are also defined by the type of services they provide and the type of operating systems that are installed on the network computers. Although they may be configured in different physical arrangements, LANs are typically either peer-to-peer networks or server-based networks when viewed by the type of service provided. Server-based networks are sometimes called client-server networks. In the next few paragraphs you will learn about the physical and logical definitions of computer networks. First, let's look at the definitions for each type of network.

Local Area Networks

Early generation networks were expensive and consisted primarily of large mainframe computers connected to "dumb" terminals. The development of the desktop personal computer dramatically changed how networking was conducted by corporations and also opened the opportunity for worldwide use of the Internet. The "smart" desktop computer changed network computing from a centralized architecture to a distributed architecture in which the processing tasks are divided between the central server machine and the client machines.

Local Area Networks (LANs) are composed of several personal computers connected together and are limited in scope to a single building or facility.

Local Area Networks (LANs)

LANs are the simplest and cheapest of the three types of networks mentioned earlier. Figure 2-2 shows a typical network connection scheme using a router. The implementation of Ethernet and its subsequent standardization by the Institute of Electrical and Electronic Engineers (IEEE) made LANs less expensive. With the rapid development of network operating system software coupled with advances in network interface cards for personal computers, networking has become an essential part of every organization throughout the world. LAN networks use media and hardware designed for relatively fast transmission speeds with very low bit error rates over short distances.

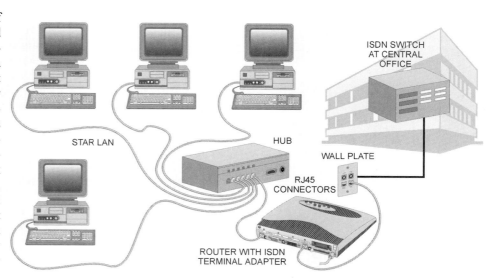

Figure 2-2: A Local Area Network

Ethernet has become the preferred technology for LANs because it has been able to evolve over several years as a standard through several "fast" Ethernet upgrades from 10 Mbps, 100 Mbps, and 1 Gbps transmission rates. Ethernet has proven to be a very scalable technology. It continues to thrive as a preferred LAN technology, whereas competing technologies have virtually disappeared from the market.

Metropolitan Area Networks

A **Metropolitan Area Network (MAN)** encompasses an area larger than a LAN. It serves larger areas within a city or, as the name implies, a major metropolitan area.

An example is fire, police, emergency response teams and 911 emergency answering locations that would individually be configured as LANs but connected together in a single metropolitan region to form a MAN. The distinction between a MAN and a WAN is narrow. MANs are generally confined to a single metropolitan area while WANs can cover much larger geographical areas.

Wide Area Networks

Several LANs or MANs may need to be connected together over longer distances. An example would be a corporate headquarters located in a city utilizing **Wide Area Network (WAN)** connections to several branch offices located in other cities or other countries.

Figure 2-3 depicts a wide area network.

LANs are usually owned and maintained by private corporations. WANs and MANs are more expensive and are commonly owned by large communications service companies since the cost to individual private corporations would be prohibitive.

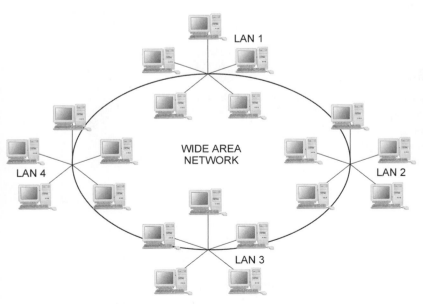

Figure 2-3: A Wide Area Network

WAN technologies use media such as fiber-optic cable, satellite networks, and fixed wireless microwave radio systems to move data over long distances. Several high-speed WAN protocols have been standardized to permit the sharing of circuits among multiple organizations. Popular WAN transmission systems include Wavelength Division Multiplexing (WDM) over fiber-optic cable, Asynchronous Transfer Mode (ATM), and Frame Relay.

> The **Internet** is actually a network of networks and is the infrastructure that makes up a large share of the WANs on a global scale.

Internet

A typical WAN is a local city- or countywide network, like the one in Figure 2-4.

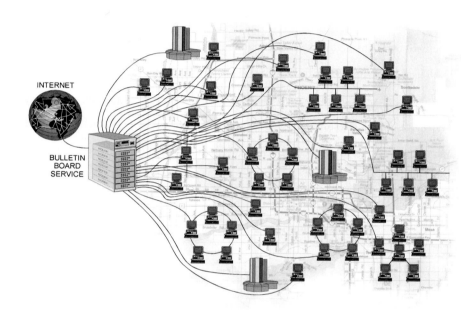

**Figure 2-4:
County-Wide Network**

> **Intranets** are networks that provide connections between different facilities within a single corporation, government organization, or university campus.

Intranets

With the availability of Internet type tools and protocols, it is feasible and economical for organizations to provide controlled private networks (intranets) using Internet protocols and standards to allow easy access to corporate resources for employees.

Peer-to-Peer Networks

As described earlier, local area networks can be categorized by the type of services they provide to the using organization. In the case of peer-to-peer vs. server-based networks, they are defined primarily by the software running on the network computers.

> **Peer-to-peer networks**, as the name suggests, is where all computers on the network are peers and have equal status concerning sharing rights and capabilities.

Peer-to-peer networks

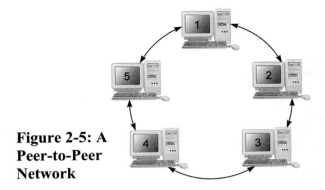

Figure 2-5: A Peer-to-Peer Network

Figure 2-5 illustrates a typical peer-to-peer network arrangement. Any individual machine can share data or peripherals with any other machine on the same local area network. No other computer on the network has a higher priority or rank than any other computer. Each computer user can act as his or her own network administrator since the degree of sharing resources is not centrally managed. A computer may act as a client or a server depending upon the specific transaction called for by each peer user. Peer-to-peer networks are suitable for small groups of 10 computers or less, but are not practical for larger groups.

Some advantages to peer-to-peer networks are:

- They are inexpensive to install and maintain.

- Each user manages all the resources on his or her computer.

- No central network administrator is required.

- The operating system software is less expensive and no server computer is required.

- Minimal training is needed for individual users to become proficient using peer-to-peer networks.

Some disadvantages to peer-to-peer networks are:

- They do not support good security management. Every access to all shared network resources is determined by individual peer users.

- They have very limited growth potential.

- Users have to keep track of many passwords as the sharing of resources expands among peers.

- Backing up critical data that is shared is required for all computers.

Server-Based Networks

> **Server-based networks** are often referred to as **client/server networks**. The popular trend has been towards server-based as the proper term since it focuses on the central component called the server.

Server-based networks

client/server networks

All other machines are clients or workstations, which obtain services and access authority from the server (or servers) on the same network. Server networks are used extensively by institutions because security has become an important issue in practically all business and government computer environments. Servers and server-based operating systems offer superior security for the protection of files when compared to peer-to-peer networks. Server-based networks divide processing tasks between the server and the client machine. Figure 2-6 depicts a typical client/server LAN configuration. Operating system software is structured to provide the division of tasks between the server and client computers on a LAN.

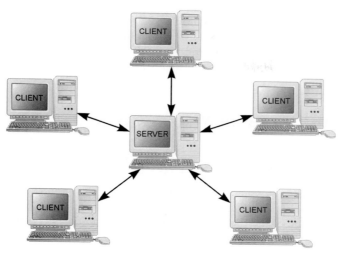

Figure 2-6: A Client/Server Network

Servers can be added as the number of users in an organization grows. Server-based networks are very scalable. They can handle tens or thousands of users as the organization expands. Specialized servers can be arranged to handle specific tasks such as communications, database management, or applications so that greater control can be maintained for critical information. Access control can be implemented on a user-by-user or group-by-group basis with a user ID and password on a hierarchical basis. As an example, users can be divided into groups such as payroll information, corporate financial data, or personnel records so that access can be managed from a server.

Some advantages of server-based networks are:

- They offer better security.

- They are very expandable.

- They provide a centralized location for backing up important files.

- They offer a wider range of services.

Some disadvantages of server-based networks are:

- They are more expensive to install and maintain.

- The operating system software for server-based networks is more expensive than peer-to-peer system software.

- Server-based networks require specially trained personnel to manage the network resources.

Network Services

All computer networks have a common objective: to provide services for the users.

As the technology for LAN software and hardware continues to mature, vendors are finding new applications to aid in improving productivity for the modern office. Let's examine some of the basic services offered by LANs.

File Sharing and File Management

This service helps ensure that files can be protected by limiting the access to authorized users and assuring that only one version of the file exists. Centralized management on a file server also protects against loss by scheduling backups. Files often represent the most valuable resource in an organization, therefore file sharing is not the only task for the network. Management and safeguarding of the file system is also an important service that is provided by properly designed local area networks.

> The purpose of file sharing on a network is to provide a common location or storage area (typically on a file server) where a number of users can have access to files.

Sharing Peripherals

In addition to sharing files, LANs provide a method for sharing peripheral devices. For example, it would not be cost effective to provide a dedicated printer for each computer on a network. But it is possible to connect a printer to a single computer on the network and share it with other members. Although printer sharing can be effective on peer-to-peer networks, a more efficient method is to have a dedicated print server on the network. This provides a higher throughput for print tasks and also allows a centralized access control to be maintained for workstations that need to share the network printer.

E-mail

Electronic mail
(e-mail)

Electronic mail (e-mail) is one of the most popular applications for computer networks. E-mail provides a way to reach users on a network when they are away from their computer. Distribution lists can be maintained so that several people in an organization can receive important bulletins at the same time. E-mail messages are transmitted to a user's e-mail address, which is a mail server located on the user's network. Members can then access their mailbox on a server by entering the correct password. This method of communication also enables members of the network to reach the mail server and read their mail even if they are away from their home location. This requires a remote access capability via a dedicated private network (or more often the public Internet) to access a member's home network mail server.

Communications Servers

Communication servers enable network users to dial in to a corporate network from a remote location while on travel status or working at home. Novell and Microsoft network operating systems provide support for **Remote Access Servers (RAS)**.

A RAS can support multiple connections to the network at the same time. With dial-in access, authorized users can utilize network resources over phone lines.

Communication servers

Remote Access Servers (RAS)

Applications Servers

Applications servers are different from the other types of servers discussed previously. Client machines generally must run client-side applications. In this type of service, the client typically generates requests and sends them to the application server.

The applications server handles all of the background processing of the request and delivers the results back to the client. The client software then formats the results and displays them for the user.

Applications servers

Basic Design Concepts

Network design is the first step an organization must undertake before proceeding with the installation of network components.

Although this sounds obvious and straightforward, a surprising number of organizations fail to adequately plan for the future. This often results in numerous add-ons and extensions to the basic architecture because the requirements for performance of the network are not documented and reviewed by qualified network design engineers. In order to recognize some of the basic design features and topologies, let's examine some of the common characteristics of network design. You will learn about network terms, topologies, and design options in the following paragraphs.

The Mesh Network Topology

A **mesh network topology** provides a multitude of connections between individual nodes, as shown in Figure 2-7.

mesh network topology

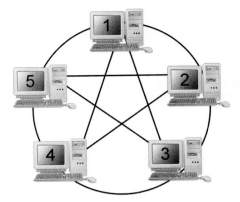

The Internet as well as most WANs are examples of the mesh topology. Mesh networks provide a high degree of redundancy that permits more than one path to exist between any two network nodes. Routers are used in mesh networks to select the best routes for data and voice circuits between source and destination addresses. Traffic can be switched to different routes in the network based upon traffic loads and outages that may occur between nodes.

Figure 2-7:
A Mesh Network

The Bus Network Topology

The **bus network topology** (also sometimes referred to as a linear bus) is formed by a simple connection between each computer.

The topology is named for the single cable or "bus" that provides a common connector for all devices. The bus forms the backbone or access point, which by definition carries all the traffic between computers on the network.

Only one computer at a time can send messages on a bus. In future sections, you will learn how protocols provide a set of rules that permit each computer to transmit information on a time-shared basis with other users on the network. Bus topologies are simple to install but suffer from a potential single point of failure since a broken cable will disable the entire network.

┌─ **TEST TIP** ─────
Know the purpose and need
for a terminator.
└────────────────

A terminator must be connected to each end of the bus to prevent reflections. Messages reflected from an improperly terminated bus can interfere with the network by allowing signals to "bounce" between the ends of the bus. The locations of the terminators and the bus architecture are shown in Figure 2-8.

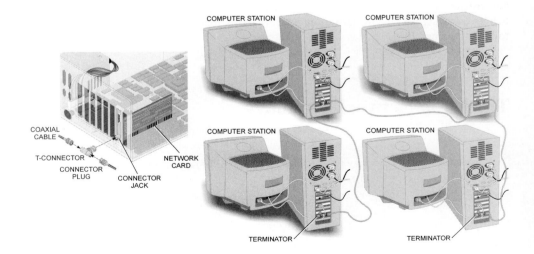

**Figure 2-8: A Bus
Network**

The bus architecture was used in an earlier generation of LANs. The bus architecture uses coaxial cable, which was later standardized as Ethernet 10Base-2 (Thinnet). Ethernet network features are covered later in Chapter 3.

The Star Network Topology

A **star network topology** as shown in Figure 2-9 connects all computers to a single central point called a hub.

star network topology

The central point is served by a hub that acts as a repeater. The star topology provides the same universal connection of each computer to all other computers, as is the case for the linear bus but with several advantages. For example, a faulty cable only affects one computer and is easier to troubleshoot than in the bus topology.

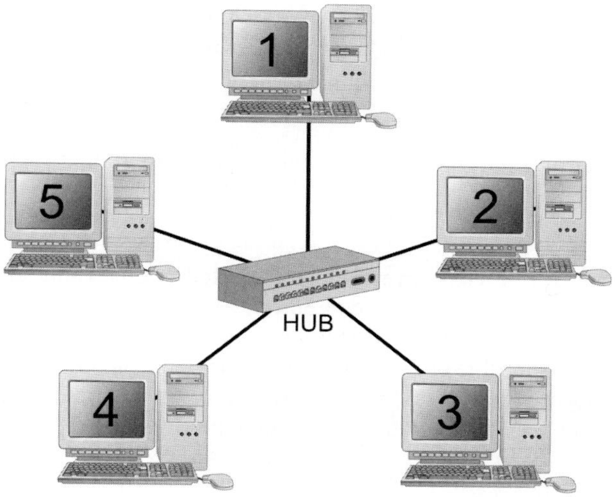

Figure 2-9:
A Star Network

The hub also provides the proper termination for each network computer, thereby eliminating the signal reflection problem associated with the bus architecture. The hub can also provide a central control point where each computer is allowed to transmit by a central hub controller. The hub may also act as only a common termination point with no specific control over network access for the network computers. Star topology became the preferred topology as twisted pair cable became the media of choice over coaxial cable. Twisted pair cable is easily configured for star topologies. (Twisted pair cable is also the basis for the 10Base-T IEEE Ethernet standard.)

The Ring Network Topology

A **ring network topology** is formed by a single conductor or backbone connecting each computer to form a circle or ring.

ring network topology

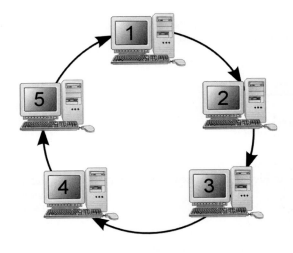

As each computer receives a message, it examines the information to determine whether it is the intended recipient, then acts on it, regenerates the signal, and passes it on to the next computer on the ring. A ring topology is shown here in Figure 2-10.

IBM developed the token ring network. It was later standardized by the IEEE, which popularized the ring topology and provided a high degree of reliability with several built-in diagnostic features.

Figure 2-10: A Ring Network

Hybrid Network Topologies

Usually, in a typical LAN installation, the topologies described above are configured as hybrids.

> The combination of a star topology and a bus topology is a popular form of **hybrid network**.

This configuration is composed of several hubs connected by a backbone circuit. Each hub is connected in a star topology to several computers, as shown in Figure 2-11. This type of network allows several star/hub combinations to exist in an office building or campus environment. The coaxial cable or fiber-optic cable forms the bus connection between each hub. Hubs may be located conveniently on separate floors of an office building with the bus running vertically between floors to form the backbone circuit.

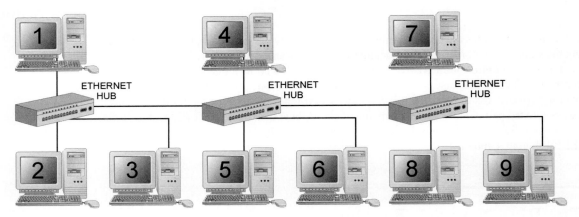

Figure 2-11:
A Hybrid Network

Limitations of Wired LANs

The preceding paragraphs have shown the basic design features of **wired networks** that are representative of most business LANs. The widespread use of computer networks on a global basis has developed very rapidly. It is almost taken for granted now that no business can succeed without modern data communications using desktop and mainframe computer connectivity between the business units.

The Internet has become a primary driving force for global communications. New technology for mobile wireless handheld data terminals in turn is pushing the fixed wired network into a new era. Wireless customers now want the features they have become accustomed to in wired networks. There is increasing demand for handheld wireless phones and wireless laptop computers.

Access to corporate assets and home networks via the Internet for mobile and home users is in demand worldwide. Network administrators also are now able to have the flexibility afforded by wireless LAN equipment. It is proper therefore to now take a closer look at how this transition to wireless LAN technology is occurring and also at the limitations of the wired network.

- Wired LANs require permanent cable runs within walls and flooring. Relocating personnel often requires expensive overhaul of the cable plant.

- Users are locked to the network point of attachment for their desktop or laptop computer. There is no mobility of computer resources within the office environment.

- Installation or relocation of wired networks is time consuming.

- Maintenance and testing of the cable plant is a recurring expense item.

- Wired LANs are limited to where cables can be installed. If a building is configured without cable conduit or cable ducts, it is expensive and time consuming to modify the building for cable installation and routing.

Advantages of Wireless LANs

Although many advantages are now available to the wireless user, wireless LANs are most effective for extensions to existing LANs.

Wireless LAN extensions may be used for specific applications such as medical or healthcare in which medical staff members need network mobility as a method of improving efficiency in a hospital or emergency response center. Warehouse personnel in a corporation may need the mobility feature for exchanging shipping and receiving information between sales or accounting personnel. The wireless network extension accesses the wired network via an access point interface device.

Other advantages of wireless LANs are:

- Reduced time for installation. Because wireless LANs have no wires or cable trays to install, they can be brought on line in much less time than wired LANs.

- Wireless LANs provide mobility for the network users. Ad hoc networks can be implemented on an as-needed basis for specific business meetings, conferences, or special projects.

- Wireless LAN hardware is rapidly becoming less expensive due to advances in technology and mass market growth.

- Wireless LANs now have the benefit of an IEEE standard for hardware (IEEE 802.11). This has accelerated acceptance of and confidence in the wireless alternative for network managers and administrators. Interoperability among various products that meet the standard is now assured.

- Training and classroom sites can be rapidly set up for distance learning and classroom conferencing.

The OSI Reference Model

The **OSI reference model** divides the problem of moving information between computers over a network into seven smaller and more manageable areas, as shown in Figure 2-12.

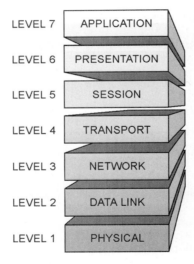

Figure 2-12: OSI Reference Model

LEVEL 7 — APPLICATION
LEVEL 6 — PRESENTATION
LEVEL 5 — SESSION
LEVEL 4 — TRANSPORT
LEVEL 3 — NETWORK
LEVEL 2 — DATA LINK
LEVEL 1 — PHYSICAL

Each of the seven smaller areas was chosen because they were reasonably self-contained and did not require any reliance on external information.

We can conclude therefore that moving information from an application running on a computer to another computer running on another network is a complex task. The OSI model simplifies greatly the tasks performed during the transfer by breaking them down into distinct and smaller subtasks. Each layer is able to perform a set of tasks independently of the layer above or below it. As a result, vendors may be able to supply software or hardware products that work independently at one or more of the layers. The lower two OSI layers are typically implemented with hardware and software; the upper five layers are generally implemented in software.

Protocols are the rules that govern the transmission and control of the flow of information between network nodes.

When referencing the tasks performed by each layer of the OSI model in this chapter, it is understood that a set of protocols (a **protocol stack**) actually perform the tasks at the various levels of the model. As a result, technicians and engineers will often refer to a "layer 4 protocol" or a "Transport layer protocol."

The OSI reference model describes how information makes its way from an application program through a network medium to another application program in another computer. In the next section, the tasks allocated to each layer are explained.

Table 2-1 shows a summary of the tasks performed at each layer of the OSI reference model.

Layer	Function
Application	The Application layer provides the interface for all applications that need access to network resources.
Presentation	The Presentation layer has the job of determining how data should be formatted between the sender and receiver.
Session	The Session layer establishes, manages, and terminates sessions between applications running on different host computers on a network.
Transport	The Transport layer provides the necessary quality of service (QoS) required by the Application layer. This layer has the flexibility to utilize either connection-oriented (CO) protocols or connectionless (CL) protocols. The Transport layer supplies the upper layers with a virtual path, from transmitter to receiver, by assigning end users an address without regard to their position in the network.
Network	The Network layer is responsible for the routing of packets completely through the network from the source to the destination.
Data Link	The Data Link layer provides point-to-point link management and access to the network.
Physical	The Physical layer defines the physical and electrical characteristics of the interface to the network. The Physical layer is the first layer of the OSI reference model.

Table 2-1: Basic Functions of the OSI Reference Model Layers

The Application Layer

The **Application layer** provides the interface for all applications that need access to network resources. It is the reason for the existence of all the other layers.

Some examples of applications are:

- A client/server transaction occurring across a network

- Downloading a file from the Internet

- Performing some type of network management application

- A client computer accessing an e-mail server

- Mail, ftp, telnet, DNS, NIS, and NFS

The Application layer identifies and establishes the availability of intended communication partners and establishes agreement on procedures for error recovery and control of data integrity. The Application layer also determines whether sufficient resources for the intended communication exist.

The Presentation Layer

Presentation layer

The **Presentation layer** provides a variety of coding and conversion functions that are applied to Application layer data.

This layer has the job of determining how data should be formatted between the sender and receiver. These tasks involve the management of various formats used in computer systems to ensure that information sent from the Application layer of one system is readable by the Application layer of another system. Some examples of Presentation layer coding and conversion schemes are:

- *Data representation formats.* The use of standard image, sound, and video formats allow the interchange of application data between different types of computer systems.

- *Conversion of character representation formats.* Conversion schemes are used to exchange information with systems using different text and data representations (such as EBCDIC and ASCII).

- *Data compression schemes.* The use of standard data compression schemes allows data that is compressed at the source device to be properly decompressed at the destination.

- *Data encryption schemes.* The use of standard data encryption schemes allows data encrypted at the source device to be properly unencrypted at the destination.

The following multimedia standards are used at the Presentation layer:

- **Joint Photographic Experts Group (JPEG):** A common graphic standard compression scheme

- **Motion Picture Experts Group (MPEG):** A coding and video compression scheme for motion

- **QuickTime:** An Apple Computer specification that runs video and audio clips

- **Graphics Interchange Format (GIF):** A standard for compression and coding of graphic images

- **Tagged Image File Format (TIFF):** A standard coding format for graphic images

- **Musical Instrument Digital interface (MIDI):** A standard used for recording and playing digital sounds and music

The Session Layer

Session layer

> As its name implies, the **Session layer** establishes, manages, and terminates sessions between applications running on different host computers on a network.

Sessions are dialogues between two or more presentation entities and consist of service requests and service responses. This layer organizes communications between two end points by offering three different modes of transmission:

- Simplex
- Half duplex
- Full duplex

The Session layer sets up a session in three phases. They are the connection establishment, data exchange, and connection release. In addition to basic regulation of conversations (or sessions), the Session layer offers provisions for data expedition, class of service, and exception reporting of Session-layer, Presentation-layer, and Application-layer problems.

The Transport Layer

Transport layer

> The **Transport layer** provides the necessary quality of service (QoS) required by the Application layer.

This layer has the flexibility to utilize either connection-oriented (CO) protocols or connectionless (CL) protocols. Whatever mode is used by the Transport layer, it is up to the lower level functions to carry out the demands, whatever they may be. Some applications encountered in networks require CO services while other applications do not. In common terminology, CO service is referred to as "reliable" transport service while CL service is referred to as "unreliable" or "best effort" service.

Transport-layer functions, when called for by the upper layers, may include some or all of the following CO or reliable functions:

- Flow control manages data transmission between devices so that the transmitting device does not send more data than the receiving device can process. The receiving node uses this feature to prevent the buffers in the receiving device from overflowing and creating the potential for lost packets.

- Multiplexing allows data from several applications to be transmitted onto a single physical link.

- Virtual circuits are established, maintained, and terminated by the Transport layer.

- Error checking involves various mechanisms for detecting transmission errors. Error recovery involves taking an action (such as requesting that data be retransmitted) to resolve any errors that occur.

The Network Layer

The **Network layer** is responsible for the routing of packets completely through the network from the source to the destination.

As you will see, the Data Link and the Physical layer are local. They are associated with moving data on a local area network or from point to point across a radio interface.

The Network layer contains the address, which may carry the packet across the Internet on a path that may involve several router hops across thousands of miles. The global scope of the Internet is often represented graphically as a cloud. The network address is a logical address (as opposed to the physical address used at the lower layers).

The Network layer, like the Transport layer above it, supports both connection-oriented and connectionless service.

The Data Link Layer

The **Data Link layer** controls access to the Physical layer.

This layer encapsulates the upper layer message with a data link frame header that includes the hardware source and destination address. It provides the point-to-point data link control on the physical medium.

The Physical Layer

The **Physical layer** consists of the bits that are transmitted from one place to another.

The Physical layer also defines electrical signaling on the transmission channel; how bits are converted to electrical current, light pulses, or some other physical form. This conversion is called encoding. Encoding of binary data into analog or other representations of the state of a one or a zero is the task performed by the Physical layer. The encoding of binary data into an analog signal involves the controlled fluctuation of the signal amplitude, frequency, phase, or a combination of these attributes. The capture and conversion of encoded signals to data bits when receiving packets of information is also performed by the Physical layer. Different encoding schemes are used by different Physical layer standards. One type of signaling is called polar signaling.

The OSI model makes no distinction concerning the actual hardware involved. The Physical layer includes any and all connectors, hubs, repeaters, transceivers, and network interfaces, as well as the cable. The Physical layer is not concerned with packets, frames, addressing, or the ultimate destination of the data. It is only concerned with the transmission of data bits on the transmission media.

The DOD Reference Model

The **DOD reference model** described earlier contains four layers compared to the OSI 7-layer model. This model was developed in the early 1970s by the developers of the ARPANET. The TCP/IP layered architecture, which is the reason we have the DOD model, is used to define the stack of protocols commonly referred to as the TCP/IP protocol suite. Table 2-2 compares the DOD model with the OSI model.

DOD reference model

DOD Reference Model	OSI Reference Model
Process/Application	Application, Presentation, and Session
Host-to-Host	Transport
Internet	Network
Network Access	Data Link and Physical

Table 2-2: Comparing the DOD Model to the OSI Model

The common features of both the DOD model and the OSI model are summarized as follows.

- The Process/Application layer is a consolidation of the OSI Application, Presentation, and Session layers.

- The Host-to-Host layer controls packet traffic on the logical link, which is the complete end-to-end connection between devices at the source and destination addresses. This layer is compatible with the OSI model Transport layer. The name of this layer is derived from the terminology used by the early ARPANET developers in which all computers were large mainframes or hosts.

- The Internet layer performs the same functions as the OSI model Network layer. It handles all of the addressing and routing functions between network segments.

- The DOD model consolidates the OSI model Data Link and Physical layers into the Network Access layer. This layer provides the physical interface to the network medium as well as the sharing of access to the medium by network devices.

Defining Frames and Packets

Some confusion and inconsistency exists in network texts and literature regarding the use of the terms packet, frames, protocol data unit (PDU), message, and segment. Each of these terms is usually associated with the logically grouped units of information at various layers in the OSI model. As data units are moved, processed, and encapsulated at the various levels, different names are often associated with the information as it travels through each layer. Throughout this book the following definitions will apply.

The term **frame** denotes an information unit at the Data Link layer. The term **packet** describes an information unit with a source and destination address that exists at the Network layer.

┌─ **TEST TIP** ─────────────────
Know the difference between frames and packets.
└────────────────────────────────

The term **message** will always denote an information unit that exists above the network layer. While the remaining terms are often used at various layers, the three layers referenced here are the most widely used.

How Packets Are Delivered on a Network

Packets can be different in size depending on the network type and their function. It is essential that data be transmitted in some form of discrete unit so that there is a measure of empty space on the network media between transmissions. (The protocol stack manages this requirement.)

By definition, every station must be allowed to transmit its own data at some time. In other words, each station must gain access to the network through some orderly process. This task, as we noted earlier, is performed and managed by the Data Link layer.

All protocols manage to restrict the length of packets to some agreed upon limit. This assures that data can be transmitted in small units with a discrete interval controlled by the Data Link layer. Routers handle packets and route them through the packet-switched network by reading the destination address information contained in the header of each packet.

Key Points Review

In this chapter, the following key points were introduced concerning the basic structure and characteristics of wired networks. Review the following key points before moving into the Review Questions section to make sure you are comfortable with each point and to verify your knowledge of the information.

- A computer network, in its simplest form, can be defined as two or more computers connected together with the ability to share data or peripheral equipment.

- Computers may be connected using various types of media. The type of connection depends on the specific environment and the needs of the organization.

- Computer networks are usually classified by the physical arrangement and distance they serve as well as the type of services they are designed to provide. Three types of networks classified according to physical size are Local Area Networks (LANs), Metropolitan Area Networks (MANs), and Wide Area Networks (WANs).

- Local Area Networks (LANs) are composed of several personal computers connected together and are limited in scope to a single building or facility.

- A Metropolitan Area Network (MAN) encompasses an area larger than a LAN. It serves larger areas within a city or, as the name implies, a major metropolitan area.

- Several LANs or MANs may need to be connected together over longer distances. An example would be a corporate headquarters located in a city utilizing Wide Area Network (WAN) connections to several branch offices located in other cities or other countries.

- LANs are usually owned and maintained by private corporations. WANs and MANs are more expensive and are commonly owned by large communications service companies since the cost to individual private corporations would be prohibitive.

- The Internet is actually a network of networks and is the infrastructure that makes up a large share of the WANs on a global scale.

- Intranets are networks that provide connections between different facilities within a single corporation, government organization, or university campus.

- Peer-to-peer networks, as the name suggests, are networks in which all computers on the network are peers and have equal status concerning sharing rights and capabilities.

- Server-based networks are often referred to as client/server networks. The popular trend has been towards server-based as the proper term since it focuses on the central component called the server.

- All computer networks have a common objective: to provide services for the users.

- The purpose of file sharing on a network is to provide a common location or storage area (typically on a file server) where a number of users can have access to files.

- Network design is the first step an organization must undertake before proceeding with the installation of network components.

- A mesh network topology provides a multitude of connections between individual nodes.

- The bus network topology (also referred to as a linear bus) is formed by a simple connection between each computer.

- A star network topology connects all computers to a single central point called a hub.

- A ring network topology is formed by a single conductor or backbone connecting each computer to form a circle or ring.

- The combination of a star topology and a bus topology is a popular form of a hybrid network.

- Although many advantages are now available to the wireless user, wireless LANs are most effective as extensions to existing LANs.

- The OSI reference model divides the problem of moving information between computers over a network into seven smaller and more manageable areas.

- Protocols are the rules that govern the transmission and control of the flow of information between network nodes.

- The Application layer provides the interface for all applications that need access to network resources. It is the reason for the existence of all the other layers.

- The Presentation layer provides a variety of coding and conversion functions that are applied to Application layer data.

- As its name implies, the Session layer establishes, manages, and terminates sessions between applications running on different host computers on a network.

- The Transport layer provides the necessary quality of service (QoS) required by the Applications layer.

- The Network layer is responsible for the routing of packets completely through the network from the source to the destination.

- The Data Link layer controls access to the Physical layer.

- The Physical layer consists of the bits that are transmitted from one place to another.

- The term frame denotes an information unit at the Data Link layer. The term packet describes an information unit with a source and destination address that exists at the Network layer.

- Packets can be different in size depending on the network type and their function. It is essential that data be transmitted in some form of discrete unit so that there is a measure of empty space on the network media between transmissions. (The protocol stack manages this requirement).

REVIEW QUESTIONS

The following questions test your knowledge of the material presented in this chapter:

1. What type of LAN provides the best security features?

2. What type of LAN topology requires the installation of terminators for proper operation?

3. Describe the features of a hybrid network.

4. Name two disadvantages for wired LANs when compared to wireless LANs.

5. What standard was developed to assure interoperability between wireless LAN devices?

6. Name two advantages offered by wireless LANs over wired LANs.

7. What type of network topology is generally associated with WANs?

8. What layer of the OSI model is responsible for point-to-point access between network devices?

9. Name the three OSI model layers contained in the DOD model Application layer.

10. What layer of the OSI model provides for end-to-end network routing?

11. Name the OSI model layer that is responsible for handling network access protocols.

12. What layer of the OSI model is responsible for establishing simplex, half duplex, or full duplex transmission modes?

13. What is the correct term to describe data units at the Data Link layer?

14. Describe the difference between connection-oriented service and connectionless service. What layers of the OSI model provide for these two types of transmission service?

15. Describe the purpose of flow control.

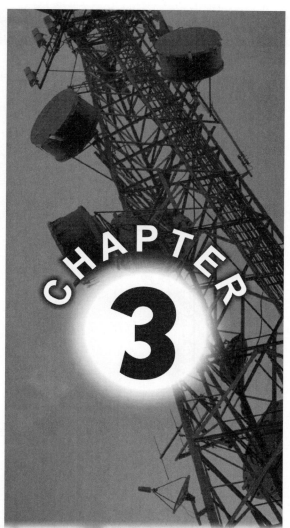

CHAPTER 3

WIRED NETWORK PROTOCOLS AND STANDARDS

LEARNING
OBJECTIVES

LEARNING OBJECTIVES

Upon completion of this chapter and its related lab procedures, you should be able to perform the following tasks:

1. Describe the features of both wired and wireless transmission media used in various types of communications networks.

2. List the advantages and limitations of the different types of media.

3. Describe the most popular and widely used types of local area network standards and protocols.

4. Compare the various characteristics, advantages, and disadvantages of wired and wireless local area networks.

Wired Network Protocols and Standards

INTRODUCTION

In the preceding chapter, the basic structure and topologies of computer networks were introduced. This chapter takes a more in-depth view of the differences between wired physical media and wireless transmission media employed in computer data network architectures. The topologies introduced in the preceding chapter are further applied here to specific Local Area Network (LAN) standards used widely throughout the world.

The purpose of including a chapter on wired LANs in this study guide for wireless networks is as follows:

1. Wireless networks are closely integrated with the largest network in the world — the Public Switched Telephone Network (the PSTN). You must understand the protocols and topologies of wired networks because information often crosses the boundaries of wired and wireless media when traveling from the source to the destination.

2. Wireless protocols, as we shall see in subsequent chapters, often work closely with wired protocols at the network interface point. This is particularly true with Local Area Networks (LANs) and cellular wireless networks. The information on wired LANs will help you understand these relationships in future chapters on advanced wireless air interfaces, networks, protocols, and standards.

Wired Local Area Networks (LANs) have emerged over several years as a natural business application of more powerful desktop and portable laptop computers arranged to provide maximum efficiency through shared resource computing. You will discover here how the more widely used LAN standards and accompanying protocols are employed. Also covered are the basics of how wireless LANs are now able to provide enhanced features and benefits to wired LANs.

Network Communications Media

The transmission of information from one location to another requires some type of medium.

The various media used to accomplish the transfer can be grouped into two basic types. The first type includes all physical wired media in which the information is contained within a guided or **bounded** area, such as a metallic wire or glass fiber-optic cable. In wired media terms, these media are referred to as **conductors**. The second type of media in which the information travels over an unguided or **unbounded** path between the sender and the receiver is referred to as **wireless**.

bounded

conductors

wireless

unbounded

Figure 3-1 shows several examples of bounded and unbounded media that are covered in more detail in this chapter.

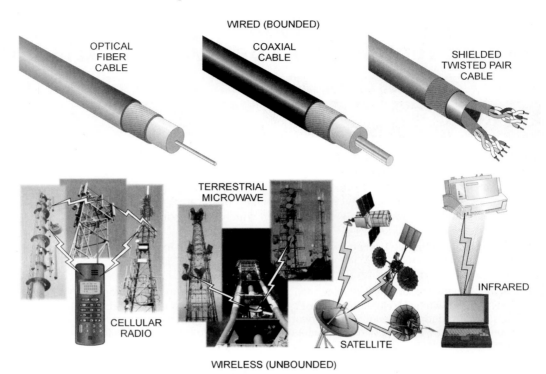

Figure 3-1: Examples of Wired (Bounded) and Wireless (Unbounded) Media

Wired Media

Wired media is widely used in most regions of the world for telephone (voice), data, and multimedia (voice, video, data, and combinations of all three). Wired media use metallic or glass material as the conducting element for electric current or light waves.

The conductors are usually covered by a protective layer of material to isolate them from the elements. The choices vary according to the distance involved, the bandwidth or amount of information the media must handle, cost constraints, errors encountered, environment, and equipment interface requirements. In most cases, the features of the media are closely related. For example, high bandwidth typically costs more, and as the distance increases, the type of media tends to become higher in cost also.

The most widely used wired media types include: twisted pair cable, coaxial cable, and fiber-optic cable.

Let's take a closer look at some of the characteristics of wired media.

Twisted Pair Cable

Twisted pair cable

Twisted pair cable (the original telephone cable) consists of two insulated copper conductors twisted around each other as shown in Figure 3-2.

balanced media

The name is derived from the fact that each insulated conductor is part of a pair in which each conductor is twisted around the other to protect against unwanted external electromagnetic interference. The twisting process also serves to maintain the electric field within the pair. Since twisted pair cable is usually manufactured with several pairs in a protective sheath, the twist for each pair minimizes interference to adjacent pairs in the same sheath or cables placed in the same cable tray during installation. The current flow or signal in each conductor of a pair is similar, but in an opposite direction, therefore twisted pair cable is considered to be a **balanced media**. As a general rule, the more twists per unit of length, the better the interference rejection of the cable pair.

Advanced computer internetworking created a need for higher quality twisted pair cabling. Voice circuits do not require as much capacity as computer-to-computer digital data circuits. In 1985, companies representing the telecommunications and computer industries were concerned with the lack of a standard for building telecommunications wiring systems. A joint industry group, the **Computer Communications Industry Association (CCIA)**, requested that the **Electronics Industries Association (EIA)** develop a standard for in-building wiring systems. Six years in the making, the EIA/TIA 568, 568A, and 568B Commercial Building Telecommunications Wiring Standard is the result of their efforts. Published in July of 1991, the EIA/TIA 568 standard sets standards for twisted pair cabling and fiber-optic cabling for local area networks.

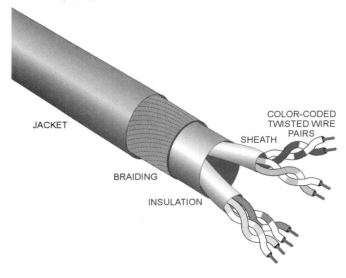

Figure 3-2: Twisted Pair Cable

Computer Communications Industry Association (CCIA)

Electronics Industries Association (EIA)

Coaxial cable

Coaxial Cable

Coaxial cable gets its name from the common axis shared by a center conductor. It contains an insulating material, a braided metal shield/outer conductor, and an outer insulating cover.

The insulating layer or dielectric (nonconductive) material isolates and maintains a common spacing between the center conductor and the outer metal shield conductor. A coaxial cable is illustrated in Figure 3-3.

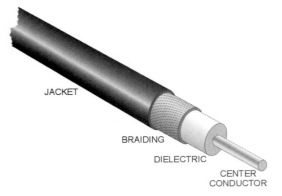

Figure 3-3: Coaxial Cable

Coaxial cable was initially used to carry television broadcast programs over long distances. It was later adapted for use in computer networking. Coaxial cable was the first media type used for Ethernet LANs prior to the development and standardization (by the EIA) of improved twisted pair cable. With coaxial cable, the center conductor carries the information while the outer conductor is used as an electrical ground to maintain a 0 voltage. Coaxial cable is therefore considered to be an **unbalanced media**.

unbalanced media

Coaxial cable is manufactured in a wide range of electrical properties, size, quality, and cost range. Standard low-cost coaxial cable is used in cable TV distribution systems and backbone circuits for computer networks. Higher quality coaxial cable with lower loss and minimum radio signal attenuation is called **hardline**. It is often used to provide the interface cable between microwave and cellular radio transmission equipment at ground level and antennas located on transmission towers.

hardline

Coaxial cable is physically larger and works well in high-capacity long distance applications. However, it also costs more than twisted pair cable and is less flexible and somewhat more difficult to install in conduit or in-building standard cable runs. In modern local area network environments, twisted pair cable is the preferred media.

Fiber-Optic Cable

Fiber optic cable

Fiber-optic cable has revolutionized long distance communications. It works on the principle that light waves traveling in a glass medium can carry more information over longer distances than copper wire media.

Fiber-optic cables contain a large number of glass fiber pairs enclosed in a protective sheath. Each fiber strand may be as small as .002 inches. The glass fiber must be free from impurities in order to pass the light waves with a high degree of efficiency. Each fiber is encased in a layer of glass cladding that has a slightly different refractive index. The cladding serves to direct the light waves back into the central core. Surrounding the cladding is a dielectric material that provides a buffer layer between the cladding and the braiding. The dielectric layer is enclosed with a braiding material called Kevlar, a patented product of DuPont, which adds strength to the cable. Figure 3-4 shows the composition of a fiber-optic cable. The final outer jacket consists of a material to protect the cable from damage. Many fibers may be enclosed in an armored sheath. Fiber-optic cable designed for installation under the ocean is typically enclosed in with multiple layers of armor for additional strength.

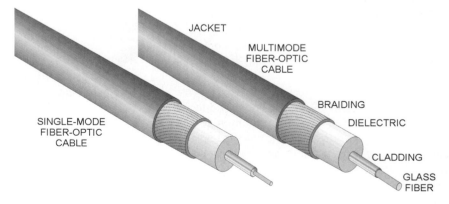

Figure 3-4: Fiber Optic Cable

Fiber-optic systems are a form of hybrid **opto-electric** transmission system. They first convert the electrical digital signal into an optical signal. The light source may be one of two types, Light Emitting Diodes (LEDs) and lasers. Light sources are required to generate the light pulses that travel down the fiber-optic cable. Light detectors are used at the receiving end to convert the light waves to electrical signals.

There are two types of fiber-optic cable. They are:

- Multimode fiber

- Single-mode fiber

Multimode fiber is the least expensive to produce but also has a lower performance figure because the inner core diameter is larger. As the light rays move down the fiber strand, they spread out due to what engineers call **nodal dispersion**. Although the light rays are directed back to the fiber inner core by the cladding, they travel different distances. Although these distances are very small, they result in the rays arriving at different times. This condition gets worse as the distance increases, and the light detector at the receiving end becomes unable to distinguish the modulation of the original light pulses. Multimode fiber, therefore, is the less expensive alternative for applications such as LANs.

Single-mode fiber, shown in Figure 3-5, is more expensive and is used in long distance, higher bandwidth applications.

Single-mode fiber-optic cable uses a very narrow (and higher cost) light source. Single-mode fiber requires fewer repeaters. It is used for long distance applications in which the added cost is able to return the higher performance required.

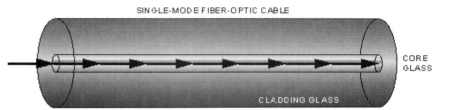

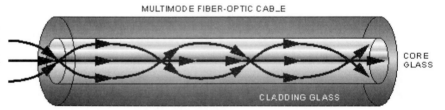

Figure 3-5: Single-Mode and Multimode Fiber

Wireless Media

Wireless media do not require a physical metallic or glass conductor to guide the flow of electrical current or light waves. This type of media transmits electromagnetic energy through space without any physical connection between the receiver and transmitting devices.

As discussed in Chapter 1, electromagnetic wave transmission (wireless) technology was not developed until near the end of the 19th century. Guglielmo Marconi developed the first wireless telegraph system and went on to prove the commercial value of wireless transmission systems. Later, when commercial broadcasting was introduced, the term "radio" was introduced. In this context, radio referred to the broadcast receiving set in homes. As the wireless industry has rapidly expanded into numerous market segments, the term **radio** is now used to refer generically to practically any wireless communications facility or wireless device capable of processing, transmitting, and receiving electromagnetic signals, as depicted in Figure 3-6.

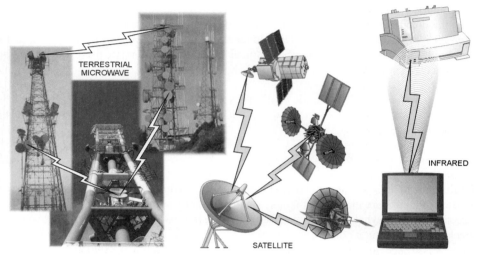

Figure 3-6: Wireless Communications

Wireless media consists of the following classifications: terrestrial microwave, satellites, cellular radio, and infrared.

Each type of wireless system has a specific market application.

Terrestrial Microwave

Terrestrial microwave

Terrestrial microwave radio systems are used to provide high-capacity communications between fixed locations.

Microwave radio

Microwave radio refers to the relatively short wavelength of the frequencies used. Due to the curvature of the earth and limitations on distance due to radio signal attenuation, microwave systems employ repeaters at 20- to 30-mile intervals. The term terrestrial (meaning "of the Earth") is used in this text to describe wireless communications services using radio frequencies above 2 gigahertz (GHz). This term is used to distinguish this type of service from satellite microwave service. Satellites use similar microwave frequency bands above 2 GHz but utilize space-based repeaters as opposed to terrestrial repeaters.

Satellites

Satellites

Satellites provide communications links between earth stations located anywhere in the world.

Satellites are functionally similar to terrestrial microwave repeater stations with the added advantage of being located very high above the surface of the Earth. They can provide distance-insensitive links between point-to-point and point-to-multipoint locations. Satellites can be placed into various orbits around the earth to support a wide range of applications. As we will point out later in Chapters 7 and 8, satellites are placed in low, medium, and geosynchronous orbits to meet the requirements of a variety of new applications.

Most satellites are deployed in **geosynchronous** (or **GEO**) orbits. At an altitude of approximately 22,300 miles, a satellite will have an orbital period of 24 hours. Since the orbital period is the same interval as the rotation of the Earth, it remains stationary with respect to an earth station and provides a constant area of coverage for approximately 1/3 of the Earth's surface.

geosynchronous (GEO)

GEO satellite networks span the oceans to provide intercontinental coverage. They also are widely used for maritime safety, emergency search and rescue, position location and tracking, international television program distribution, direct-to-home television program service, high-speed Internet access, and numerous military applications. Satellites are also opening a new segment of the wireless industry called the Mobile Satellite Service, which is discussed in more detail in Chapters 7 and 8.

Cellular Radio

Cellular radio is a special application of wireless media that has enjoyed phenomenal growth.

Cellular radio

Cellular wireless communication provides users with complete mobility for voice, fax, data, and Internet connectivity throughout metropolitan areas and some rural areas. Subscribers use portable handset radios or portable data terminals to connect to fixed wireless base stations. Base stations are located throughout the service provider's coverage area to form a series of interconnecting cells. The base stations are connected by terrestrial or microwave circuits called **backhaul** circuits to a mobile switching center (MSC). The MSC manages the cellular base station network and also serves as the gateway between the wireless cellular network and the Public Switched Telephone Network (PSTN). The cellular radio network can be viewed as a wireless extension to the PSTN. Mobile cellular subscribers have the same access to the PSTN as wireline subscribers. All calls between mobile subscribers are also routed through the MSC and the PSTN even if they are located in the same cellular area of coverage. Details of cellular wireless networks will be covered in Chapters 5 and 6.

backhaul

Infrared

Infrared is a wireless media that can be used for short-range transmission in local area network applications.

Infrared

Infrared devices use light waves just below the frequencies of visible light. The most widely used application for infrared transmission is the remote control unit for a television set. Infrared network devices modulate the beam of infrared light in a room or area where it can be detected by an infrared receiver.

Similar to wireless LAN radio devices, infrared can be used to connect computers to peripheral devices or access points between separate wired LANs. Infrared is a low-cost solution for connecting network components in areas where wired connections may be difficult or impractical to install. Infrared transmission has been incorporated as one of the wireless media standards in the IEEE 802.11 wireless LAN standard.

Hybrid Media Networks

Over the past several years, communications networks have migrated to a hybrid variety of wired and wireless transmission systems.

Figure 3-7 illustrates one well-known example of how the cellular telephone system provides mobile subscribers with a wireless interface to the wired Public Switched Telephone Network (the PSTN). A radio interface exists between the handset used by the wireless subscriber and the cellular base station. The base station may use a wireless or wired media connection to the Mobile Switching Center (MSC). The MSC is the network connection between the PSTN (wired) network and the cellular (wireless) network.

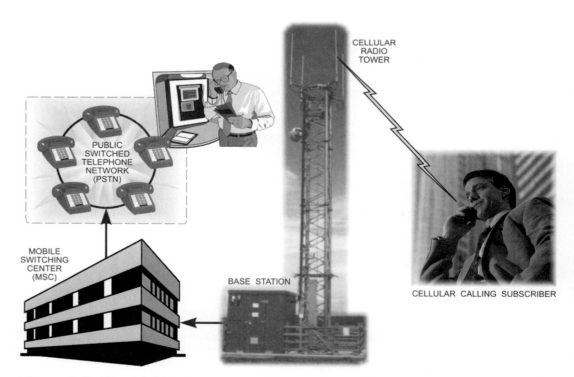

Figure 3-7: The Mobile Services Switching Center Provides Wireless Cellular Network Access to the Wired Public Switched Telephone Network

Combination or hybrid networks are widely used.

Wired Local Area Network Architectures

Today, local area networking is a **shared access** technology. This means that all of the devices attached to the LAN share a single communications medium, usually a coaxial, twisted pair, or fiber-optic cable. Figure 3-8 illustrates this basic concept of local area networking.

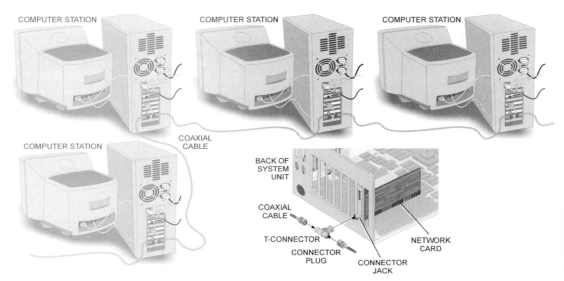

**Figure 3-8:
Basic LAN
Concept Using
a Shared Media**

Several computers are connected to a single cable that serves as the communications medium for all of them. The physical connection to the network is made by putting a **network interface card** (**NIC**) inside each computer and connecting it to the network cable. Once the physical connection is in place, network software manages communications between stations on the network. When a station sends a packet to another station on the LAN, the packet passes by all of the stations connected to that LAN. On the bus network illustrated in Figure 3-8, the electrical signal representing the packet travels away from the sending station in both directions on the shared cable. All stations will see the packet, but only the station it is addressed to will pay attention to it.

A Local Area Network (LAN) is by definition a short distance, packet network for computer-to-computer communications.

LANs may use a variety of media types and are physically implemented using one or more of the network topologies you have read about in this chapter. They are arranged to serve a community of users such as a department within a company, a workgroup, or a single floor in an office complex. Several LANs can be connected within a campus or office building to extend the range of each LAN. LANs are installed for the purpose of sharing access to peripheral equipment, applications, files, and databases.

This has provided an industry with many vendors, each able to provide hardware and software that can fulfill the requirements of the IEEE standards as well as some ANSI standards.

Several LAN standards have been developed over the past several years. Ethernet has clearly emerged as the favorite local area network standard although other types are mentioned in this chapter for reference.

Ethernet

Ethernet

Bob Metcalfe and David Boggs of the Xerox Palo Alto Research Center (PARC) in California developed the first experimental **Ethernet** system in the early 1970s. It interconnected Xerox Alto computers and laser printers at a data transmission rate of 2.94 Mb/s. (This data rate was chosen because it was derived from the system clock of the Alto computer.) In 1979, Digital Equipment Corporation (DEC), Intel, and Xerox joined for the purpose of standardizing an Ethernet system that any company could use. In September 1980 the three companies released Version 1.0 of the first Ethernet specification called the "Ethernet Blue Book", or "DIX standard" (after the initials of the three companies). The first Ethernet controller boards based on the DIX standard became available in 1982. The second and final version of the DIX standard, Version 2.0, was released in November 1982.

Early Industry Standard

The original IEEE standard and most popular version of Ethernet supported a data transmission rate of 10 Mb/s. Newer versions of Ethernet called "Fast Ethernet" and "Gigabit Ethernet" support data rates of 100 Mb/s and 1 Gb/s (1000 Mb/s).

An Ethernet LAN may use coaxial cable, special grades of twisted pair wiring, or fiber-optic cable. "Bus" and "star" topology wiring configurations are supported. Ethernet Network Interface Cards (NICs) installed in desktop workstations and servers compete for access to the network using a protocol called Carrier Sense Multiple Access with Collision Detection (CSMA/CD). Released in 1983, the formal title of the standard was IEEE 802.3 Carrier Sense Multiple Access with Collision Detection (CSMA/CD) Access Method and Physical Layer Specifications. This specification was also known by the large coaxial cable called for in the standard. It was referred to as **thicknet**.

thicknet

In 1985, IEEE 802.3a defined a second version of Ethernet called "thin" or **thinnet** Ethernet, or **cheapnet**, or more correctly as the IEEE standard 10Base-2. It used a thinner, cheaper coaxial cable that simplified the cabling of the network. Although both the thick and thin systems provided a network with excellent performance, they utilized a bus topology, which made implementing changes in the network difficult, and also left much to be desired in regard to reliability.

thinnet

cheapnet

Also released in 1985 was the IEEE 802.3b 10Broad36 standard that defined transmission of 10 Mb/s Ethernet over a "broadband" cable system.

In 1987, two standards were released. The IEEE 802.3d standard defined the Fiber Optic Inter-Repeater Link (FOIRL) that used two fiber-optic cables to extend the maximum distance between 10 Mb/s Ethernet repeaters to 1000 meters. IEEE 802.3e defined a "1 Mb/s" Ethernet standard based on twisted pair wiring. This 1 Mb/s standard was never widely used.

> In 1990, a major advance in Ethernet standards came with introduction of the IEEE 802.3i 10Base-T standard. It permitted 10 Mb/s Ethernet to operate over simple Category 3 Unshielded Twisted Pair (UTP) cable.

The widespread use of UTP cabling in existing buildings created a high demand for 10Base-T technology. 10Base-T also permitted the network to be wired in a "star" topology that made it much easier to install, manage, and troubleshoot. These advantages led to a vast expansion in the use of Ethernet.

All of the Ethernet IEEE specifications use a designation indicating the transmission wire speed in megabits per second. The second term indicates that the information is transmitted as unmodulated baseband information. The last term is used to indicate the maximum segment length allowed. As an example, 10Base-5 is the 802.3 initial specification for a 10-megabit transmission speed using data at baseband with a maximum segment length of 500 meters.

In order to understand how Ethernet controls access to the transmission media, we need to look closer at wire speed and the **Carrier Sense Multiple Access with Collision Detection (CSMA/CD) protocol**. Collision detection is shown in Figure 3-9. A workstation waiting to transmit must first determine whether the medium is being used by another workstation. (This is what is meant by "carrier sense" in the term CSMA/CD.) If it finds that the medium is not being used, it begins transmitting on the wire at a rate of 10 Mb/s (or 100 Mb/s or 1000 Mb/s at the Fast Ethernet speeds). If another workstation started transmitting at exactly the same time, the information contained in each of the frames would collide and both messages would be lost. The Ethernet protocol has the ability to detect a collision and provide a random **backoff**, or programmed delay time, whereby one workstation has a statistical probability of capturing the medium and locking out the other stations during transmission. The term **wire speed** refers to the actual transmission rate once the workstation starts transmitting. There are obviously some quiet times when stations are recovering from a collision or when stations are waiting for the medium to be unused. This gets us to the concept referred to as throughput in a LAN. Throughput is a measure of the actual amount of information that gets transmitted on a per second basis. By definition therefore, throughput will always be less than the wire speed. As more stations are added to an Ethernet LAN, throughput is degraded. Ethernet enjoys some benefit from the fact that the CSMA/CD is an elegant and simple protocol that regulates itself without a central controlling device. However, its utilization is limited to networks with limited traffic demands.

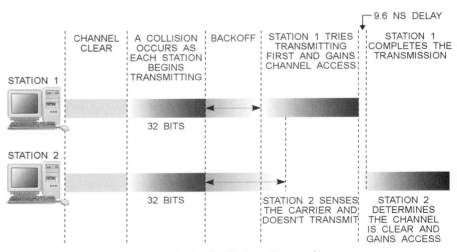

Figure 3-9: Collision Detection

The Ethernet IEEE 10Base-5 Specification

The 10Base-5, thicknet topology uses an external transceiver to attach to the network interface card.

vampire tap

The NIC attaches to the external transceiver by a connector on the card. Some external transceivers clamp to the thicknet cable with metal points that resemble a vampire's teeth (**vampire tap**), and others connect with standard BNC *or* N-series barrel connectors. As with thinnet, each network segment must be terminated at both ends, with one end using a grounded terminator. The components of a thicknet network are shown in Figure 3-10.

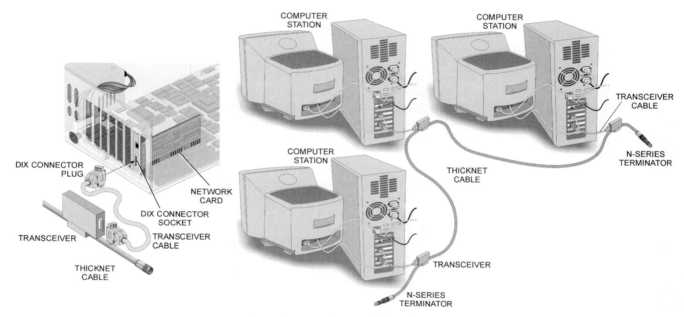

Figure 3-10: Ethernet 10Base-5 Thicknet Standard Network Configuration

There are 10Base-5 specification rules that must be followed. These rules ensure that the CSMA/CD protocol will be able to detect collisions and recover correctly. This results in some physical distance constraints on network segments. The rules are:

- The minimum cable distance between transceivers is eight feet or 2.5 meters.

- You may not go beyond the maximum network segment length of 1,640 feet or 500 meters.

- The entire network cabling scheme cannot exceed 8,200 feet or 2,500 meters.

- One end of the terminated network segment must be grounded.

- The maximum number of nodes per network segment is 100. (This includes all repeaters.)

The Ethernet IEEE 10Base-2 Specification

The 10Base-2 specification calls for RG-58 A/U coaxial type cable, 50-ohm terminators, and BNC T-connectors that directly attach to the connector on the NIC. A grounded terminator must be used on only one end of the network segment.

The components of an Ethernet 10Base-2 network topology are shown in Figure 3-11. Each thinnet NIC is connected to the coaxial cable using a BNC-T connector.

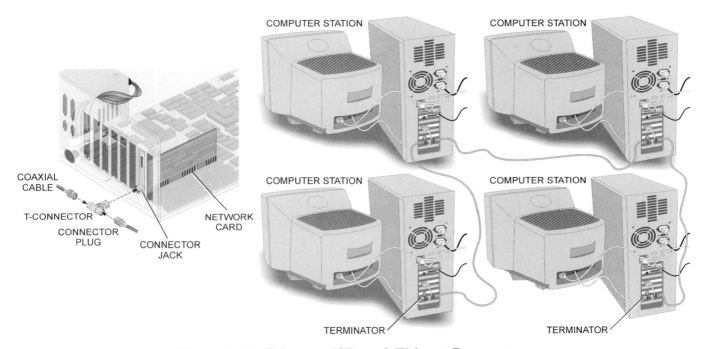

Figure 3-11: Ethernet 10Base-2 Thinnet Parameters

The Ethernet 10Base-T Specification

Unshielded Twisted Pair (UTP) cable has become the favorite type of media for Ethernet LANs.

Unshielded Twisted Pair (UTP)

UTP cable has a lower cost than coaxial cable. UTP is also smaller than coax, which relieves congestion of wiring conduits. 10Base-T Ethernet is wired in a star topology. UTP cable uses RJ-45 type connectors. Each network interface card has a RJ-45 jack included in the back panel.

Figure 3-12 shows 10Base-T Ethernet topology using twisted pair and an Ethernet hub.

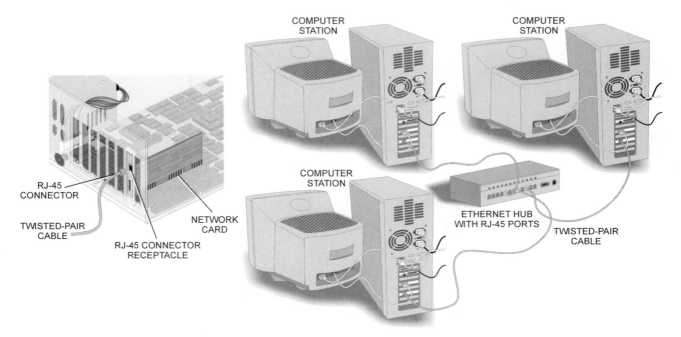

Figure 3-12: Ethernet 10Base-T Configuration

Later Ethernet Standards Evolution

In 1993, the IEEE 802.3j standard for 10Base-F (FP, FB, & FL) was released, which permitted attachment over longer distances (2000 meters) via two fiber-optic cables. This standard updated and expanded the earlier FOIRL standard.

In 1995, IEEE improved the performance of Ethernet technology by a factor of 10 when it released the 100 Mb/s 802.3u 100Base-T standard. This version of Ethernet is commonly known as **Fast Ethernet**. Three media types were supported:

Fast Ethernet

1. 100Base-TX operates over two pair of Category 5 twisted pair cable.

2. 100Base-T4 operates over four pair of Category 3 twisted pair cable.

3. 100Base-FX operates over two multimode fibers.

In 1997, the IEEE 802.3x standard became available, which defined "full-duplex" Ethernet operation. Full-duplex Ethernet bypasses the normal CSMA/CD protocol to allow two stations to communicate over a point-to-point link. It effectively doubles the transfer rate by allowing each station to concurrently transmit and receive separate data streams. For example, a 10 Mb/s full-duplex Ethernet station can transmit one 10 Mb/s stream at the same time it receives a separate 10 Mb/s stream. This provides an overall data transfer rate of 20 Mb/s. The full-duplex protocol extends to 100 Mb/s Ethernet and beyond. Also released in 1997 was the IEEE 802.3y 100Base-T2 standard for 100 Mb/s operation over two pairs of Category 3 balanced cabling.

In 1998, IEEE once again improved the performance of Ethernet technology by a factor of 10 when it released the 1 Gb/s 802.3z 1000Base-X standard. This version of Ethernet is commonly known as **Gigabit Ethernet**. Three media types are supported:

1. 1000Base-SX operates with a 850 nm laser over multimode fiber.

2. 1000Base-LX operates with a 1300 nm laser over single- and multimode fiber.

3. 1000Base-CX operates over short haul copper "twinax" shielded twisted pair (STP) cable.

Also released in 1998 was the IEEE 802.3ac standard that defines extensions to support Virtual LAN (VLAN) tagging on Ethernet networks.

ARCnet

ARCnet is a pioneering technology originally developed and marketed by the Datapoint Corporation.

ARCnet is an acronym for Attached Resource Computer NETwork. ARCnet operates at a wire speed of 2.5 Mbps throughput and can be connected using RG-62 A/U coax cable or unshielded twisted pair (UTP) wiring. Although ARCnet can support up to 255 node numbers on a single network, systems of this size are not practical.

Typical ARCnet coax cable specifications are:

*   RG-62 A/U, solid conductor, CL2, 90%+ braided shield, PVC jacket, nominal 93 ohm impedance, 15.5 nominal capacitance/ft.

*   RG-62 A/U, solid conductor, CL2P, 90%+ braided shield, plenum jacket, nominal 93 ohm impedance, 12.5 nominal capacitance/ft.

ARCnet was one of the pioneering topologies for local area networking. It is rarely used as the topology of choice in new LAN environments. It continues, however, to have strong support from the ARCNET Trade Association (ATA).

Token Ring

The **Token Ring** network was originally developed by IBM in the 1970s.

As you may assume from the name, Token Ring uses a ring topology. An information packet travels around a closed loop, passing through each workstation NIC. The information packet is regenerated and passed on to the next workstation. Although the information travels in a closed loop, the central control unit, called the **Multistation Access Unit** (**MSAU**), provides a connection path for each Token Ring workstation.

Although the network physically appears as a star configuration, internally signals travel around the network from one station to the next. The cabling configurations and the addition or removal of workstations must ensure that the logical ring is maintained. All workstations connect to central hubs called **MSAUs**. Multiple hubs are connected together to create large multistation networks. The hub itself contains a "collapsed ring." If a workstation fails, the MSAU immediately bypasses the station to maintain the ring of the network. The MSAU, workstations, and associated cabling are shown in Figure 3-13.

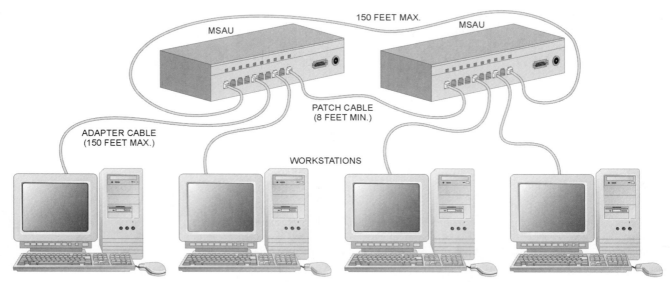

Figure 3-13: Token Ring Network Wiring Configuration

Token Ring was eventually adopted as the 802.5 standard by the IEEE.

The term Token Ring generally refers to both IBM's original Token Ring network and IEEE 802.5. On a Token Ring network, a special frame called a token controls the right to transmit and is passed from one station to the next in a physical ring.

If a station has information to transmit, it captures the idle token, marks it as being in use, and inserts the information. The token is now considered by all computers on the network to be a "busy" token because it is carrying the message to a destination computer. The message is then passed around the circle, copied when it arrives at its destination, and eventually returned to the sender to verify error-free reception by the receiving station. The sending station removes the attached message and then passes the free token to the next station in line.

Token Ring and IEEE 802.5 Comparison

IBM Token Ring and IEEE 802.5 networks are compatible, although the specifications differ in relatively minor ways. IBM's Token Ring network specifies a star topology, with all
end stations attached to a MSAU, whereas IEEE 802.5 does not specify a topology (although
virtually all IEEE 802.5 implementations also are based on a physical star topology). Other
differences exist, including media type (IEEE 802.5 does not specify a media type, whereas
IBM Token Ring networks use twisted-pair wire).

The MSAU is the central cabling component for IBM Token Ring networks. The 8228
MSAU was the original wiring hub developed by IBM for Token Ring networks. Figure 3-14
shows an example of a network cabling several workstations and 8228 MSAUs. Each 8228
has ten connectors, eight of which accept cables to workstations or servers. The other
connectors are labeled RI (ring in) and RO (ring out). The RI and RO connectors are used to
connect multiple 8228s to form larger networks.

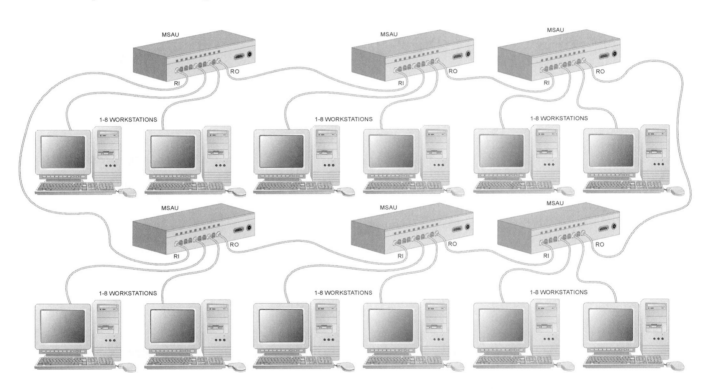

Figure 3-14: Network Cabling Several Workstations and 8228 MSAUs

IBM Token Ring networks use two (or three) types of connectors. NICs are equipped with a female (DB-9) nine-pin connector (only four pins are used) and/or an RJ-45 connector. MSAUs, repeaters, and most other equipment use a special IBM Type-1 unisex data connector. (Newer Token Ring cards and MSAUs use RJ-45 connectors *and* CAT 5 UTP cabling). IBM specifies a special class of shielded twisted pair cabling for Token Ring networks. The five types and their use in the network are shown in the Table 3-1.

Table 3-1: IBM Shielded Twisted Pair Types

Type	Pairs	AWG	Use
1	Solid, twisted	22	Data
2	Type 1 plus(4) voice pairs	22	Data and Voice
6	Stranded, twisted	26	Patch
8	Solid, parallel	26	Under carpet
9	Solid, twisted	26	Plenum

The following rules apply to Token Ring networks:

- The minimum patch cable distance between two MSAUs is eight feet.

- The maximum patch cable distance between two MSAUs is 150 feet. Patch cables come in standard lengths of 8, 30, 75, and 150 for Type 6.

- The maximum patch cable distance connecting all MSAUs is 400 feet.

- The maximum patch cable distance between a MSAU and a node is 150 feet.

IBM continues to support the Token Ring 802.5 specification and heads the IEEE 802.5 standards committee. In order to maintain its competitive position in the industry, IBM led the development of the Token Ring 100 megabits per second IEEE 802.5t specification.

Wireless Local Area Networks — An Alternative Solution

In this chapter, you have reviewed the most widely used wired LAN standards, topologies, and performance issues. With continuing research and development by the wireless industry, costs continue to come down, and performance is increasing for **wireless** LAN technology. Here, we will briefly review some of the advantages of wireless LANs compared to wired LANs. (Wireless LANs will be covered in more detail in Chapter 6.)

wireless

The term wireless is somewhat misleading, since most wireless LANs interconnect with wired networks. The bulk of the distance between a wireless node and another node may well be over wires or fiber. Nevertheless, it is possible to build a network that is completely wireless. In such an instance, the physical size of the network is determined by the maximum reliable propagation range of the radio signals. Networks such as these are referred to as **ad hoc networks**, and are well suited for temporary situations such as meetings, conferences, and sporting events.

In 1985, the FCC approved the use of unlicensed radio spectrum for wireless network devices. This was a major incentive for the wireless industry to move ahead with wireless networking standards.

This FCC action authorized the use of the industrial, scientific, and medical (ISM) frequency bands for all wireless devices. Wireless equipment could be used without obtaining a FCC license if the products met the limited transmit power specified in the FCC order. The authorized bands are as follow:

- 902 to 928 MHz

- 2.40 to 2.4835 GHz

- 5.725 to 5.850 GHz

You will more likely encounter what is called an **infrastructure network**, where your wireless LAN connects to an existing wired LAN, as shown in Figure 3-15.

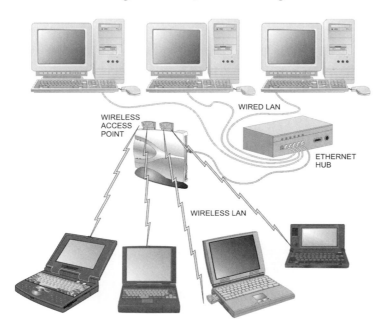

Figure 3-15: Wireless LAN Access Point Provides the Wireless Interface to a Wired LAN

In this case you will need an access point that effectively bridges wireless LAN traffic onto your LAN. This function may be handled by software in a workstation that houses both a wireless card and a wired (e.g., Ethernet) card. But most wireless LAN vendors recommend dedicated hardware called an access point for this function. The access point can also act as a repeater for wireless nodes, effectively doubling the maximum possible distance between nodes.

Wireless Local Area Networking Advantages

Wireless LANs provide services that are not practical for wired local area networks. Some examples are:

- **Scalability:** Wireless LAN systems can be configured in a variety of topologies with no constraints on the location. Configurations are easily changed and range from peer-to-peer networks suitable for a small number of users to full infrastructure networks of thousands of users that enable roaming over a broad area.

- **Mobility:** Wireless LAN systems can provide LAN users with access to real-time information anywhere in their organization. This mobility supports productivity and service opportunities not possible with wired networks.

- **Installation Speed and Simplicity:** Installing a wireless LAN system is fast and easy and eliminates the need to pull cable through walls and ceilings.

- **Installation Flexibility:** Wireless technology allows the network to go where wire cannot go.

- **Reduced Cost-of-Ownership:** While the initial investment required for wireless LAN hardware can, in some cases, be higher than the cost of wired LAN hardware, overall installation expenses and life-cycle costs can be significantly lower. Long-term cost benefits are greatest in dynamic environments requiring frequent moves and changes.

Lab 1 – Installing and Testing a NIC and a Modem Card in a Laptop Computer

Lab Objectives

Install the Network Interface Card (NIC) in the laptop.

Verify the functionality of the NIC.

Verify the functionality of the laptop's modem.

Materials Needed

- Laptop computer w/ internal modem

- Windows XP compatible NIC with RJ-45 adapter

- NIC documentation and drivers

Instructions

In this lab we will set up and install a network card with Windows XP. We will then verify the functionality of the card on your local area network. We will also verify the functionality of the computer's internal communication with the onboard modem.

NIC Installation

1. Gather the materials needed and boot the laptop to Windows XP.

2. Record the manufacturer and the model number of your NIC in Table 3-2.

NIC Manufacturer	NIC Model

Table 3-2: Manufacturer and Model Number

3. Connect the RJ-45 adapter to an available RJ-45 cable on your wired LAN. Connect the other end of the adapter to the NIC PC Card.

4. Plug the NIC into an available PC Card slot on the laptop. You may hear an audible alert from the laptop indicating that a card has been inserted. You will see a pop-up window similar to Figure 3-16.

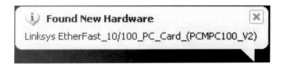

Figure 3-16: Found New Hardware 1

5. Windows will automatically install the drivers for the NIC and display a series of pop-up windows similar to Figures 3-17 and 3-18. Wait a few moments and the card will be installed. If Windows doesn't have the drivers "built in" you must insert the manufacturer's CD (containing the drivers) and follow the prompts.

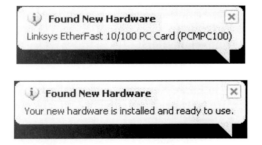

Figure 3-17: Found New Hardware 2

Figure 3-18: Found New Hardware 3

6. A pop-up window similar to Figure 3-19 will finally appear. Click in the window to start the Network Setup Wizard.

Figure 3-19: Network Setup Wizard Prompt

7. Click on **Next** to begin the Network Setup Wizard.

8. Choose the default option **This computer connects to the Internet through another computer on my network or through a residential gateway**, unless otherwise directed by your instructor. Click on **Next**.

9. Leave the "Computer name" the same and click on **Next**.

10. Enter a Workgroup name like **Marcraft** and click on **Next**. If all the computers in your classroom have the same workgroup, it will be easier for them to access each other.

11. You will now see a window similar to Figure 3-20. It shows a summary of your settings. Click on **Next**.

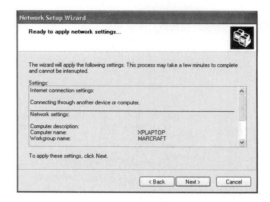

Figure 3-20: Ready to Apply Network Settings

12. On the subsequent window that opens, select the last option **Just finish the wizard...**, click on **Next**, and then click on **Finish**.

13. You will be prompted to restart the computer. Select **Yes**.

Network Testing

14. After Windows has restarted, log on, and click on **Start/My Network Places**.

15. Click on **View workgroup computers**.

16. You can now view other computers on your network. They will be listed directly or in their own workgroup in Entire Network. Click on the drop-down address menu and you can select **Entire Network**. The window will look similar to Figure 3-21.

17. Browse through the network for a minute and close all windows.

NOTE

You will need to set up an Internet connection for later use. This procedure assumes that your school's network uses DHCP (Dynamic Host Configuration Protocol) to set up an IP address and gateway automatically. If not, you will need to set up a manual IP address and gateway per your instructor.

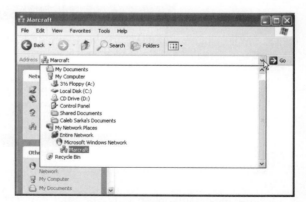

Figure 3-21: Network Places

Modem Testing

18. Assuming that you have a modem "built in" and installed in the laptop, click on **Start/Control Panel**.

19. Double-click on **Phone and Modem Options**.

20. Click on the **Modems** tab.

21. Select the modem from the list and click on **Properties**.

22. Click on the **Diagnostics** tab.

23. Click on **Query Modem**. The window will look similar to Figure 3-22. The computer will send a series of commands to the modem to find out its characteristics.

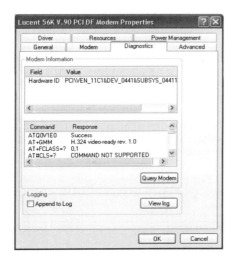

Figure 3-22:
Modem Properties

24. Click on the **View Log** button.

25. Record the value of "ATI3" in Table 3-2. This should be the model of the modem.

ATI3	

Table 3-3:
Modem Model

26. Close all windows and shut down the computer.

> **— NOTE —**
>
> Since the computer has issued commands and the modem has responded, you can assume that the modem is functioning correctly. You now could connect the modem to a phone line and dial up to the Internet or another computer.

Lab 1 Questions

1.   Did your NIC installation require the manufacturer's CD-ROM?

2.   How many times was a restart necessary to install the NIC in Windows XP?

Key Points Review

In this chapter, the following key points were introduced concerning wired networks, common protocols, and wired local area network standards. Review the following key points before moving into the Review Questions section to make sure you are comfortable with each point and to verify your knowledge of the information.

- The media used to accomplish the transfer can be grouped into two basic types. The first type includes all physical wired media where the information is contained within a guided or bounded area such as a metallic wire or glass fiber-optic cable. In wired media terms, these media are referred to as conductors. The second type of media is referred to as wireless, where the information travels over an unguided or unbounded path between the sender and the receiver.

- Wired media is widely used in most regions of the world for telephone (voice), data, and multimedia (voice, video, data, and combinations of all three). Wired media consists of metallic or glass material as the conducting element for electric current or light waves.

- The most widely used wired media types include: twisted pair cable, coaxial cable, and fiber-optic cable.

- Twisted pair cable (the original telephone cable) consists of two insulated copper conductors, each twisted around the other.

- Coaxial cable gets its name from the common axis shared by a center conductor. It contains an insulating material, a braided metal shield/outer conductor, and an outer insulating cover.

- Fiber-optic cable has revolutionized long distance communications. It works on the principle that light waves traveling in a glass medium can carry more information over longer distances than a copper wire medium.

- Wireless media do not require a physical metallic or glass conductor to guide the flow of electrical current or light waves. This type of media uses electromagnetic energy that is transmitted through space without any physical connection between the receiver and transmitting devices.

- Wireless media consists of the following classifications: terrestrial microwave, satellites, cellular radio, and infrared.

- Terrestrial microwave radio systems provide high-capacity communications between fixed locations.

- Satellites provide communications links between earth stations located anywhere in the world.

- Cellular radio is a special application of wireless media that has enjoyed phenomenal growth.

- Infrared is a wireless media that can be used for short-range transmission in local area network applications.

- Over the past several years, communications networks have migrated to a hybrid variety of wired and wireless transmission systems.

- Combination or hybrid networks are widely used.

- Today, local area networking is a shared access technology. This means that all of the devices attached to the LAN share a single communications medium, usually a coaxial, twisted pair, or fiber-optic cable.

- A Local Area Network (LAN) is by definition a short distance, packet network for computer-to-computer communications.

- LAN architectures have been standardized largely through the efforts of the Institute for Electrical and Electronic Engineers (IEEE).

- The original IEEE standard and most popular version of Ethernet supported a data transmission rate of 10 Mb/s. Newer versions of Ethernet called "Fast Ethernet" and "Gigabit Ethernet" support data rates of 100 Mb/s and 1 Gb/s (1000 Mb/s).

- In 1990, a major advance in Ethernet standards came with introduction of the IEEE 802.3i 10Base-T standard. It permitted 10 Mb/s Ethernet to operate over simple Category 3 Unshielded Twisted Pair (UTP) cable.

- The 10Base-5, thicknet topology uses an external transceiver to attach to the network interface card.

- The 10Base-2 specification calls for RG-58 A/U coaxial type cable, 50-ohm terminators, and BNC T-connectors that directly attach to the connector on the NIC. A grounded terminator must be used on only one end of the network segment.

- Unshielded Twisted Pair (UTP) cable has become the favorite type of media for Ethernet LANs.

- In 1993, the IEEE 802.3j standard for 10Base-F (FP, FB, & FL) was released, which permitted attachment over longer distances (2000 meters) via two fiber-optic cables. This standard updated and expanded the earlier FOIRL standard.

- ARCnet is a pioneering technology originally developed and marketed by the Datapoint Corporation.

- The Token Ring network was originally developed by IBM in the 1970s.

- Token Ring was eventually adopted as the 802.5 standard by the IEEE.

- The term Token Ring is generally used to refer to both IBM's original Token Ring network and IEEE 802.5. On a Token Ring network, a special frame called a token controls the right to transmit and is passed from one station to the next in a physical ring.

- In 1985, the FCC approved the use of unlicensed radio spectrum for wireless network devices. This was a major incentive for the wireless industry to move ahead with wireless networking standards.

REVIEW QUESTIONS

The following questions test your knowledge of the material presented in this chapter:

1. List the two basic types of transmission media.

2. What type of media is used in opto-electric transmission systems?

3. What type of media is the most popular and widely used for wired LANs?

4. Explain the rationale for twisting the conductors in twisted pair cables.

5. List the primary advantages offered by satellites compared to terrestrial wireless media.

6. What are the two basic types of fiber-optic cable?

7. What is the name of the facility that serves as a gateway between the PSTN and the cellular (wireless) network?

8. What Ethernet standard allowed the use of unshielded twisted pair cable as the network media?

9. What is the maximum network segment length permitted in the 10Base-5 Ethernet specification?

10. What type of cable is specified in the 10Base-2 Ethernet standard?

11. Token Ring was originally developed by what corporation?

12. What type of media is required for IBM Token Ring networks?

13. List the three corporations that developed the first experimental Ethernet system.

14. List two main advantages offered by wireless LANs.

15. What frequency bands are used for wireless LAN devices?

CHAPTER
4

RADIO COMMUNICATIONS FUNDAMENTALS

Upon completion of this chapter, you should be able to perform the following tasks:

1. Describe a transverse electromagnetic wave.

2. Describe the relationship between frequency and wavelength.

3. Explain the purpose of using powers of ten to express the value of large and small numbers.

4. Describe the procedure for converting frequency to wavelength values.

5. Describe the procedure for converting wavelength to meters and feet.

6. Describe the ITU standard names used for portions of the radio spectrum.

7. Identify three types of radio waves based upon their propagation characteristics.

8. Provide a definition of Line of Sight (LOS) wireless communications links.

9. Briefly list the most important propagation problem areas.

10. Describe the use of the decibel.

11. Explain how interference disrupts radio signals. Define path loss.

12. Provide a definition of the inverse square law as used in communications theory.

13. Describe the problems associated with fades and multipath.

14. Explain how ground waves can degrade a wireless communications link.

15. Describe the mathematical relationship between distance to the horizon and required antenna height.

16. Identify and explain the seven basic functions of a radio system.

17. Explain the purpose and use of an antenna polar plot.

18. Briefly describe how to calculate antenna gain.

19. Describe the three main types of microwave antenna feed assemblies.

20. Explain the relationships between channel spacing and channel bandwidth.

21. Prepare a definition describing the difference between simplex, half duplex, and full duplex modes of communication.

22. List the purpose and use of Forward Error Correction (FEC) in wireless communications networks.

23. Describe the three basic methods of modulation of a radio carrier wave.

24. Describe the fundamental concept of Amplitude Shift Keying (ASK).

25. Describe the advantages of Frequency Shift Keying (FSK).

26. Describe the theory of Phase Shift Keying (PSK).

27. Briefly explain the relationship between bit rate and symbol rate.

28. Describe the theory and advantages for Quadrature Amplitude Modulation (QAM).

29. Identify the purpose and use of digital voice compression.

30. Explain how voice quality relates to voice compression bit rates.

CHAPTER 4

Radio Communications Fundamentals

INTRODUCTION

The German scientist Heinrich Hertz, shown in Figure 4-1, confirmed the existence of electromagnetic waves in 1887 through experiments carried out in a laboratory. His work confirmed the earlier theories of electromagnetic radiation published in 1865 by James Clark Maxwell, shown in Figure 4-2. Electromagnetic waves consist of periodic variations created by the motion of electric charges. A moving charge gives rise to a magnetic field, and if the motion is changing, then the magnetic field varies and in turn produces an electric field. These variations travel at right angles to each other and also to the direction of propagation of the energy of the waves.

Electromagnetic waves are measured by the frequency of the variations and the distance between crests of the waves. Maxwell's equations predicted their travel in free space at the speed of light. The electromagnetic spectrum covers wavelengths from many thousands of meters to micrometers.

Figure 4-1: Heinrich Hertz

Figure 4-2: James Clark Maxwell

The number of **oscillations** or reversals of the fields per second is called the **frequency of the wave**. The constant velocity of the wave therefore creates an inverse relationship between frequency and wavelength, that is, as the frequency of oscillations increases, the wavelength of the resulting wave decreases.

A convenient way to express large and small wavelength and frequency values of the radio spectrum is by using powers of ten.

The frequency range of the radio spectrum is enormous. It is therefore more practical to refer to portions of the spectrum by common naming conventions. Our emphasis for wireless applications will concentrate on the standard frequency band designations adopted for use by the International Telecommunications Union (ITU), the Federal Communications Commission, and the IEEE.

oscillations

frequency of the wave

Electromagnetic waves do not require a material medium. They can travel in a vacuum or the earth atmosphere. The use of electromagnetic waves for wireless communications, however, requires an in-depth understanding of the propagation problems that can occur between two or more cooperating wireless transmitting and receiving locations within a wireless network. Various terms are used by engineers and technicians to describe the problems associated with the wireless environment, such as fading, interference, multipath, and path loss. It is important for you to study these terms and become familiar with their origin and relevance to wireless communications.

The bel and decibel are logarithmic mathematical tools used by wireless technical personnel to express relative power ratios and power levels in wireless communications link analysis and network design.

Energy levels for electromagnetic waves can vary widely as they travel through space and transmission feeder systems such as antennas and transmission lines. Power levels for various points in a network are often referenced to a standard logarithmic value rather than decimal arithmetic. Like the powers of ten, the bel provides a convenient mathematical shortcut when working with wireless systems.

The basic purpose of a wireless network is to move information from the source to the destination using radio waves as the medium. The information is added to the radio waves using a process called modulation. Modulation can be implemented by varying one or more of the attributes of a radio carrier wave. They are the frequency, amplitude, or phase. Amplitude modulation has been used for many years with the widespread use of standard radio and TV broadcasting. Digital communications, however, uses advanced modulation methods more appropriate for data communications and offers the potential for wider use of digital signal processing technology. Digital modulation also provides higher tolerance to the propagation problems encountered in the wireless environment.

All basic digital wireless transmitter systems contain as a minimum the following functional subsystems:

1. A source encoder

2. A channel encoder

3. A multiplexer

4. A modulator

5. A radio frequency source

6. A radio frequency amplifier

7. A transmitter antenna system

The receiver system contains the following basic subsystems:

1. A receiver antenna

2. A radio frequency amplifier

3. A demodulator

4. A radio frequency source

5. A demultiplexer

6. A channel decoder

7. A source decoder

Radio transmission modes include simplex, half duplex, and full duplex. Simplex and half duplex use a single radio channel. Full duplex requires two radio channels for simultaneous two-way communications.

The use of digital communications techniques requires the conversion of analog voice source information to a digital representation of the original analog form suitable for digital modulation on the radio carrier wave. Advances in digital voice processing and compression have led to rapid and dramatic improvements in cellular, microwave, and satellite voice networks. The wireless world is converging on all digital voice and data radio transmission technology made possible in part by rapid advances in digital signal processing technology.

In this chapter, you will gain new insights into the basics of wireless communications. Each fundamental concept covered here will aid in your understanding of specific wireless applications in future chapters.

Understanding the Electromagnetic Spectrum

The **electromagnetic spectrum** is the entire range of wavelengths or frequencies of electromagnetic radiation.

electromagnetic spectrum

The range of wavelengths is very large, as shown in Figure 4-3. They extend from many kilometers for radio waves, to 0.4 to 0.7 millionths of a meter for visible light, to a billionth of a meter for X rays and even less for gamma rays.

Electromagnetic waves travel through space at the speed of light and, as the name suggests, consist of periodic variations of electric and magnetic fields. These two fields are displaced in free space from each other by 90 degrees, as shown in Figure 4-4.

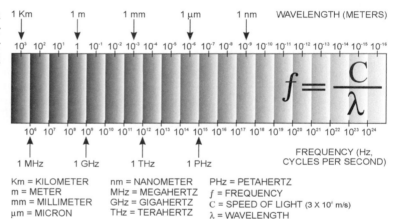

$$f = \frac{C}{\lambda}$$

Km = KILOMETER
m = METER
mm = MILLIMETER
μm = MICRON

nm = NANOMETER
MHz = MEGAHERTZ
GHz = GIGAHERTZ
THz = TERAHERTZ

PHz = PETAHERTZ
f = FREQUENCY
C = SPEED OF LIGHT (3 X 10⁸ m/s)
λ = WAVELENGTH

Figure 4-3: Range of Frequencies

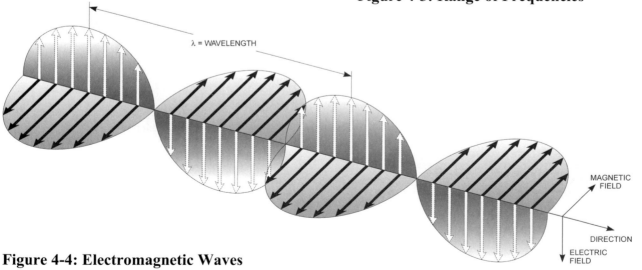

Figure 4-4: Electromagnetic Waves

transverse wave

An electromagnetic wave is therefore a **transverse wave**.

wavelength

The distance from one crest of the wave to the next crest is the **wavelength**. Frequency is measured by the number of oscillations that occur in one second.

The unit of measurement for frequency is hertz (Hz), named after the German scientist Heinrich Hertz discussed earlier, who confirmed the existence of electromagnetic waves experimentally in 1887.

You may have concluded from the figures that frequency and wavelength are related in a special way.

lambda (λ)

Since we know that electromagnetic waves travel at the speed of light, which is approximately 300,000,000 meters per second or 186,000 miles per second, we can determine the frequency of the wave by dividing the speed of light (*c*) by the wavelength, which is represented in radio theory by the Greek letter **lambda (λ).**

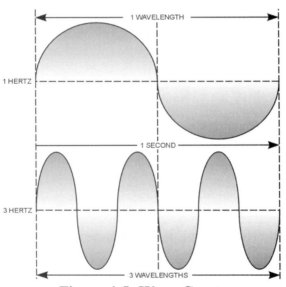

Figure 4-5: Wave Crests

Because all waves propagate through space at a constant speed, the higher frequencies have a shorter distance between the wave crests, as shown in Figure 4-5. In the example, a frequency of one hertz is compared with a frequency of 3 hertz. Notice that both of the waves have traveled through space in the same time period (1 second), but the higher frequency has traveled a shorter distance during one oscillation. The lower frequencies would have a longer distance between crests.

We may conclude therefore, that the wavelength is inversely proportional to frequency.

A frequency of 1 hertz is used here to simplify the comparison. In real applications for wireless systems, the frequencies used are much higher, as indicated in Figure 4-3. The following examples illustrate the method for calculating frequency or wavelength when one of the two terms is unknown.

Example 1

A radio wave has a wavelength of 2 meters. Calculate the frequency in hertz.

— TEST TIP —
Know how to convert between frequency and wavelength.

$$f = \frac{c}{\lambda}$$

$$f \text{ (hertz)} = \frac{300,000,000 \text{ (meters / second)}}{2 \text{ meters (wavelength or } \lambda)}$$

$$f = 150,000,000 \text{ hertz}$$

Example 2

A radio wave has a frequency of 10,000,000 hertz. Calculate the wavelength in meters. Since we know the formula for calculating frequency, we can solve it for the wavelength as follows:

$$\lambda = \frac{c}{f}$$

$$\lambda \text{ (meters)} = \frac{300,000,000 \text{ (meters / second)}}{10,000,000 \text{ (hertz)}}$$

$$\lambda = 30 \text{ meters}$$

Powers of 10

We frequently deal with large numbers in wireless calculations of frequency and wavelength. It is easier to express large numbers as a **power of 10.**

power of 10

You will typically use powers of 10 to describe the unit of frequency in kilohertz (kHz) or megahertz (MHz). It is also convenient to express wavelength in meters as a power of 10 instead of using the actual base number. Table 4-1 lists the prefixes we will be using in this book to describe the operating frequencies or wavelengths of various wireless systems.

Table 4-1: Operating Frequency or Wavelength Prefixes

Prefix	Definition of the Prefix	
micro(μ)	$\dfrac{1}{1,000,000,000}$	10^{-6}
milli	$\dfrac{1}{1,000}$	10^{-3}
kilo	1,000	10^{3}
mega	1,000,000	10^{6}
giga	1,000,000,000	10^{9}

In the previous example, we calculated frequency and wavelength using the units of hertz and meters. Using powers of 10, we can adjust the numbers in the equation to simplify the calculations as follows:

$$f \text{ (Hz)} = \frac{3 \times 10^8}{\lambda(m)} \quad \text{or} \quad f \text{ (kHz)} = \frac{3 \times 10^5}{\lambda(m)} \quad \text{or} \quad f \text{ (MHz)} = \frac{3 \times 10^2}{\lambda(m)} \quad \text{or} \quad f \text{ (GHz)} = \frac{3 \times 10^{-1}}{\lambda(m)}$$

The following calculations indicate how powers of 10 can be used to simplify the equations.

Example 3

A radio wave has a wavelength of 2 meters. Calculate the frequency in megahertz.

$$f \text{ (MHz)} = \frac{c}{\lambda}$$

$$f \text{ (MHz)} = \frac{300}{2}$$

$$f = 150 \text{ MHz}$$

Example 4

A radio wave has a frequency of 10 megahertz. Calculate the wavelength.

$$\lambda \text{ (m)} = \frac{300}{f \text{(MHz)}}$$

$$\lambda \text{ (m)} = \frac{300}{10}$$

$$\lambda = 30 \text{ meters}$$

Example 5

It is also possible to use English measurements for calculating wavelength. The following equation can be used for determining the approximate wavelength in feet.

Calculate the wavelength in feet for a frequency of 2 GHz.

$$\lambda \text{ (ft)} = \frac{1}{f \text{(GHz)}}$$

$$\lambda = \frac{1}{2} = .5 \text{ feet}$$

Thus the approximate wavelength for a frequency of 2 GHz is .5 feet or 6 inches.

We can obtain a more accurate answer with the previous formula for calculating the wavelength in meters as follows:

$$\lambda \text{ (m)} = \frac{.3}{f \text{(GHz)}}$$

$$\lambda = \frac{.3}{2} = .15 \text{ meters}$$

Converting meters to feet:

Since 1 meter = 39.37 inches, then .15 meters × 39.37 inches = 5.9055 inches for the more accurate calculation of the wavelength of a frequency of 2 GHz.

Frequency Band Designations

Various portions of the electromagnetic spectrum are categorized using different names.

The naming conventions have changed over the past several years with the introduction of new radio technologies. In dealing with the hundreds of names, application areas, and categories for the electromagnetic spectrum, you will need to become familiar with the vocabulary of the wireless world. The following represents the most widely used categories in the wireless industry.

The ITU Band Designations

The most popular method for naming portions of the spectrum is the International Telecommunications Union – (Radio Sector) standard names.

These standard names are generally accepted by all countries. This standard for spectrum bands divides the spectrum into subsets, each bounded by wavelength increments of a power of 10. In the plan in Figure 4-6 you will note that the High Frequency (HF) band extends from a wavelength of 100 meters (10^2) to 10 meters (10^1). Another example would be the Super High Frequency (SHF) band with wavelengths between 10^{-1} (100 cm) and 10^{-2} (10 cm). The bottom scale of Figure 4-6 gives the related frequency band limits. From the earlier examples on wavelength versus frequency, you could calculate the frequency limits for the HF band as 3 MHz and 30 MHz.

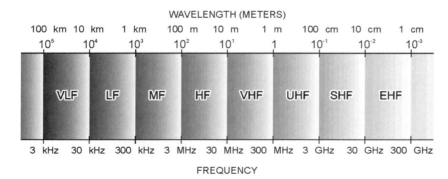

Figure 4-6: Wavelength Frequencies

The IEEE Radar Band Designations

A unique nomenclature exists for a category of frequency bands that use the letters L, S, C, X, Ku, K, Ka, V, and W. There seems to be no logic to the letters and the actual bands they represent, and for a good reason. These terms originated during World War II as a secret code so that scientists and engineers could talk about frequencies that were being considered for use with early radar systems without divulging them.

These codes were later declassified and remain in popular use today. The nomenclature is widely used today in radar, satellite, and microwave terrestrial communications. The frequency bands above 40 GHz, designated as V and W, were added later. These last two additional bands are still in the experimental category. The Radar Band designations were eventually adopted as a standard by the IEEE. Table 4-2 shows the IEEE Radar Bands and the associated letter designation for each band of frequencies.

Table 4-2:
IEEE Military Radar Frequency Bands

Frequency	Wavelength	Designation
1-2 GHz	30-15 cm	L Band
2-4 GHz	15-7.5 cm	S Band
4-8 GHz	7.5-3.75 cm	C Band
8-12 GHz	3.75-2.5 cm	X Band
12-18 GHz	2.5-1.67 cm	Ku Band
18-27 GHz	1.67-1.11 cm	K Band
27-40 GHz	1.11 cm -7.5 mm	Ka Band
40-75 GHz	7.5-4 mm	V Band
75-110 GHz	4-2.7 mm	W Band

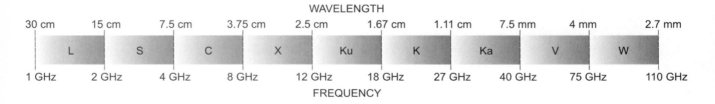

The Microwave Band

Frequencies at the higher end of the radio spectrum have been the subject of continuing research for a wide variety of reasons. The need for additional spectrum to meet the explosive demand for new wireless applications has been the major incentive. A general term used to describe the high end of the spectrum of frequencies (and also the term used for the electronic oven) is **microwave**. Various sources give varying definitions of where the microwave band begins and ends. Similar inconsistencies exist for the use of the term **short wave** as well. Some textbooks describe microwave frequencies very broadly as everything above 300 MHz.

microwave

short wave

> However, the wireless industry generally defines microwaves as all wavelengths between 1 GHz and 300 GHz.

So called short waves are generically described as frequencies between 5 MHz and 30 MHz and are the domain of international broadcasting stations and the amateur radio service.

Wireless Allocations

The selection, licensing, and management of frequency bands for U.S. wireless applications is administered by the Federal Communications Commission. Other international agencies described in Chapter 3 cooperatively work to coordinate the use of the radio spectrum between countries. The international cooperation is based upon the need to avoid the potential for radio interference between nations. Certain types of applications, such as satellite communications, maritime safety, and international broadcasting, require frequency planning and cooperation on an international scale. Much work remains to be done to provide international roaming for cellular telephone users through cooperative frequency planning. This is a complex issue for international standards bodies to deal with effectively due to the uncoordinated frequency bands that have evolved over time in different countries for domestic services.

Radio Frequency Propagation

The behavior of each portion of the radio spectrum in different environments has been studied and documented over a long period of time.

> Research has shown that different frequency bands have different **propagation** characteristics.

propagation

Wireless design engineers select appropriate frequencies from the different frequency bands for wireless applications that fit the objectives and goals of the system they are designing.

The FCC also allocates the frequencies for different services based upon the propagation limitations and features of the various radio bands.

Radio frequency bands can categorized into the following three major groups based upon their fundamental propagation characteristics:

1. Ground waves

2. Sky waves

3. Line of sight waves

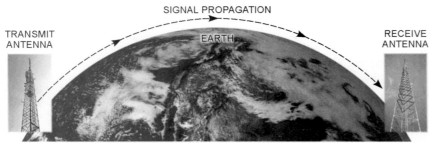

(a) GROUND-WAVE PROPAGATION (BELOW 2 MHz)

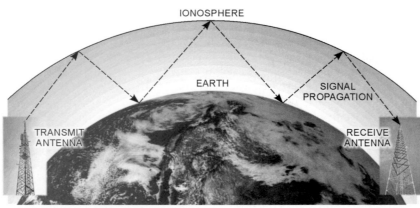

(b) SKY-WAVE PROPAGATION (2 TO 30 MHz)

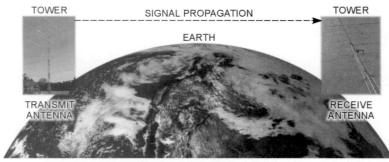

(c) LINE-OF-SIGHT PROPAGATION (ABOVE 30 MHz)

Figure 4-7: Radio Frequency Bands

As an example, it would be impractical for satellite communications systems to operate in the HF band, mostly due to propagation features. Conversely, the microwave bands do not have the proper propagation characteristics to serve long distance applications such as international broadcasting. Frequencies below 30 MHz in the HF, LF, and MF bands are primarily used for long range terrestrial communications. Frequencies in the UHF band work well for mobile and cellular applications but would not be suitable for long distance ship-to-shore communications.

We now will examine how these radio propagation features affect wireless communications across the different frequency bands and why we need to know more about them.

Figure 4-7 illustrates how each radio frequency is named according to its basic propagation path features.

Table 4-3 shows the frequency bands associated with each of the three groups.

Ground Waves	3 kHz – 2 MHz
Sky Waves	2 MHz – 30 MHz
Line of Sight Waves	30 MHz – 300 GHz

Table 4-3: Radio Propagation Groups

Ground Wave Propagation

Ground waves travel close to the surface of the earth, as illustrated in Figure 4-7. They occupy the region of the spectrum below approximately 2 MHz.

The advantages of ground waves are:

1. They can travel very long distances.

2. They are used by the military for communicating between land-based stations or aircraft and submarines.

3. Ground waves are dependable. They are relatively immune to atmospheric interference or propagation variations.

The disadvantages of ground waves are:

1. They require very large antenna structures.

2. They are expensive.

3. They are limited in the amount of information (bits per second) they can carry.

4. Outside of military and government applications, there is no practical commercial wireless application for ground wave frequencies below 300 kHz.

> **NOTE**
>
> The frequency spreads of the three categories have some overlap. As an example, frequencies between 500 kHz and 1.5 MHz are classified as MF, are used for standard commercial broadcasting, and exhibit some of the characteristics of both ground and sky waves under certain conditions.

Sky Wave Propagation

Sky waves

ionosphere

skip

Sky waves are frequencies between approximately 2 MHz and 30 MHz. Sky waves travel in a straight line until they encounter a layer of the atmosphere called the **ionosphere**. At that point they are refracted by the ionosphere and eventually reach a point on the earth at some distance from the transmitting station. The distance or **skip** of the propagation path, shown in Figure 4-7, depends to a great extent on the height of the ionosphere. The ionosphere provides the means for long distance communication. Typical applications include international broadcasting, amateur radio service, and ship-to-shore communications and maritime weather alerts.

The ionosphere is not a stable medium. It obtains all of its energy from the sun. It varies in its ability to refract radio waves over time. It can change throughout the year, or even over the day-night transition. The ionosphere also varies over the eleven-year solar cycle. A frequency that provides successful propagation at a specific time may not do so an hour later. There are reliable techniques, however, for predicting the optimum usable sky wave frequencies between any of the earth's regions on a day-by-day basis.

Electrons are produced in the upper regions of the atmosphere when solar radiation collides with uncharged atoms and molecules at altitudes from 50 km to approximately 500 km above the earth. Figure 4-8 shows the process of free electron production, which creates the layer called the ionosphere. Since this process requires solar radiation, production of electrons only occurs in the daylight hemisphere of the ionosphere. During daylight hours, the sun is constantly recharging the ionosphere with electrons. When a free electron combines with a charged ion a neutral particle is usually formed, as illustrated in Figure 4-8. Some loss of electrons occurs continually, both day and night.

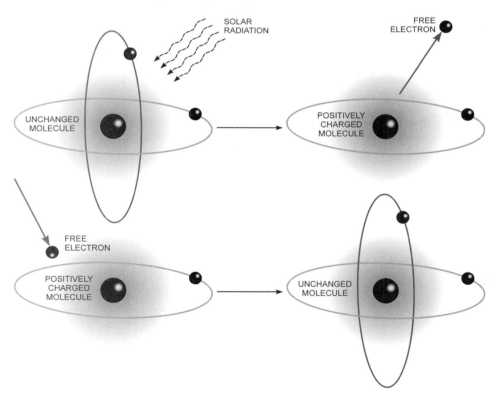

Figure 4-8: Neutral Particle Formation

The advantages of sky waves are:

1. They support long distance communications with relatively modest transmitter power and antenna requirements.

2. They can be used for point-to-multipoint voice service or low-speed radio teletype services on a global scale.

3. They support an economical maritime safety communications service.

4. Sky wave propagation frequencies extend over a relatively large portion of the spectrum (3 MHz to 30 MHz or approximately 27 MHz of bandwidth).

The disadvantages of sky waves are:

1. Sky wave propagation depends upon the ionosphere, which is not a stable medium.

2. The frequencies for sky waves are not suitable for carrying high-capacity data transmission circuits.

3. The sky wave frequencies are not suitable for most emerging wireless technologies such as cellular and wireless broadband applications.

Line of Sight Propagation

Radio waves above 30 MHz can penetrate the ionosphere and continue on in space without returning to a point on the earth. They travel in straight lines with characteristics similar to visible light waves. These frequencies are therefore limited primarily to **line of sight (LOS)** communications links between the transmitter and receiver, as illustrated in Figure 4-7. LOS radio wave propagation is in the range of the radio wavelengths less than 10 meters designated as the VHF, UHF, and SHF bands. Intuitively, LOS waves would appear to be of minimal value for wireless communications when compared to ground waves and sky waves. However, as you will discover in future chapters, LOS propagation is the primary domain for the new wireless revolution in communications. Propagation advantages, applications, and problems associated with this region of the spectrum will be the main area of interest for the remainder of this chapter.

line of sight (LOS)

Propagation Problems

A radio wave is also referred to as the **carrier wave** or the **signal**. As the names imply, some type of information is carried by the medium between a radio transmitter and receiver.

carrier wave

signal

Loss of transmitted power as it propagates through the medium is called attenuation (reduction in the magnitude of the transmitted power) and occurs with respect to both frequency and distance.

Propagation loss can be explained by considering the forces that degrade the communications link between the transmitter and receiver as two main types:

1. Path loss as a function of frequency and distance. A transmitted radio wave observes certain laws of physics. The radio signal loses energy at a constant rate as it travels through the air medium. Higher frequencies are absorbed by the atmosphere at a progressive rate that is nearly linear as the frequency is increased. Path loss is well understood and to a large extent predictable.

2. All other types of propagation problems are random with respect to time and intensity. They are more complex to deal with and are not predictable. As you will discover later, this type of propagation degradation is more severe in the mobile wireless network environment because the communications channel conditions are more dynamic due to movement.

The Bel

Before we investigate specific wireless propagation problems, we need to become familiar with the **bel**. The bel is a an exponential tool that uses logarithms for expressing the ratio of power transmitted and power received. This tool is very useful when calculating gain and loss in wireless radio links. The numbers associated with the ratio of power transmitted and power received can become large. It is easier to express the ratios exponentially. Another reason for using logarithms for the expression of power ratios is that the human ear processes sound intensity on a logarithmic scale. As an example, the perception of doubling the loudness is actually the result of a power increase of a factor of 10.

For those of you who need a refresher on logarithms, you need to recall that the logarithm to the base 10, expressed as **$\log_{10}$**, of a number is equivalent to how many times 10 is raised to a power to equal the number. As an example, $\log_{10} 100$ is 2 and $\log_{10} 1000$ is 3. Using the bel in a formula for expressing a power ratio gives:

$$B = \log_{10} \frac{P_o}{P_i} \text{ where } P_o \text{ is power out and } P_i \text{ is power in}$$

Another important feature of the bel equation is that the power out (received power) or numerator is often smaller than the power in (transmitted power). The result can be negative in this case. To illustrate this example, assume that the received power is one one-thousandth of the transmitted power. Then the equation becomes:

$$B = \log_{10} \frac{1}{1000} = -3$$

When using logarithms, a negative number indicates a power loss and a positive number illustrates a power gain.

bel

$\log_{10}$

Using the Decibel

A decibel is a more precise measure of power ratios because it is one-tenth of a bel. If we are using the decibel as the standard measure, we would revise the equation as follows:

$$dB = 10\log_{10} \frac{P_o}{P_i}$$

TEST TIP

Know how to express power levels in dBW and dBm.

The decibel has 10 times the resolution of the bel and provides a better measure of power ratios.

Using the previous example for calculating dB:

$$dB = 10\log_{10} \frac{1}{1000} = -30$$

Decibels are, as we can observe here, a measure of power ratios. Another form of power measurement is needed to express power as a fixed reference. One milliwatt (mW) is used as the fixed reference and is expressed as dBm. The term dBm is a reminder that we are using the absolute reference measurement value of 1 mW as opposed to power ratios. Every power level can now be expressed as a value greater than or less than 1 mW. A value of 10 watts would now be expressed as 40 dBm (**40 decibels above one milliwatt**) since it is 10,000 times larger than 1 mW.

Similarly, a reference level of 1 watt may be used. A value of 10 watts would then be expressed as 10 dBW (**10 decibels above one watt**).

The decibel and dBm are useful tools for calculating the gains and losses in a transmission path between a transmitter and receiver. Path losses can occur due to the impairments described in the following paragraphs.

Interference

> Interference refers to any electromagnetic disturbance that contains energy on or near the frequency being used for wireless communications.

Interference is often used in technical literature to define almost anything in the spectrum that disrupts the communications channel. The definition used here is restricted to those sources of electromagnetic energy from transmitters operating on or near the same radio frequency as well as electrical disturbances occurring in the atmosphere from lightning and electrical systems. This energy can cause the receiver to encounter errors when processing the transmitted information on the desired carrier wave. **Electromagnetic interference (EMI)** can originate from a variety of natural or man-made sources as follows:

Electromagnetic interference (EMI)

1. Lightning
2. Ignition noise from autos and trucks
3. Electrical motors and motor speed controllers

cochannel interference

4. Reception at the receiver of other transmitters operating on the same frequency. This is called **cochannel interference**. Cochannel interference is usually mitigated by proper separation in distance between transmitters and receivers using the same band of frequencies. This is managed through licensing of the use of the various radio frequencies by the FCC.

5. Adjacent channel interference caused by spillover of some of the transmitted energy between two or more radio frequency carrier waves operating on adjacent channels. This is also avoided by proper frequency planning and site survey in multiple carrier wireless networks.

6. Emissions of radio energy from faulty power lines and improperly installed TV cable systems.

Cellular and mobile wireless systems must deal with cochannel interference as the dominant interference factor. This topic is discussed in more detail in Chapter 5.

Path Loss

free space

Path loss is the measure of loss of the radio energy in a transmission path between the transmitter and the receiver. It is normally considered to be the loss that would be expected in **free space**.

inverse square law

Another way of thinking about it is to consider the loss that would occur if there were no objects in the region that could absorb or reflect the radio waves. This is the ideal situation, which we hope to achieve in the real world, but in fact, this is rarely the case. Free space loss is the starting point for engineers to determine the amount of power required and the types of antennas needed when designing wireless systems. Free space path loss observes a rule called the **inverse square law**. Path loss is not caused by the air having resistance such as the ohmic resistance you have encountered in electrical circuits. Path loss occurs because radio waves "spread out" as the distance increases. The inverse square law tells us that as the distance from a transmitting source to the receiver doubles (x2), the received power density decreases by the inverse of the square of 2 or $1/2^2$.

Assume, as in Figure 4-9, that p is a radio wave source that spreads out equally in all directions. We could then view the energy received at distance d as being the same at all points on the surface of an imaginary sphere. Reviewing high school geometry for a moment, the area of a sphere with radius R can be found with the formula $A = 4\pi R^2$. From the formula we can see the area increases in direct proportion to the square of the distance. If we doubled the radius R, the area covered would increase in direct proportion to the square of the radius in the outer sphere in Figure 4-9. The power density on any point on the sphere would now be spread out over a larger area relative to the square of the radius. The power density for a given area on the sphere would now be less relative to the inverse square of the distance. As explained above, this is expressed mathematically as $1/d^2$. Free-space loss, therefore, indicates that the power of the signal at any given point on the sphere diminishes as the inverse function of the distance of the receiver from the transmitter.

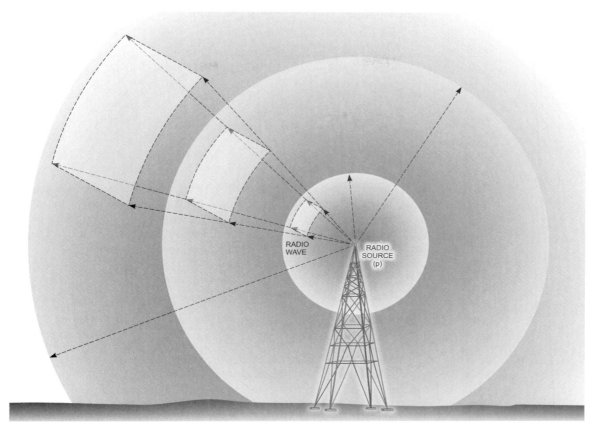

RADIO
WAVE

RADIO
SOURCE
(p)

**Figure 4-9:
Inverse Law**

The following example shows how the inverse law works:

Example 1

In free space, if the received signal is 100 watts at a distance of 1 mile, it will be 25 watts at 2 miles ($1/d^2$ or 1/4 of the power). At 4 miles it would be 6.25 watts ($1/d^2$ or 1/16 of the power), at 5 miles 4 watts ($1/d^2$ or 1/25 of the power), and at 10 miles 1 watt ($1/d^2$ or 1/100 of the power).

The inverse square law assumes a free space environment. The free-space path loss is the starting point for planning a microwave radio installation. Free-space path loss can be calculated much more easily with a formula using our decibel terms mentioned earlier. We will use decibels to calculate the gains and losses later when looking at the total link features.

The formula for calculating free-space path loss between two isotropic antennas is:

$$Lp = (96.6 + 20\log_{10} f) + (20\log_{10} d)$$

where:

Lp = free-space path loss between antennas (in dB)
f = frequency in GHz
d = path length in miles

Sample problem:

Calculate the free-space path loss for a microwave installation where the antenna separation between the two sites is 5 miles and the operating frequency is 18 GHz.

Lp = (96.6 + 20 × log *f*) + (20 log *d*)
Lp = (96.6 + 20 × log 18) + (20 log 5)
Lp = 96.6 + (20 × 1.2553) + (20 × .699)
Lp = 96.6 + 25.106 + 13.98
Lp = 135.68 dB

In the real world of wireless communications, total path loss may be much greater than free-space path loss, particularly in mobile systems. In some cases, engineers use the inverse of a cube ($1/x^3$) and even higher inverse powers, such as the 4th, 5th, or 6th inverse power, because of the other effects of the transmission path. Some of the causes of degradation beyond the ideal inverse square law are explained in the following paragraphs.

Atmospheric Effects

Weather and gases in the atmosphere can cause absorption of the energy in a radio carrier wave. The absorption is also related to frequency. For example, rain has a pronounced effect on radio waves at microwave frequencies. However, rain hardly affects radio waves in the HF range and below. Attenuation may be caused by absorption where a raindrop absorbs power from the radio wave and dissipates the power by heat loss or by scattering. Figure 4-10 shows the effect of raindrops on propagation. Raindrops cause greater attenuation by scattering than by absorption at frequencies above 100 megahertz. At frequencies above 6 GHz, attenuation becomes progressively more severe and peaks at 23 GHz. Oxygen in the atmosphere also causes absorption. It has a peak attenuation at approximately 60 GHz. The latter is not important to wireless propagation since most of the current services exist at frequencies below 30 GHz. Rain intensity or rain rate must be compensated for in the design of microwave and satellite networks because they use spectrum in the range of frequencies where rain affects propagation.

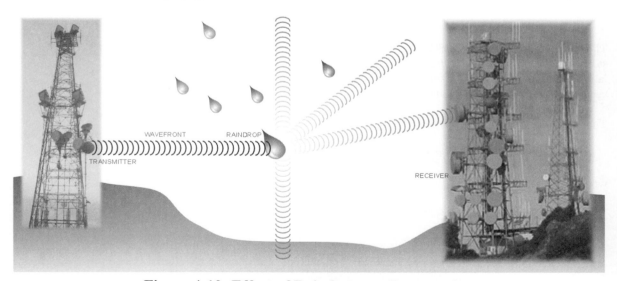

WAVEFRONT   RAINDROP

TRANSMITTER

RECEIVER

Figure 4-10: Effect of Raindrops on Propagation

Fading and Multipath

Path loss for the wireless link would be simple to calculate if no obstacles were present in the vicinity of the transmission path. Unfortunately, this is not the case. Signals used in fixed and mobile wireless systems in the UHF and SHF bands have wavelengths less than one meter. Radio signals in this part of the spectrum are subject to reflection from surrounding buildings, mountains, and other obstacles. A graphical representation of a typical mobile user environment, such as a cellular user, is shown in Figure 4-11. The radio propagation between the mobile user and the transmitting station is composed of several different paths. This condition creates a multipath environment that is often encountered in mobile wireless networks between a fixed base station and a mobile subscriber. Some signals may follow a direct line of sight path, shown as path 1 in Figure 4-11.

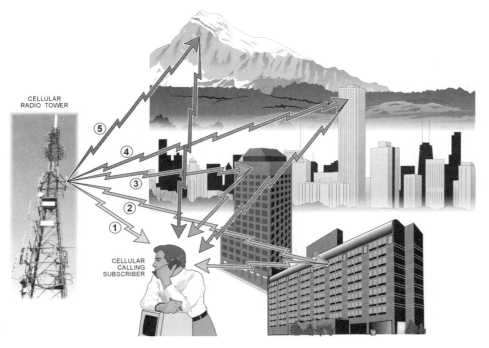

CELLULAR
RADIO TOWER

CELLULAR
CALLING
SUBSCRIBER

Figure 4-11: Typical Mobile User Environment

Other signals (paths 2 through 5) may travel over longer paths due to reflections from objects. The reflected signals may therefore arrive at the mobile station with a different phase relationship when compared to the direct path signal. If one of the reflected signals is delayed sufficiently to cause the phase to be 180 degrees ($\lambda/2$) out of phase with the direct signal represented in Figure 4-12, the two signals cancel each other, reducing the received signal power available at the mobile subscriber's location.

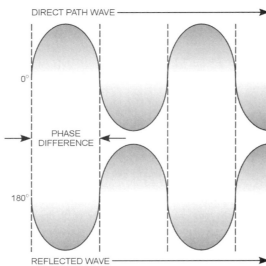

DIRECT PATH WAVE

0°

PHASE
DIFFERENCE

180°

REFLECTED WAVE

Figure 4-12: Effect of Raindrops on Propagation

As the mobile station moves with respect to the fixed cellular transmitting station, the received power at the mobile station becomes more complex. The received power will fluctuate depending upon the combined phase relationship of the multipath signals. Cancellation would occur between waves that arrive at multiples of each odd number of half wavelengths ($\lambda/2$, $3\lambda/2$, $5\lambda/2$, $7\lambda/2$, etc). The plot on a time scale of the received signal power is shown in Figure 4-13. The deep nulls called **fades** typically occur when the mobile station moves at intervals of half a wavelength at the operating frequency being used. Fading and multipath are particularly difficult problems for the wireless network designer to deal with in a mobile wireless environment.

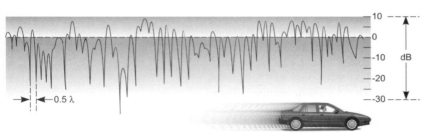

Figure 4-13: Received Signal Power

Fresnel Zones

The radio signal in a point-to-point microwave link occupies space between the transmitter source and the receive site. Although the transmission path is usually illustrated as a straight line for simplicity, in reality the signal components are contained in an ellipsoid, as illustrated in Figure 4-14. This region is called the Fresnel zone. The regions are divided into zones with each zone being concentric with the first. The first zone, illustrated in Figure 4-14, is a surface containing all points for which the sum of the distances from that point to the two ends is exactly ½ wavelength longer than the direct path. Each zone surrounding the first will be some multiple of ½ wavelength either out of phase or in phase with the direct wave, thereby producing alternate constructive and destructive interference zones due to multipath signals.

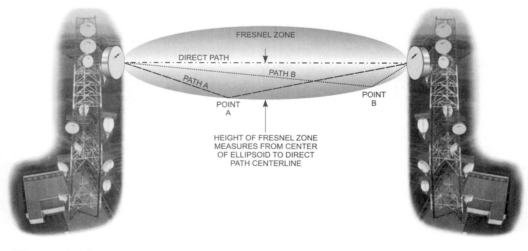

Fresnel zone 1 is a surface containing every point for which the sum of the distances from that point to the two ends of the path is exactly ½ wavelength longer than the direct end-to-end path. Point A and Point B are examples where the sum of each path is points ½ wavelength longer than the direct path.

As covered earlier, wavelength is inversely proportional to frequency, therefore the size and shape of the Fresnel zones are frequency dependent. When planning a microwave link, the parameters of the Fresnel zone must be known in order to plan the tower height to provide clearance from the earth surface and objects between the towers. As long as 60 percent of the first Fresnel zone is clear of the earth and obstructions, the radio transmission path behaves essentially the same as a clear free-space path.

Figure 4-14: Fresnel Zone

The formula in Table 4-4 is used to calculate Fresnel zone 1 that incorporates the 60 percent factor mentioned earlier.

$$H = 43.26 \sqrt{\frac{D}{4F}}$$

Where:

H = height of the first Fresnel zone (feet)
D = distance between the antennas (miles)
F = frequency (GHz)

Example:

Find the height of the Fresnel zone where the distance between the towers is 16 miles and the frequency is 2 GHz.

$$H = 43.26 \sqrt{\frac{16}{8}} = 43.26\sqrt{2} = 61.18 \text{ ft}$$

Table 4-4: Calculation for Fresnel Zone

Remember, the height of the Fresnel zone in the formula refers to the distance from the first zone edge to the direct path centerline, and not to the height above the earth.

Ground Reflections

The Earth may act as a reflector for radio waves. Under the proper conditions, the reflected earth wave may cause fades similar to multipath signals from other surrounding objects.

Ground reflections occur over broad, flat areas rather than at a fixed point, as indicated in Figure 4-15. In metropolitan areas, ground reflections may be blocked by buildings or trees. Ground reflections can be diminished between fixed locations by adjusting the antenna height or by using narrow beam antennas that focus more of the radio signal energy into the direct path. When surveying for a microwave link, it is important to look at the ground profile between the antennas to see if problems may be potentially present. Remember also that the reflection point may not be at the midpoint if the antennas are not equal in height or if the ground is sloped between the towers.

Figure 4-15: Ground Reflections

Earth Curvature

space waves

line of sight (LOS)

As stated earlier, radio frequencies in the upper portion of the radio spectrum (UHF and higher) are classified as **space waves**. They generally follow a straight line path and are often referred to as **line of sight (LOS)** wavelengths because they require the transmitter and receiver to be above the horizon, otherwise the Earth will block the line of site path. When planning for a link that is greater than 7 miles, the curvature of the Earth may become a factor.

> Earth curvature limits the distance between two fixed LOS transmitting and receiving stations.

The rate of the curvature of the Earth establishes the radio horizon. A similar phenomenon occurs for ships at sea when they appear and disappear on the line of sight horizon. The line of sight can obviously be extended by raising the antenna height to overcome the curvature of the Earth. The earth curvature as a function of distance to the horizon can be calculated as follows:

$$\frac{D^2}{2} = H$$

where:

D = distance to the horizon
H = antenna height required to compensate for earth curvature (feet)

Examples are listed in Table 4-5.

Table 4-5: Total Antenna Height Formula

Tower Height (Feet)	Distance to Horizon (Miles)
100	14.14
150	17.32
300	24.5
400	28.3

When it is necessary to calculate the antenna height at each end of the transmission path, the following simple formula may be used:

$$H = \frac{D^2}{8}$$

where:

D = total path length between the antennas (miles)
H = height of each antenna required to compensate for the earth curvature (feet)

(These calculations assume that both antenna sites are on near level terrain and are approximately the same height above mean sea level.)

Example:

What height would be required for each antenna site when the total distance between the towers is 22 miles?

$$H = \frac{D^2}{8} = \frac{(22)^2}{8} = 60.5 \text{ miles}$$

Calculating the Total Antenna Height

We now must plan for an antenna height at both ends of the transmission path that includes the clearance for the first Fresnel zone as well as the additional height required for earth curvature. The formula for both now becomes:

D = distance between the antennas (miles)
H = antenna height required to compensate for the earth curvature and first Freznel zone (feet)
F = frequency in GHz

$$H = 43.26\sqrt{\frac{D}{4F}} + \frac{D^2}{8}$$

Example:

Calculation the total antenna height for two sites located 16 miles apart. The operating frequency is 2 GHz.

$$H = 43.26\sqrt{\frac{16}{8}} + \frac{256}{8}$$

$$H = 43.26(1.414) + 32 = 61.18 + 32 = 93.18 \text{ feet}$$

Link Budgets

Link budgets are used as a planning tool for microwave transmission installations for terrestrial networks. They are also used to plan the installation of satellite earth stations. Link budgets are derived by simply adding together all of the gains and losses that are present between the transmitter location and the receiver site. Since all systems typically operate full duplex, the link budget is applicable to the path in both directions.

The purpose of the link budget is to determine whether the signal power delivered to the receiver is sufficient to "close the link." All microwave radios have a threshold that determines the minimum amount of received signal power necessary to amplify the signal. If the received signal is below this value, the system will not work.

Some additional link margin must be allowed in the design over and above the threshold value for the receiver. Interference from transient events such as rain, high wind, electrical interference from storms, man-made random interference, and other disturbances takes up some of the margin or "headroom" for the receiver. Most plans for terrestrial microwave systems include link margins of 20–25 dB. These values will vary due to rain rate tables and geographical location. For example, additional margins would be needed in tropical regions where rain rate attenuation may be high. In the following exercise, you will work with some examples for calculating link budgets and link margins.

Lab 2 – Planning a Microwave Network Site Installation

<u>Lab Objectives</u>

Prepare a planning document for the installation of a microwave wireless network link.

Work with the formulas and planning considerations for installing microwave towers, antennas, connectors, and cabling.

Calculate the values required to complete a link budget work sheet.

<u>Materials Needed</u>

- This textbook contains the formulas and data you will need to complete this lab.
- Tables and worksheet examples are shown in the following lab portion.
- Hand calculator (scientific)

<u>Instructions</u>

This lab exercise will take you through the steps necessary to calculate the values for planning a microwave radio site installation. The plan covers two microwave transmitter sites operating full duplex over a distance of 13 miles. The overall system plan is illustrated in Figure 4-16. Values in actual site planning may vary due to terrain, manufacturer specifications, and climate. The values in this exercise are selected to show the process necessary to develop a link budget.

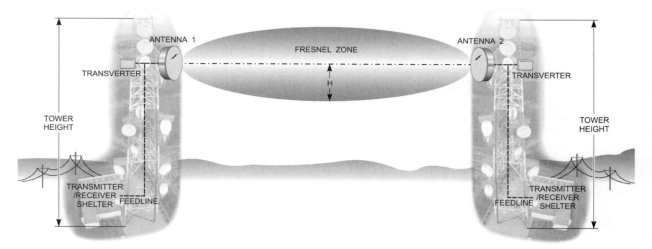

Figure 4-16: Fresnel Zone

The transverters shown in the figure are mounted on the towers near the antenna to reduce cable loss at the higher operating frequency. The transverter changes (up converts) the feed line frequency from the transmitter of 700 MHz to 11 GHz. On receive, the transverter changes (down converts) the incoming frequency from 11 GHz to 700 MHz.

The links are licensed to operate in the 11 GHz band. The FCC rules for licensing stations in this band are contained in Title 47, Code of Federal Regulations (CFR), Part 101.

TIP: Part 101 of the FCC rules state that a maximum EIRP of 55 dBW is permitted in the 11 GHz band.

The survey team has determined that the path between the two sites is free from obstructions and the terrain is reasonably flat. Your task is to perform a link analysis of the two sites. You will calculate the required tower height, feed cable loss, and free-space path loss. When you complete the computations, you will need to determine whether the link margin can be satisfied with the values you have selected.

1. Make a worksheet to record the values as you progress through each step using Table 4-6 as a model, or complete the answer by recording them in the book if you wish. Be sure to check your work after each step.

TIP: The transmitter power for each site and the antenna gain has been specified in the FCC license. The values are shown in Table 4-6.

Item		Value
1.	Transmit Power (dBW)	9 dBW
2.	Antenna Gain (dBi)	38.9 dBi
3.	Fresnel Zone Height (Feet)	
4.	Antenna Height	
5.	Total Height (Sum of items 2 and 3)	
6.	EIRP (dB)	
7.	Cable Loss Transmitted (dB)	
8.	Free Space Path (dB)	
9.	Receive Antenna Gain (dBi)	38.9 dBi
10.	Cable Loss (Receive dB)	
11.	Total Path Loss (dB)	-79 dB
12.	Minimum Receiver Signal Threshold (dB)	
13.	Link Margin (dB)	
14.	Link Margin Minimum (dB)	20 dB

Table 4-6: Link Budget Work Sheet

2. Calculate and record the Fresnel zone height using the formula:

$$H = 43.26\sqrt{\frac{D}{4F}}$$

where:

D = distance between the antennas (miles)
F = frequency (GHz)

3. Calculate and record the required antenna height using the formula:

$$H = \frac{D^2}{8}$$

TIP: The sum of the values in steps 2 and 3 is the required tower height for the two sites. You may assume that the two sites are the same height above mean sea level.

4. Calculate the Effective Isotropic Radiated Power (EIRP) by adding the antenna gain (dBi) and the transmit power (dBW). The values for the antenna gain and the transmit power are specified in Table 4-6.

5. Calculate the cable loss for the total tower height. (The transmit frequency from the shelter to the transverter is 700 MHz.) Use the values shown in Figure 4-17 to calculate cable loss. Use the total tower height to calculate the loss. Do not forget the two extra feet at the transverter shown in Figure 4-16. The cable specifications list cable loss in dB/100 feet, so you will need to calculate the loss on the tower height in proportional amounts of 100 feet. Remember to include the waveguide loss at the transverter.

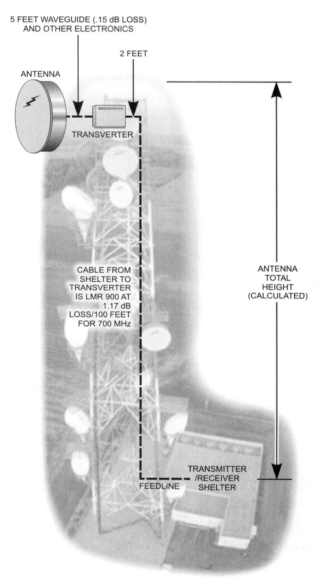

Figure 4-17: Cable Attenuation

TIP: Remember, combined waveguide and cable loss is a negative number or loss in the link budget.

6. Calculate and record the free-space path loss using the following formula discussed earlier in the text.

$$Lp = (96.6 + 20 \log_{10} f) + (20 \log_{10} d)$$

where:

Lp = free-space path loss between antennas (in dB)
f = frequency in GHz
d = path length in miles

TIP: Remember, free-space path loss is a negative number or loss in the link budget.

7. Record the receiver antenna gain given in Table 4-6.

8. Record the waveguide loss from the antenna to the transverter and the loss from the transverter to the shelter in item 10. (These values will be the same numbers calculated at the transmitter site. The two sites operate full duplex and are essentially the same installation at both sites.)

9. Add the sum of the values in your worksheet. Record this value in your worksheet in line 11 to obtain the total loss and gain summary. This represents the amount of signal power remaining at the input to the receiver.

TIP: Item 12 lists the minimum signal threshold for the receiver used in this installation.

10. Compare the total signal power available at the receiver to the required minimum signal threshold. Record the difference on line 13. This represents the margin for this installation.

11. Compare the link margin on line 13 with the minimum selected link margin for this installation on line 14.

Lab 2 Questions

1. Will the plan using the values you have computed in this exercise provide the minimum (or better) value for the link margin?

2. What is the purpose of establishing a link margin for a microwave transmission system such as the example shown in this exercise?

3. What was the calculated EIRP of the two sites? Were these values in compliance with the maximum allowable EIRP for this frequency band as stated in part 101 of the FCC rules?

Basic Radio System Architecture

The basic architecture applicable to all wireless systems can be shown as consisting of the following basic elements:

The Radio Transmitter Processing Chain

1. A source encoder
2. A channel encoder
3. A multiplexer
4. A modulator
5. A radio frequency source
6. A radio frequency amplifier
7. An antenna

The Radio Receiver Processing Chain

1. An antenna
2. A radio frequency amplifier
3. A demodulator
4. A radio frequency source
5. A demultiplexer
6. A channel decoder
7. A source decoder

The functional flow of information in a basic digital radio system is illustrated in Figure 4-18.

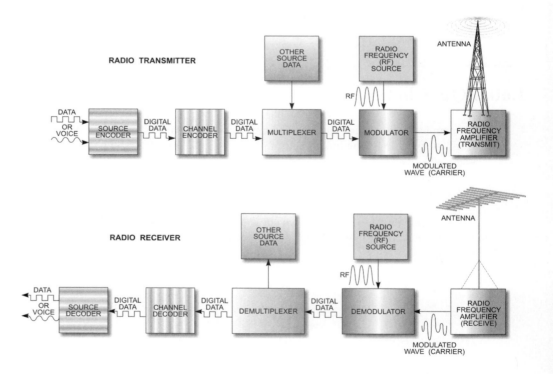

Figure 4-18: The Functional Flow of Information

The Transmitter Functional Diagram

The following is a functional description of the tasks performed by the transmitter modules.

Source Encoder

The **source encoder** accepts input digital data and analog voice information for processing prior to the modulator and channel coding functions. Source coding converts the different formats received by the wireless transmitter into the correct data rates and format required for transmission over the wireless link. The processing tasks may involve one or more of the following depending upon the type of wireless application:

source encoder

1. Voice sampling and quantization

2. Voice waveform conversion to **Pulse Code Modulation (PCM)**

Pulse Code
Modulation (PCM)

3. Digital voice compression to a lower bit rate

4. Data input formatting and data compression

5. Encryption

Channel Encoder

Channel encoding adds extra bits to the information bit stream. The additional redundant coded bits are called **convolutional codes**. They are designed to improve the performance of the channel and reduce the bit error rate on the radio link. This is done at the expense of using more bandwidth. The added bits are structured such that the receiver is able to use them to detect and correct errors. This process is widely used in wireless communications. It is called **Forward Error Correction (FEC)**.

Channel encoding

convolutional codes

Forward Error
Correction (FEC)

Multiplexer

Multiplexing refers to the sharing of a fixed communications channel. In the digital processing model shown earlier, the multiplexer adds bits from other source encoders to the information bit stream to form alternating frames of data. Individual frames are passed to the modulator serially to form a composite modulated signal. This allows multiple information sources to share a common radio channel. The net effect is a reduction in the number of radio channels required since several sources are time sharing the radio channel although at a higher composite bit rate.

Multiplexing

Modulator

The **modulator** receives the digital bit stream and modulates a radio carrier signal. The modulator can be viewed as a mixer of two frequencies. It multiplies the carrier frequency and the bit stream frequency to produce new radio frequencies adjacent to the radio carrier wave. This additional sideband information plus the radio carrier wave is the direct output product of the modulator.

Radio Frequency Source

The **radio frequency** or **RF source** provides the radio carrier wave at a specified assigned frequency. It is applied to the modulator as the component for modulating the information bit stream. This radio frequency source is developed by an oscillator circuit that is tunable on command from the internal digital controller. Cellular handsets, for example, contain radio frequency control circuits to automatically tune the subscriber's phone to the frequency required by the wireless cellular network.

Radio Frequency Amplifier

The modulated radio wave is amplified by the **radio frequency amplifier** to boost the output of the modulator to a higher power level. The amount of amplification depends upon the power required to produce an acceptable signal above the noise and interference at the input of the target receiver. As an example, a power amplifier of several hundred watts may be required for a satellite earth station, but a cellular phone may require one watt or less at the output of the radio frequency amplifier.

Antenna

The radio transmitter **antenna** provides the interface between the radio frequency amplifier and open space. Antenna designs vary widely depending upon the directivity and the range of frequencies required for the specific application. Antenna design features are covered in more detail later in this chapter.

The Receiver Functional Diagram

The following is a functional description of the tasks performed by the receiver modules.

The Antenna

The receiving antenna captures the transmitted radio frequency carrier and couples it to the radio frequency amplifier. The transmitting and receiving antennas are often the same device. The antenna may receive on one frequency and also transmit energy from the radio transmitter at the same time. This mode of operation is called full duplex and requires a duplexer to isolate the transmitter output from the receiver input (optional depending on the mode of operation).

The Radio Frequency Amplifier

The received modulated carrier wave is extremely weak and requires an amplifier to increase the amplitude of the received signal. The receiver radio frequency amplifier increases the sensitivity of the receiver by boosting weak signals to a higher level prior to processing in the receiver circuits.

The Demodulator

The **demodulator** performs the reverse function of the modulator. Several tasks are performed by the demodulator for the purpose of retrieving the original digital data bit stream. The inbound modulated carrier is mixed with a local radio frequency source called the local oscillator. The mixing reduces the radio carrier and sidebands to a lower intermediate frequency, which is processed further by a detector circuit that filters out the radio carrier frequency and captures the original data bits.

demodulator

The Radio Frequency Source

The radio frequency **oscillator** is tuned to the correct frequency to mix with the desired inbound carrier wave. Since many other carriers closely spaced to the desired carrier may enter the demodulator, the radio frequency source value is chosen such that the desired input carrier wave is tuned similar to the tuning of a broadcast receiver to the desired frequency. All other carrier waves operating on other frequencies are rejected by the detector circuits by a filtering process since they do not fall into the band of frequencies called the **bandpass** of the receiver input circuits. The local radio frequency source can therefore act as a selective tuner for aligning the modulator to process the desired received radio signal.

oscillator

bandpass

The Demultiplexer

NOTE

The output from the multiplexer for other source data will be processed by another external channel decoder and source decoder.

The **demultiplexer** captures the data frames that were originally combined (multiplexed) in the transmitter. The frames are examined for routing and address labels for proper routing to the destination.

The Channel Decoder

The **channel decoder** uses the structured redundant bits to perform error correction on the received bit stream. Using a complex error correction algorithm, the channel decoder can correct bit errors without assistance or repeat message requests from the transmitter. This is the basis for correcting errors at the "forward" received location (from which the term forward error correction or FEC is derived).

The Source Decoder

The **source decoder** is the final processing function designed to reformat and restore the data bits to their original condition as presented to the transmitter. Depending upon the protocols used, the source decoder may include one or more of the following:

1. Conversion of the compressed digital voice data stream using decompression algorithms and digital to analog (D to A) conversion, thereby restoring the voice signal to the analog format as was applied to the transmit source encoder.

2. Data input formatting and data decompression

3. Decryption antenna (if required)

Antenna

As shown in the radio system diagram in Figure 4-18, the antenna is a device for radiating or receiving electromagnetic waves. It plays a very important role in a wireless communication system. You will need to become familiar with the characteristics, performance, and design issues of the types of antennas used for UHF and SHF frequency bands. These bands are the primary domain for the systems covered in this study guide. They include cellular, satellite, and microwave system applications.

There is little fundamental difference between transmitting and receiving antennas, since the same type of antenna is used for both purposes with only minor variations. The following is a summary of the characteristics of both receiving and transmitting antennas.

Polar Plots

By far, the most important property of an antenna is its radiation pattern or polar diagram. In the case of a transmitting antenna, this is a plot of the power field strength radiated by the antenna in different angular directions. **Polar plots** can be obtained for the vertical and horizontal planes, and they are called the vertical and horizontal patterns respectively.

Figure 4-19 shows the two polar plots. Notice that the horizontal plot shows, in this case, an antenna considered to be omnidirectional. In other words, it radiates and receives power equally well in all directions.

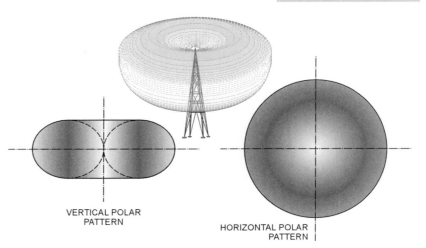

VERTICAL POLAR
PATTERN

HORIZONTAL POLAR
PATTERN

**Figure 4-19:
Two Polar Plots**

The Gain of an Antenna

The merit of an antenna, especially one designed to focus transmissions in a particular direction, is expressed in terms of the **antenna gain**. The gain is generally based upon a comparison with a reference antenna. The reference antenna is an **isotropic radiator** that theoretically radiates the same power in all directions.

The gain of any antenna is therefore based on the amount of power supplied to the reference antenna to the power that would be required to the antenna under test in order to produce the same field strength in the desired direction of the receiving antenna. An example of calculating gain is as follows:

Example 1

During an installation test, 1 kW of power is required at the feed point of a parabolic microwave antenna to produce 20 microvolts per meter of flux density at a receiving antenna located on a transmission tower 22 miles away. An omnidirectional antenna at the same transmitter location requires 18.8 kW to produce the same received signal strength at the receiving antenna. The gain for the test antenna is calculated as:

$$\text{Gain(dB)} = 10 \log \frac{P_2}{P_1} = 10 \log \frac{18.8}{1} = 12.7$$

Microwave Antennas

microwave antennas

At microwave frequencies, antenna dimensions that are convenient to handle can reach several wavelengths in size. It is therefore possible to apply some optical approaches to **microwave antennas** where the beam of radio energy can be focused like a visible light beam. The main application of microwave dish-type antennas for the frequency range of 1 GHz and more is in terrestrial point-to-point communications, satellites, and radar. In such systems, high-gain antennas with very narrow beam widths in one or more planes are required and may be achieved with antennas of reasonable size. Parabolic and horn-type antennas are used for point-to-point microwave communications. Horn-type antennas are often used for parabolic reflector feed systems or as stand-alone antennas.

> Parabolic reflector and horn-type antennas are efficient at microwave frequencies.

A typical array of horn-type antennas used for frequencies in the microwave region is shown in Figure 4-20.

**Figure 4-20:
Horn-Type Microwave
Antennas**

A parabolic surface has the useful property of being able to convert a diverging spherical wavefront into a parallel plane wavefront, producing a highly focused and narrow beam. Figure 4-21 illustrates this feature. Notice that all angles formed on the parabolic surface converge at the focus of the antenna.

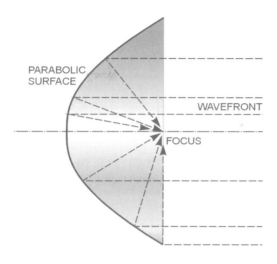

**Figure 4-21:
Converting a
Wavefront Feature**

Phased Array Antennas

As illustrated in Figure 4-22, phased array antennas are used in **smart antenna** cellular applications.

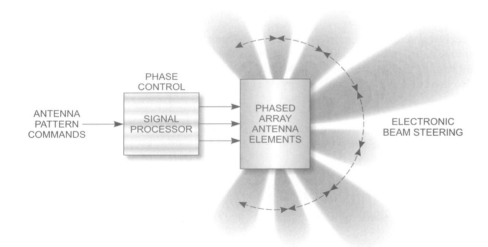

**Figure 4-22:
Smart Antenna
Cellular Applications**

By dynamically altering the phase shift between successive elements in an array antenna, the pattern and directivity of the antenna array may be steered electronically without physically moving the antenna structure.

Modulation

> **Modulation** is the process of combining the information to be transmitted with a radio carrier wave.

This requires a **modulator**, as shown in the radio system diagram earlier. All wireless devices, including cellular phones, wireless pagers, and satellite transmitting stations, require some type of modulator to combine the **baseband** source encoded information with a radio frequency transmitter. The modulated wave is passed to the radio frequency amplifier, which increases the power level to the required level to overcome the propagation problems discussed in the preceding paragraphs. The transmitter also couples the amplified modulated carrier wave to an antenna. The transmitter antenna is designed to provide the maximum transfer of power in the direction of the intended receive station. Radio transmitter power levels are regulated through licensing by the FCC to avoid interference to other licensed users sharing the same spectrum. There are also unlicensed radio bands for special low-power wireless applications, such as cordless telephones and wireless LAN devices, that are discussed in the next chapter.

The receiving station is the mirror image of the transmitter. It consists of a receiving antenna, a receiver that is tuned to receive the required frequencies, and a **demodulator**. The demodulator performs the reverse function of the transmitter and extracts the baseband information from the received radio carrier.

The receiver and transmitter circuits are combined into a single device to facilitate two-way communications. Modulation and demodulation functions are often combined into a single assembly, thus the basis for the name **MODEM** for **MOD**ulator – **DEM**odulator. Modems are also used over wired circuits to provide signaling between computers or other digital systems.

Modulation can be accomplished by altering one or more of three parameters of a radio carrier wave with the baseband information as follows:

1. Changing the amplitude of the radio carrier (**Amplitude Modulation** or **AM**)
2. Changing the frequency of the carrier wave (**Frequency Modulation** or **FM**)
3. Changing the phase of the radio carrier wave (**Phase Modulation** or **PM**)

Figure 4-23 illustrates the three types of modulation employed in wireless communications systems. There are numerous variations of these three basic types of modulation for different applications.

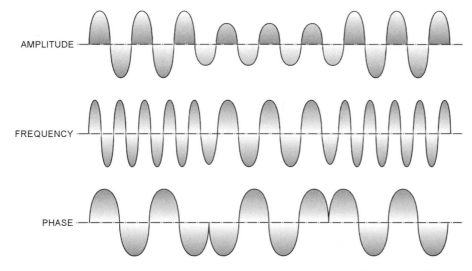

AMPLITUDE

FREQUENCY

PHASE

Figure 4-23: Three Types of Modulation

The information sources may be in the form of an analog signal or a digital signal. The modulator treats each of them in a similar but also different manner as we shall see next.

First, a definition of both analog and digital information is in order. You may have heard about both types when working with wired networks, so they are probably familiar terms to most of you.

Analog information has (nearly) an infinite number of values and is not predictable. This means that the receiver must receive and demodulate an exact replica of the transmitted analog information, which can have a wide variety of values from time to time with no specific prior indication of what the transmitter is going to send. This is the type of information transmitted by commercial broadcasting and television stations using AM and FM to carry voice, music, and video. Analog voice information is also transmitted and received over first generation cellular systems using FM. AM and to a lesser extent FM are subject to electromagnetic interference and gradual degradation along the transmission path. The receiver demodulates the analog wave, as shown in Figure 4-24, with all the problems of multipath, atmospheric, and natural interference. If the wave is corrupted, the receiver has no method for separating the distortions from the transmitted analog information since the intended signal may have an infinite set of amplitude values.

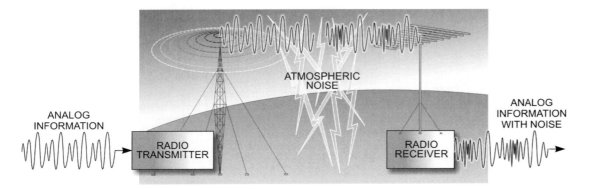

ATMOSPHERIC NOISE

ANALOG INFORMATION

RADIO TRANSMITTER

RADIO RECEIVER

ANALOG INFORMATION WITH NOISE

Figure 4-24: Demodulating the Analog Wave

The Radio Channel, Bandwidth, and Channel Spacing

A modulated radio carrier wave uses **bandwidth**; therefore, each radio carrier frequency must be assigned to a specific channel that provides enough space between frequencies to avoid interference or "spillover" of power into the adjacent channel.

When modulation is imposed upon a radio carrier wave, additional frequencies adjacent to the carrier wave with additional power content are introduced at the output of the transmitter.

The modulated radio wave is best observed by plotting the power versus the frequency of the transmitter output in Figure 4-25. The new power components are called sidebands. The amount of bandwidth required depends upon several factors:

1. For digital modulation, the bit rate of the source data determines the bandwidth created by the modulated carrier wave. Higher bit rates consume more bandwidth. The ratio of the bandwidth to the bit rate of a modulated carrier is expressed in bits per second per hertz (b/s/h).

2. The higher level modulation techniques such as QAM are more efficient because they result in the transmission of many bits per symbol. Therefore, they are more "bandwidth efficient."

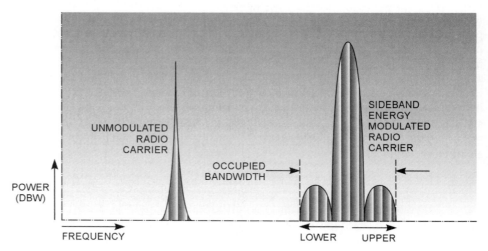

Figure 4-25:
Power vs Frequency of the Transmitter Output

3. Commercial wireless systems are allocated specific frequency bands for licensed operation by the FCC. The frequency bands are further divided into **channels** or bandwidth for each radio carrier wave. Upper and lower limits of the frequency band are defined for each channel. Different wireless applications use different amounts of radio spectrum. For example, a television channel uses 6 MHz channels. A standard analog cellular radio channel assigns each call to a 30 kHz channel for voice, but an FM radio broadcast channel is 200 kHz wide. Figure 4-26 illustrates some concepts of channel frequencies and channel spacing.

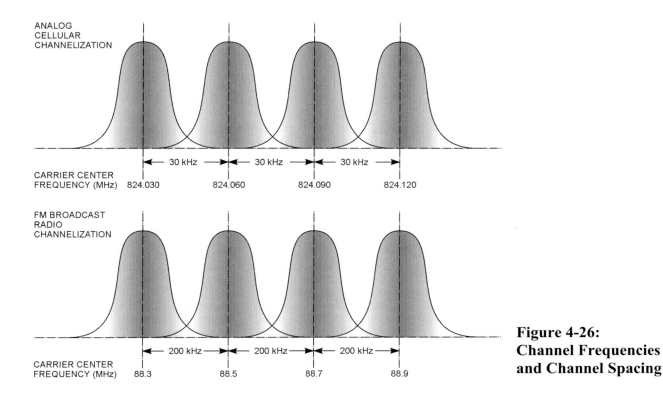

ANALOG CELLULAR CHANNELIZATION

← 30 kHz → ← 30 kHz → ← 30 kHz →

CARRIER CENTER FREQUENCY (MHz)   824.030   824.060   824.090   824.120

FM BROADCAST RADIO CHANNELIZATION

← 200 kHz → ← 200 kHz → ← 200 kHz →

CARRIER CENTER FREQUENCY (MHz)   88.3   88.5   88.7   88.9

**Figure 4-26:
Channel Frequencies
and Channel Spacing**

Transmission Modes

Wireless transmission systems operate in one or more of the following three modes, as shown in Figure 4-27.

1. **Simplex** mode, where the transmission always travels in one direction from a transmitter to a receiver.

2. **Half-duplex** mode, where communications travel in both directions, but not at the same time.

3. **Full-duplex** mode, where the information travels over a radio link in both directions at the same time. Full-duplex mode requires two separate radio carriers operating on different frequencies.

┌─ **TEST TIP** ──────────────┐

Be able to describe the three modes of transmission systems.

└────────────────────────────┘

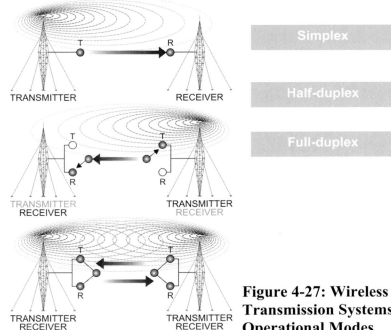

**Figure 4-27: Wireless
Transmission Systems
Operational Modes**

Digital Advantages

The main reasons for going to a digital transmission format are as follows:

signal-to-noise ratio

1. The main challenge for a wireless radio link is to deliver to the receiver a modulated radio carrier wave that will survive the major challenges of propagation losses with as few errors as possible. The receiver's ability to properly demodulate an analog received signal depends upon the **signal-to-noise ratio**, which is a measure of the original radio carrier signal strength compared to the radio noise (static) present at the receiver antenna. If an analog transmission becomes corrupted with noise, there are only limited options available for improving performance at the receiver. The receiver is often unable to distinguish between the near infinite number of analog values being transmitted compared to random noise.

forward error correction (FEC)

2. Powerful error detection and correction algorithms are available for handling errors on a digital wireless channel. By adding a specific pattern of redundant bits to the transmitted bit stream, errors can be corrected "on the fly" at the receiver. This process is called **forward error correction** or **FEC**. The receiver can therefore process a received digital signal by making the best guess between the probable reception of a binary 1 or 0. It is not concerned with the distortion or signal-to-noise corruption of the received signal that occurs with analog signal processing. Perfect reproduction of the received bits is possible at the receiver since the measure of quality is now based upon making the correction decision between two binary values.

FEC is the basis for the human perception of "good" quality when using a cellular phone that employs digital processing with FEC. The perceived improvement in quality is attributable to a much improved bit error rate for a digitally encoded voice transmission. This processing enhancement is not possible with analog FM or AM voice transmissions.

Digital data

3. **Digital data** transmissions can use compression algorithms to reduce the amount of information transmitted over a radio link. This results in conservation of radio spectrum and much greater throughput of information per unit of time. The compression algorithm compresses the information through a complex mathematical process and the receiver is able to decompress the information and restore it to its original form. Compression of digital channels is the underlying technology that supports digital direct broadcast satellite TV service.

4. Digital data can share the same radio channel in the time domain with other similar information channels by packaging information in small clusters called packets. This is similar to wired packet switched networks. Packets can then be transmitted in specific time slot intervals on the radio channel. This process is called multiplexing. Unlike analog information, digital bit streams can be segmented into packets and interleaved with other bit streams. The bit streams can be demultiplexed at the receiving location. This results in more efficient use of radio spectrum.

Digital Modulation

Digital modulation is the act of changing one or more of the three attributes of the radio carrier such that the instantaneous state of the carrier wave can be used to represent a binary encoded value of a logical 1 or a 0. Each type of modulation has some advantages that favor specific applications. For example, PSK is widely favored for wireless applications. Other types such as QAM are more efficient because they transmit multiple bits for each state change; however, they require a higher quality radio link condition to maintain a reasonable bit error rate performance. Every choice has a tradeoff between complexity and spectral efficiency. Popular types of digital modulation will now be described.

Amplitude Shift Keying

Amplitude Shift Keying (ASK) is the simplest form of modulation whereby the radio carrier is turned on and off. The amplitude on and off changes represent binary 1s or 0s. **Keying** is the signaling concept of the telegraph, which has remained in the vocabulary of digital transmission terms over many years. **On-Off Keying (OOK)** is a form of ASK accomplished very simply by turning the carrier off and on at predetermined synchronized times or using different intervals of on and off times to represent binary 1s or 0s, as indicated in Figure 4-28.

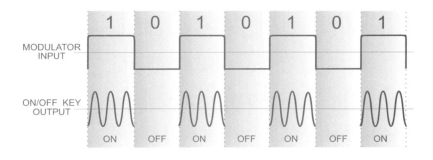

Figure 4-28:
On-Off Keying

Frequency Shift Keying

Frequency Shift Keying (FSK) is accomplished by shifting the radio carrier frequency between two discrete values to represent a 1 or a 0. FSK is simple to generate and detect in the receiver. Notice also that unlike ASK, it has a constant amplitude and is therefore not disturbed by the fluctuations in received amplitude that are often encountered by the receiver in noisy radio circuits.

Each transition from one state to another in this case represents the value of a 1 or a 0. Another name for the number of state changes per unit of time in digital communications terminology (typically one second) is the **symbol rate**.

Example 1

The transmission symbol rate for a FSK modulator that shifts the frequency 3000 times per second would be 3000 symbols per second (sps). Since each symbol represents one bit of binary information, the transmission bit rate and the symbol rate are the same. This seems to have some redundancy in this example; however, as we shall see with other types of digital modulation, one symbol or state change may represent more than one bit of information. In future examples the symbol rate and the effective bit rate may be different.

It is possible that a phase jump or discontinuity could occur when the FSK modulator changes from on frequency to another. This can cause the radio carrier to use more spectrum than is desirable due to the harmonics that are generated by the switching in standard FSK modulation. If the modulator is designed such that a radio carrier frequency shift is accomplished without loss of the phase continuity, as shown in Figure 4-29, the spectrum used by the carrier is much less, improving the overall quality of the transmission. This type of FSK is called Continuous Phase Frequency Shift Keying (CPFSK). This is also an example of **Minimum Shift Keying** (**MSK**), which is widely used in digital transmission systems.

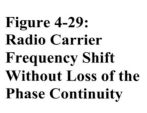

Figure 4-29: Radio Carrier Frequency Shift Without Loss of the Phase Continuity

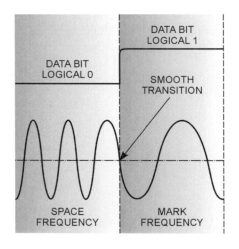

Phase Shift Keying

Phase Shift Keying (**PSK**) is a type of digital modulation where the phase of the carrier wave is switched between two states to represent a binary 1 or 0. The instantaneous phase of the carrier is usually compared with a reference phase developed in the receiver to determine whether the transmitted carrier is in phase or shifted 180 degrees, as shown in Figure 4-30. This type of modulation is called **binary phase shift keying** (**BPSK**). You may have concluded that BPSK has two possible states (and 180 degrees) and therefore has a symbol rate that is the same as the transmitted bit rate. It is also possible to represent a binary value by the phase difference between consecutive symbols. In this case, the value of the bits is determined by a change from the previous phase rather than the absolute phase. This variation of PSK is called **Differential PSK** (**DPSK**). DPSK has the advantage over regular PSK because there is no requirement for a phase reference to be present in the receiver. DPSK is also capable of transmitting two or three bits per symbol by using multiple phase offsets. This variation of PSK is called 4PSK, better known as **Quadrature PSK** (**QPSK**), which provides 2 bits per symbol and 8-DPSK which yields a higher efficiency of 3 bits per symbol.

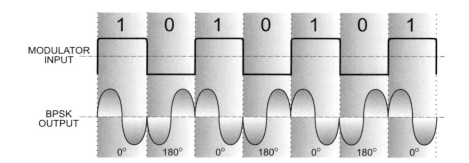

Figure 4-30:
BPSK Modulation

Quadrature Amplitude Modulation

Quadrature Amplitude Modulation (QAM) uses a combination of amplitude modulation and phase modulation. By combining the amplitude and phase modulation of a carrier signal, we can increase the **number of states** and thereby transmit more bits per every state change. A state refers to the instantaneous value of either the carrier phase or the carrier amplitude. QAM can use several states. When referring to the number of states that can exist for the carrier, the expression m-ary is used where m = the number of possible states of the carrier wave. As an example, 8-PSK has eight possible states. In this example $m = 8$.

With any modulation scheme, the number of bits that can be transmitted for each state is dependent upon the number of possible states used. QAM is implemented using 8-QAM, 16-QAM, and so on to 256-QAM. Table 4-7 shows the relationship between the number of states and the bits transmitted per symbol.

Quadrature Amplitude
Modulation (QAM)

number of states

Number of States (m-ary)	Bits Transmitted Per Symbol
2	1
4	2
8	3
16	4
32	5
64	6
128	7
256	8

Table 4-7:
QAM Modulation

One convenient way to represent the possible states for QAM is to use a constellation pattern diagram, a truth table, and a phasor diagram. For 8-QAM where $m = 8$, there are 8 possible combinations of 3 bit patterns from 000 to 111, as shown in Figure 4-31. The states are present at different amplitudes and phases. The truth table in Figure 4-31 shows the relationship between the 8 possible states of the carrier wave phase and amplitudes. Dots at plus and minus 45 and 135 degrees with two possible amplitudes yield eight possible states. With eight unique states, it is possible to transmit 3 bits in every state. For example, if the modulated signal is of amplitude 1 at −135 degrees, three zeros (000) are transmitted with only one symbol. In other words, the bit rate is 3 times the symbol rate for 8-QAM.

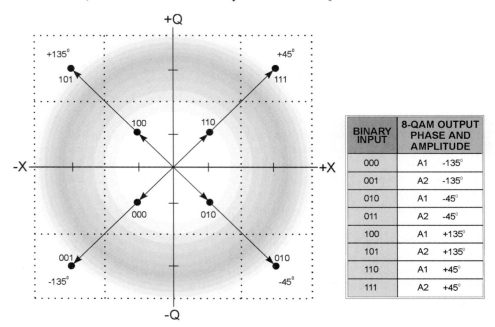

Figure 4-31: QAM

BINARY INPUT	8-QAM OUTPUT PHASE AND AMPLITUDE	
000	A1	-135°
001	A2	-135°
010	A1	-45°
011	A2	-45°
100	A1	+135°
101	A2	+135°
110	A1	+45°
111	A2	+45°

16-QAM is another m-ary system where $m = 16$. The input bits to a 16-QAM modulator are in groups of four. With 16-QAM, the bit rate is 4 times the symbol rate. Similar to 8-QAM, 16-QAM can be represented by the phasor diagram, truth table, and constellation diagram shown in Figure 4-31.

Other Modulation Options

The modulation types used in wireless communications are all extensions of the basic approach of using the phase, amplitude, or frequency to represent information content. The list is extensive and is included here in Table 4-8 to show a few of the popular extensions of PM, AM, and FM.

Acronym	Modulation Type
AM	Amplitude Modulation
ASK	Amplitude Shift Keying
OOK	On-Off Keying
PAM	Pulse Amplitude Modulation
PCM	Pulse Code Modulation
PPM	Pulse Position Modulation
PTM	Pulse Time Modulation
PWM	Pulse Width Modulation
FM	Frequency Modulation
FSK	Frequency Shift Keying
CPFSK	Continuous Phase Frequency Shift Keying
MSK	Minimum Shift Keying
PSK	Phase Shift Keying
QPSK	Quadrature Phase Shift Keying
OQPSK	Offset Quadrature Phase Shift Keying
QAM	Quadrature Amplitude Modulation

**Table 4-8:
Modulation Types and
Terminology**

Digital Voice Processing

Converting analog voice information to a digital format was implemented for wired telephone networks in the 1950's. The advantages offered to the wired network were numerous. The primary reason was to minimize the number of wires connecting switching centers in the telephone hierarchy. Digitizing the voice information allowed high-speed multiplexing of several voice channels on a single copper path. This led to the standard Pulse Code Modulation (PCM) standard for encoding analog voice information into samples of 64 kilobits per second. With PCM, the inbound analog voice signal is processed by an analog-to-digital converter shown in Figure 4-32. With a sample rate of 8,000 samples per second and a binary word length of 8 bits per sample, the result is a 64,000 bit per second digital encoded replica of the analog voice information. The digital-to-analog converter performs the reverse process by decoding the digital voice information into the original analog voice information. PCM provides an efficient digital encoding scheme for analog voice; however, it does not include any attempt to compress (lower) the bit rate through additional processing.

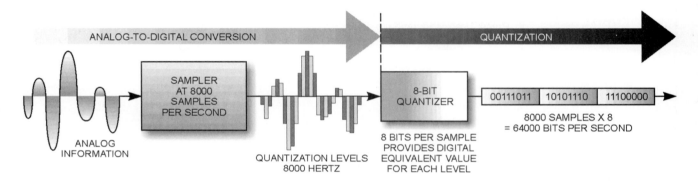

Figure 4-32:
An Analog-to-Digital
Converter

The PCM digital encoding scheme may be used for wireless network transmission as well with the attendant advantages offered by wireless digital transmission. Radio channels, however, have some additional constraints that must be considered. They are as follows:

1. Converting the analog voice signal to a digital format will require more channel bandwidth than would be the case for the equivalent analog modulated radio carrier.

2. Radio spectrum is a valuable resource. Digital cellular radio networks in particular would be very inefficient if cell phones used a PCM digital format.

3. Digital encoded transmissions can tolerate the severe propagation problems of the radio environment much better than analog transmissions. Forward error correction techniques dramatically improve the quality and performance of wireless links when applied to digital encoded transmissions.

4. Digital compression is used to reduce the bandwidth required for transmitting information while increasing the effective bit rate. It is used extensively in digital cellular radio systems to reduce the bandwidth required for each voice channel. Voice compression is a relatively new technology. It continues to improve with the development of better algorithms and more efficient digital signal processing integrated circuits.

The challenge for voice compression is to maintain reasonable voice quality with the lowest practical bit rate output.

The tradeoff between greater compression and voice quality is shown in Figure 4-33. The figure shows the comparison of compression rates compared to **64 kbps wireline quality** voice. Reasonable quality voice compression algorithms can bè accomplished with compression output rates of between 8,000 and 13,000 bits per second.

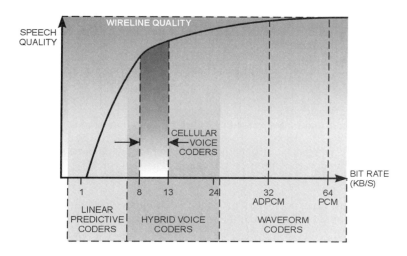

Figure 4-33: Tradeoff between Greater Compression and Voice Quality

Key Points Review

This chapter has provided an overview of the fundamentals of radio communications. Review the following key points before moving into the Review Questions section to make sure you are comfortable with each point and to verify your knowledge of the information.

- The electromagnetic spectrum is the entire range of wavelengths or frequencies of electromagnetic radiation.

- Electromagnetic waves travel through space at the speed of light and, as the name suggests, consist of periodic variations of electric and magnetic fields. These two fields are displaced in free space from each other by 90 degrees.

- The distance from the crest of one wave to the next crest is the wavelength. Frequency is measured by the number of oscillations that occur in one second.

- The unit of measurement for frequency is the hertz (Hz), named after the German scientist Heinrich Hertz who confirmed the existence of electromagnetic waves experimentally in 1887.

- Radio waves travel through space at a speed of approximately 300,000,000 meters per second.

- The frequency of a radio wave is inversely proportional to its wavelength.

- It is easier to deal with large numbers by expressing them as a power of 10.

- The bel and the decibel are logarithmic tools used to express relative power ratios in wireless communications link analysis and network design.

- Microwaves are generally defined as all frequencies between 1 GHz and 300 GHz.

- Various portions of the electromagnetic spectrum are categorized using different names.

- The most popular method for naming portions of the spectrum is the International Telecommunications Union – (Radio Sector) standard names.

- A unique nomenclature exists for a category of frequency bands that uses the letters L, S, C, X, Ku, K, Ka, V, and W. These designations were used during WW II as radar band designations.

- Research has shown that different frequency bands have different propagation characteristics.

- The FCC allocates the radio frequencies for different services based upon the propagation limitations and features of the various radio bands.

- Radio frequency bands can categorized into the following three major groups based upon their fundamental propagation characteristics: ground waves, sky waves, and line of sight waves.

- Loss of transmitted power as it propagates through the medium is called attenuation (reduction in the magnitude of the transmitted power).

- Interference refers to any electromagnetic disturbance that contains energy at or near the frequency being used for wireless communications.

- Cochannel interference is caused by two transmitters operating in close proximity using the same frequency.

- Adjacent channel interference is caused by spillover of some of the transmitted energy from a radio carrier operating on a frequency on an adjacent radio channel.

- Path loss is the measure of loss of the radio energy in a transmission path between the transmitter and the receiver. It is normally considered to be the loss that would be expected in free space.

- LOS radio signals are affected by nearby objects. Reflections create a multipath problem that can cause the received signal strength to fade when one or more stations are in motion.

- Earth curvature limits the distance between two fixed LOS transmitting and receiving stations.

- Ground reflections can cause an out-of-phase interfering multipath signal to be present at the receiver location.

- All basic digital wireless transmitter systems contain as a minimum the following seven basic functional subsystems: a source encoder, a channel encoder, a multiplexer, a modulator, a radio frequency source, a radio frequency amplifier, and a transmitter antenna system.

- All basic digital wireless receiver systems contain as a minimum the following seven basic subsystems: a receiver antenna, a radio frequency amplifier, a demodulator, a radio frequency source, a demultiplexer, a channel decoder, and a source decoder.

- The merit of an antenna, especially one designed to focus transmissions in a particular direction, is expressed in terms of the antenna gain or performance. The gain is generally based upon a comparison with a reference antenna. The reference antenna is an isotropic radiator that theoretically radiates the same power in all directions.

- Parabolic reflector antennas are efficient at microwave frequencies.

- By dynamically altering the phase shift between successive elements in an array antenna, the pattern and directivity of the antenna array may be steered electronically without physically moving the antenna structure.

- Modulation is the process of combining the information to be transmitted with a radio carrier wave.

- Modulation can be accomplished by altering the amplitude, frequency, or phase of the radio carrier.

- A modulated radio carrier wave uses bandwidth, therefore, each radio carrier frequency must be assigned to a specific channel that provides enough space between frequencies to avoid interference or "spillover" of power into the adjacent channel.

- For digital modulation, the bit rate of the source data determines the bandwidth created by the modulated carrier wave. Higher bit rates consume more bandwidth. The ratio of the bandwidth to the bit rate of a modulated carrier is expressed in bits per second per hertz (b/s/h).

- The higher level modulation techniques such as QAM are more efficient because they result in the transmission of many bits per symbol. Therefore, they are more "bandwidth efficient."

- The challenge for voice compression is to maintain reasonable voice quality with the lowest practical bit rate output.

REVIEW QUESTIONS

The following questions test your knowledge of the material presented in this chapter:

1. Describe the method for calculating the frequency in MHz of a radio wave when the wavelength is given in meters.

2. Name two disadvantages relative to the use of ground waves for wireless communications.

3. What is the wavelength in meters for a frequency of 175 MHz?

4. A radio transmitter is rated as having an output power of 50 dBm. What is the output of the transmitter in watts?

5. What is the distance in miles to the horizon from the top of a 125-foot microwave antenna tower?

6. List the functions performed by the source encoder in a wireless transmission system.

7. What is the definition of cochannel interference?

8. What three parameters of a radio carrier wave are altered to accomplish modulation?

9. What is the major cause of rapid fluctuations in received signal strength in a mobile subscriber terminal?

10. What is the main advantage of the use of data compression in wireless networks?

11. What is the major challenge of voice compression in wireless applications?

12. How many bits per symbol are transmitted when using 8-QAM modulation?

13. What range of frequencies is generally accepted as the proper definition for microwave frequencies?

14. What is the main advantage of phased array antennas?

15. What are the three types of antennas most widely used for microwave applications?

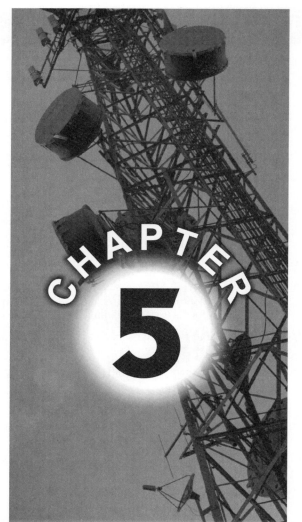

CHAPTER 5

CELLULAR RADIO

**LEARNING
OBJECTIVES**

LEARNING OBJECTIVES

Upon completion of this chapter and its related lab procedures, you should be able to perform the following tasks:

1. State the difficulties that delayed the introduction of cellular radio service in the U.S.

2. Draw a diagram of a typical cellular radio system.

3. Explain the goals for a cellular radio system.

4. Provide an explanation of the rules for implementing a cellular reuse pattern.

5. Explain the relationship between cell size and carrier-to-interference ratio.

6. Construct a diagram explaining the concept of a 7-cell reuse factor.

7. State the basis for using hexagon-shaped cells when describing cellular reuse factors.

8. Provide a definition and explain the basic principles of the three main multiple access protocols used in cellular radio systems.

9. Describe the three identification numbers associated with a cellular handset and list the purpose and function of each.

10. Explain the theory of operation for the signaling formats used in the AMPS.

11. Create a flow chart showing the sequence of events associated with placing a cellular telephone call.

12. Provide a definition and rationale for paired channels and frequency offset for the cellular radio 800 MHz band.

13. Explain the fundamental difference between cellular and Personal Communications Services (PCS).

14. Describe the nature of the problems associated with global roaming.

15. Provide the rationale and definition of dual-mode phones.

16. List the four members of the GSM family. Explain the attributes of each.

17. Explain how the gross and net bit rates are calculated for a GSM frame.

18. Explain the basic difference between the CDMA and TDMA air interfaces.

19. Describe the difference between chip rate and information rate as applied to CDMA.

20. Prepare a list of the four spread spectrum techniques.

21. Describe the channels associated with the forward and reverse CDMA links.

22. Explain rate adaptation in CDMA and why it is used.

23. Describe the purpose of power control in a CDMA cellular system.

24. Describe the difference between soft handoff and hard handoff in cellular radio systems.

25. Explain how the chipping bits are altered during modulation of a CDMA carrier.

Cellular Radio

INTRODUCTION

Cellular radio is one of the fastest growing segments of the wireless industry. Since the introduction of commercial cellular telephone service in 1983, its market demand has increased exponentially.

Early mobile radio systems were primitive compared to modern cellular radio systems. Law enforcement agencies, dispatch, and delivery services were the first enterprises to recognize the advantages of mobile radio technology. Later systems would allow mobile users to access the Public Switched Telephone Network thereby increasing the commercial value of radiotelephony.

In the U.S., the development of cellular radio was delayed for several years due to political arguments over the priorities for the use of scarce radio spectrum. There was also a shift in the regulatory attitude for telecommunications due to the breakup of AT&T in the 1970s.

Engineers at the Bell Laboratories formalized the idea for cellular radio as early as 1947. Cellular radio was a solution to the limitations faced by early systems planners concerning the acute shortage of radio frequencies suitable for mobile radio applications. Cellular radio proposed to reuse the radio frequencies over and over in a geographic metropolitan setting. Although logical and straightforward on paper, the testing and fielding of cellular mobile radio systems is a serious technical challenge. In Chapter 4, you learned about some of the general problems faced by a mobile radio system technician and designer. In this chapter, you will gain further knowledge about how these challenges are met and solved in the real world for both analog and digital cellular radio systems.

You will gain insight on why we have so many different cellular radio standards in the world. The U.S. and Europe embarked upon different paths toward selecting standards for mobile cellular radio. In Europe, the Global Systems for Mobile Communications (GSM) standard, a totally digital system, evolved early as a cooperative effort of the telecommunications governing agencies in the European marketplace. It has been enormously successful. The U.S. started with an analog cellular system and later opted for a digital radio system using a Time Division Multiple Access (TDMA) protocol that was constrained to some extent by the requirement to be backward compatible and interoperable with the analog legacy system. The proliferation of standards was again extended in 1993 with the emergence of a spread spectrum–based Code Division Multiple Access (CDMA) cellular radio standard, developed by Qualcomm. In this chapter, you will learn how these competing systems work. You will also be able to understand the terms and "jargon" associated with the technical side of cellular radio.

Radio spectrum for cellular use has not been coordinated very well on an international scale. The International Telecommunications Union and others have coordinated spectrum planning for satellites and international short-wave broadcasting, whereas little attention has been paid to the unification of cellular radio frequency bands worldwide. This has complicated the issue that now faces international planners who have new technology and marketing issues to worry about concerning global roaming with cell phones. You will learn about the cellular standards and radio bands used by the different countries. The issues and proposed standards for global wireless device roaming in various national wireless networks will be further explained in Chapters 6 and 9.

The procedures for establishing and maintaining a call from a mobile handset (called a mobile station by the standards world) are complex due to the validations, security checks, link integrity checks, and mobile unit tracking required to maintain effective communications with a power-limited cell phone that is moving from location to location. You will learn all the facts about mobility management and the integrity checks required in cellular radio standards and protocols.

Chapter 5 also provides the basic understanding and groundwork you will need later for study of the advanced cellular third and fourth generation systems explained in Chapter 9.

Early Mobile Radio Systems

The first radio communications between a fixed radio base station and a mobile station began with ship-to-shore radiotelephone service in the late 1920's.

Interest quickly grew among law enforcement agencies that realized the benefits of mobile radio. Although it was initially a one-way simplex system, the Detroit police department built and tested a broadcast system for sending messages to police cars as early as 1928. Many other cities would later follow with mobile networks modeled after the Detroit system.

The primary difficulty during this early period was developing a radio that would work reliably in a vehicle. Radios used vacuum tubes and discrete components, were large and heavy, and required a bulky power supply. Much work and research would be required to advance to the shirt pocket digital cellular phone of the late 1990s.

The invention and development of frequency modulation by Edwin Armstrong, discussed in Chapter 1, proved to be particularly useful for mobile radio applications.

trunking

mobiles

frequency agile

Another development in early mobile radio systems was **trunking**. In order for several stations to share a number of available radio frequencies, it was necessary for the **mobiles** to be **frequency agile**. Instead of mobile stations being assigned a fixed frequency shared by other groups, trunked systems allowed mobile stations to scan the available channels and use a pooled set of channels. If the transmitter could be tuned to a vacant channel, greater efficiencies were possible than was the case for fixed frequency assignments. The concept of trunked radio networks is shown in Figure 5-1. This was also a prelude to one of the concepts of operation for cellular radio systems that were to follow.

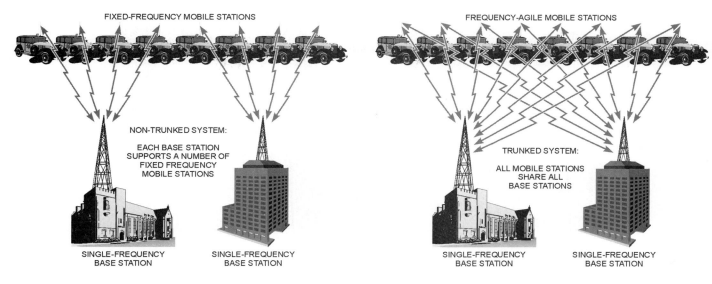

Figure 5-1: Non-trunked vs. Trunked Mobile Radio System

The Mobile Telephone Service

The first U.S. commercial radio system designed to provide access to the Public Switched Telephone Network for wireless mobile subscribers was introduced in St. Louis, Missouri on June 17, 1946. This inaugural system was called the **Mobile Telephone Service (MTS)**. MTS was a joint effort between AT&T and Southwestern Bell.

The Mobile Telephone Service Operational Overview

A model of the St. Louis MTS system is illustrated in Figure 5-2. The MTS operated on only six channels in the 150 MHz band with 60 kHz channel spacing. The base station was located at a central location and operated with a power output of 250 watts for the base-to-mobile link. The higher power was required to provide quality signal strength to the mobile stations throughout the metropolitan area. The mobile stations were designed to operate with lower power for economic reasons and therefore were not capable of providing sufficient transmit power for the return link to the base station.

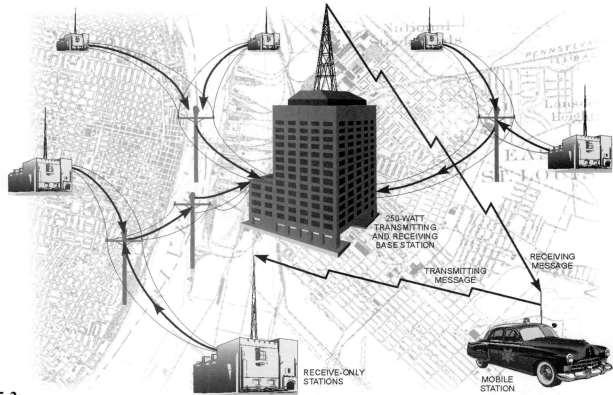

250-WATT
TRANSMITTING
AND RECEIVING
BASE STATION

RECEIVING
MESSAGE

TRANSMITTING
MESSAGE

RECEIVE-ONLY
STATIONS

MOBILE
STATION

**Figure 5-2:
First Bell System Mobile
Telephone Service -
St. Louis, MO**

The mobile-to-base radio link was provided by a series of receive-only sites located at several locations throughout the city. Each receive site relayed the mobile return link back to the base station, thereby allowing the mobile transmitters to use lower power. The base station used a manually operated switchboard. Mobile subscribers were required to contact the mobile operator in order to place a call.

The MTS operated in the half-duplex mode. You pushed a handset button to talk, then released the button to listen.

The Improved Mobile Telephone Service

The FCC licensed the first non-AT&T Radio Common Carriers (RCCs) to provide radiotelephone service for other cities following the trials in St. Louis.

On March 1, 1948 the first fully automatic radiotelephone service began operating in Richmond, Indiana, eliminating the operator needed to place most calls.

AT&T followed in 1964 with the Improved Mobile Telephone Service (IMTS). It operated in the full-duplex mode, which made the system much easier to use and more closely paralleled the convenience of the landline telephone system. IMTS, however, required two radio channels; one for the base to mobile link, and a second channel for the return link. IMTS is still available in some rural areas where cellular is not available.

The Improved Mobile Telephone Service Limitations

The IMTS was a major improvement over the initial MTS service; however, it also had serious limitations.

1. The IMTS required two dedicated radio channels for each call placed by a mobile subscriber. In New York City, AT&T could support only twelve channels at one time. The waiting list for potential subscribers in large urban areas soon reached a year or more. The centerpiece of the IMTS relied on a high-power base station reaching out to a radius of approximately 50 miles. While this provided good coverage, it also prevented the frequencies from being reused within the coverage area in a radius of approximately 50 miles or more.

2. The demand for both private and public mobile radio service grew dramatically during the 1950s. It stayed well ahead of the spectrum capacity to satisfy the demand.

3. Mobile radio dispatch systems not connected to the telephone network, such as taxi, commercial vehicles, police and fire, search and rescue, and railroads, were rapidly gaining popularity. Unlike mobile telephone services, dispatch systems consisted of short transmissions. The telephone system transmissions typically lasted much longer per call. Dispatch systems were therefore able to support more users in a given amount of radio spectrum without blocking. Commercial mobile radiotelephone service was terrible. Blocking, caused by insufficient capacity, often made the service totally unacceptable for widespread commercial use.

4. Early systems did not have the ability to perform frequency shifting of the transmitter when scanning to locate an available idle channel. Each mobile used a single frequency that was shared among several users like a party line. The trunked radio systems discussed earlier would solve this problem as mobile radio technology advanced after WWII.

Cellular Radio Communications

The solution to the IMTS spectrum saturation problem was proposed by Bell Laboratory engineers as early as 1947. It was called **cellular radio**.

cellular radio

The cellular concept took a completely different approach to providing mobile radiotelephone service. The cellular approach abandoned the high-power broadcast in favor of small cells (thus the origin of the term **cellular**) each with low-power transmitters. Using lower power transmitters allowed frequencies to be reused if sufficient separation between cells could be maintained.

cellular

The arrangement for allocating sets of frequencies among dispersed cellular locations was called the cellular frequency reuse plan.

In theory at least, the reuse plan seemed to be a way out of the spectrum shortage problem. However, there were several problems that had to be worked out before mobile cellular telephony would become a reality.

> It would require over 35 years, or approximately from 1947 to the mid 1980s, before the first cellular telephone system would become operational.

The problems summarized below, as you will see, were both political and technical.

1. Studies by Bell Laboratory engineers concluded that the cellular concept would need additional radio spectrum before it could become a practical commercial mobile telephone service. During this developmental era, the FCC was strongly biasedt oward the TV broadcasters for available spectrum targeted by the cellular community. The FCC was attempting to promote the expansion of UHF television broadcasting. A large amount of spectrum was set aside for UHF channels 14 to 83, more than 400 MHz of spectrum between 470 MHz and 890 MHz! It would take approximately 20 years for the regulatory climate to abandon the promotion of UHF television and reallocate channels 70 to 83 in the 800 MHz band for cellular telephony.

2. The original plan was for AT&T to operate the first cellular system as a monopoly. This was justified by the complexity and cost of building a cellular network. And after all, AT&T invented the whole idea. This was not to be. The Radio Common Carriers lobbied heavily for competition and won in a lengthy court battle.

3. During the 1980s, the competitive environment changed dramatically. The U.S. Justice department, in a dramatic antitrust action, divided AT&T into eight separate companies. Cellular development would again be delayed for several years.

4. There were substantial technical problems to solve. Mobile users were now going to be moving between small cells during a call. Adjacent cells would be assigned different frequencies to avoid interference, therefore mobile stations would be required to change frequencies when crossing cellular boundaries without dropping the call in progress. This would result in one cell base station handing off a mobile station to an adjacent cell base station. The cellular system would require its own dedicated central switch to manage the handoff based upon relative dynamic changes to signal strengths of mobile stations.

Cellular Radio Architecture

A diagram of a typical cellular radio system is shown in Figure 5-3. The main elements and their functions are explained in the following paragraphs.

Mobile Switching Center (MSC)

- The **Mobile Switching Center** (**MSC**) is the controller for all base stations located in a service area. (In some older documents and standards the MSC is called the Mobile Telephone Switching Office or MTSO). It handles all of the mobility management functions for the cells. It also provides the interface functions between the cellular network and the Public Switched Telephone Network (PSTN). It also contains transmission control equipment connecting each base station controller over the backhaul circuits. The backhaul link may be fixed microwave radios or a T1 transmission circuit.

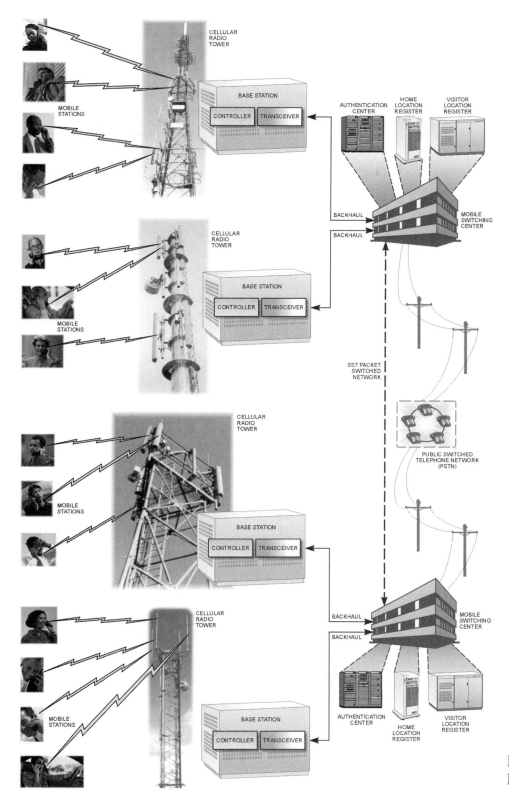

Figure 5-3: Cellular Radio System

- The **Home Location Register (HLR)** is a database that maintains all information about a subscriber's profile. It is accessed each time a subscriber requests service such as making a call. All subscribers have a home location. The home location is the MSC that covers the local area code where the subscriber's phone is registered. The home location register also keeps an updated file on the location of all home subscribers so that calls can be processed when the subscribers roam outside the home location serving area.

Home Location
Register (HLR)

- The **Visitor Location Register** is also a database that keeps track of all visiting subscribers when they are roaming in an area served by an MSC that is not in their home area. When a subscriber turns on the phone it "logs in" and registers its ID number, home telephone number, and carrier identification with the VLR. The VLR then exchanges information with the subscriber's HLR. The HLR then is able to keep track of all home subscribers for call forwarding, billing, and processing of calls.

- The **Authentication Center** is a database that maintains all secret authentication ID numbers and a secret key associated with each subscriber. When a mobile station requests service, it registers with the HLR (or VLR if it is roaming). The HLR (or VLR if outside the HLR area) issues a challenge to the mobile station to determine whether the phone is a valid phone with the proper ID and stored secret code. The authentication center is the only location in the network where the "correct answer" to the secret code challenge is stored. It is asked to verify the secret code answer associated with the subscriber's phone. If the registration information received from the subscriber's phone (or any authorized wireless device) is correct, it is validated (*authenticated* if you wish) and the HLR honors the subscriber request and processes the call. If the answer is incorrect, the caller is denied service. If the phone is outside the home area, the HLR sends a service authorization "OK" message to the VLR to permit service in the foreign HLR area. This exchange can take place over a very long distance in a few milliseconds.

- Cells are the geographical areas covered by a base station radio system. The base station may be located at the center of the cell area or, with a directional sector antenna, may be located at the edge of the cell. Cells vary in size based upon the service provider's design for providing service to urban or rural areas.

- A **Cellular Base Station Transceiver** is the radio equipment serving a single cell area. It is usually collocated with the base station controller and is located in a shelter adjacent to the antenna and tower. It is capable of transmitting and receiving all paired channel frequencies pre-assigned to the individual cell.

- The **Base Station Controller** is the equipment that provides the interface between the base station transceiver and the mobile switching center. It performs the switching and control of all cell radios to different frequencies on commands transmitted by the mobile switching center. It also houses the transmission equipment for the backhaul link to the MSC.

- The **Mobile Stations** are the cell phones or other authorized data terminal deices that contain the transceiver, microphone, earphone, and keypad. Mobile stations contain stored secret codes and electronic ID numbers to prevent fraud and provide authentication information to the cellular network authentication center.

It is not possible to build perfect cellular systems, therefore they are designed to be sufficiently tolerant of interference. In order to accomplish this goal, the frequency reuse plan is an important design issue. Let's examine how frequency reuse works.

> It is important to understand the primary goal for a cellular radio system. The goal is to provide the maximum subscriber density for a given amount of radio spectrum and geographical area while maintaining an acceptable quality of service. This is one of the fundamental principles driving all cellular system design.

Mobility Management

Cellular mobile networks are more complex than fixed wireline networks.

The wireline network connects telephone subscribers to fixed locations. The mobile network, however, must manage mobile stations that may be constantly changing locations. The location may change for a subscriber when a call is in progress or when the user is moving with the phone on standby. In both cases, the mobile station may cross several cell boundaries. The mobile station location must be monitored in order for inbound calls to be properly routed.

When a call is in progress, the mobile station must be transferred to the neighboring cell without dropping the call. This is referred to as **handoff** in U.S. standards or **handover** in some international standards. Handoff is managed by the MSC and is initiated when the signal strength falls below a specified threshold. The MSC monitors the mobile signal strength of adjacent cells and determines which cell can provide the best signal for the handoff decision. There are basically two types of handoff: Network handoff where the MSC makes the decision based upon received signal strength in neighboring cells and Mobile Assisted Handoff (MAHO) where the mobile station actively participates in the handoff decision by supplying signal strength measurements to the MSC.

handoff

handover

All cellular subscribers have a home service area dictated by the mobile telephone area code number. When a mobile station enters a service area outside the home service area, the subscriber is in the roaming mode.

Advantages of Frequency Reuse

Cellular radio uses low-power radio base stations to bring the geographical coverage area down to smaller areas called "cells."

Frequency reuse is accomplished by assigning a subset of the total frequencies available to each cell.

Each cell is able to reuse the frequencies over a wide geographical area, thereby using the total allocated spectrum more efficiently. A reuse pattern is implemented in such a way that no adjacent cells are assigned the same set of frequencies.

Reuse patterns are also implemented to provide equal distance separation between cells using the same frequencies.

For example, cell size in a rural area would be larger than a busy metropolitan area.

There are, however, other factors that must be considered in the layout of cellular networks. Cellular reuse is a completely different approach for mobile radio system design. The IMTS depended upon brute force radio power to cover the largest area possible, whereas cellular radio uses low power in many cells placed adjacent to each other. With MTS, the challenge was to provide enough power to overcome any fades, eliminate any holes in the coverage area, and maintain an acceptable signal-to-noise ratio (S/N_o) for the base to mobile radio and path. With a cellular system, the radio transmitters must be balanced against each other. The main issue with cellular is to provide an **acceptable carrier to interference plus noise ratio ($C/I+N_o$)**.

> The primary element the cellular engineer must deal with is the interference from other cells using the same frequency, called **cochannel interference**, and cells using the adjacent frequency, called **adjacent channel interference**.

This is a more dominant issue than the noise on the radio channel. Cellular systems, even with geographic separation of the cells using the same frequency, often operate within the same radio horizon. In analog systems using FM the level of C/I for an acceptable quality of service is about 17 to 18 dB, or approximately 55 to 1. Digital modulated systems using forward error correction can withstand C/I values in the range of 9 to 13 dB or approximately 15 to 1. This is a major advantage of digital over analog modulation. As a result, digital systems can tolerate smaller cells with closer spacing between cochannels. This results in more mobile subscribers for a given geographical area.

With these design issues identified, the following general features of cellular frequency reuse can be summarized as follows:

1. Cellular reuse of frequencies results in a more efficient use of radio spectrum than is the case with the MTS.

2. Cellular radio allows many mobile stations to share the same frequencies within a cellular network coverage area.

3. Cell diameter governs the distance between cells using the same frequency. In other words, as the cell diameter decreases, the distance between cells using the same frequency will decrease. This results in greater cochannel interference.

4. Analog systems using FM can tolerate carrier-to-interference ratios of 17 to 18 dB. This dictates a limit on minimum cell size. Digital systems are more tolerant to interference and can operate satisfactorily with C/I values of 9 to 13 dB.

The Cellular Frequency Reuse Factor

Different reuse factors are used in analog and digital cellular systems throughout the world.

> North America and several other regions of the world use a 7-cell reuse plan.

Figure 5-4 shows a reuse pattern with 7 cells arranged in a cluster. Each number represents a cell with an assigned set of frequencies. Within each cluster, 7 cell groups are arranged using the familiar hexagon-shaped pattern. Cellular coverage areas are actually a series of overlapping circles, as shown in Figure 5-5. The hexagon pattern is used as a planning tool only and is a convenient method for illustrating frequency reuse patterns.

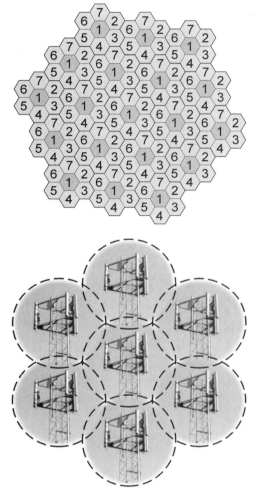

**Figure 5-4:
7-Cell Reuse Pattern**

**Figure 5-5: Typical
Cell Overlap**

Table 5-1 illustrates how frequency reuse can increase the spectrum efficiency for a given geographical area. It would appear from this table that the number of channels can be increased substantially by progressively decreasing cell size. Unfortunately this is true only for a limited scale of values. The limits are based upon the amount of cochannel interference that can be tolerated among the mobile and base stations as the cell size is reduced. As cell size is decreased, the transmit power for each base station must be decreased, but there is a ultimately a limit where cochannel interference overwhelms the system. Analog modulation systems are limited in their tolerance for cochannel interference. Digital cellular systems, as we shall see later, are designed to tolerate a higher value of interference and therefore can exist in a smaller cell separation environment.

┌─ NOTE ─────────────────────────────

With the 7-cell cluster arrangement, all numbered cells are separated by an equal distance. Assuming a total number of available frequencies of 420, the 7-cell reuse plan would result in 420/7 or 60 radio frequencies per cell. Each cluster of 7 cells could be repeated throughout a given geographical coverage area.

└────────────────────────────────────

Table 5-1:
Frequency Reuse vs.
Cell Size Ratio

ASSUMING 2800 SQUARE MILE AREA
(RADIUS = ~ 30 MILES)

EXAMPLE 1	EXAMPLE 2

2.5 MILE CELL RADIUS	4 MILE CELL RADIUS
$A = \pi r^2$	$A = \pi r^2$
$A = \pi 6.25$ SQ. MILES	$A = \pi 16$ SQ. MILES
$A = 19.63$ SQ. MILES	$A = \sim 50$ SQ. MILES

CLUSTER SIZE	CLUSTER SIZE
19.63 x 7 = 137.4 SQ. MILES	50 x 7 = 350 SQ. MILES
$\frac{2800}{137.4} = \sim 20$ CLUSTERS	$\frac{2800}{350} = \sim 8$ CLUSTERS

SYSTEM CAPACITY	SYSTEM CAPACITY
420 CHANNELS x 20 CLUSTERS = 8400 CHANNELS	420 CHANNELS x 8 CLUSTERS = 3360 CHANNELS

Cell Sizing and Sectoring

Cell size may vary from .5 miles in a metropolitan area to 10 miles in a rural area.

> The size of the cell can be controlled by antenna height, antenna tilt, and the transmitted power output of the base station.

The cell size is also dictated by the economics of population density, traffic statistical studies, and surrounding terrain.

> Cell splitting is often used by cellular service providers to increase capacity in areas with high subscriber growth.

This does not provide additional channels. It only allows more channels to be placed in a given geographic area, thereby increasing subscriber density and reducing the possibility of blocking. This comes at the expense of more handoffs and more base stations, which increases the cost to the cellular service provider. The tradeoff decision for cell splitting is always driven by an attempt to balance quality of service against equipment cost.

> Cell sectoring is another technique used to increase the capacity of a cellular network.

With sectoring, the base station uses a set of directional antennas to establish the transmit and receive patterns that favor a specific direction. Figure 5-6 illustrates a 7-cell reuse pattern with 120° sectors in each cell. In this 3-sector-per-cell system, the frequencies in each cell have been divided into 3 subsets. As an example, with a 60-channel cell, each sector would be allocated 20 channels each. Notice in Figure 5-6 that each sector with the same subset of frequencies is pointing the directional pattern in the same direction, thereby limiting the potential interferers to a much smaller number than was the case in the omnidirectional antenna example shown in Figure 5-5. Recalling a fundamental principle of cellular systems discussed previously, any reduction in interference from cochannel interferers in other cells results in the system having more capacity. This does not mean that more channels have been created, but that the cells can be smaller with sectorization with no additional increase in cochannel interference.

The objective of sectorization is to provide more subscriber channels for a given area while maintaining an acceptable carrier-to-interference ratio (C/I). An example of the physical arrangement of a 3-sector antenna is shown in Figure 5-7. The triangular platform provides a mounting platform for the antennas with the flat face forming the 120-degree sector. The center antenna is a phased array antenna for the base station transmitter. The two antennas at the corners of the platform are used for receiving the mobile reverse channels. These two antennas provide diversity reception. The receive antennas are typically spaced at odd-number multiples of half wavelengths to provide the best diversity reception for the mobile stations.

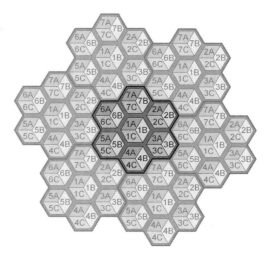

Figure 5-6: 7-Cell Reuse Pattern with 120-Degree Sectors in Each Cell

Figure 5-7: 3-Sector Antenna Configuration

Multiple Access Protocols for Cellular Radio

Multiple access protocols are used in cellular radio systems to provide a structured method of sharing radio channels among a number of mobile stations.

An access protocol sets the rules for how channels are assigned to a user community. Three basic access protocols are defined as follows:

- **Frequency Division Multiple Access (FDMA)** divides the entire radio spectrum into channels. Each user is assigned a radio channel during a call. Users are separated in the frequency domain, as shown in Figure 5-8. This is similar to broadcast radio, television, and cable TV systems in which each transmitting station uses a separate radio channel.

- **Time Division Multiple Access (TDMA)** allows more than one user to use a radio channel at the same time by dividing the channel into non-overlapping time slots. TDMA is a multiplexing type of access protocol. It assigns time slots to different mobile stations and multiplexes each of the transmissions at the proper time into one of the time slots, as illustrated in Figure 5-8. TDMA can be characterized as an extended FDMA protocol. In cellular radio, TDMA further divides each FDMA channel into separate time slots, therefore users are segregated in both the time domain and the frequency domain.

- **Code Division Multiple Access (CDMA)** allows all users to share the same radio channel at the same time. Users are separated by different codes, as shown in Figure 5-8. Although all stations operate on the same carrier frequency, the receiving station knows in advance the specific code assigned to the desired transmitting station. It can extract the desired transmitter information from the desired station and reject all other stations that are using other codes. The codes are digital sequences of binary ones and zeros. In a CDMA access system, the codes are selected so that they are different. In CDMA terminology, the codes are said to be **orthogonal**. A channel, in CDMA terms, is therefore a unique code assigned to a base station and a mobile station rather than a separate frequency. CDMA has become a common term used synonymously with spread spectrum communications technology. In specific terms, CDMA is an access protocol used in cellular radio. Spread spectrum communications is a much broader subject, and is used in many applications beyond the scope of this chapter.

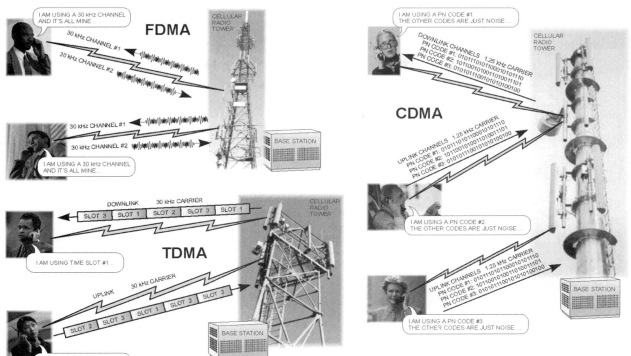

Figure 5-8: Multiple Access Protocols

Cellular Standards, Access Protocols, and Acronyms

With so many acronyms to remember, you need to understand how terms, standards, and acronyms are often misused. So let's clear up a few issues before you read further about cellular network terminology. FDMA, TDMA, and CDMA are **access protocols** for cellular systems used worldwide. They are generally referred to in the industry as **air interfaces** because they define the fundamental method used by the radio equipment to gain access to the wireless portion of a large composite network. There are also cellular standards developed by the Telecommunications Industry Association and the Electronics Industry Association (TIA/EIA) in coordination with the American National Standards Institute (ANSI) in the U.S. and the European Telecommunications Standards Institute (ETSI) in Europe. The three air interface protocols are not stand-alone standards, but they are the core multiple access technologies used in the various cellular standards documents. For example, the specification for the North American Advanced Mobile Phone System (AMPS), which defines the total architecture for the analog cellular standard ANSI-553, specifies FDMA as the access protocol. The most widely used cellular standards and terms are outlined in Table 5-2.

access protocols

air interfaces

Table 5-2: North American Cellular Standards

Cellular Standard	Name	Description
ANSI-554	Analog Cellular	The original version of the analog standard.
ANSI-88	Narrowband Analog Cellular	A Motorola system using 10 kHz analog channels. Referred to as N-AMPS.
ANSI-91	Analog Cellular	Consolidation of ANSI-94 and ANSI-88.
ANSI-94	In-building Cellular	Standard for analog low-power system for buildings.
ANSI-732	Cellular Digital Packet Data	A packet radio system for data using an analog voice channel.
ANSI-54	TDMA (D-AMPS)	Evolution of ANSI-91 to a digital system using the original ANSI-91 signaling standard but incorporating digital voice channels.
ANSI-627	D-AMPS	Finalized as TIA/EIA-627. Replaced ANSI-54.
ANSI-136	TDMA Digital Cellular System	Upgrade to ANSI-54 using an advanced signaling protocol on a digital control channel.
ANSI-95	CDMA Digital Cellular System	A digital cellular system originally developed by Qualcomm using spread spectrum technology.
ANSI-41	Intersystem Operation	A specification for handling intersystem roaming, authentication, accounting, and authorization between cellular networks.

Frequency Division Duplex and Time Division Duplex

Two different multiplexing schemes are used by cellular radio systems. Duplex is a term used to describe how a transmission between two locations is handled. As discussed in previous chapters, full duplex refers to a connection where information is exchanged in both directions at the same time, similar to a landline telephone connection. Some radio systems, such as aircraft traffic control radios, operate in half-duplex mode where the channel is alternately shared with transmission in one direction at a time.

Cellular radio uses the terminology "Frequency Division Duplex (FDD)" and "Time Division Duplex (TDD)." As illustrated in Figure 5-9, FDD uses a separate channel in each direction between two parties at the same time. The channels are therefore separated in the frequency domain. TDD uses a single radio channel that is used alternately between two parties on an alternate or "ping pong" type scheme. The two channels are therefore separated in the time domain.

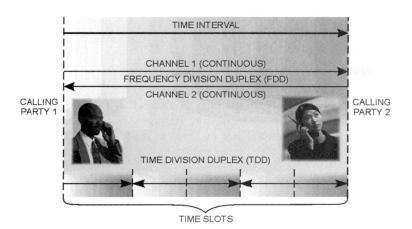

TIME INTERVAL

CHANNEL 1 (CONTINUOUS)

FREQUENCY DIVISION DUPLEX (FDD)

CHANNEL 2 (CONTINUOUS)

CALLING PARTY 1

CALLING PARTY 2

TIME DIVISION DUPLEX (TDD)

TIME SLOTS

Figure 5-9: Frequency Division Duplex (FDD) and Time Division Duplex (TDD)

Both FDD and TDD utilize transmission paths to and from the base stations and mobile stations. TDD is not used in current second generation cellular systems.

The radio transmission channel path from the base to the mobile is called the **forward channel**. The radio transmission channel path from the mobile station to the base station is called the **reverse channel**.

forward channel

reverse channel

The Advanced Mobile Phone Service Standards

Advanced Mobile Phone Service (AMPS) was the pathfinder system for cellular communications.

In March 1977, the FCC authorized the Illinois Bell Telephone company to install and test a prototype version of AMPS in the Chicago area. This was followed by the first commercial service in Chicago in 1983. AMPS was originally an AT&T proprietary service name but later came into common use when the AMPS was adopted as the TIA/EIA-553 standard. It is often referred to as the first generation of cellular communications systems. As we shall see later, the transition from AMPS analog to all-digital cellular technology is referred to as the second generation (2G).

As you can see from Table 5-2, AMPS is covered by several standards documents. AMPS uses FDMA as the access protocol. It operates using FDD for the forward and reverse link and 30 kHz channels in each direction. The forward and reverse channels are always selected in pairs with a constant separation in the frequency domain of 45 MHz. AMPS uses FM as the modulation method for the voice channels using a peak deviation of 12 kHz.

We will now examine how signaling protocols and call initiation procedures are used in the AMPS TIA/EIA-553 standard.

Establishing the Identity of the Mobile Handset

As illustrated in Figure 5-10, three identification numbers are used to establish the identity and provide a deterrent to fraud for the mobile station handset. They are the **Electronic Serial Number (ESN)**, the **Mobile Identification Number (MIN)**, and the **System Identification Number (SID)**. Their purpose and function are described as follows:

- The *Electronic Serial Number* is burned into the handset programmable read-only memory at the factory by the manufacturer. It is used as a permanent 32-bit identification number for the mobile station. Any attempt to alter the ESN makes the phone unusable.

- The *Mobile Identification Number* is programmed from the keypad into the handset memory by the service provider when service is initiated. It is a 34-bit number that is governed by the handset 10-digit phone number.

- The *System Identification Number*, programmed into the handset, is a 15-bit number that is assigned to all cellular service providers. It enables the network to know the cellular carrier your phone is registered with (i.e., Verizon, Sprint, Cingular, etc.). It is used by the network to identify whether a subscriber is operating in its "home" system or if it is roaming in another service provider's network.

**Figure 5-10:
Handset Identification
Numbers**

Advanced Mobile Phone Service Channel Utilization

As indicated in Figure 5-11, AMPS uses both **voice channels** and **control channels** to establish and maintain communications in a cellular radio system. The control channels are dedicated to call setup, paging, and monitoring between the base and mobile stations. Control channels are used when a voice call is not in progress but the phone is powered on. The voice channels are used to conduct the full-duplex voice communications between the mobile station and the base station after a call has been established.

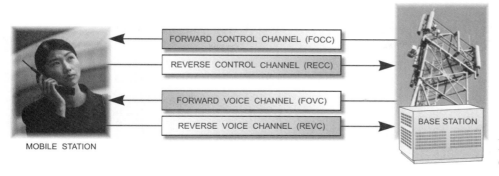

Figure 5-11: AMPS Channel Utilization

The control channels as well as the voice channels are used for full-duplex service. The base station to mobile links are called the forward control channels (FOCC) and the forward voice channels (FOVC). Similarly, the mobile station to base station links are called the reverse control channels (RECC) and the reverse voice channels (REVC).

The Control Channels

Control channels are broadcast-type channels that send continuous digital data streams using frequency shift keying (FSK) modulation with a deviation of +/− 8 kHz and a data rate of 10 kbps. All mobile stations monitor the FOCC for essential information when not engaged in a call. The mobile station seeks the strongest signal and locks its frequency to that signal. If the mobile subscriber is moving, the control channel signal may fade, but the mobile station will always seek and lock on the strongest signal to be able to make and receive calls when necessary from any cell.

The mobile station can also issue requests on the Reverse Control Channel (RECC). However, since the RECCs are used by more than one mobile station, collisions are possible since base stations are not aware of other mobiles that may be contending for the RECC. To handle this situation, the base station transmits a busy/idle status on the FOCC to alert the mobile stations when the RECC may be seized by a mobile for a request to make a call. Both the FOCC and the RECC operate at a digital data rate of 10 kbps.

The Voice Channels

Voice channels are allocated to authorized mobile stations by the mobile switching center following the initialization steps carried out on the control channels. Voice channels are referred to as traffic channels in the second generation cellular standards documents. Voice channels are only used during the duration of the call. For full duplex, one 30 kHz analog channel is used for each direction. 60 kHz of radio spectrum is therefore utilized for each call. These channels are called the forward voice channels (FOVC) and the reverse voice channels (REVC). They are released for other users when the mobile station call is terminated.

Channel Signaling on the Voice Channels

After a call has been established on a voice channel frequency pair, there is still a requirement to transmit command information to the mobile station, such as handoff and power change commands.

Because the mobile station cannot receive on two channels at the same time, the base station uses a special signaling format on the voice channel to send commands without disrupting the call in progress. This is called **blank and burst encoding**. (It occurs too fast for you to hear.)

It is implemented by turning the FM voice transmission off for approximately 100 ms and inserting a "burst" of digital data that commands the mobile station to perform specific tasks.

Two additional signaling formats are used by the AMPS. They are called the Supervisory Audio Tone (SAT) and the Signaling Tone (ST).

SAT

SAT Color Code

These signaling protocols are used to verify the integrity of the link when a call is in progress. First, let's look into the details of how SAT works.

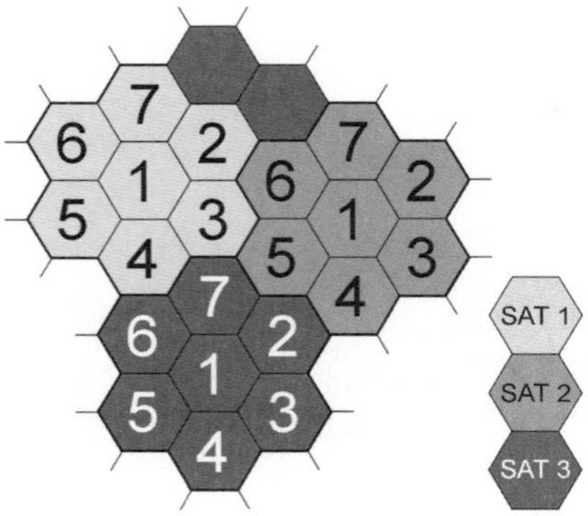

Figure 5-12: Supervisory Audio Tone (SAT) Cluster Distribution

The **SAT** is an audio tone that is transmitted on the Reverse Voice Channel (REVC) and the Forward Voice Channel (FOVC). After a call has been established, the base station transmits a selected continuous SAT on the FOVC. The mobile turns the signaling tone around and sends it back to the base station. This is the confirmation to the base station that the mobile station radio link is still connected and has not been lost due to a fade or other anomaly. There are three SATs that can be used by a base station. If the tone disappears for more than one second, the base station considers the call to be lost and terminates the call. Also, when a handoff occurs, the base station sends a special **SAT Color Code** notifying the mobile station what SAT to expect in the new cell. This tone is picked up and returned on the new voice channel by the mobile after changing to a new assigned voice channel. This is the confirmation to the base station that the handoff was executed correctly. During a call, the mobile station continuously transmits the SAT assigned to the cluster within which the mobile phone is located. The other two SAT frequencies are assigned to adjacent clusters, as shown in Figure 5-12.

The SATs and their corresponding color codes are shown in Table 5-3.

Table 5-3: AMPS Supervisory Audio Tones (SAT) and Corresponding SAT Color Codes (SCC)

SAT Audio Frequency	SAT Color Code
5970 Hz	00
6000 Hz	01
6030 Hz	10

The **ST** or **Supervisory Tone** is the second signaling format. The ST is transmitted only on the REVC. The ST performs four tasks and is recognized not by a tone frequency, but by the duration of a 10 kHz tone. The burst lengths for each of the mobile station to base station commands are as follows:

- The 50 ms burst is sent to the base station by the mobile station confirming the receipt of a handoff message.

- The 400 ms burst is used as a request to send signal. This is a tone that allows the subscriber to input data from the keypad.

- The 1.8-second tone is used to terminate a call (the "hang up" signal). When the subscriber is finished with a call and presses the "end" key, the base station terminates the call. The call can also be terminated on the PSTN side by a caller. In this case, the PSTN alerts the MSC concerning the call termination. The MSC forwards this information on the FOVS. The mobile station responds with a 1.8-second ST tone on the REVC and shuts off the mobile station radio carrier.

- The last use for the ST is a continuous tone. It is used by the mobile station as a confirmation to the base station that the subscriber has answered the call. When the cell phone is ringing, the ST is transmitted to the base station continuously. When the subscriber answers by pressing the send key, the ST tone stops. The base station knows that the mobile station has "picked up" the call.

AMPS Call Processing

With all of the preceding AMPS signaling protocols and channel names defined, it is now time to take a tour of the events supporting a cellular phone call initiated on an AMPS network. Figure 5-13 shows the sequence of events for the initialization phase of a mobile initiated call.

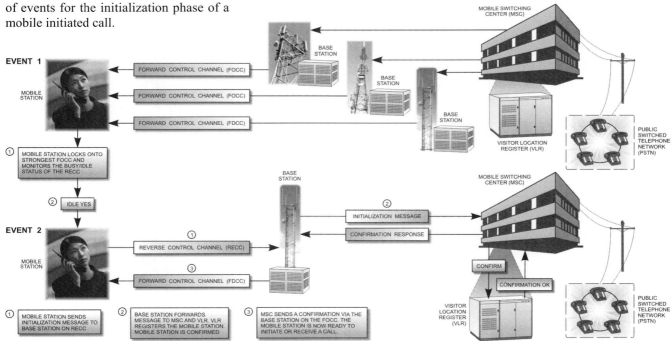

Figure 5-13: Mobile Station Initialization Phase

The Initialization Phase

Event 1. The mobile station is powered on. The mobile station scans, seeks, and locks on to the strongest forward control channel (FOCC) broadcast by the base stations. The FOCC provides several required parameters, including the number of available control channels that can be scanned and information concerning the identity of the cellular service provider called the System Identification Number or SID. This alerts the mobile station if it is in a cell operated by its "home" SID or in a "roam" system. It also indicates the busy/idle status of the reverse control channel (RECC). Remember, the RECC is a resource shared among the mobile stations, therefore the mobile stations within the cell must monitor the busy/idle bit setting to know when it is clear to transmit on the RECC to avoid collisions.

Event 2. When the RECC becomes idle, the mobile station identifies itself to the network by sending an initialization message to the base station on the RECC. The initialization message includes the mobile station ESN, MIN, and SID.

The base station forwards the identification information for proper validation to the Mobile Switching Center (MSC). The MSC and its associated Visitor Location Register (VLR) confirm the mobile station identity as a safeguard against fraud and also determine the mobile station's "home" location. The confirmation is returned to the mobile station as a control message on the FOCC. The mobile station is now ready to receive or originate a call.

The Call Initiation Phase

Event 1. As indicated in Figure 5-14, the mobile station subscriber keys in the called number and presses the send key. The mobile station again attempts to send a message to the base station by searching for an idle control channel (the mobile is "listening" to the FOCC busy status). It may return to the original channel if it is not busy. When a vacant RECC is found, this triggers an origination message on the RECC. This repeats the validation of the mobile station identity as in event 2 above. The origination message also includes the called party number.

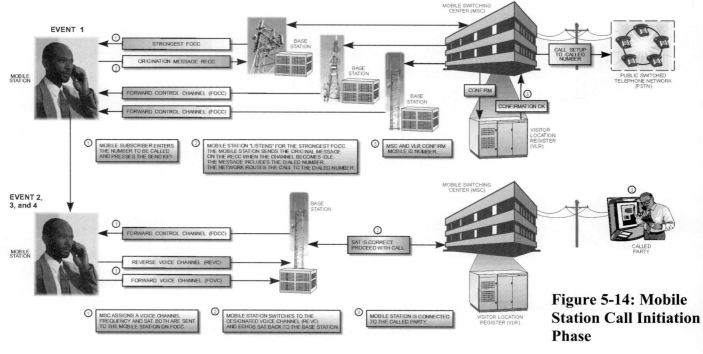

Figure 5-14: Mobile Station Call Initiation Phase

Event 2. The base station receives the call origination message with the called party number and passes it to the MSC as before during the previous initialization phase. When the validity of the mobile station is confirmed by the MSC and VLR, the MSC forwards a control message back to the base station and on to the mobile station on the RECC. The network also contacts the called party location via network call routing facilities. This includes the assigned voice channel frequency (REVC) and the SAT to be used for the call. The mobile station hears the ring signal indicating that call initiation is occurring.

Event 3. The mobile station receives the control message and switches from the RECC frequency to the assigned voice channel (REVC) frequency. (This event assumes that the called party has answered the outbound call.) The mobile returns (transponds) the SAT that was sent on the FOCC back to the base station on the newly assigned reverse voice channel (REVC). This is the link validity check for the air interface to ensure that the mobile received the control information.

Event 4. This link integrity check can now be confirmed by the base station, which "listens" for the correct SAT to be echoed back on the assigned voice channel (REVC). The base station now knows it has a mobile station assigned on the correct frequency. The called party is now connected to the mobile station on the REVC. Full-duplex conversation now proceeds on the REVC and the FEVC. The SAT will remain for the duration of the call as a continuous link integrity check.

The Handoff Phase

Event 5. The mobile station may complete the call without leaving the cell where the call originated. The mobile subscriber may also be moving and eventually leave the originating cell. This would require a handoff to the new cell, as shown in Figure 5-15, and a new frequency assignment for the mobile station. The handoff is accomplished by some clever signaling on the voice channel, which normally does not interfere with the call in progress. When the base station is notified by the MSC that the mobile signal strength is changing between adjacent cells, it orders the mobile station to switch to an adjacent cell frequency. To do this, the base station employs the **blank and burst feature** discussed previously. The FM carrier is muted for 100 ms when the digital "burst" message is sent commanding the mobile station to switch to a new voice channel (REVC) frequency and to use a new SAT. The confirmation from the mobile to the base station is returned on the FOVC using the signaling tone (ST) feature. The mobile sends a 50 ms, 10 kHz tone, notifying the base station that it has received the command. The base station and MSC then monitor the new voice channel (REVC) frequency for the transponded SAT to confirm that the handoff to the new cell frequency was accomplished correctly. If it hears no tone or the wrong tone, the call is dropped.

blank and burst feature

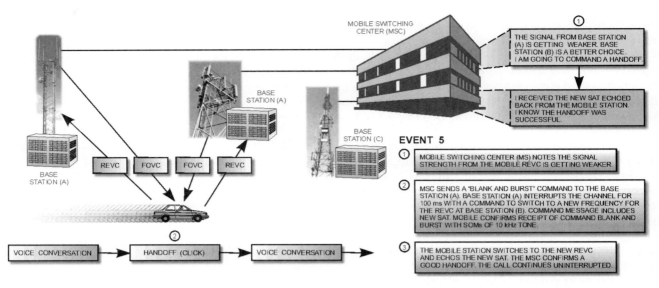

Figure 5-15: Mobile Station Handoff Phase

The Cellular Radio Spectrum Plan

FCC Docket 18262, a FCC proposed rule-making notice published in 1968, was a major milestone in the development of cellular service. It proposed a reallocation of the UHF television channels 70 through 83 in the 800 MHz band for cellular operation. It would require over 12 years for all of the legal issues to be worked out before cellular wireless telephone service became a reality. Figure 5-16 shows the frequency plan and paired channels for both the "A band," or non-wireline carrier, and the "B band," or wireline carrier. In the interest of generating competition, the FCC ruled that two licensed operating companies would be authorized for each metropolitan area. The "A" licensee would be legally restricted to not have a financial interest in the local telephone operating company. The "B" operating company would be the local telephone company or "wireline" company. Currently, the regulatory environment is different. The "A" and "B" carrier designations are no longer relevant.

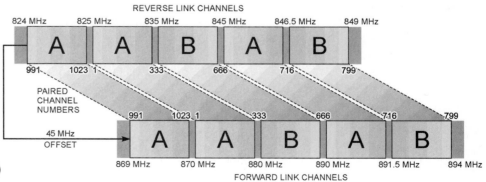

Figure 5-16: Cellular 800 MHz Band Plan

Any company with sufficient resources can apply for a license in any area. In the current North American cellular system, 50 MHz of spectrum is allocated for cellular use. This is broken into two segments of 25 MHz each. The 824 MHz to 849 MHz portion is used for the mobile transmit to base link or **reverse channels**, and the 869 MHz to 894 MHz portion is used for the base station to mobile station link or the **forward channels**. Each transmit and receive carrier is allocated a **30 kHz channel**. The bands are arranged such that all forward and reverse channels are paired and separated by 45 MHz. As an example, the reverse channel of 824.030 would be paired with the forward channel of 869.030. 45 MHz of separation allows the transmit and receive functions for full-duplex operation to be separated into different bands.

The original allocation provided a total of 40 MHz of spectrum between 825 to 845 MHz for the reverse channels and 870 to 890 MHz for the forward channels. The bands were then divided into numbered 30 kHz channels. Channels 1 to 333 were assigned to the "A band" carrier and channels 334 to 666 to the "B band" carrier.

The later addition of 10 MHz of spectrum added channel space both above and below the original 40 MHz of spectrum. This resulted in new channels 667 to 799 and 991 to 1023. You may be wondering what happened to channels 800 to 990. The answer is that the additional 10 MHz of spectrum was allocated after the original 666-channel plan was put in place. When the expanded cellular plan was established, channel numbers 880 through 990 had been assigned to other radio services, so the expanded channel numbers picked up at 991 instead of 800.

Doing the math to clear things up a bit, we can see the evolution of the 800 MHz cellular band plan as follows:

- **Original allocation of 40 MHz supporting 666 channels:**

    666 channels @ 30 kHz each (x2 for forward and reverse links) = ~ 40 MHz

- **Expanded allocation of 10 MHz supporting 166 channels:**

    166 channels @ 30 kHz each (x2 for forward and reverse links) = ~ 10 MHz

The original plus expanded spectrum resulted in 666 + 166 channels = 832 channels.

Summing things up on the cellular 800 MHz band channel numbers, it is important to remember the following points.

1. The AMPS cellular system operates in **full-duplex mode**. That is, the mobile and base stations transmit in both directions at the same time.

2. The base station transmit channels (base to mobile) are referred to as the **forward link channels**. There are 832 forward channels. Each voice channel is centered on the channel frequency. Each channel is 30 kHz wide, therefore channel separation is 30 kHz.

3. The forward link channels are allocated in the 869 to 894 MHz portion of the cellular band.

4. Each forward link channel number is paired with the equivalent channel number for the reverse link. Forward and reverse channel pairs are offset by 45 MHz.

5. The mobile station transmit channels (mobile to base) are referred to as the **reverse link channels**. There are also 832 reverse channels of 30 kHz each.

reverse channels

forward channels

30 kHz channel

full-duplex mode

NOTE

Refer back to Figure 5-9 for an example of full-duplex mode with forward and reverse link channels.

forward link channels

reverse link channels

6. The reverse link channels are allocated in the 824 to 849 MHz portion of the cellular band.

In 1981, the FCC finally completed the rules and procedures for a commercial cellular radio-telephone service for the United States. American commercial cellular development began in earnest only after AT&T's breakup in 1984. The United States government decided to license two carriers in each geographical area. One license went automatically to the local telephone companies, in telecom parlance, the local exchange carriers or LECs. The other went to an individual, a company, or a group of investors who met a long list of requirements, properly petitioned the FCC, and, perhaps most importantly, won the cellular lottery. Since there were so many qualified applicants, operating licenses were ultimately granted by the luck of a draw, not by a spectrum auction as they are today.

The local telephone companies were called the **wireline carriers**. The others were the **non-wireline carriers**. Each company in each area took half the spectrum available in what's called the "A band" and the "B band." The non-wireline carriers usually got the A band and the wireline carriers got the B band. There's no real advantage to having either one. It's important to remember, though, that depending on the technology used, one carrier might provide more connections than a competitor does with the same amount of spectrum.

Personal Communications Services

Personal Communications Services (PCS) has many definitions. For our purposes in this study guide, I will begin with the FCC definition. The FCC defines PCS as a broad range of radio communication services that free people from the constraints of the regular public switched telephone network — so that you can communicate when you are away from your home or office.

The **PCIA (Personal Communication Industry Association)** defines PCS as a broad range of individualized telecommunications services that enable people or devices to communicate independent of location. PCS represents the latest advance in wireless services consisting of voice, data, and video applications that allow users to communicate virtually anytime and anywhere. PCS empowers the user to customize these services to meet his or her individual lifestyle.

You have probably concluded by now that these definitions sound similar to cellular services already defined. The FCC intent was to make the definition as broad as possible. It was a dramatic change from the previous regulatory approach. Technology was changing at such a rapid pace, the FCC decided to let the industry define PCS. That is how we ended up with so many variations for a standard definition. The FCC specifically recognized the advantages of this approach when it established the rules for PCS. Frequencies within the new PCS spectrum could be used for almost anything the licensee wanted to do other than broadcasting.

In future references in this book, we will follow a general rule of thumb in the wireless industry of referring to PCS not as a set of specific services, but as the radio frequency bands allocated to new PCS services in the 1900 MHz band. The 800 MHz band is still called the cellular band for no other reason than it was the original service defined for that frequency band. PCS follows the cellular design with a few exceptions as follows:

- The PCS responds to the repeated wireless industry demand for more spectrum to keep up with explosive growth in the cellular market.

- PCS uses a higher frequency band than the cellular band (cellular is 800 MHz versus the PCS 1900 MHz band). In North America, both cellular and the PCS follow a frequency reuse factor of 7 cells per cluster.

- Due to the propagation attenuation for the PCS frequency bands, the cells can be smaller, allowing a higher subscriber density.

- The PCS frequency band was intended to encompass a new generation of digital technologies. Although the licensee was not directed to use any specific radio air interface standard, it was assumed that the economics of digital systems would drive the market in this direction.

- The "personal" in PCS was intended to allow each subscriber to have a phone number that would follow them wherever they go.

- The cellular market focused on voice service, whereas PCS was designed to open a wide range of caller services such as caller ID, messaging, e-mail, high-speed data services, FAX, paging, specialized mobile radio, in-building systems, and almost anything not yet invented.

- PCS was aimed at increasing competition in the wireless market. No longer would the "A" and "B" carriers be licensed in each area. The licensing would hopefully encourage a new generation of wireless service providers.

- The PCS frequencies were to be auctioned off to the highest bidder. For the first time in the history of telecommunications, the radio spectrum would be sold by the government.

- PCS licensing would be allocated on a population distribution plan. This was to establish a pricing structure for large and medium sized metropolitan areas. Unlike cellular licensing, PCS spectrum would be allocated on a statistical population center formula.

- PCS would be divided into three categories called narrowband, broadband, and unlicensed. It was expected initially that service providers would offer different services in each of these areas.

> **TEST TIP**
>
> Know the three categories of PCS.

Broadband, Narrowband, and Unlicensed Personal Communications Services

Figure 5-17 shows the allocated spectrum for PCS. Broadband services in the band from 1850 to 1990 MHz serve the voice market and provide the required expansion of cellular and a wide range of other enhanced services. Narrowband services are 3 MHz of spectrum allocated in the 900 MHz band and are intended for two-way paging and messaging. Unlicensed spectrum provides radio spectrum for low-power applications such as cordless phones. Figure 5-17 also shows the PCS markets. PCS licenses were auctioned for both large Metropolitan Trading Areas (MTA) and smaller Basic Trading Areas (BTA). The total amount of broadband spectrum is evenly divided between MTAs and BTAs. These divisions in economic areas were drawn from work performed by the Rand McNally Corporation.

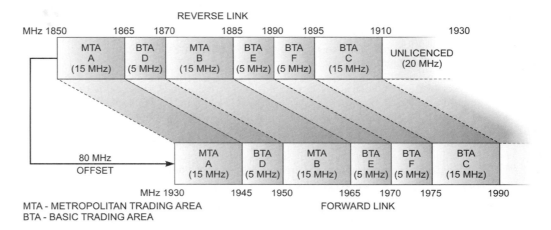

Figure 5-17: PCS Frequency Bands

Global Cellular and PCS Frequency Bands

The frequency bands used in different countries have evolved with little consideration for global roaming. As a result, very few cellular systems are compatible. Air interfaces as well as frequency bands are different in various countries of the world. Efforts were initiated by the ITU in the year 2000 to work with the different standards groups throughout the world to establish a common set of standards and spectrum plans called the International Mobile Telecommunications (IMT) 2000 plan. Table 5-4 lists the global frequency bands for the second generation cellular and data network standards. This is an example of the scope of the problem faced by IMT-2000 participants working toward global standards. (More on the IMT-2000 third generation wireless effort will be covered in Chapter 9.)

International Wireless Telecommunications	Frequency Range
Advanced Mobile Phone System (AMPS)	869-894 MHz 824-849 MHz
C450 (Germany)	461-466 MHz 451-456 MHz
Cordless Telephone (CT-2)	864-868 MHz
Digital Cellular System (DCS-1800) Also called GSM 1800.	1805-1880 MHz 1710-1785 MHz
Digital Cordless Telephone (DCT)	1900 MHz
Digital European Cordless Telephony (DECT)	1880-1900 MHz
Enhanced Total-Access Communications System (ETACS)	916-949 MHz 871-904 MHz
Global System for Mobile Communications (GSM) Also called GSM 900	935-960 MHz 890-915 MHz
IEEE 802.11 Wireless Local-Area Networks (North America/Japan)	2400-2483 MHz 2470-2499 MHz (Japan)
Industrial, Scientific, Medical (ISM) - Unlicensed	902-928 MHz. 2400-2483 MHz 5725-5850 MHz
ANSI-54/ANSI-136 Time-Division Multiple Access (TDMA)	869-894 MHz 824-849 MHz 1930-1990 MHz 1850-1910 MHz
ANSI-95 Code-Division Multiple Access (CDMA)	869-894 MHz 824-849 MHz 1930-1990 MHz 1850-1910 MHz
Japanese Cordless Telephone (JCT)	254-380 MHz
MCS (Japan)	870-885 MHz 925-940 MHz
Personal Communications Services - Narrowband (N-PCS)(U.S.)	901-902 MHz 930-931 MHz 940-941 MHz

Table 5-4:
International Wireless Telecommunications Frequency Bands

Table 5-4: International Wireless Telecommunications Frequency Bands (Continued)

International Wireless Telecommunications	Frequency Range
Narrowband Total-Access Communications System (NTACS)	860-870 MHz 915-925 MHz
Nordic Mobile Telephone (NMT-450)	463-468 MHz 453-458 MHz
Nordic Mobile Telephone (NMT-900)	935-960 MHz 890-915 MHz
Personal Communications Services Wideband (W-PCS)	1930-1990 MHz 1850-1910 MHz
Personal Handyphone System (PHS) (Japan)	1895-1918 MHz
RCR-27 Japanese Digital Cellular (JDC)	810-826 MHz 940-956 MHz 1429-1441 MHz 1477-1489 MHz 1453-1465 MHz 1501-1513 MHz
Total-Access Communications System (TACS)	935-960 MHz 890-915 MHz

* Although the U.S. Digital standards ANSI-136 (TDMA) and ANSI-95 (CDMA) may use the 800 MHz band, the PCS 1900 MHz band is the preferred radio band for second generation digital cellular systems.

The Digital Advanced Mobile Phone Service Standard

The AMPS analog system was a tremendous success with sustained growth through the late 1980s. But customer growth was rapidly exceeding the capacity of the AMPS system. The solution was either to seek additional radio spectrum for AMPS or to examine new technologies to use the radio spectrum more efficiently. The industry and the FCC opted for new technology solution.

Digital Advanced Mobile Phone Service (D-AMPS)

The **Digital Advanced Mobile Phone Service (D-AMPS)** was the introduction in North America of the second generation of cellular communications.

The feature that established the basis for a new second generation (2G) was the transition from an analog to a digital system. D-AMPS was the result of study and subsequent design by the Telecommunications Industry Association (TIA) for an improved service for the AMPS analog standard, ANSI-553. The TIA activity led to the publication of the **D-AMPS standard ANSI-54** in 1990 and ANSI-54B in 1991. They offered several improvements over AMPS (ANSI-54B was later finalized as TIA/EIA-54B, which was subsequently rescinded in 1996 and replaced by **TIA/EIA-627**). The key features were:

- Proceeding with a backward-compatible **dual-mode** approach in which phones would be capable of operating in the analog mode or the digital mode. This would allow a progressive conversion of the cells from analog to digital modulation. Subscribers would be able to use the new digital system but could also have coverage for AMPS where the digital service was not yet available. The goal was to minimize the impact on the AMPS design. ANSI-54 specifies the dual mode as support for analog traffic channels (AMPS), digital traffic channels, and the same analog control channels that are used for AMPS.

- D-AMPS would use the same 30 kHz AMPS channels in the same 800 MHz cellular band.

- D-AMPS would use digital modulation with encoding and compression of the analog voice information. Digital modulation would allow improved error correction techniques and overall improved quality of service.

- Time Division Multiple Access (TDMA) was selected as the radio air interface. The 30 kHz carrier, by using TDMA with time slots with digital voice compression, would provide the ability for three subscribers to share a 30 kHz carrier instead of one subscriber with AMPS. This provided a 3 to 1 increase in capacity with the same amount of spectrum.

- ANSI-54 was designed to have 6 time slots per carrier. Each time slot supports 8 kbps of bandwidth. At the time of this study guide publication, the voice coder with overhead bits for D-AMPS uses 16 kbps. This requires two time slots for each call. Future improvements in voice compression algorithms will offer the possibility of one call for each time slot and a capacity improvement of 6 to 1 over AMPS.

The TIA/EIA-136B Standard

The ANSI-136B specification is a newer TDMA specification incorporating a digital control channel in place of the AMPS analog control channel.

The ANSI-136B standard is a migration to new services from the earlier ANSI-54 specification. Most frame structures remain the same; however, the Digital Control CHannel (DCCH) provides increased signaling capacity. The frame structure for the ANSI-136B is shown in Figure 5-18. The figure shows that the frame length of a TDMA frame is 40 milliseconds. Frames are divided into 20-millisecond blocks that are equal to three time slots. Each time slot is 6.67 milliseconds in length. Each carrier has a bandwidth of approximately 48 kbps. Each time slot therefore provides 48/6 or 8 kbps. The voice coder plus overhead bits equals 13.3 kbps, therefore two time slots are required for each call.

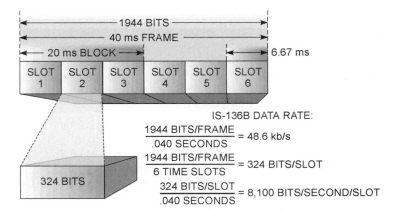

Figure 5-18: TDMA ANSI-136B Frame Structure

The Global System for Mobile Communications Standard

The Global System for Mobile Communication (GSM) is the most widely used form of digital wireless technology for mobile telephony and data communications in the world.

The GSM acronym is derived from the original French name, Groupe Spécial Mobile.

GSM was developed as a second generation digital standard in Europe. Unlike the evolutionary nature of the second generation ANSI-136 TDMA standard in the U.S., GSM is not a follow on to any first generation system in Europe. It is called "second generation" or "2G" because it is a digital wireless standard with characteristics similar to the U.S. ANSI-136 second generation standard. The GSM standard is maintained by the European Telecommunications Standards Institute (ETSI). GSM does not set specific requirements for hardware. It only defines interfaces and leaves the specific implementation to the hardware manufacturers and service providers.

The GSM standard does not have a numeric designation (with the exception of the U.S. TIA J-STD-007 standard). It is known simply as GSM in most countries. The GSM standard has historically been referred to as GSM 450, DCS 900, 1800, 1900, PCS1900, and several hybrid names based upon the frequency band being used by GSM service providers in different countries. For simplicity and to avoid confusion with U.S. PCS digital technologies, I will refer to the GSM "family" as follows:

- **GSM 400** is the GSM standard adopted for use in some European countries, principally in Germany and the Nordic Countries. Although the radio band is actually in the 450 MHz region, ETSI refers to this version as GSM 400, probably for consistency with the other generic band references.

- **GSM 900** is the original GSM standard used in Europe on the 900 MHz band. It is widely deployed in over 120 countries. There is no 900 MHz spectrum allocated for GSM use in North America.

- **GSM 1800** is an "upbanded" version of GSM 900 with minor modifications. It is also used worldwide wherever cellular spectrum is available in the 1800 MHz band. The frequency plan for both GSM 900 and 1800 is illustrated in Figure 5-19.

- **GSM 1900** is the version of GSM adopted by the U.S. standards bodies for use in the current PCS 1900 MHz band in the U.S.

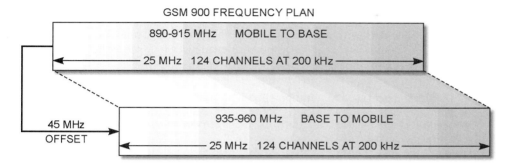

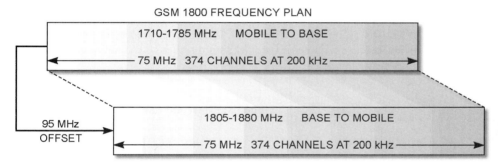

Figure 5-19: GSM 900 and GSM 1800 Frequency Plan

GSM uses a combination of FDMA and TDMA as the air interface, similar to the U.S. ANSI-136 TDMA air interface. The similarities end there, however. GSM uses 8 time slots per carrier with a bandwidth of 200 kHz. Eight calls are supported for each 200 kHz of bandwidth. The time slots can support voice or data. The standard specifies FDD operation with an 8-slot frame length of 4.615 ms, as indicated in Figure 5-20, for both the uplink and downlink. GSM uses Gaussian Minimum Shift Keying (GMSK) modulation that yields 270.8 kbps for each 200 kHz carrier. These values translate into a spectral efficiency of approximately 1.35 bits/hertz.

> **TEST TIP**
> Know the bandwidth and number of time slots per carrier of a GSM system.

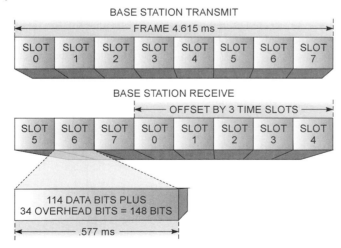

Figure 5-20: GSM System Overview

GSM Channels and Frames

It is easy to misunderstand the GSM terms "channels" and "frames." Previously, in analog systems, it was explained that a single radio carrier supported a single call. This was referred to as a channel. With a GSM TDMA access protocol, a single radio carrier supports 8 calls per radio carrier. Within this context, a **channel** is a time slot burst where the transmitter at the base and mobile station is turned on and off during a prescribed time. Keep in mind that the channel (now a time slot) is still the resource allotted to a single subscriber call. In GSM there are two types of channels:

- Traffic channels used to transport speech and data information

- Control channels used for network management messages and some channel maintenance tasks

Traffic channels are defined using a group of 26 TDMA **frames** called a 26-Multiframe. The 26-Multiframe lasts 120 ms. Note that Figure 5-21 shows how frames contain groups of 8 time slots. In this 26-Multiframe structure, the traffic channels for the downlink and uplink are separated by 3 bursts, as illustrated in Figure 5-20. As a consequence, the mobiles will not need to transmit and receive at the same time, which simplifies considerably the electronics of the mobile station.

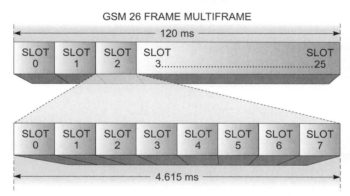

The frames that form the 26-Multiframe structure have different functions:

- Twenty-four frames are reserved to traffic.

- One frame is used for the Slow Associated Control Channel (SACCH).

The last frame is unused. This idle frame allows the mobile station to perform other functions, such as measuring the signal strength of neighboring cells.

Figure 5-21: GSM Frame to Multiframe Hierarchy

Dividing each frame into 8 time slots (channels) results in a gross bit rate of $270.8/8 = \sim 33.9$ kbps per time slot. As indicated previously, only 114 bits are used for data after subtraction of the timing and synchronization bits. This results in a net bit rate of 22.8 kbps. Summarizing the bit rate calculation for a GSM full-rate single-channel data rate:

$$\frac{114 \text{ bits per time slot} \times 24 \text{ traffic frames per multiframe}}{120 \text{ ms per multiframe}} = 22.8 \text{ kbps of bandwidth per time slot}$$

Figure 5-21 illustrates the frame and multiframe relationship. Each multiframe contains 24 frames reserved for traffic. These frames contain traffic channels plus one SACCH frame (#12) and one idle frame (#25).

The GSM System Overview

The GSM system components are similar in functionality to the ANSI-136 network elements; however, they have different names. Figure 5-22 shows the typical configuration and names for the GSM system. The main components are as follows:

- The **Mobile Services Switching Center** (**MSC**) provides all of the control, switching, and mobility management features and the interface to the PSTN. The GSM MSC functions are similar to the ANSI-136 MSC except the GSM MSC can connect directly to other MSCs without the interconnection of the PSTN.

- The **Base Station Controller** (**BSC**) manages the radio resources for one or more BTSs. The BSC is the connection between the mobile station and the Mobile Services Switching Center.

- The **Base Transceiver Station** (**BTS**) houses the radio tranceivers that define a cell and handles the radio-link protocols with the Mobile Station.

- The **Home Location Register** (**HLR**) is a database used to provide information about the subscriber's identity. The information includes the subscriber's profile, the home location MSC, and all services purchased by the subscriber. It is placed in the HLR by the service provider when a customer signs up for service.

- The **Visitor Location Register** (**VLR**) receives information from a subscriber's HLR in order to provide the correct list of subscribed services to visiting mobile subscribers. When a subscriber enters the area of a new MSC, the VLR associated to this MSC will request information about the new subscriber to its corresponding HLR. The VLR will then have enough information in order to authorize the subscribed services for the visiting mobile subscriber.

- The **Mobile Station** provides the radio interface for the reverse link radio transmitter and the forward link receive functions. The mobile unit sends bursts of information by turning on the transmitter for short periods of time when commanded by the base transceiver station (BS).

- The **Authentication Center** (**AuC**) is a protected database that stores a copy of the secret key stored in each subscriber's SIM card, which is used for secure authentication of every subscriber on the network.

- **Equipment Identity Register** (**EIR**) is a database that contains a list of all valid mobile equipment on the network, where each mobile station is identified by its International Mobile Equipment Identity (IMEI). An IMEI is marked as invalid if it has been reported stolen or is not type approved.

Mobile Services Switching Center (MSC)

Base Station Controller (BSC)

Base Transceiver Station (BTS)

Home Location Register (HLR)

Visitor Location Register (VLR)

Mobile Station

Authentication Center (AuC)

Equipment Identity Register (EIR)

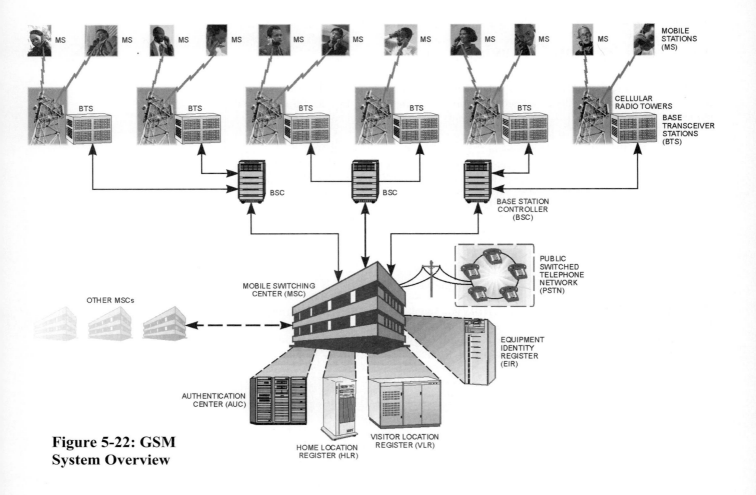

Figure 5-22: GSM System Overview

The Code Division Multiple Access Standard

The ANSI-95 Code Division Multiple Access (CDMA) Standard is another second generation digital cellular technology that has gained considerable market share and growth in the U.S. It is also the leading air interface technology for third generation cellular systems as we will see later in Chapter 9.

The CDMA standard, referred to in this text as ANSI-95, requires some clarification on how it evolved. In this case, we will depart momentarily from the "rule of thumb" of using the simple reference to "ANSI" for the wireless cellular standards.

In 1991, Qualcomm completed field tests in San Diego for a commercial digital multiple access cellular technology called CDMA. Although used extensively by military organizations for many years as an anti-jam (AJ) stealth communications technology, CDMA was deemed by many experts to be too complex and expensive for any practical application as a commercial mobile wireless multiuser environment. While it worked well for many years in point-to-point communications applications and was resistant to intentional jamming, it was not considered practical for cellular use. Moreover, at that time, the U.S. standards groups and industry already had a new second generation digital standard in place called TDMA (ANSI-136). TDMA was well understood, relatively low risk, and considered by experts to be a mature technology. CDMA was not well understood outside of military circles. It also had numerous problems that seemed to place it in the high technical risk category. The Qualcomm tests along with other participating carriers and manufacturers established CDMA as a viable technology for cellular radio. The tests confirmed that CDMA could meet the original CTIA goals for a second generation digital cellular standard. In July 1993, the TIA officially voted on and accepted IS-95 as the official air interface specification for CDMA operating in the 800 MHz cellular band. Several improvements and iterations resulted in the release of IS-95A. Later features not included in IS-95A were subsequently published as TSB-74 and, finally, the ANSI-J-STD-008 brought forth CDMA standards for operation in the new 1900 MHz PCS band. In 1998, the revised standard was released as TIA/EIA-95 (no longer an "interim standard") and incorporated all the features of IS-95, J-STD-008, TSB-74, and additional features. As this scenario illustrates, standards can become complex due to the nature of a fast paced technology for wireless communications. (As a footnote, we are up to ANSI-95B already!) CDMA has been well tested and proven in wide commercial deployment since the early tests.

In the next section, you will learn how CDMA works in a cellular radio environment and how it compares with the TDMA air interface.

CDMA and Multiple Access Codes

Code Division Multiple Access is a multiple access technique for sharing a radio spectrum resource among several users. This definition sounds similar to TDMA and FDMA multiple access methods; however, the technology used in CDMA is quite different. With FDMA and TDMA, mobile stations were assigned different radio frequencies and time slots for the duration of a call. Frequency reuse in the U.S. was a 7 cell per cluster arrangement with separate sets of frequencies allocated within each cell. Care was taken to avoid using the same frequencies within a cluster of adjacent cells in surrounding clusters. With CDMA, subscribers in a single coverage area use the same carrier frequency at the same time. CDMA uses one radio carrier for all mobiles and another carrier for all base stations. It therefore operates in the FDD mode, similar to AMPS and TDMA. The reuse pattern is therefore said to be 1. This simplifies the reuse pattern, but the mystery of CDMA is how users can operate on the same frequency without intolerable interference and total chaos. CDMA separates each caller by using separate codes for the duration of the call. Each CDMA channel is therefore defined as a unique code used temporarily between a base station and a mobile station. Each mobile station responds only to its assigned code and ignores all other codes in use by other mobile/base station channels.

The different codes used with CDMA ANSI-95 are binary sequences that are said to be **orthogonal** to each other.

The codes are referred to as **pseudonoise (PN)** codes because the codes appear as noise to the other mobile stations that do not have the correct PN code sequence to decode the voice signal. Mobile stations ignore the incorrect codes, as shown in Figure 5-23.

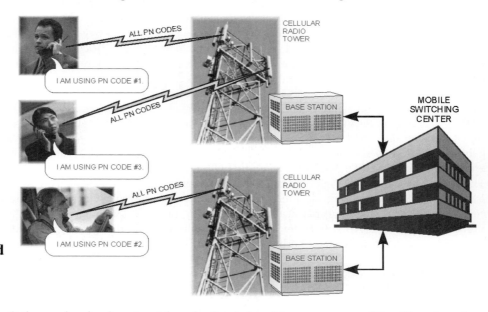

Figure 5-23: CDMA Mobile Stations Respond Only to Their Assigned Codes

Orthogonal codes do not match each other in long binary sequences. We will see how these codes are applied to the radio carrier along with the digitally encoded voice information later in this section. A PN code is not purely random since it can be duplicated electronically and repeats after a designated number of bits. The "pseudo" in PN tells us that the code is **deterministic** and not totally random. Looking closer at a PN sequence, the rules say that the number of 1s and 0s within a code sequence must differ by no more than 1. The sequence relates to the coin flipping experiment shown in Figure 5-24 where the number of 1s and 0s follow a run length pattern called the Bernoulli sequence.

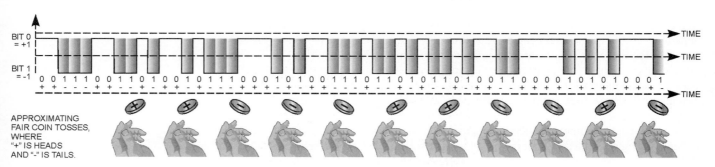

Figure 5-24: PN Code Sequence

Figure 5-25 illustrates the idea of orthogonal codes. This indicates that when two PN orthogonal codes that are bipolar (positive and negative) for equal numbers of bit intervals over a given run length are multiplied together the product is zero. This explains how mobile stations can decode the correct PN code with the correct replicate of an assigned code. The other codes would multiply out to 0 in the receiver, thereby allowing the receiver to accept only the code that matches the assigned known code. Keep in mind that the **randomizing** of these codes makes incorrect codes appear as random noise to the mobile station. Only the correct replicated code in the mobile station will match the base station transmitted code when properly time-aligned in the mobile station receiver.

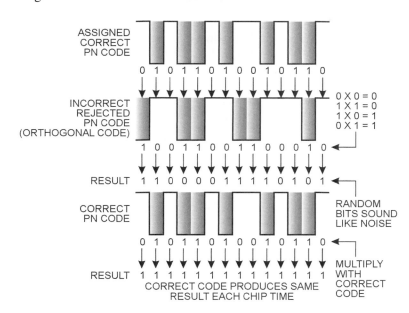

Figure 5-25:
Orthogonal Codes

CDMA and Spread Spectrum Basics

CDMA is a multiple access technique. The term CDMA has become, to some degree, interchangeably used with the term spread spectrum. The two terms have different definitions. Although CDMA defines the "code division" concept of a resource allocation algorithm, spread spectrum communications has a broader set of applications and implementation options. Spread spectrum communications can be implemented in four ways as follows:

- **Direct Sequence Spread Spectrum (DSSS)** is the type of spread spectrum used with ANSI-95. In this case the carrier wave is modulated with a binary bit stream whose bit rate is much larger than the information bit rate. This code is called the spreading code. As an example, ANSI-95 uses a spreading code of 1.2288 megabits per second. The information rate from the voice coder is in the range of 13 kbps. In this case the spreading code to information bit rate ratio is shown as:

$$\frac{1,228,000 \text{ bits / seconds}}{13,000 \text{ bits / seconds}} = \sim 100$$

This indicates that the spreading code will result in a modulated radio carrier wave with a channel over **100 times wider** than would be the case for a 13 kbps modulating bit rate applied to the carrier wave. Remembering our rule from the basic radio chapter, a modulating signal of 1.2288 megabits per second with a 1 bit/hertz modulator would produce a carrier wave 1.2288 MHz wide.

chipping code

megachips

┌─ TEST TIP ─┐
Be able to define
processing gain.
└────────────┘

processing gain

With DSSS, we need to have separate names for the two modulating digital bit streams to keep from confusing the two signaling sources. The spreading code is called the **chipping code** or simply the chipping rate (in this case 1.2288 **megachips**/second). The 13 kbps coded voice signal is called the information bit rate. As you can see from this example, the chipping code contains no information and is used only to create the "spreading" of the radio carrier wave. The difference between the modulated DSSS carrier and the bandwidth that would result from a 13 kbps modulated carrier is shown in Figure 5-26. You may ask the obvious question at this point as to why we would want to use an inordinate amount of spectrum to transmit a 13 kbps digital information rate.

As will be explained later, the ratio between the chipping code and the information code rate, called the **processing gain**, is the feature that provides CDMA with its inherent robustness and tolerance for interference from other mobile stations using the same carrier frequency.

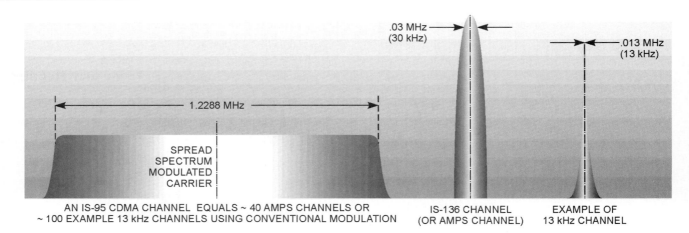

.03 MHz (30 kHz)

.013 MHz (13 kHz)

1.2288 MHz

SPREAD SPECTRUM MODULATED CARRIER

AN IS-95 CDMA CHANNEL EQUALS ~ 40 AMPS CHANNELS OR ~ 100 EXAMPLE 13 kHz CHANNELS USING CONVENTIONAL MODULATION

IS-136 CHANNEL (OR AMPS CHANNEL)

EXAMPLE OF 13 kHz CHANNEL

Figure 5-26: CDMA and AMPS Bandwidth Comparisons

CHANNEL BANDWIDTH COMPARISONS:

- CDMA CARRIER OCCUPIES 1.2288 MHZ OF SPECTRUM FOR A 13 kHZ VOICE CHANNEL, BUT THE CHANNEL IS SHARED AMONG MANY USERS.

- CDMA CARRIER FREQUENCY IS REUSED IN <u>ALL</u> CELLS AND <u>ALL</u> CLUSTERS.

- ANSI-136 AND AMPS CHANNELS USE 30 kHZ OF SPECTRUM FOR AN ANALOG VOICE CHANNEL, BUT THE SAME CHANNEL CANNOT BE SHARED IN THE SAME CELL OR THE SAME CLUSTER OF CELLS.

- ANSI-136B USES 30 kHZ OF SPECTRUM FOR 3 EA 13 kHZ VOICE CHANNELS OR 48 kB/S OF BANDWIDTH.

- EXAMPLE 13 kHZ VOICE CHANNEL @ 1 BIT/HERTZ IS ~100 TIMES WIDER THAN A CDMA CHANNEL.

- **Frequency-Hopping Spread Spectrum (FHSS)** is a method where the radio carrier frequency is shifted in selected increments across a given frequency band. The PN code is again used to provide a pattern for the frequency shifts. A PN code provides the randomness to the shifts. Codes can be selected that are orthogonal and, with proper design, can result in multiple stations hopping across the same band of frequencies with minimal mutual interference. FHSS systems depend upon the statistical probability that other stations will not land on the same frequency at the same time. A graphic of how a FHSS spectrum plot would appear is shown in Figure 5-27. FHSS is used extensively in wireless LAN applications.

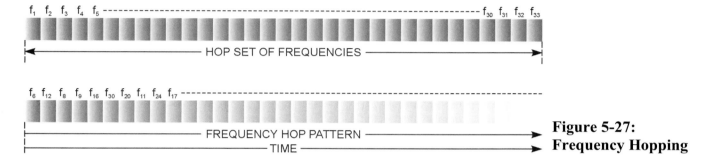

**Figure 5-27:
Frequency Hopping**

- FREQUENCY-HOPPING SPREAD SPECTRUM (FHSS)

- RANDOM HOPS OVER SELECT HOP SET

- RANDOMNESS DETERMINED BY PN CODE SEQUENCE

- **Time-Hopped Spread Spectrum (THSS)** is the third type of spread spectrum communications. With THSS, the transmitter is pulsed on and off. The period and the duty cycle are varied by the PN code. The advantage of a pulsed THSS system is the very low duty cycles that are used. THSS is often used in hybrid systems with FHSS.

- **Chirped FM Pulsed Spread Spectrum** is also a pulsed system where the carrier is turned on and off in a fixed period and fixed on-off time interval. When the transmitter is pulsed on, the frequency is varied (swept) over a fixed range, creating an additional spreading of the carrier. Chirped FM systems are used primarily in radar systems to obtain better resolution in range measurements.

CDMA Channels

Three forward link channels, called the pilot, sync, and paging channels, are used for control functions and carry no data or voice traffic. The other forward channels are called traffic channels.

Traffic channels are used to transmit voice and data information. The channels for CDMA are listed here for both the forward and reverse links.

pilot channel

- The **pilot channels** are transmitted by the base station continuously. Mobile stations scan for the strongest pilot channel when power is first applied. Because all base stations use the same frequency, base stations use a time offset that is different for each base station. The offset is a specific number of bits from a code reference point. This establishes an identity in the network for each pilot channel and associated base station. The amplitudes of the pilots must be carefully controlled because their relative amplitudes control handoff boundaries between stations.

  All base stations use the same short code, and thus have the same pilot waveform. They are distinguished from one another only by the offset of the pilot channel. The short period of the short code, 2^{15}, facilitates rapid pilot searches by the mobiles for the nearest pilot carrier.

Pilot Offset

  The specification indicates that pilot phases are always assigned to stations in multiples of 64 chips, giving a total of $2^{15-6} = 512$ possible assignments. The 9-bit number that identifies the pilot phase assignment is called the **Pilot Offset**. As an example, the short code is 2^{15} bits in length or 32,768 bits. With multiples of 64 bits for each base station offset, this gives a total of:

$$\frac{32{,}768 \text{ bits for the short code}}{64\text{-bit increments for each base station}} = 512 \text{ possible offset assignments}$$

sync channel

- The **sync channel** carries a repeating message that identifies the station, and the absolute phase of the pilot bit sequence. The data rate is always 1200 bits/second. The sync channel allows the mobile station to establish a bit timing reference between the base station and the mobile station. All CDMA base stations are time synchronized. They use the U.S. Global Positioning System (GPS) to provide a common time reference.

paging channel

- The **paging channel** is used to communicate with the mobile station when the mobile is not using a traffic channel. The mobile station will begin monitoring the paging channel after it has set its timing to the system time provided by the sync channel. After a mobile station has been paged and acknowledges that page, call setup and traffic channel assignment information is passed on this channel to the mobile unit.

Traffic channels

- **Traffic channels** are assigned dynamically, in response to mobile station accesses, to specific mobile stations. The mobile station is informed which code channel it is to receive via a paging channel message. The traffic channel always carries data in 20 ms frames.

The following channels are used by the reverse link.

traffic channel

- The **traffic channel** is used as the other half of the FDD link from the mobile station to the base station. It also carries the mobile power control information from the mobile unit to the base station.

access channel

- The **access channel** is used by the mobile stations for originating calls and responding to a page. Information transmitted on the reverse channel is contained in 20 ms frames.

Walsh Codes

Each channel on the forward link is modulated by a Walsh code. Each base station uses 64 different Walsh codes to provide separation (orthogonality) for each channel. Walsh codes are also used on the reverse link but for another purpose.

> It is the Walsh code that provides the separation or "code division" feature for the forward channels in CDMA.

The pilot channel is always assigned to Walsh code #0. The sync channel is assigned to Walsh code #32. There are 7 paging channels that are given Walsh codes 1 through 7. The remaining Walsh codes are used for forward link traffic channels. Walsh codes are orthogonal codes if they are all adjusted to the same start time. Walsh codes do not provide orthogonal code separation on the reverse channel because different mobile stations have different propagation delays, so they will arrive shifted in time relative to each other. Walsh codes are used to provide a more rugged "64-ary" encoding on the reverse channel.

CDMA Modulation and Rate Adaptation

Both the forward and reverse traffic channels use a similar control structure consisting of 20-millisecond frames. For the system, frames can be sent at either 14,400, 9600, 7200, 4800, 3600, 2400, 1800, or 1200 bits/second.

For example, with a traffic channel operating at 9600 bits/second, the rate can vary from frame to frame, and can be 9600, 4800, 2400, or 1200 bits per second. When operating 14,400 bits per second, the rate can vary from 14,400, 7200, 3600, or 1800 bits per second. The receiver detects the rate of the frame and processes it at the correct rate. This technique allows the channel rate to dynamically adapt to the speech or data activity. For speech, when a talker pauses, the transmission rate is reduced to a low rate. When the talker speaks, the system instantaneously shifts to a higher transmission rate. This technique decreases the interference to other CDMA signals and thus allows an increase in system capacity.

> Any feature that reduces interference in a CDMA system results in a potential increase in subscriber capacity.

CDMA starts with a basic data rate of 9600 or 14,400 bits per second. This is then spread to a transmitted bit rate, or chip rate (the transmitted bits are called chips), of 1.2288 MHz. The spreading process applies digital codes to the data bits, which increases the data rate while adding redundancy to the system.

The chips are transmitted using a form of QPSK (quadrature phase shift keying) modulation, which has been filtered to limit the bandwidth of the signal.

Demodulation and Despreading of the CDMA Carrier

Figure 5-28 illustrates the process used to capture the information at the receiver from a modulated CDMA carrier. The top line represents the baseband data stream carried on the CDMA carrier. The second line in the figure represents the PN code used for this traffic channel example. The third line shows the actual PN code transmitted after combining with the information bits. Notice the inversion of the PN traffic channel code bits when the data bit is a logical zero. The fourth and fifth lines in the figure show the results in the receiver when the transmitted bits are multiplied by the correct PN code. The result is that the baseband information bits are recovered by the multiplication with the correct PN code in the receiver. All other incorrect codes being transmitted by the base station on the same carrier frequency would not match the correct traffic channel assigned and the time-aligned PN code in the receiver and would appear as noise.

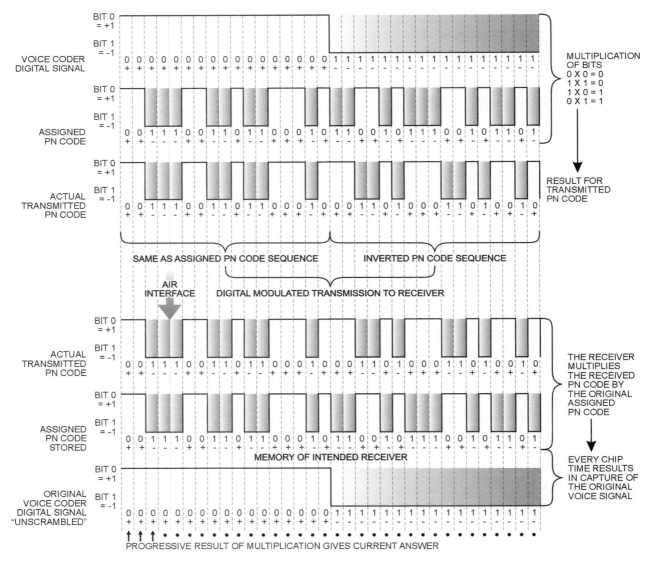

Figure 5-28: Unscrambling the CDMA Voice Channel

CDMA Handoff and Control Functions

Several features are used in CDMA to enhance performance and overcome some of the challenges that were previously believed to be difficult, if not impossible to solve. Other processing functions are used to enhance the performance of the network. These features are summarized below.

The "Near/Far" Problem

> The characteristics of CDMA require a constant (or nearly constant) received power at the base station from mobile stations.

┌─ TEST TIP ─
Be able to describe the
CDMA near/far problem.
└

This is because CDMA is an interference limited system. Since all mobiles transmit on the same frequency and are moving around at varying distances from the base station, it would be possible for those mobile stations nearest to the base station to create interference that prevents the base station from adequately receiving mobile stations far away near the periphery of the cell (the "near/far" problem situation). In the early planning phase for CDMA, this problem was believed so severe as to make CDMA unusable for a mobile environment. The base station has a more severe problem to deal with than is the case for the mobile stations, therefore some method of power control is required. The objective of power control is to limit the radiated power on the forward and reverse links while still maintaining adequate quality of service.

> As indicated in Figure 5-29, power control for CDMA is called adaptive power control since it responds to the dynamic link conditions as necessary to maintain a minimum power level for both the near and far mobile stations.

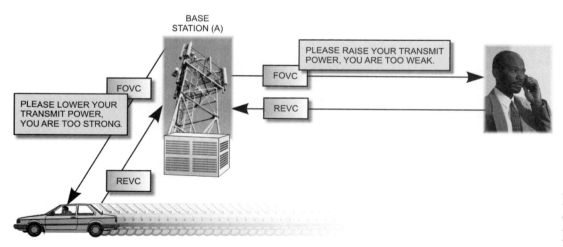

**Figure 5-29:
CDMA Dynamic
Power Control**

Two types of power control are:

- Open loop control
- Closed loop control

Open loop control is where the mobile station determines its transmit power from the strength of the received signal from the base station. When the mobile detects that the signal is getting weaker, it assumes it is moving away from the base station and therefore needs to increase power. This provides a slow, coarse power control.

Closed loop power control is where the mobile station is commanded in the forward link traffic channel to increase or reduce power based upon the received signal strength at the base station. A power control bit is sent to each mobile station at a rate of 800 times per second to either increase or decrease power. This closed loop control provides a fast, efficient fine control. A side benefit to dynamic power control is that CDMA cell phones typically use less transmitter power than TDMA and AMPS phones, thereby obtaining on average a longer battery endurance time.

Soft Handoff

The advantage with soft handoff is fewer dropped calls and better performance on the channel when the mobile station travels across cell boundaries.

As shown in Figure 5-30, CDMA takes advantage of multipath signals, which are common in a mobile environment. A RAKE receiver is used to separately detect multipath components.

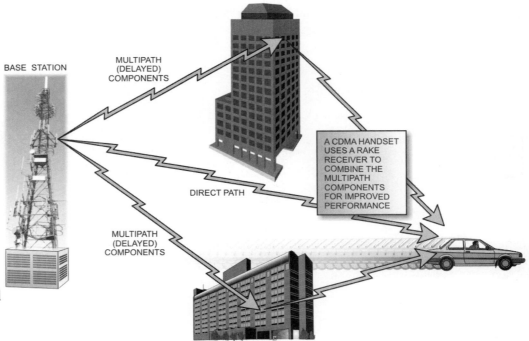

**Figure 5-30:
Multipath Signal
Processing with
CDMA**

In the CDMA system where adjacent cells use the same carrier frequencies, it is possible for a mobile station to receive transmissions from two or more base stations at the same time. It is also possible for multiple base stations to receive transmissions from a mobile station at the same time. Using this advantage, it is possible to process a handoff from one base station to another using a "make before break" strategy. With CDMA, a mobile can be in contact with as many as three base stations during handoff. When the MSC determines that one or more of the cells are not receiving the mobile with sufficient signal quality after handoff, the connections are dropped in favor of the best signal from a cell. This feature is called **soft handoff**.

┌─ TEST TIP ─┐
Know the advantages for soft handoff.

soft handoff

RAKE Receivers

Combining is performed in a **correlator** where the received CDMA carrier is added to the delay-compensated copies of the signal that result from reflections in an urban environment. This combining is called **diversity reception**. It aids performance in a fading radio environment. The term RAKE is not an acronym but is named for the appearance of the tapped delay line circuit in the receiver shown in Figure 5-31. Multiple RAKE receivers have the ability to process more than one delayed multipath component. Each RAKE path in the receiver is called a finger. RAKE receivers are also used during soft handoff where multiple base station signals are received and processed.

correlator

diversity reception

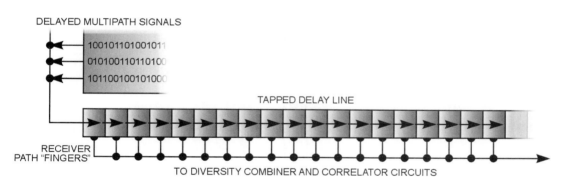

- RAKE RECEIVER PROCESSES DELAYED, MULTIPATH SIGNALS FOR INCREASED PERFORMANCE.

- TYPE OF DIVERSITY RECEPTION.

- CIRCUIT DIAGRAM RESEMBLES A GARDEN RAKE. 'HANDLE' IS OPTIONAL.

Figure 5-31: RAKE Receiver

Selecting a Cell Phone and Service Provider

Selecting the best cell phone and the best wireless subscriber plan can become a complicated task. There are many options to choose from; therefore, conducting an objective evaluation of cell phones and various service plans is a good way to begin the selection process. The number of service providers and geographical coverage will vary with each geographical area in the U.S. If you already own a cell phone, the exercise at the end of this chapter may help you review the process you used in selecting your service plan and phone. Figure 5-32 shows various cell phone models.

Figure 5-32:
Cell Phone Models

Some of the specifications for the cell phone you purchase are governed by the service provider's technical standards. As you learned in this chapter, the three digital air interface standards are TDMA, CDMA, and GSM using digital modulation. The analog first generation cellular systems use FDMA with frequency modulation. The two frequency bands used in North America are the 800 MHz band and the 1900 MHz PCS band. When selecting a service provider for your cellular service, the phones that are compatible with the air interface and frequency bands are described in the service plan. It is important to know the terms and functions described for each feature when completing the evaluation checklist in this lab exercise. Remember, there are no "stronger" or "weaker" frequency bands or air interface standards listed in sales literature or brochures.

First generation analog cellular service does not offer the robustness and interference rejection of the later generation digital service, but it does fill the gaps where digital mode service may not be built out with cell site coverage by your primary service provider. You would want to select a service plan that provides the best coverage with both modes when roaming in different areas. If you never plan to use a wireless phone outside your home area, this is not a choice you need to worry about.

Figure 5-33 shows a typical circuit board and integrated circuit layout of a digital cellular phone.

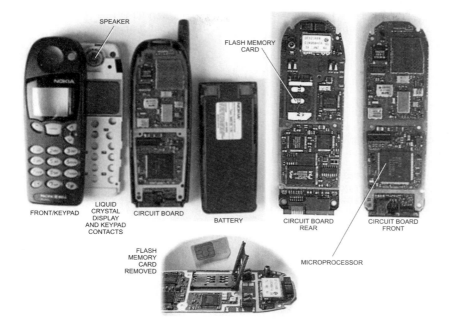

SPEAKER

FLASH MEMORY CARD

FRONT/KEYPAD

LIQUID CRYSTAL DISPLAY AND KEYPAD CONTACTS

CIRCUIT BOARD

BATTERY

CIRCUIT BOARD REAR

CIRCUIT BOARD FRONT

FLASH MEMORY CARD REMOVED

MICROPROCESSOR

**Figure 5-33:
Cell Phone Parts**

Most carriers offer two ways to connect to the World Wide Web. The phone can act as a data modem when connected to a laptop or PDA. A smart phone with a Wireless Access Protocol (WAP) microbrowser can be used to connect to text-only web sites. Web surfing is not for everyone. If you seriously need to check up on financial news, news headlines, e-mail, or the latest flight information on your way to the airport, this may be your best option, but remember to ask where the service is available and how much extra it will cost. Expect slow download speeds in the range of 14.4 Kbps due to the limited bandwidth of second-generation cellular voice channels.

In Japan, NTT DoCoMo has introduced a proprietary access protocol called I-mode service. Some wireless carriers in the U.S. may offer it in the future.

You may want to consider a cell phone that will allow you to connect to the Internet when you are away from home or the office. Several wireless service providers began wireless Internet service in 2000. A typical wireless Internet menu is shown on a Sprint Internet-capable phone in Figure 5-34. The web site information will be displayed in a condensed form since web page information including graphics is not available with this type of service in second generation digital phones due to the limited screen size and bandwidth limitations.

When selecting a cell phone the selection guidelines listed in Table 5-5 should be considered.

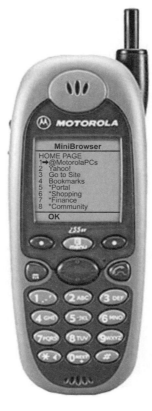

**Figure 5-34:
Internet-Capable
Phone**

Table 5-5: Cell Phone Selection Guide

Vendor Name/ Model Number	When selecting a vendor, be certain they have the type of phone that meets your overall needs. A comparison matrix will help you decide on the model number after you review the features in the Cell Phone Comparison Chart.
Price	Select a phone and the price range that meets your needs and budget.
Dual-Mode and Dual-Band Phones	Many vendors have different marketing names for dual-mode and dual-band phones. Most popular brand name suppliers have phone products that support dual band/dual mode of operation. Dual mode means that the phone supports both analog and one digital access mode (CDMA, TDMA, or GSM). Carriers operate in one of the digital modes and also have roaming agreements with many analog operators to provide roaming coverage where no digital mode access service is yet available. Dual-band phones operate in both the 800 MHz and the 1900 MHz bands and will automatically search for the frequency band and strongest signal that is available in your primary or roam service area. If your needs involve frequent international travel, you may want to consider GSM phones, which operate with European GSM standards in the 900 MHz or 1800 MHz bands. Nextel is an international wireless service provider that offers products and plans that will allow you to use your phone in the U.S. and other selected countries outside the U.S.
Phone Talk Time and Standby Time	Cell phones consume more power when conducting a call than when in standby mode. In standby mode, the phone will automatically maintain communications with the network as you move from one location to another. Both standby time and talk time are functions of phone design and battery efficiency. Cell phones use two primary battery technologies: NiMH (nickel-metal hydride) is a high-capacity battery that provides extra power for extended use. They are widely used and available for most phones on the market. Li-ion (lithium ion) is a lightweight package in a smaller physical space with good efficiency but typically at higher cost. Lithium batteries have a charging control circuit to prevent overcharging. They are available for nearly all phones on the market.
Voice Quality	Different wireless carriers offer various claims about the quality of the voice service. Digital standards for voice compression are very similar for each digital mode. Voice quality evaluation is a subjective process. You may need to use the phone and service selected in the lab to make your personal evaluation. Ask your cellular dealer for a demonstration.
Display Size and Memory	An important item if you are planning to use your phone for Internet access or one of the text messaging services. Small screens reduce the size of the phone but often at the expense of display area. Memory slots allow you to store phone numbers for "one button" speed dialing.
Hands-Free Kit and Headset	The hazards associated with using your cell phone while driving have been widely publicized. If you are going to initiate and receive calls while driving, consider selecting a phone that has provisions for a hands-free kit. Technology is advancing in this area for voice recognition software that permits dialing using voice commands. Headsets can also free up your hands to conduct work on your laptop in an airport while waiting to board. Consider both options if they fit your user profile.
Automatic Redial	This allows you to redial a number without keying in the numbers a second time.

Table 5-5: Cell Phone Selection Guide (Continued)

Incoming Call Number Storage	This feature stores the numbers for the inbound calls you have received in memory.
Digital Mode Access	Digital modes come in three types, as discussed in the text of this chapter. Carriers in the U.S. have adopted digital access standards that are not compatible or interoperable with each other. The debate about which standard is better is a long ongoing debate. It should not be a factor in your selection process. The cellular service is very competitive. Each carrier strives to provide good service with good voice quality; however, you may want to know about the TDMA, CDMA, or GSM mode of service and how extensive the digital service area is across the U.S.
Service Plan Comparison Chart	The Service Plan Comparison Chart contains a sample of most of the items offered by various cellular service providers. Ask yourself how you now use or intend to use your cell phone. Are you going to use it only for emergencies? If so, you may want a prepaid service plan with a limited number of minutes. Do you plan to travel outside of your primary home number service area? If you do, you may want to focus on the coverage area and free long distance calling plans. If you spend a lot of your business travel time outside of the U.S., you need to evaluate the GSM service plans that are available in the U.S. and Europe and other countries. Talk to a service provider that supports international roaming.
Access Mode	Ask about the type of digital access used by the carrier (TDMA, CDMA, or GSM). Use the same selection process outlined in the cell phone selection criteria.
Local Coverage and National Coverage	Obtain a map that shows each carrier's local coverage as well as national coverage. Ask about roaming charges for calls outside the home coverage area. Many carriers offer different plans based upon how far your travels may take you across the country. Look for local, regional, and national rates with no roaming fees.
Annual Contract	Annual contracts fix the rate for a specified period. The plan usually offers a fixed amount of air time minutes. Carriers often offer perks such as free phones to get you to sign up for 1 to 3 years. These plans offer good value but be aware that canceling out early may cost you.
Prepaid Plan	Research each carrier to determine whether they offer a pre-paid wireless plan as an alternative to long-term contracts. Like pre-paid phone cards used by long distance wire line carriers, these plans let you buy wireless minutes that you can draw from until they run out. You can buy service in $25, $50, and $100 blocks. There is no monthly fee.
Minutes Included	Check the number of minutes available with each plan. Note the cost and availability of anytime minutes and weekend or night-time minutes in your preferred plan.
Cost per Minute Beyond Plan	Determine the penalty for going over the plan's allowable number of minutes.
Free Long Distance and Roaming	Check to see if there are extra costs for calls outside your home area.
Activation Fee	Some carriers include activation fees to program the setups in your phone such as the MIN and SID codes supplied by a service representative.

Table 5-5: Cell Phone Selection Guide (Continued)

Caller ID	This feature works like the wireline service available in your home or office. Incoming numbers are identified on the phone display.
Web Access	Remember that surfing the Internet can consume minutes in your basic service plan.
Voice Mail	A nice feature available in most plans. You will need this feature to record missed calls.
Call Forwarding	Allows calls to be forwarded to any other number when you know you will be unable to answer your cell phone. When activated, calls are immediately forwarded to the number you have programmed. Note the additional cost if available.
Short Message Service	Short message service (SMS) allows you to send and receive short text messages, usually less than 110 alphanumeric characters, to and from other mobile phones.

Lab 3 – Selecting the Best Service Plan Among Cellular Carriers

Lab Objectives

Given a cell phone comparison table, perform an analysis of different cell phone features and associated costs.

Given a service plan comparison table, perform an analysis of different cellular subscriber service plans and associated costs.

Materials Needed

- Computer access to the Internet

- Wireline telephone access for obtaining cost quotations for service plans from sales representatives as required

Instructions

1. Select at least three cellular wireless service providers in your area. Using the Internet or direct phone contact, collect the information necessary to complete the two comparison charts in this lab exercise.

2. Start with the cellular phone comparison chart shown in Table 5-6. Narrow your search to three phones based upon your preference for size, color, and general physical features. The selection criteria for cell phones are outlined in Table 5-5. Different vendors may have features not shown in the matrix; however, the items listed are the most important features you need to review when selecting a cell phone in this lab exercise.

Cell Phone Feature	Vendor 1	Vendor 2	Vendor 3
Vendor Name/Phone Model	_____	_____	_____
Price	$_____	$_____	$_____
Dual Mode/Dual Band	Yes __ No__	Yes __ No__	Yes __ No__
Talk Time/Standby Time	_____	_____	_____
Sound quality	Good __ Poor__	Good __ Poor__	Good __ Poor__

Battery Type:
  Nickel-Metal-Hydride (NiMH)
  Li-ion (Lithium-ion)
	Yes __ No__
	Yes __ No__
		Yes __ No__
		Yes __ No__
			Yes __ No__
			Yes __ No__

Display size	_____	_____	_____
Memory size/Phone directory	_____	_____	_____
Hands-free kit	Yes __ No__	Yes __ No__	Yes __ No__
Headset available	Yes __ No__	Yes __ No__	Yes __ No__
Car adapter/charger	Yes __ No__	Yes __ No__	Yes __ No__
AC wall charger	Yes __ No__	Yes __ No__	Yes __ No__
Weight	_____	_____	_____
Wireless Web Access	Yes __ No__	Yes __ No__	Yes __ No__
Volume adjust	Yes __ No__	Yes __ No__	Yes __ No__
Ringer/silent/vibrate mode	Yes __ No__	Yes __ No__	Yes __ No__
Automatic redial	Yes __ No__	Yes __ No__	Yes __ No__
Incoming number storage	Yes __ No__	Yes __ No__	Yes __ No__

Digital Mode Access:
  TDMA
  CDMA
  GSM
	_____
	_____
	_____
		_____
		_____
		_____
			_____
			_____
			_____

Table 5-6: Cellular Phone Comparison Chart

TIP: Think about how you plan to use the phone. Will it mostly be a car phone? Or do you plan to carry it in a pocket all day long? This will help you determine whether a cheaper phone with the same features that weighs 12 ounces is a better deal than the one that weighs only 6 ounces but costs 20 percent more. Size and shape should match how you intend to use your phone. Try making a call before you buy. Figure 5-32 shows some examples of various size and shapes of phones as well as display areas.

3. The Internet is an excellent source to obtain photos and technical specifications for cell phones. You may try starting with a general search for wireless cell phones to obtain the most recent information on technical specifications and cost.

4. Evaluate any free cell phone offers from service providers. Determine whether any free phone offers meet all the criteria in the cellular phone comparison chart. Also remember that free phone offers are often tied to an annual contract.

5. Based on your results from the cellular phone comparison shown in Table 5-6, select a phone that provides a digital access mode (CDMA, TDMA, or GSM) that is compatible with the service provider's digital mode of operation. Your service provider will answer this question for you. (Very few customers ask for this information.) Phones are often marketed with the brand name of the service provider on the phone (Sprint, Verizon, Voice Stream, etc.).

6. Complete the service plan comparison chart in Table 5-7. Make your overall evaluation-ranking estimate after gathering information. This selection should be based on the factors in this chart that meet **your individual needs**. Pay particular attention to the number of minutes you plan to use and the areas beyond the primary home area (if any) where you will be using your cell phone.

Table 5-7: Service Plan Comparison Chart

Cell Phone Feature	Vendor 1	Vendor 2	Vendor 3
Vendor Name	_____	_____	_____
Price	$_____	$_____	$_____
Access Mode	_____	_____	_____
Local area coverage	Good __ Poor__	Good __ Poor__	Good __ Poor__
U.S. coverage	Good __ Poor__	Good __ Poor__	Good __ Poor__
Prepaid plan	Yes __ No__	Yes __ No__	Yes __ No__
Minutes included	_____	_____	_____
Monthly plan	Yes __ No__	Yes __ No__	Yes __ No__
Minutes included	_____	_____	_____
Cost per minute beyond plan	$_____	$_____	$_____
Free long distance	Yes __ No__	Yes __ No__	Yes __ No__
Free roaming outside home area	Yes __ No__	Yes __ No__	Yes __ No__
Long distance cost/minute	$_____	$_____	$_____
Activation fee	$_____	$_____	$_____
Caller ID	Yes __ No__	Yes __ No__	Yes __ No__
Web access	Yes __ No__	Yes __ No__	Yes __ No__
Voice mail	Yes __ No__	Yes __ No__	Yes __ No__
Call forwarding	Yes __ No__	Yes __ No__	Yes __ No__
Short message service (SMS)	Yes __ No__	Yes __ No__	Yes __ No__
Overall preference (1, 2, or 3)	_____	_____	_____

Lab 3 Questions

1. What type of service plan did you find that met most of your personal requirements?

2. Were the costs between different carriers for similar plans within the same range?

3. Describe the advantages you believe your first choice had over the two other service plans.

4. What features of the cell phone influenced you the most in your selection?

Key Points Review

In this chapter, the following key points were introduced concerning analog and digital cellular radio. Review the following key points before moving into the Review Questions section to make sure you are comfortable with each point and to verify your knowledge of the information.

- The first radio communications between a fixed radio base station and a mobile station began with ship-to-shore radiotelephone service in the late 1920s.

- The invention and development of frequency modulation by Edwin Armstrong, discussed in Chapter 1, proved to be particularly useful for mobile radio applications.

- The first U.S. commercial radio system designed to provide access to the Public Switched Telephone Network for mobile wireless subscribers was introduced in St. Louis, Missouri on June 17, 1946. This inaugural system was called the Mobile Telephone Service (MTS).

- The MTS operated in the half-duplex mode. You pushed a handset button to talk, then released the button to listen.

- On March 1, 1948 the first fully automatic radiotelephone service began operating in Richmond, Indiana, eliminating the need for an operator to place most calls.

- The IMTS was a major improvement over the initial MTS service; however, it also had serious limitations.

- The solution to the IMTS spectrum saturation problem was proposed by Bell Laboratory engineers as early as 1947. It was called cellular radio.

- The arrangement for allocating sets of frequencies among dispersed cellular locations was called the cellular frequency reuse plan.

- It would require over 35 years, or approximately from 1947 to the mid 1980s, before the first cellular telephone system would become operational.

- Studies by Bell Laboratory engineers concluded that the cellular concept would need additional radio spectrum before it could become a practical commercial mobile telephone service.

- Cellular radio uses low-power radio base stations to bring the geographical coverage area down to smaller areas called "cells."

- During the 1980s, the competitive environment changed dramatically. The U.S. Justice Department, in a dramatic antitrust action, divided AT&T into eight separate companies. Cellular development would again be delayed for several years.

- It is important to understand the primary goal for a cellular radio system. The goal is to provide the maximum subscriber density for a given amount of radio spectrum and geographical area while maintaining an acceptable quality of service.

- Cellular mobile networks are more complex than fixed wireline networks.

- All cellular subscribers have a "home" service area dictated by the mobile telephone area code number. When a mobile station enters a service area outside the home service area, the subscriber is in the roaming mode.

- Frequency reuse is accomplished by assigning a subset of the total frequencies available to each cell.

- Reuse patterns are also implemented to provide equal distance separation between cells using the same frequencies.

- Cells can vary in size based upon the anticipated traffic density for mobile users.

- The primary element the cellular engineer must deal with is the interference from other cells using the same frequency, called cochannel interference, and cells using the adjacent frequency, called adjacent channel interference.

- Analog systems using FM can tolerate carrier to interference ratios of 17 to 18 dB.

- Digital systems are more tolerant to interference and can operate satisfactorily with C/I values of 9 to 13 dB.

- North America and several other regions of the world use a 7-cell reuse plan.

- The size of the cell can be controlled by antenna height, antenna tilt, and transmitted power output of the base station.

- Cell splitting is often used by cellular service providers to increase capacity in areas with high subscriber growth.

- Cell sectoring is another technique used to increase the capacity of a cellular network.

- Multiple access protocols are used in cellular radio systems to provide a structured method of sharing a radio channel among a number of mobile stations.

- Frequency Division Multiple Access (FDMA) divides the entire radio spectrum into channels.

- Time Division Multiple Access (TDMA) allows more than one user to use a radio channel at the same time by dividing the channel into non-overlapping time slots.

- Code Division Multiple Access (CDMA) allows all users to share the same radio channel at the same time.

- Cellular radio uses the terminology "Frequency Division Duplex (FDD)" and "Time Division Duplex (TDD)."

- FDD uses a separate channel in each direction between two parties at the same time. The channels are therefore separated in the frequency domain.

- The radio transmission channel path from the base to the mobile is called the forward channel.

- TDD uses a single radio channel that is used alternately between two parties on an alternate or "ping pong" type scheme. The two channels are therefore separated in the time domain.

- The radio transmission channel path from the mobile station to the base station is called the reverse channel.

- Advanced Mobile Phone Service (AMPS) was the pathfinder system for cellular communications.

- AMPS uses both voice channels and control channels to establish and maintain communications in a cellular radio system.

- Because the mobile station cannot receive on two channels at the same time, the base station uses a special signaling format on the voice channel to send commands without disrupting the call in progress. This is called blank and burst encoding.

- Two additional signaling formats are used by the AMPS. They are called the Supervisory Audio Tone (SAT) and the Signaling Tone (ST).

- The Digital Advanced Mobile Phone Service (D-AMPS) was the introduction in North America of the second generation of cellular communications.

- The ANSI-136 (now ANSI-136B) specification is a newer TDMA specification incorporating a digital control channel in place of the AMPS analog control channel.

- The Global System for Mobile Communication (GSM) is the most widely used form of digital wireless technology for mobile telephony and data communications in the world.

- GSM 400 is the GSM standard adopted for use in some European countries, principally in Germany and the Nordic Countries.

- GSM 900 is the original GSM standard used in Europe on the 900 MHz band.

- GSM 1800 is an "upbanded" version of GSM 900 with minor modifications.

- GSM 1900 is the version of GSM adopted by the U.S. standards bodies for use in the current PCS 1900 MHz band in the U.S.

- The ANSI-95 Code Division Multiple Access (CDMA) Standard is another second generation digital cellular technology that has gained considerable market share and growth in the U.S.

- In July 1993, the TIA officially voted on and accepted IS-95 as the official air interface specification for CDMA operating in the 800 MHz cellular band.

- The different codes used with CDMA ANSI-95 are binary sequences that are said to be orthogonal to each other.

- Three forward link channels, called the pilot, sync, and paging channels, are used for control functions and carry no data or voice traffic. The other forward channels are called traffic channels.

- It is the Walsh code that provides the separation or "code division" feature for the forward channels in CDMA.

- Any feature that reduces interference in a CDMA system results in a potential increase in subscriber capacity.

- The characteristics of CDMA require a constant (or nearly constant) received power at the base station from mobile stations.

- Power control for CDMA is called adaptive power control since it responds to the dynamic link conditions as necessary to maintain a minimum power level for both the near and far mobile stations.

- The advantage with soft handoff is fewer dropped calls and better performance on the channel when the mobile station travels across cell boundaries.

- CDMA takes advantage of multipath signals, which are common in a mobile environment. A RAKE receiver is used to separately detect multipath components.

REVIEW QUESTIONS

The following questions test your knowledge of the material presented in this chapter:

1. What was the first U.S. commercial radio system designed to provide access to the Public Switched Telephone Network for wireless mobile subscribers?

2. What is "trunking" as applied to mobile radio service?

3. Describe the major goal of any cellular radio system.

4. Describe the two types of handoffs used by cellular systems.

5. What is cochannel interference?

6. Describe the major advantage of digital cellular radio over analog cellular radio systems.

7. Describe the two techniques used by service providers to increase the capacity of a cellular system.

8. Briefly describe the three multiple access protocols used for cellular radio systems.

9. Where and when was the first commercial cellular service offered in the U.S.?

10. How many channels are authorized by the FCC for cellular operation in the 800 MHz band?

11. Explain what is meant by "dual-mode" phones.

12. Describe the "GSM Family."

13. What are orthogonal codes? Where are they used?

14. What is the bandwidth used for each channel as specified in the ANSI-95 CDMA standard?

15. Briefly describe how soft handoff is used. What are the advantages offered by soft handoff?

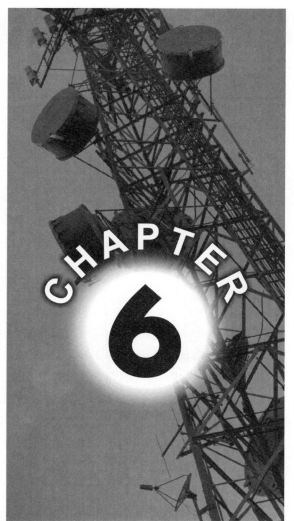

CHAPTER 6

WIRELESS NETWORKS

Upon completion of this chapter and its related lab procedures, you should be able to perform the following tasks:

1. Identify the standards that support cellular radiocommunications intersystem operation.

2. Label a block diagram showing the cellular network elements.

3. Describe the functional capabilities of a mobile switching center.

4. Describe the content of the home location register database.

5. Identify the functions performed by the visitor location register.

6. List the main tasks performed by the authentication center.

7. Describe the equipment and functional elements of a cellular base station.

8. Describe the processing sequence for registration of a mobile station.

9. State the two main steps in a CAVE authentication sequence.

10. Explain how identity codes stored in the handset prevent fraud.

11. Analyze the procedure for authenticating a mobile station.

12. List the six steps required for successful call completion.

13. Describe the limitations of the TCP/IP protocol suite in a mobile environment.

14. Describe the process of obtaining a temporary IP address for a portable user.

15. Analyze and list the differences between a permanent IP address and a "care of" address.

16. List the six steps involved in the registration of a mobile IP station.

17. Explain the concept of tunneling and encapsulation in network operation.

18. Describe the spectrum allocations used for all unlicensed WLAN devices.

19. Describe the features of HIPERLAN.

20. List the general features of the HomeRF specification.

21. Describe the general characteristics of the Bluetooth specification.

22. Create a chart showing 802.11b channel separation.

23. Describe how directional antennas can be used to minimize interference in WLANs.

24. Explain the difference between ad hoc mode and infrastructure mode.

25. Describe the two processes involved in WEP data encryption.

WIRELESS NETWORKS

INTRODUCTION

In the preceding chapter, you learned the basic concepts of the cellular radio network with the emphasis on the radio side of the network. In this chapter, we will look at the network side and the functional elements that collectively perform all tasks dealing with mobile subscriber management.

Cellular Network Intersystem Operation

Cellular networks perform many tasks prior to establishing a call connection between the calling parties. In this section, we will take a closer look at how the cellular network manages the multitude of tasks necessary to provide such services as registration, authentication, call routing, intersystem roaming, subscriber location tracking, and security for cellular subscribers. You will be able to follow the sequence of events and tasks that occur during cellular call initiation and completion. Although most of the tasks are performed without any interaction with the mobile subscriber, it is necessary to take a closer look at the functions supporting network intersystem operation.

Cellular Network Intersystem Operation Standards

There are many cellular service providers in North America. Each provider may use different air interface standards such as AMPS, GSM, TDMA, and CDMA, as discussed in Chapter 5. Different vendors also supply cellular equipment such as base stations, mobile switching centers, and network interface entities.

Without a set of standards, the capability of cellular carriers to exchange information concerning roaming subscribers would be severely limited.

Standards also facilitate the ability to confirm the identity of a cellular mobile station as a legal subscriber, thereby protecting carriers from fraud and cloning of cell phones. This feature adopted in the network standards is called authentication.

As stated in Chapter 3, the Telecommunications Industry Association (TIA) subcommittee TR45.2 is responsible for the development of wireless network standards, including the ANSI-41 standard for intersystem operation. ANSI-41 has been revised several times since its original publication in January 1991.

> The ANSI-41-C has been replaced with the current version called **Cellular Radiotelecommunications Intersystem Operations (ANSI/TIA/EIA-41-D-97)**.

The TIA describes the purpose of this standard as follows:

The purpose of this document is to identify those cellular services that require **intersystem cooperation**, to present the general background against which those services are to be provided, and to summarize the principal considerations that have governed and directed the particular approaches taken in the procedural recommendations. This latest document replaces TIA/EIA/ANSI-41-D.

The Cellular Telecommunications and Internet Association (CTIA) reported that in 1995, the wireless industry lost $650 million in cloning fraud. Because of the potential for fraudulent cloning of cell phone Electronic Serial Numbers (ESN) and Mobile Identification Numbers (MIN), the standards bodies developed standards and security procedures to help cellular service providers avoid revenue loss through fraud.

Authentication is considered by cellular engineers as the ultimate solution to cloning fraud. However, few cellular phones originally supported authentication and many cellular service providers did not see its implementation as cost-effective. Now, however, authentication is possible in all cellular mobile stations based on the latest TIA standards including AMPS, TDMA, and CDMA.

Network Elements

The cellular system architecture shown in Figure 6-1 illustrates the interfaces and cellular network entities described in ANSI/TIA/EIA-41-D-97. The model is referred to in the TIA standards as the TR-45/TR-46 Reference Model. Each element and the reference points shown in Figure 6-1 provide a common set of protocols to allow different vendors and different service providers to exchange information using a common set of rules for intersystem operation. These operations are described in more detail in the following sections. Figure 6-1 illustrates the network elements and reference points we are concerned with for network intersystem operation. Each element and related interface definitions are described in the following paragraphs.

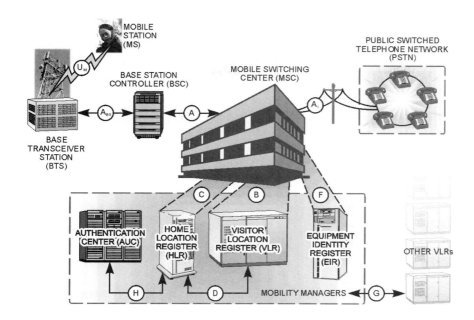

**Figure 6-1:
Cellular Network
Reference Model**

The Mobile Switching Center

The **Mobile Switching Center** (**MSC**) is the main control entity in the network that provides the interface between the PSTN and the base station.

The MSC contains an Electronic Switching System (ESS), a power supply system, computer resources, and alarm monitoring equipment. It controls the handoff functions, channel assignments, and all mobility management of a group of cellular base stations. The MSC is also connected to databases that contain specific information relative to the identity and location of mobile subscribers being served by the MSC.

The Home Location Register

The **Home Location Register** (**HLR**) database contains all specific subscriber service data, including the profile of services that each subscriber is authorized to use on the network.

Every mobile station has a "home" location determined by the service area of the service provider. When a subscriber is using the cellular network, they are being served in one of two possible configurations. They are being served (1) by the MSC associated with their home area phone number or (2) by a MSC that is not located in the home subscriber area. An example would be a subscriber who has a cellular number associated with a Chicago MSC but is "roaming" in the Denver area. The Denver area databases would notify the Chicago area HLR concerning the location of the Chicago subscriber. Depending upon the network service provider's implementation plan, a HLR may be shared among several MSCs in a metropolitan area.

The HLR is therefore the database that also holds information concerning the location of each subscriber whose phone number is served by an associated MSC. This assumes that the subscriber has registered by turning the power on the mobile station handset. If the subscriber is currently being served in a distant MSC that is not the "home" MSC, the HLR is always notified and essentially tracks the location of the subscriber.

The Visitor Location Register

The **Visitor Location Register (VLR)** is a database associated with an HLR. It contains dynamic information about subscribers *who are currently active* in the area served by the associated MSCs.

The term "visitor" used in the ANSI-41 standard is sometimes confusing because the VLR contains information about active home subscribers who are active as well as visitors from another service area.

The VLR database content varies due to the dynamics of active mobile stations. In contrast, the HLR is a database with a relatively stable set of entries about current home subscribers, whether active or inactive. Depending upon the service provider's implementations, the VLR and HLR may be the same physical database with entries maintained in each category in a single mass storage device. Switch manufacturers often implement the MSC, VLR, and HLR together; however, the trend for modern network architectures is toward separate implementations. A simplified scenario of exchanges between VLRs and HLRs is shown in Figure 6-2.

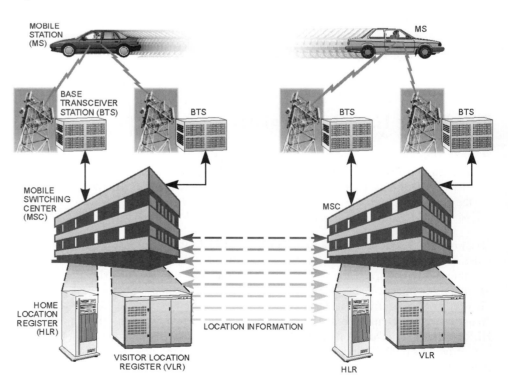

Figure 6-2: HLRs and VLRs Track Location of Mobile Subscribers

The Authentication Center

The **Authentication Center** (**AC**) is a database that manages authentication and encryption information of subscribers. The AC may be collocated with the MSC or HLR or it may be located at another location on the network. An AC may serve several MSCs in the network.

It is a secure database that contains secret information. This data is used to issue a challenge message to a mobile station using a set of codes. The AC is the only element in the network where the secret code information about each mobile station is stored. The data is used to determine whether the mobile station responds with the correct challenge response information known only to the authentication center and the mobile station.

The Equipment Identity Register

The **Equipment Identity Register** (**EIR**) is an optional database that contains the list of all valid mobile equipment on the network. The equipment in this database is listed by the International Mobile Equipment Identity (IMEI).

If a mobile station shows up on the network that is reported stolen or is not approved for use, it can be marked as invalid and will be denied service.

The Base Station Controller

The **Base Station Controller** (**BSC**) handles all mobile management tasks for one or more Base Transceiver Systems (BTS).

It is one of the two elements that are contained within the Base Station. The BSC provides commands to the BTS for frequency selection, handoff, and call management.

The Base Transceiver System

The **Base Transceiver System** (**BTS**) contains the radio transmitter, receiver, antenna, and tower that define a cell site.

The BTS is the element that communicates across the air interface with the Mobile Stations (MS).

The Mobile Station

The **Mobile Station** (**MS**) is the radio communications device that is designed to handle the protocols on the air interface.

A MS may be equipped to work with both digital and analog air interface protocols. The radio in a MS may be integrated with a laptop computer, data terminal, fax machine, or standard voice handset.

Network Interfaces

The reference model contains the designated interfaces shown in Figure 6-1. Each interface provides a standard protocol or set of protocols for the handling of data and commands between network elements. Almost all of the interface specifications shown in the reference model are described in TIA/EIA-41D. The "A" interface is covered in TIA/EIA-634.

Registration and Authentication

Registration is the process that occurs when a cell phone or any cellular wireless device classified as a mobile station informs the network of its presence in the cellular system.

Registration may be viewed as "logging on" to the wireless network. Re-registration is also required under specific scenarios discussed later in this section.

Authentication is a more elaborate step in the call processing routine of a cellular network.

Authentication is the processing executed between a mobile station and the network to confirm that the mobile station is who it claims to be prior to honoring a request for service.

In other words, it is a process designed to prevent fraud. The successful authentication of a mobile station is achieved when the mobile station and the network demonstrate through equal processing steps that each possess the same shared secret data. The secret data, however, is never transmitted in the clear over the air interface but may be shared within the network between the VLR and the AC.

Unlike the wired telephone network where the identity of the subscriber is a fixed physical location, a cellular subscriber may show up anywhere at any time within a region or country served by the home carrier or another service provider.

The network must be able to verify the identity of a subscriber for billing purposes as well as detecting illegal cloned phones that may be using stolen phone numbers or subscriber identities. In addition, location tracking of a mobile station must be performed in order to route calls to the roaming mobile station. In addition to the initial registration, a mobile station may perform one of several registration type messages based on events that may occur subsequent to the initial power-on registration.

- **Power-off registration** is performed when the power-off switch is pressed. This notifies the network that the phone is no longer available for receiving calls. Most service providers provide voice mail service when the mobile station is not registered.

- **Location-based registration** occurs when the mobile station has moved beyond a specified threshold of distance since the previous registration.

- **Zone or area based registration** occurs when the mobile station travels to a new area within the same network. This reduces the number of cells that need to be paged to notify the subscriber of an inbound call.

- **Time-elapsed registration** is performed by the mobile station when a time interval since the last registration is exceeded. This allows the network databases to be cleared if a mobile station does not re-register after a time established by the service provider.

After the mobile station has been registered, it may be challenged by the network to authenticate its identity using a secret code exchange that is unique to each mobile station. We will now look at this process in more detail.

Power-off registration

Location-based registration

Zone or area based registration

Time-elapsed registration

Cellular Phone Subscriber Identity Codes

Cellular phones have several identification codes stored in memory that are used in the registration and authentication process. Some of the codes are added at the time of manufacture and cannot be changed in the field. Others are stored in programmable ROM. The mobile station stored codes are outlined in Table 6-1.

Table 6-1: Mobile Station Identity Codes

Code Identity	Description
Electronic Serial Number (ESN)	A 32-bit number that is permanently stored in the handset during the manufacturing. It uniquely identifies the mobile unit and cannot be changed in the field.
Mobile Identification Number (MIN)	The MIN is divided into two subsets. MIN1 is a 24-bit transformation of the 7-digit handset phone number. MIN2 is a 10-bit transformation of the handset 3-digit area code.
System Identification Number (SID)	A 15-digit number that identifies the cellular operator associated with the handset. Each operator in the U.S. is assigned a SID number. It is used by the network to determine whether the handset is in the same service area as the SID or is roaming in another service provider area.
A-Key	The A-Key is a semipermanent 64-bit secret code that is used during the authentication of the handset's validity as a fraud prevention measure.
Shared Secret Data (SSD)	SSD is composed of SSD_A (64-bits) and SSD_B (64-bits). These codes are derived from the CAVE processing algorithm during the authentication and registration process. The values are used to form the authentication response message from the handset to the network.

Authentication Procedures

mobile station and
the authentication
center

The procedures for authentication involve similar processing routines in the **mobile station and the authentication center**. The results are then compared in the authentication center to confirm that the answer to the algorithm is the same in both cases.

mobile station

The two steps for authentication processing in the **mobile station** are shown in Figure 6-3. The mobile station may initiate the authentication under the conditions listed earlier such as initial registration, call or power-off registration, or each time it initiates a request for service called origination registration. The mobile station may also perform the same processing steps when it is challenged by the network to confirm its identity under certain conditions. This network-initiated authentication challenge/response sequence is performed if the initial authentication fails for any reason or if the mobile station moves to a new section of the network.

Mobile Station Authentication Sequence

Cellular
Authentication and
Voice Encryption
(CAVE)

The sequence and processing flow for authentication by the mobile station is shown in Figure 6-3. The A-Key is a semipermanent secret key stored in the mobile station memory. It is used to generate the 128-bit shared secret data (SSD_A and SSD_B). The A-Key is known only to the mobile station and the AC and is never transmitted over the air interface. The algorithm used to generate the SSD is the **Cellular Authentication and Voice Encryption (CAVE)** algorithm. The mobile station ESN is also an input to CAVE. The output of the first step, SSD_A, is processed by a second step using CAVE to produce a unique AUTHR 18-bit sequence. SSD-B may be used for voice encryption and/or Signaling Message Encryption (SME). Voice encryption and SME are supported in the specification and are required for implementation in all handsets but are optional for implementation on the network. The second processing step also uses the MIN, the ESN, and conditionally the dialed number as inputs. The Random Number (RAND) is broadcast by the MSC to all mobile stations in the cell as an added security feature. The SSD_A and SSD_B may be updated from time to time by the network using the RAND_SSD update message to all mobile stations in the network, as shown in Figure 6-3.

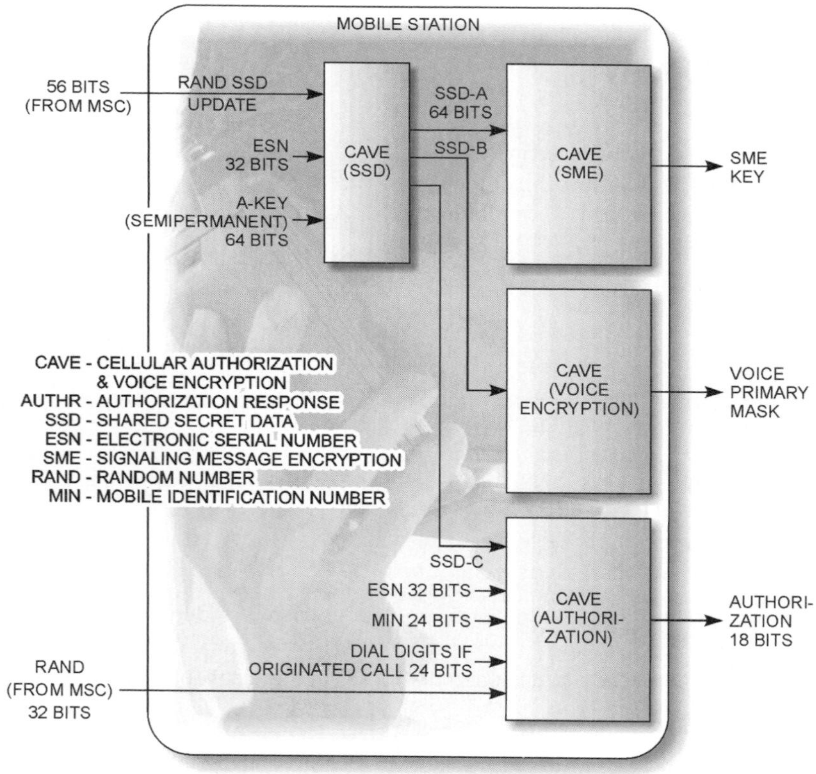

Figure 6-3: Mobile Station Authentication Calculations

Authentication Center and Mobile Station Processing

The combined mobile station and authentication center processing steps are shown in Figure 6-4. The same secure A-key for the mobile station is contained in a secure file on the AC computer and is used with the same parameters as the mobile station to produce a matching SSD and AUTHR result with the CAVE algorithm. The mobile station AUTHR parameter is transmitted through the cellular base station to the MSC. The MSC then forwards the mobile station AUTHR parameters to the AC where the comparison is performed. If the mobile station is in possession of the correct A-Key that is in the AC database, the results of the comparison step shown in the AC should match. If they do not match, the network attempts through a second authentication challenge to verify the mobile station identity. When it fails to compare, the mobile station will not be able to access the network. In addition the mobile station and the authentication center each keep a cumulative value called COUNT of the recent significant events such as registrations and calls. The COUNT value goes up to 64 and then rolls over to 0. The counts are transmitted by the MS and also compared with the equivalent COUNT in the AC. This added security feature assumes that a fraudulent phone user that has cracked the authentication keys would be less likely to guess the right count.

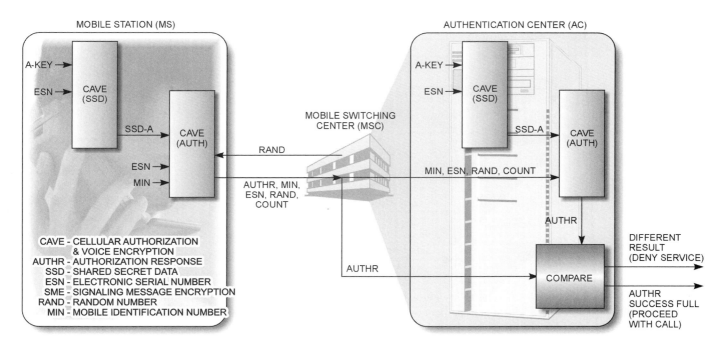

Figure 6-4:
MS and MC Authentication Processing

Cellular Network Call Processing

Voice call processing and call delivery proceeds after the authentication of the originating mobile station.

A similar authentication sequence occurs for the destination mobile station. Figure 6-5 illustrates the voice circuit path through the Public Switched Telephone Network (PSTN). The Common Channel Signaling System 7 (SS7), a packet-switched network that is used nationwide for call setup, is also shown in the figure. SS7 is a global standard for telecommunications published by the International Telecommunications Union. The standard defines the procedures and protocols by which network elements in the PSTN exchange information over a digital signaling network to effect wireless (cellular) and wireline call setup, routing, and control. The SS7 network and protocols are used in support of the following services:

- Basic call setup, management, and tear down

- Wireless services such as personal communications services (PCS), wireless roaming, and mobile subscriber authentication

- Local number portability (LNP)

- Toll-free (800/888) and toll (900) wireline services

- Enhanced call features such as call forwarding, calling party name/number display, and three-way calling

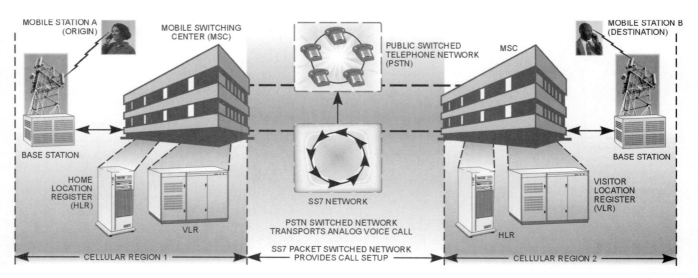

**Figure 6-5:
Mobile-to-Mobile
Phone Call**

The following example describes the sequence of events within the SS7 packet-switched signaling network and the PSTN circuit switched network (Figure 6-5) when a mobile station in a cellular network places a call to another mobile station in a distant cellular region.

Step 1 – Origination Mobile Authentication and Channel Assignment

As shown in Figure 6-6, mobile station A keys in the phone number (MIN) for station B and presses send. Mobile station A automatically sends an ORIGINATION registration message on a control channel that includes the AUTHR, ESN, MIN, and the MIN for mobile station B to the base station. The MSC passes the information to the region 1 HLR and AC. If the AUTHR information is valid, the HLR sends a registration confirmation to the MSC. The MSC establishes an active file in the region 1 VLR for mobile station A and sends a CHANNEL ASSIGNMENT message to mobile station A. Mobile station A shifts to the designated traffic channel frequency.

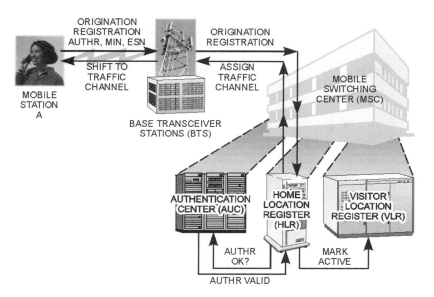

Figure 6-6: Origination Authentication and Traffic Channel

Step 2 – Locating the Destination Mobile Station

In Figure 6-7, the MSC in region A examines the 10 digits in the dialed MIN to determine the specific HLR in the network that serves as the "home" for the destination mobile station. A packet message is sent over the SS7 network to the HLR serving the destination mobile phone number (dialed MIN). Mobile station B may be registered in the home HLR serving region 2 or roaming in another region outside the home area for mobile station B. In either case the region 2 MSC will have information on the last registered location for the destination mobile station B. In the example, mobile station B is presently located in a cell served by its home HLR. At this time, the originating MSC has no knowledge of where the destination mobile station B is located. Call routing to the destination mobile is handled entirely by the HLR that is the home database for the destination mobile station. The SS7 network now has sufficient routing information to set up the switched path through the PSTN for the voice call from the origination serving MSC to the destination MSC.

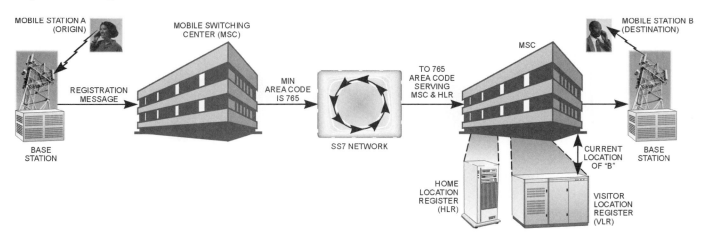

Figure 6-7: Origination Locating the Destination HLR

Step 3 – Setting up the Voice Path through the PSTN

In Figure 6-8, the MSC in region 1 sends a setup message over the SS7 network to the MSC serving the destination mobile station B. In this step, the SS7 network connects all of the switching center offices and required switching paths between the two MSCs. The destination serving MSC is now prepared to notify mobile station B concerning the incoming call.

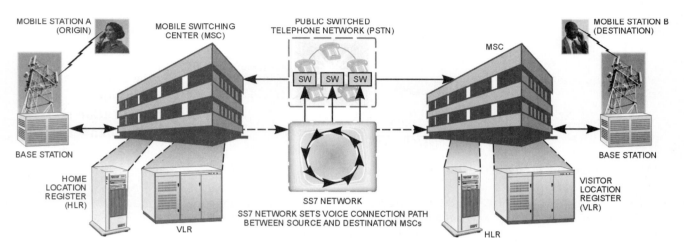

Figure 6-8: Call Setup Message

Step 4 – Paging the Destination Mobile Station

In Figure 6-9, the MSC sends a page message to the cellular paging area where mobile station B is currently registered. The page is transmitted by the base stations over a shared control channel that is monitored by all mobiles not involved in a call. The number of cells in a paging varies based upon the density of cells. A rural area would contain a smaller number of cells in a paging area than would be the case for a large city.

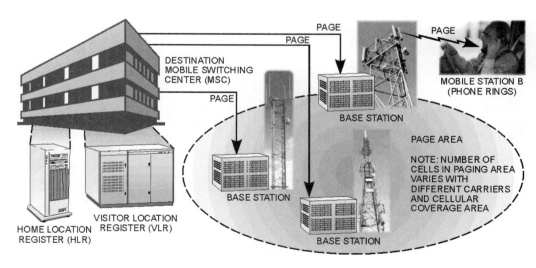

Figure 6-9: Page Message

Step 5 – Completing the Re-registration for the Destination Mobile Station B

As discussed previously, the destination mobile station must be challenged again for registration/authentication before the call can be completed. This assumes that mobile station B answers the call. Figure 6-10 shows that the paging message on the shared control channel starts the audible ring. When the subscriber using mobile station B answers by depressing the SEND key, it simultaneously sends a page response back over the shared paging radio channel with its authentication information. As before, the AC must confirm the validity of the identity of mobile station B. The destination MSC sends the authentication information to the VLR, which updates the location file and sends the information to the HLR. Then the HDR asks the AC to authenticate mobile station B (re-registration). If successful, the destination MSC sends a channel assignment message over the radio channel to mobile station B. Mobile station B switches to the designated control channel.

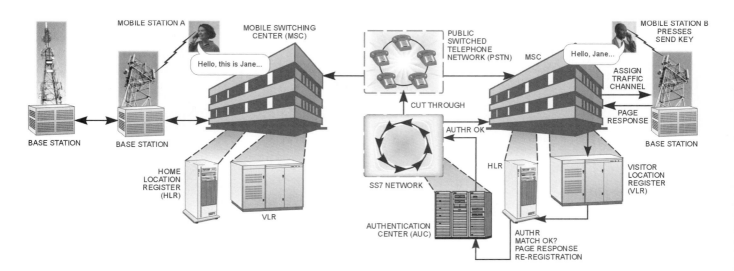

Step 6 – Call Completion

Figure 6-10: Call Completion

The authentication process for mobile station B occurs very rapidly. The logic in step 5 indicates that the authentication and channel assignment for the inbound call to mobile station B do not occur until the subscriber responds by answering the page. Premature traffic channel assignment and authentication for non-answering mobiles would waste **cellular** resources. In Figure 6-10, when mobile station B responds with an answer, the authentication and channel assignment are the final steps for the air interface elements of the network. This triggers the SS7 network to "cut through" the PSTN network to complete the path between the source and destination MSCs. Both mobile stations are now connected through the air interface traffic channels and the PSTN switches. The parties are now connected and the call can proceed with the mobile station B response "HELLO".

Mobile IP

The Internet provides access to a rapidly growing number of information sources worldwide.

As shown in Figure 6-11, Mobile IP enables corporations to provide connectivity to employees who are away from the office. Internet Service Providers (ISP) provide access to e-mail and a number of services targeted for the traveler such as airline schedules, sports information, and news headlines. Because of the global coverage and toll-free access to the Internet, the value of the Internet to mobile wireless computing is enormous.

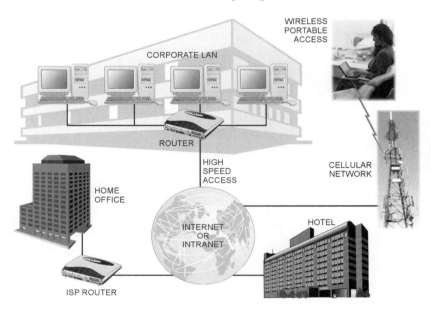

Figure 6-11: Internet Access

The problems associated with connecting the mobile wireless user with the Internet are summarized as follows:

- The Internet uses the TCP/IP suite of protocols. These protocols do not support complete mobility. The Internet Protocol (IP) address and supporting protocols were developed prior to the emergence in the 1990s of wireless mobile network computing. The TCP/IP suite assumes that a network user has a **fixed point of attachment** that does not change during a session. Packets are delivered to a network address that is based upon a fixed location known by the routers that build routing tables of network addresses. Although the IP allows access from various locations, the user must stop and connect to a fixed access point via a PSTN connection such as a hotel room. Connection can be made via an Internet Service Provider (ISP) or direct private network to a home office location using dial-up service and an ISP **borrowed temporary** IP address.

- Considerable corporate investments have been made for establishing TCP/IP as the preferred commercial standard protocol suite for business networking in most of the world. It is the established standard transport protocol suite for the Internet and most intranets. Every attachment device accessing the Internet or a corporate TCP/IP network must have a unique address, whether locally attached to a LAN or using a dial-up service, and appropriate security measures to log on to a corporate or government TCP/IP based LAN.

fixed point of attachment

borrowed temporary

- Increased popularity of handheld, portable wireless devices such as PDAs, digital cellular phones that provide access using the Wireless Access Protocol (WAP), and other wireless Internet access devices has demonstrated the need to provide true access to the Internet on an "anytime, anywhere, always connected" basis. This is a problem for standard IP because it does not support roaming in a wireless mobile network during a session for the reasons stated above.

- The Internet is not secure. It also provides no guaranteed quality of service for network users.

- The standard IP addressing protocol does not support full mobility when the attachment point changes during a session. Enhancements to TCP/IP are needed to support full mobility. These enhancements are referred to as **Mobile IP**.

Mobile IP

Portable vs. Mobile IP

The distinction between portability and mobility in wireless network computing is often unclear. Portability does not require any enhancements to the IP protocol.

Portable computing is typically defined as connecting to an office LAN from a remote location when away from home using an Internet Service Provider (ISP), as illustrated in Figure 6-12.

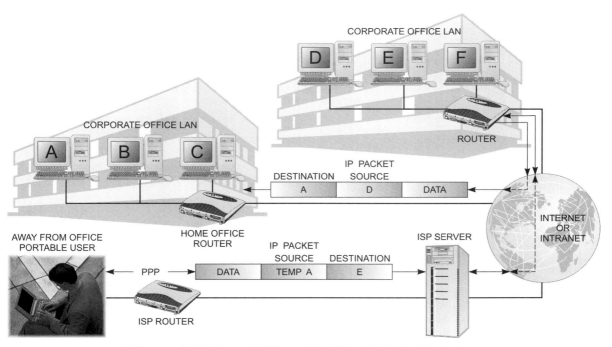

Figure 6-12: Portablility with Simple IP Addressing

The TempA must not change the point of attachment shown because the *ISP router would not know where to deliver the packets*. The router remembers the location of IP address TempA. Similarly, if host computer A were to move to another location using its permanent IP address A, the router would not be able to recognize the fact that host A was no longer located on the home LAN. With simple IP, a host computer is not able to move to a new location using its permanent IP address. *Each change in the point of attachment would require a new IP network address.*

In the example shown, the portable user wishes to send a message to workstation host E located on a corporate LAN. All IP addresses contain the unique source and destination fields for routing IP packets as illustrated in the example. The Point-to-Point Protocol (PPP) uses the IP Control Protocol (IPCP) to request that a temporary IP address be assigned to the user away from home. The address assigned for the portable user (TempA) can now be used to send and receive IP packets to IP workstation E or any other IP address. Replies will be routed to TempA as long as the connection is not terminated.

Portability in a Wireless IP Environment

Portable computing is possible for many wireless Internet access applications where cellular service providers support data services for portable wireless computing platforms. Simple IP works in the example outlined in Figure 6-13. In this digital cellular environment, the host is assigned a temporary IP address as before and may initiate data calls from anywhere in the cellular service area shown. The user will be handed off from cell to cell during a session as long as the user does not move outside of the area served by the same Inter-Working Function (IWF) and the associated router that serves as the point of attachment. The serving area MSC and the associated IWF router would recognize the TempA address as before and maintain the connection as long as the mobile station did not roam outside of the initial service area.

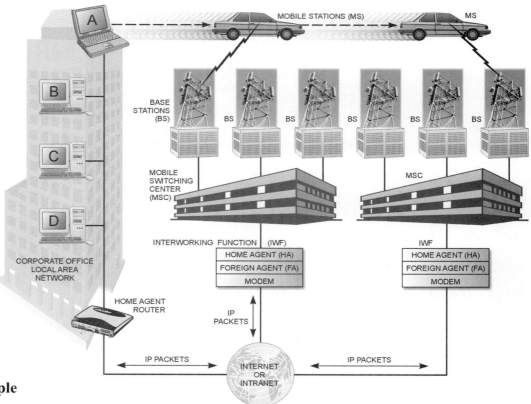

Figure 6-13: Mobile IP Example

Full mobility requires the network to maintain the connection when the computer changes the point of attachment by entering another area served by a different MSC and IWF.

As you can see from the portable operation example, a new IP address would need to be assigned when the mobile station changed its point of attachment. This would create havoc with the router tables and protocols. The solution to this problem was developed by the Internet Engineering Task Force (IETF). The specifications allow routers with enhanced software to provide mobile IP routing.

Mobile IP is published as RFC 2002 and the associated documents RFC 2003 through 2006 by the IETF.

It allows a *host to maintain its permanent IP* address while away from home by hiding it from routers. The following examples will show how the extensions to IP addressing provide this capability for mobile wireless users.

Mobile IP Network Features

Figure 6-13 shows a simplified wireless network supporting mobile IP. Special software that extends the IP standard with enhanced features is required for routers and mobile IP computing platforms (in our example, a laptop with a wireless modem).

Mobile IP routers with the enhanced software are called Home Agents (HA) and Foreign Agents (FA).

The Inter-Working Function (IWF) is a combination of routers, computers, and modems that provides the interface between the Internet or private corporate intranets that are packet-switched networks and the wireless cellular network supporting data services. Mobile IP allows the mobile user to maintain his/her IP address A and receive or send messages in both the LAN and the wireless environment. In order to accomplish this feature the routers perform a special task under the title of foreign agents and home agents.

Home Agent and Foreign Agent Registration

Mobile IP provides features that allow the mobile station and the FA to exchange information. The events outlined in Figure 6-14 show the registration process.

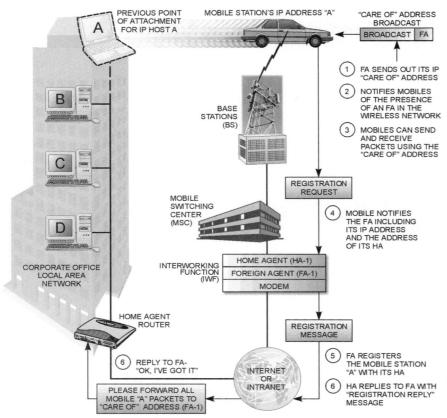

Step 1 – The FA advertises its presence in a cellular area served by an MSC. In a large city, this could include several square miles. All mobile stations in the MSC serving area can receive the FA broadcast message on a control channel. The FA sends its "care of" address. The care of address is analogous to the post office using a temporary address to forward mail to a temporary residence. This is the IP address of the FA that can be used by all mobile stations in the area to receive and transmit IP packets. This saves IP addresses by allowing the single FA to share the same address among all mobile stations.

Step 2 – The broadcast message is received by the mobile stations in the coverage area. The mobile is thereby alerted to the capability for mobile IP service in the cellular area.

Step 3 – Packets can be received and transmitted to the FA using the care of address of the FA. The IWF serves as the interface between dissimilar networks. The FA will be able to identify each mobile station by its permanent IP address. With mobile IP, the mobile retains its permanent IP address (the "A" address is used in the example in Figure 6-14). The "care of" address is used as a temporary forwarding IP address.

**Figure 6-14:
Home Agent and
Foreign Agent
Registration**

registration request

Step 4 – The mobile station sends a **registration request** message on a shared control channel on the air interface. This advises the FA of its presence in the network. The message includes the permanent mobile station IP address and the IP address of its HA. The MSC passes the registration message to the FA.

Step 5 – The FA notes the location of the HA for the mobile station. This process associates all mobile stations active in the MSC serving area with their respective HA addresses. The FA forwards the registration request message to the appropriate HA router. The HA router stores the "care of" address for the mobile station. The HA now knows the location of mobile station A in the wireless network. The registration request is a notification to the HA to forward all messages to the designated FA "care of" address.

Step 6 – The HA responds with a "registration reply" message to the FA. This is the acknowledgment that the HA has modified its routing tables and will forward all messages inbound for host A to the FA care of address. Similarly, mobile station A can send IP packets into the network with the same addressing protocols used in simple IP.

IP Encapsulation and Message Forwarding

As shown in Figure 6-15, an inbound packet with the IP header showing host A as the destination has arrived from a host with a source address F. Mobile IP software in the HA router encapsulates the inbound packet with a new header.

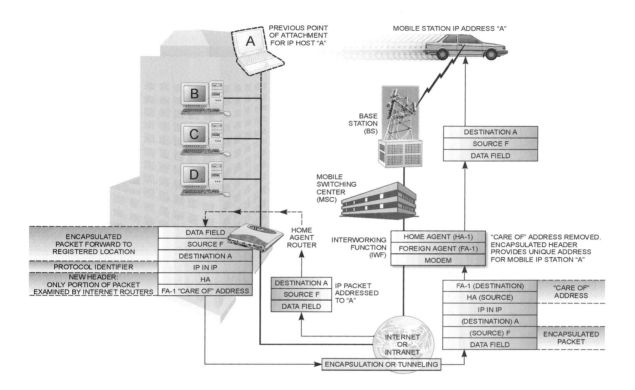

The new address header includes the "care of" address. The "care of" address is followed by the field that identifies the **IP-in-IP protocol**, as specified in RFC 2003. The Internet includes many routers that are used to connect the two sites. Each router will only examine the first source and destination fields in the IP packet. The hidden or encapsulated packet now passes through the Internet or intranet with the secondary permanent address hidden. This is analogous to passing the packet through a tunnel where the ultimate destination address is hidden or "tunneled" through the Internet. The FA removes the "care of" address and delivers the IP packet using the mobile station permanent IP address. In IP-in-IP the entire original IP header is preserved as the first part of the payload of the tunnel header. Therefore, to recover the original packet, the foreign agent merely has to eliminate the tunnel header and deliver the rest to the mobile node. The FA knows that the true permanent address is contained in the field following the IP-in-IP protocol identifier field. Handoffs will occur in the wireless network as long as the cells are served by the same IWF.

Figure 6-15:
IP Packet Encapsulation

IP-in-IP protocol

Paging and Messaging Networks

Paging and messaging networks are cellular networks designed to deliver text-only messages to small handheld terminals.

Early text delivery networks were called paging networks and provided one-way wireless messages to subscribers, usually with a number to call to reach the paging party. The person originating the page did not know if the message was received since paging networks had no method of acknowledging the page message.

The paging networks have evolved into two-way messaging networks with text delivery protocols called Short Message Service (SMS), e-mail, Internet access, and a host of other wireless services.

One of the popular messaging network systems is the BellSouth Wireless Data (BSWD) network, which is now part of the Cingular Wireless consortium of companies. The BSWD network is based on a Mobitex network architecture originally developed by Ericisson. The BSWD network architecture is illustrated in Figure 6-16. Each handheld data terminal operates in the 900 MHz band. The data rate is approximately 8 kbps including overhead bits. The base stations are arranged in cells similar to cellular voice networks. Users can originate text messages or receive messages through the cell site base stations. The network hierarchy consists of a series of packet switching nodes connected to an X.25 packet-switched network. Routing is performed via Level 1 and Level 2 Main Hierarchical Exchange Switches, as shown in Figure 6-16. Billing and accounting functions for the network are maintained by the two Network Control Centers.

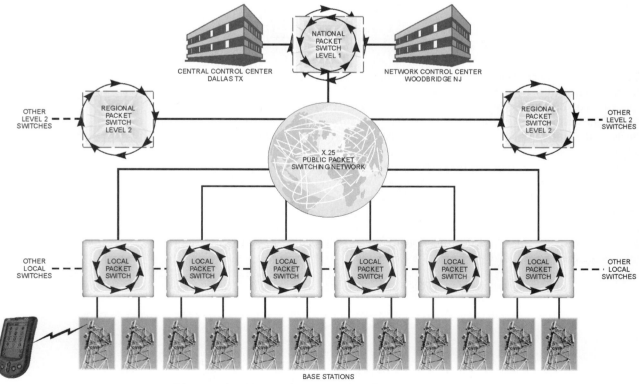

Figure 6-16: BSWD Network Architecture

Other wireless service providers have established partnerships with BSWD to offer additional services. One popular type of wireless data network is called PalmNet by Palm Inc. The PalmNet service uses the BSWD network to connect subscribers to their servers. The network provides connections to e-mail servers, instant messaging servers, and the Internet where web sites can be accessed using "web clipping" applications. The PalmNet architecture is shown in Figure 6-17. Palm computing developed a technique called "web clipping", which is based on a simple query and response instead of a system of hyperlinks. The comparison between web browsing and web clipping is illustrated in Figure 6-18. The query portion of the application is stored on the Palm handheld. (Palm literature refers to this portion of the web clipping application as Palm Query Applications (PQAs).) The access is greatly simplified. A query typically consists of 50 bytes and less than 500 bytes are returned (compressed), thereby fitting the information on a small display screen. In the lab that follows later, you will become familiar with a two-way messaging network and a handheld data terminal using the web clipping applications.

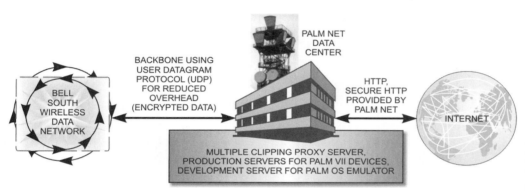

Figure 6-17: PalmNet Architecture

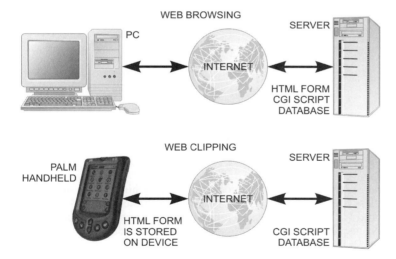

Figure 6-18: Web Browsing and Web Clipping

Wireless Local Area Networks

Chapter 3 covered the standards, topologies, and types of media for Local Area Networks (LANs). Wired LANs must have a tether connecting the subscriber to the LAN with a physical connection such as a copper wire cable.

Wireless Local Area Networks (WLAN) offer additional capabilities for mobile networking using radio technology and portability for a wide variety of applications.

A general overview of WLAN applications is illustrated in Figure 6-19.

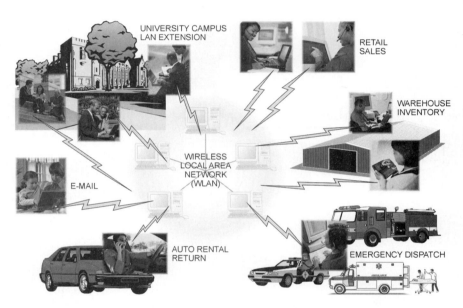

**Figure 6-19:
WLAN Applications**

Some basic differences between WLANs and the cellular networks discussed previously are as follows:

- Data rates for the second generation of cellular technology are lower at the air interface (64 kbps or less). Third generation (3G) mobile cellular data rates will offer substantially higher data transmission speeds. (Third generation wireless technology will be explained in Chapter 9.)

- Wireless LANs support data rates of 11 Mbps or higher over shorter distances. The wireless LAN can be operated as a standalone wireless network or as an extension to a wired LAN using an interface bridging device called an access point. Wireless LANs usually are designed to access a central database using handheld data terminals, barcode scanners, printers, or other low-power radio terminals.

- Cellular radio networks use **licensed** radio spectrum. Wireless LANs use special **unlicensed** radio bands.

- Cellular data networks are operated by licensed service providers offering wireless services on a public network. Wireless LANs are typically privately owned institutional networks.

You should also be aware that some similarities exist between WLANs and cellular networks. Most of the similarities are due to trends as follows:

- Cellular networks are migrating towards high data rate capabilities with 3G upgrades. The higher rates will eventually approach the speed of WLANs as the emphasis on multimedia applications and Internet access grows.

- WLANs and cellular are both converging on the use of the Internet for the transport of digital voice communications. This technology is called Voice over IP (VOIP).

- WLANs and cellular networks will migrate towards use of the mobile IP protocol for wireless Internet and intranet access.

You may have observed wireless local area networking at work when returning a rental car at the airport where the attendant gave you a receipt using a wireless LAN data terminal. Most retail stores use handheld bar code scanners to maintain inventory and price tagging. Medical information can be accessed by doctors and nurses using wireless terminals in hospitals. Numerous hotels, airports, and businesses are equipped with wireless LAN networks using mobile IP for access to the Internet for mobile computer users.

Wireless Local Area Network Standards

Several competing wireless local area network standards have evolved over the years since the release of the IEEE 802.11 standard in 1997. Some of the standards are supported by wireless industry groups. At this date, not all of the current standards are compatible or interoperable. Some may survive while others will not become commercially successful. In this text, I will focus on the IEEE 802.11 series of standards since they have wide support among industry groups and world standards bodies. You should, however, become familiar with the characteristics and claims published by the wireless LAN standards and technologies discussed briefly in the following paragraphs. Several of the competing technologies such as Bluetooth, HIPERLAN/2, and HomeRF are attempting to gain a share of the wireless LAN market.

> In the U.S. most wireless LAN products are designed for use in one of the ISM (Industrial, Scientific, and Medical) license-free bands. The FCC provides no protection from interference among the users of the ISM band.

Unlicensed Radio Spectrum Allocations

The FCC has set aside portions of the radio spectrum for unlicensed operation. Unlicensed spectrum rules allow the use of certain radio emission devices such as network interface cards for PCs, cordless phones, bluetooth chips, and so on, without an FCC license. Limitations on the amount of radiated power are established for a wide variety of such unlicensed devices. Cordless phones and wireless LAN products are manufactured for use in the unlicensed radio bands.

Mutual interference among the many emerging unlicensed products described later is a concern among many radio engineers who are developing products that share unlicensed spectrum. The ISM band allocations are:

- 902-928 MHz
- 2.4-2.483 GHz
- 5.15-5.35 GHz
- 5.725-5.875 GHz

For a wireless LAN product to be sold in a particular country, the manufacturer of the wireless device must ensure its certification by the appropriate agency in that country.

High Performance Radio Local Area Network

High Performance
Radio Local Area
Network (HIPERLAN)

The **High Performance Radio Local Area Network (HIPERLAN)** was developed in Europe by the European Telecommunications Standards Institute (ETSI).

It is designed to operate between mobile terminals or in conjunction with fixed wired LANs using radio access points. It is a technically sophisticated standard offering data transmission rates up to 54 Mbps. It uses a special form of multicarrier modulation called Orthogonal Frequency Division Multiplexing (OFDM) similar to the technology used in the IEEE 802.11a standard. OFDM is spectrally efficient and lends itself to a variety of modulation parameters and associated data rates, shown in Table 6-2. HIPERLAN operates in the 5 GHz and 17 GHz frequency bands.

Table 6-2: OFDM Data Rates and Modulation

Data Rate (Mbps)	Modulation
6	BPSK
9	BPSK
12	QPSK
18	QPSK
24	16-QAM
36	16-QAM
48	16-QAM
54	64-QAM

HIPERLAN is not likely to take market share from IEEE 802.11 products except perhaps in Europe.

HomeRF

HomeRF is a wireless LAN standard supported by the HomeRF Working Group (often referred to as the HomeRF WG).

The founding members of the HomeRF consortium include, among others, Microsoft, Intel, HP, Motorola, and Compaq. The purpose of the HomeRF initiative is to develop a standard for economical wireless home communications including data and voice.

HomeRF proposes to go beyond sharing printers and PC resources. For example, it could support a home entertainment center where music and video information could be shared throughout the house with remote TVs or speaker systems. Cordless phones could be shared on PC-based answering systems. The standard incorporates the Digital Enhanced Cordless Telephone (DECT) standard. Another example would be multiple PCs sharing a single high-speed ISP connection to the Internet. A laptop computer with a wireless modem card could be used to share the ISP connection with other computers in a small office or household. Corporations supporting the HomeRF standard envision a number of home automation features, including appliances and home security systems managed from a central computer.

Bluetooth

The **Bluetooth** wireless technology provides the capability for the replacement of cables that connect one device to another with a short-range radio link.

The name Bluetooth has its origin in early world history. It comes from the name of a Danish Viking king, Harald Blatand, who ruled in the 10th century. He is credited with uniting Norway and Sweden.

Early work and engineering studies by Ericsson Mobile Communications in 1994 focused on the feasibility of a low-power/low-cost radio interface between mobile phones and their accessories. The goals were to produce a very small chip-sized radio and signal processor that could be fitted into cell phones and other wireless terminal equipment. The chip would be required to handle both data and voice communications. Ericsson went on to establish a Special Interest Group (SIG) in 1998. The SIG members agreed to produce a formal specification for the radio interface and a control module called the baseband processor.

The Bluetooth specification defines two power levels: a lower power level that covers the shorter personal area within a room, and a higher power level that can cover a medium range, such as within a home. Software controls and identity coding built into each microchip ensure that only those units preset by their owners can communicate.

The Bluetooth wireless technology supports both point-to-point and point-to-multipoint connections. The radio interface operates in the 2.4 GHz ISM band and employs Frequency Hopped Spread Spectrum (FHSS) modulation.

With the current specification, up to seven slave devices can be set to communicate with a master radio in one device. Several of these piconets can be established and linked together in ad hoc scatternets to allow communication among continually flexible configurations. All devices in the same piconet have priority synchronization, but other devices can be set to enter at any time. The topology is best described as a flexible, multiple piconet structure.

The 802.11 Wireless LAN Standard Series

In June 1997, the IEEE completed work on the initial standard for wireless LANs, defined as 802.11. This standard specified a 2.4 GHz operating frequency with data transmission rates of 1 and 2 Mbps.

Since the initial publication date, the IEEE has released two revisions: 802.11a (40 Mbps in the 5.8 GHz band) and 802.11b (11 Mbps in the 2.4 GHz band). Most new installations for wireless LANs include products based on the later 802.11b standard.

Maintaining interoperability between different vendors' equipment is important to the success of any standard. Within the industry, the Wireless Ethernet Compatibility Alliance (WECA) was organized to certify and assure interoperability of devices manufactured to comply with the 802.11 standard. Testing for compliance is conducted by Silicon Valley Networking Laboratories. Equipment that meets the testing criteria receives the right to use the logo called Wireless Fidelity (Wi-Fi).

The 802.11b Standard

The 802.11b standard was the result of a study by the IEEE that looked at ways to achieve a higher speed that would make the technology more competitive with other emerging standards.

It was also their objective to maintain interoperability with the older 1 and 2 Mbps 802.11 legacy systems. One of the challenges in the 802.11b effort was to achieve a higher bit rate without using additional bandwidth. This design goal was achieved with the adoption of a new modulation scheme called **complementary code keying (CCK)** modulation that supports data rates up to 11 Mbps.

complementary code
keying (CCK)

802.11b specifies the use of CCK modulation and spreading of the spectrum using Direct Sequence Spread Spectrum (DSSS) techniques similar to the CDMA ANSI-95 waveform for digital cellular networks.

802.11b Spectrum Channelization

The standard calls for use of DSSS modulation in the 2400 – 2483.5 MHz band on up to 14 channels with 5 MHz spacing between each channel. DSSS systems operating in the 2.4-GHz band feature channels that occupy approximately 22 MHz of bandwidth. This channel bandwidth scheme was chosen to allow three independent, non-interfering WLAN networks across 83.5 MHz of spectrum. An example of the channel spacing and shaping of the waveform for DSSS operation is illustrated in Figure 6-20.

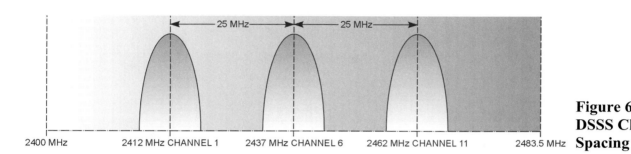

2400 MHz   2412 MHz CHANNEL 1   2437 MHz CHANNEL 6   2462 MHz CHANNEL 11   2483.5 MHz

Figure 6-20: DSSS Channel Spacing

> The ability to achieve higher data rates in the 802.11b specification was made possible in part by using a higher level modulation.

The symbol rate for CCK is 1.375 Mbps and 8 bits per symbol, which provides a throughput rate of 11 Mbps.

The 802.11b Radio Interface

Early IEEE 802.11 systems employed either a Frequency-Hopping Spread Spectrum (FHSS) or a Direct Sequence Spread Spectrum (DSSS) modulation. Both types of spread spectrum waveforms were excellent solutions for delivering data rates in the 1- to 2-Mbps range.

It became clear to the IEEE after release of the original 802.11 specification that higher data rates would be required to assure a viable position in the market for WLANs. The radio interface was the prime area of investigation to open up the possibility of supporting higher data rates using the same spectrum allocations that were available in the ISM bands. This led to the development and release of the 802.11b WLAN specification.

Legacy IEEE 802.11 DSSS systems employ Differential Binary Phase-Shift Keying (DBPSK) and Differential Quadrature Phase-Shift Keying (DQPSK) modulation techniques for delivering data packets at data rates of 1 and 2 Mbps. In DSSS systems, data is spread using an 11-bit Barker word prior to transmission. The IEEE 802.11b protocol supports data rates of 1, 2, 5.5, and 11.

Limits on Radiated Transmitter Power

Directional antennas can be used as illustrated in Figure 6-21 to improve performance over a desired area for a campus or corporate site. The FCC has actually written into the rules some language that encourages the use of directional antennas since they may reduce interference with other unlicensed services in a surrounding area. Specifically, the FCC rules that apply to spread spectrum radio transmitters in the ISM band are contained in Part 15, Subpart B, paragraph 15.247. The rules limit the transmitted power to 1 watt but provide for a minor increase in the effective radiated power using a formula for directional antennas. Part 15 of the FCC rules governing intentional radiators in the ISM band can be downloaded free of charge from the FCC web site *www.fcc.gov/oet/info/rules/*.

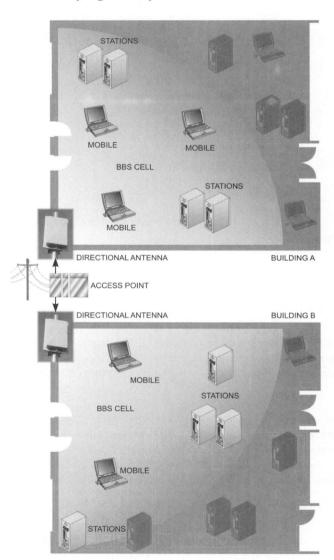

Figure 6-21: Directional Antennas Provide Improved Performance in Desired Coverage Area

802.11b Network Architecture

An 802.11b standard supports the following two topologies:

- Independent Basic Service Set (IBSS) networks

- Extended Service Set (ESS) networks

The IBSS network is illustrated in Figure 6-22. An IBSS network contains a basic building block referred to in the standard as a Basic Service Set (BSS). Stations in the BSS stay connected as long as they remain in the coverage area. If they leave the BSS area of the IBSS network, they can no longer communicate with the other stations. This type of network is referred to as an ad hoc network because it can be assembled quickly with minimum planning.

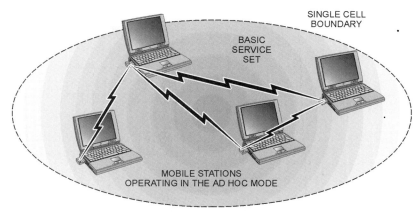

Figure 6-22: Independent Basic Service Set (IBSS) Network

When planning the layout for or troubleshooting a wireless LAN, you need to know the FCC rules governing the limitations on radiated transmitter power and the use of directional antennas.

The ad hoc mode indicates that the mobile stations are communicating among peers and can be easily configured without advanced planning or dependence upon any external devices. Ad hoc networks are very flexible and can be used for applications where rapid communications between mobile PCs is sufficient without access to a wired network server.

An ESS network is shown in Figure 6-23. This type of network has much broader capabilities because it allows communications between IBSS networks and wired LANs. Two terms used in the standard are a **Distribution System (DS)** and an **Access Point (AP)**. The distribution system may already be in existence when planning for a wireless LAN. The standard does not specify the type of network required for the distribution system although it is typically an Ethernet IEEE 802.3 LAN. The distribution system is used for connecting access points and providing address-to-destination mapping service and connection for multiple BSSs.

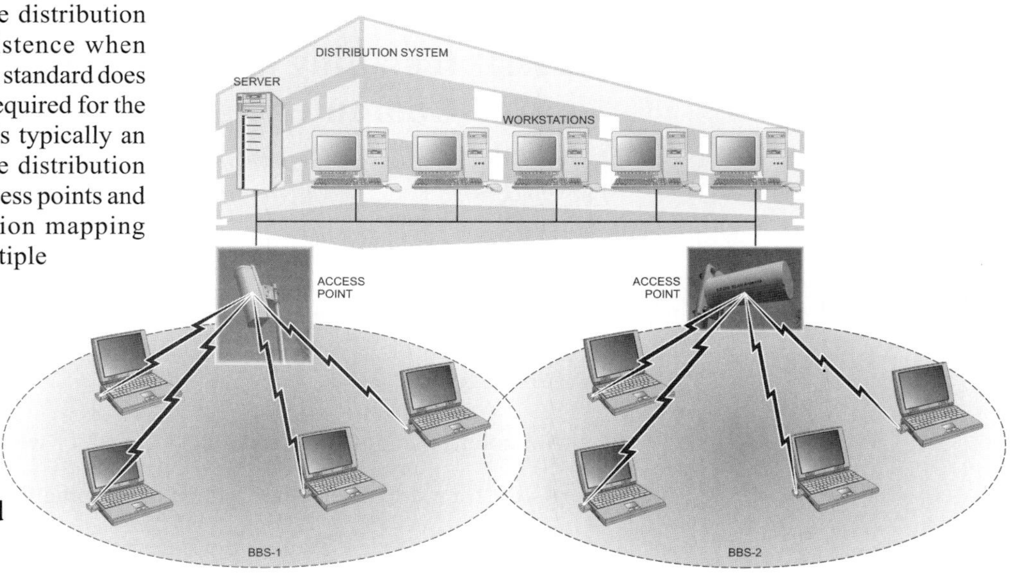

Distribution System (DS)

Access Point (AP)

Figure 6-23: Extended Service Set (ESS)

Infrastructure mode

An access point acts as a wireless bridge to the wired LAN.

An access point is similar in function to a base station in a cellular network. All communications in an ESS network must go through the access point. Access points are not mobile and are part of the wired LAN. The ESSs shown in Figure 6-20 are operating in what is referred to as the **infrastructure mode**.

802.11b Security

The authors of the 802.11b Wireless LAN Standard were concerned about security. Wireless LANs transmit information over radio channels. Any transmission may be intercepted.

It is possible for transmissions in the 2.4 GHz band to radiate outside the immediate area where an office or campus network would be located.

Transmissions could be intercepted in a parking lot or by any unauthorized mobile terminal in the vicinity outside a security perimeter. The IEEE took steps to make wireless LAN transmissions secure.

Wired Equivalent Privacy (WEP)

The IEEE 802.11b standard provides a security feature called **Wired Equivalent Privacy (WEP)**. WEP provides a means for encrypting data transmission and authenticating the nodes.

WEP uses secret keys to encrypt and decrypt the data. Two processes are involved in data encryption. One process encrypts the data, and the other protects the information from unauthorized modification.

Lab 4 – Installing a Wireless LAN Card in a Laptop Computer

Lab Objectives

Install and configure a wireless Ethernet adapter in a laptop computer.

Configure each computer and the network as an ad hoc wireless configuration.

Materials Needed

- AmbiCom model WL1100B-PC/PCI 802.11 wireless LAN card
- Two or more client laptop computers running Windows XP
- Network interface card vendor documentation

Instructions

1. Unpack the Wave2Net wireless LAN card and installation CD.

2. Prior to installing the wireless LAN card, insert the installation CD into the CD-ROM drive. The menu shown in Figure 6-24 will appear on the screen.

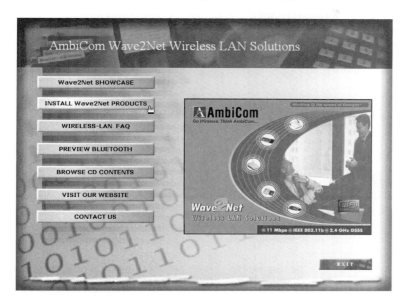

Figure 6-24: AmbiCom Flash Screen

TIP: If the display does not appear automatically, you can access the installation program by clicking on the **Start** button on the Windows desktop and selecting **Run**. In the drop-down menu box, type **D:\Setup.exe**.

TIP: The D drive here represents the CD-ROM drive. Select the appropriate drive letter for the PC you are using.

3. From the Main Menu, click on the **INSTALL Wave2Net PRODUCTS** to show the display in Figure 6-25.

4. Select **Install WL1100B-PC for WinXP (beta)**.

─ NOTE ─────────

When these instructions were written only the beta version of the driver was available. It will be assumed that you have the same version of the AmbiCom CD as used here. If the final version of the driver for XP has been released, you can use it with your installation.

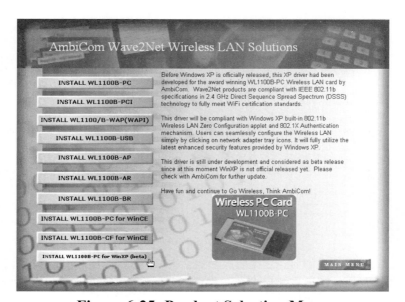

Figure 6-25: Product Selection Menu

5. The next screen, shown in Figure 6-26, appears indicating that Windows is ready to continue the installation. Click on the **Next** button to continue.

Figure 6-26: Driver Setup

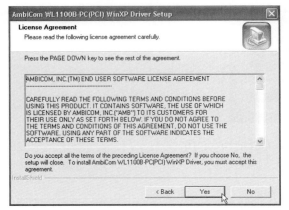

Figure 6-27: License Agreement

6. The next display, shown in Figure 6-27, is the license agreement. Click on **Yes** if you agree or **No** if you disagree.

7. You may see a warning similar to Figure 6-28. This is a warning about the beta version of the driver. Click on **Next**.

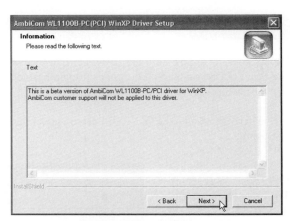

Figure 6-28: Beta Warning

8. The driver will finish installing. Click on **Finish** when you are prompted.

9. Click on **Main Menu** and Exit in the AmbiCom screens.

10. Insert the wireless card in the laptop. The card will be recognized. You will see a window similar to Figure 6-29. Click on **Next**.

11. The wizard will search for software for the device. It may give you a compatibility warning. Click on **Continue Anyway**.

12. The driver will be installed. Click on **Finish** to the Found New Hardware Wizard.

Figure 6-29: Local Area Connection Properties

13. Click on **Start/Control Panel** and open Network Connections.

14. Right-click on the connection with the wireless network card and select **Properties**. You will see a window similar to Figure 6-30.

**Figure 6-30:
Local Area Connection
Properties**

15. Click on the **Configure** button and click on the **Advanced** tab.

16. Select **Network Type** under Property. Select **802.11 Ad Hoc Mode** under **Value** as shown in Figure 6-31. This will allow you to connect to other computers without using an access point.

17. Click on **OK**. The network card is now set up.

TIP: You can now connect to other computers in your area that have done the same procedure. If the other computers have any shared files you can do so by browsing through My Network Places. Type their computer name in any Explorer window preceded by two backslashes, e.g., \\COMPUTERNAME and you will be able to browse shared files.

Figure 6-31: Ad Hoc Mode

Lab 4 Questions

1. What two types of network configurations are possible within the 802.11 specification?

2. What external visual indication told you that the cards were powered up and operating?

Lab 5 – Installing and Configuring a Wireless Access Point

Lab Objectives

Install and configure an access point in a wireless network.

Configure portable computers to operate in the infrastructure mode.

Materials Needed

- Requires access to an existing wired Ethernet network with two or more workstations configured for file sharing and a 10/100Base-T Hub. The two laptop computers with wireless network interface cards installed from Lab 4 are also required.

- AmbiCom Model 1100B-AP access point

- Vendor documentation.

- Power cord

- Universal Serial Bus (USB) interface cable

- Category 5 Cable

- Installation CD

Instructions

1. This lab requires the network component configuration shown in Figure 6-32. The two laptop computers should be configured for the ad hoc mode.

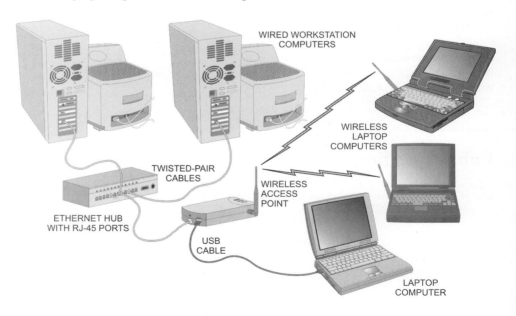

Figure 6-32: Network Configuration

2. Unpack the following items that are part of the AmbiCom installation kit.

- Model 1100-AP access point
- Power cord
- USB interface cable
- Quick installation guide
- Installation CD

3. Install the access point near the center of the area where you will be using the mobile laptop computers in the lab. The access point may be mounted on the wall or placed on a desk. Allow the location to be accessible to a 110-volt power outlet. Connect the power cord to the access point and the wall outlet or power strip. Note the "power on" red LED light on the access point.

4. Select an empty twisted pair port on the hub and connect a Category 5 network cable with RJ-45 connectors between the hub and the access point. Put the AmbiCom CD in the drive and exit any menus that open.

5. Connect the USB cable between the access point USB port and the computer. Your lab should now be configured in accordance with Figure 6-32. The USB device will be detected and you will be prompted to install a driver.

TIP: Any computer may be used to install and configure the access point. It may be more convenient to use one of the two mobile station laptop computers to accommodate the USB cable length.

6. Select **Install from a list or specific location (Advanced)** and click on **Next** twice. When you are prompted select the driver in the *\wl1100b-ap\win2000* folder on the CD-ROM and click on **Next** and **Finish**.

7. Re-insert the CD-ROM. When the CD loads, you will see the menu shown in Figure 6-33. Click on **INSTALL Wave2Net Products**.

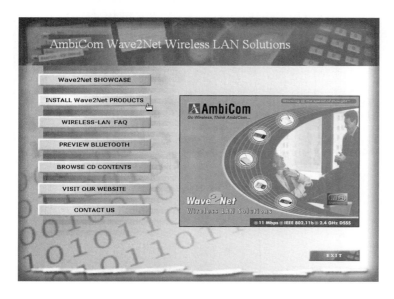

Figure 6-33: AmbiCom Flash Screen

8. Click on the button labeled "Install WL1100B-AP".

9. The next screen, shown in Figure 6-34, indicates that the install wizard is preparing to install the two utilities used to configure the access point. The *Simple Network Management Protocol (SNMP)* utility can be used to configure the access point over a network. The *USB* utility is used to upload and configure changes to the access point via a USB connection. The CD installs both utilities on your computer. For this lab, you will configure the access point using the *USB* utility.

Figure 6-34: Install Shield

10. Figure 6-35 shows the display, which provides the option of selecting the drive and folder for the installation of the utilities. Make any selection necessary for your computer configuration and click on **Next** to continue.

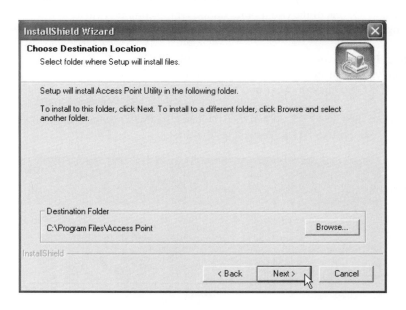

Figure 6-35: Destination Location

11. Restart your computer when prompted. The *Access Point USB Configuration Utility* shortcut will appear on the desktop. Double-click on it. A window similar to Figure 6-36 will appear.

Figure 6-36:
Access Point
Configuration Utility

12. Select the **Basic Setting** tab. The screen shown in Figure 6-37 will appear. Select the channel window and change the setting to **Channel 11**.

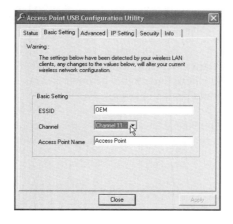

Figure 6-37:
Basic Setting Tab

TIP: The *Access Point* utility can be used to configure the access point for many different network options. Any channel can be used; however, in this lab we will use channel 11. In the following lab, we will use the network tester to identify access point channels.

13. After making the channel selection, click on the **Apply** button. The next screen prompts you to download the changes you have made over the USB connection to the access point. Click on the **Yes** button. The screen will now pause while the utility configures the access point with your changes. The access point is now operational. Close the *Access Point* dialog box.

The final step for this lab is to configure each mobile laptop computer to operate with the access point. In the previous lab, each laptop was configured in the ad hoc mode and was permitted to function as a peer network, sharing data without using an access point. You will now proceed to enable each wireless network computer to communicate with each other through the access point. This will establish a bridging function between the wired network workstations and the wireless mobile laptop stations.

14. Enter into the Advanced configuration of the wireless network interface card on the laptop and choose the **Advanced** tab. In the Property box select **Network Type**. In the Value window, select **Infrastructure Mode**. The window will look similar to Figure 6-38. Verify that the Channel is set to **11**. Click on the **OK** button. Close all dialog boxes.

**Figure 6-38:
Advanced Properties**

15. The access point will now recognize all computers throughout the wired and wireless infrastructure network. You can now browse other computers through My Network Places.

TIP: You may need to configure the workgroups and computer names for each computer to establish identities of workstations and mobile computers. Make sure that each computer is configured for file sharing and printer sharing.

The lab is now fully configured using an access point and portable computers with access to a wired Ethernet network. In the next lab, you will test the network to determine some basic performance parameters.

========

Lab 5 Questions

1. Were you able to "see" all of the network computer names in the My Network Places box?

2. What are the differences between the Ad Hoc Mode and the Infrastructure Mode?

3. What is the purpose of the USB cable connection between a computer and the access point?

4. What observation did you make to determine whether the wireless cards were properly inserted into the expansion slot?

Lab 6 – Measuring the Performance of a Wireless LAN

Lab Objectives

Use the Grasshopper wireless LAN tester to evaluate network performance.

Optimize the location of the access point for best performance.

Evaluate and determine limits of the coverage area for a wireless LAN.

Conduct a survey of a wireless LAN network for possible sources of interference.

Materials Needed

- Grasshopper wireless LAN tester and accessories kit

- AmbiCom Model 1100B-AP access point

- Network lab configuration setup used in Lab 5

- Grasshopper user manual

Instructions

1. The equipment configuration required for this lab is shown in Figure 6-39. Verify that each workstation and wireless mobile station has been configured for file sharing. Check to verify that each computer on the network is properly connected and recognized by browsing **My Network Places** for each computer.

    Make a record of the channel numbers used when you installed the access point and PC wireless cards in each laptop computer.

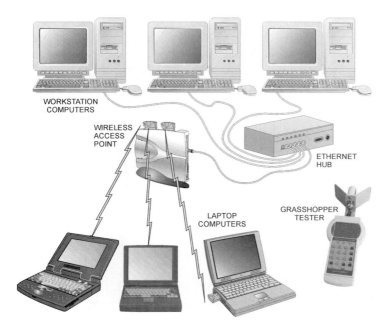

WORKSTATION COMPUTERS

WIRELESS ACCESS POINT

ETHERNET HUB

GRASSHOPPER TESTER

LAPTOP COMPUTERS

**Figure 6-39:
Lab Layout**

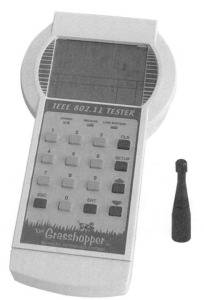

The Grasshopper tester will be used in this lab to help you obtain optimum performance from the wireless access point and wireless portable computers installed in the previous labs. Unpack the Grasshopper tester from the carrying case. Install the antenna provided as shown in Figure 6-40. Find a convenient location to place the battery pack charger used in the Grasshopper tester.

Figure 6-40: Grasshopper and Antenna

3. Take a few moments to review the Grasshopper user manual. You will refer to it occasionally in the following steps.

4. Apply power to the Grasshopper unit by turning the power switch on at the top of the unit. You will see the main menu display shown in Figure 6-41. This screen is used to monitor the performance of the access points and wireless LAN stations in the network. The window at the upper right shows the received radio multipath signal of the selected access point used later.

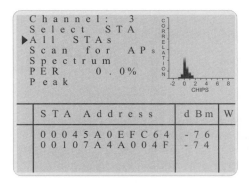

Figure 6-41: Main Menu

5. Using the up and down scroll arrow keys, scroll down to *Scan for APs* and select **ENT**. The Grasshopper tester will now scan all of the 14 radio channels in the ISM 802.11 band searching for APs, as illustrated in Figure 6-42. The display should now show the access point in the lab. The MAC 802.11 Ethernet address for the AP will appear in the left column. The other column shows the channel number for the AP. This channel number should match the number you recorded earlier at the beginning of this lab. Make a record of the AP address for future reference. Pressing the **ENT** key will repeat the scan. Note that in this step, only valid APs in the network will be detected.

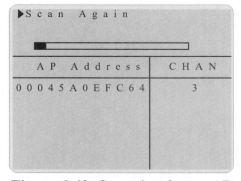

Figure 6-42: Scanning for an AP

6. Press the **ESC** key to return to the Main Screen display. Push **SETUP** and scroll to the channel for the mobile stations you recorded earlier at the beginning of the lab. You should see the display shown in Figure 6-43. Press the **ENT** key to return to the Main Screen. Scroll to *APs Only* and press **ENT**, as shown in Figure 6-44. This will toggle the selection to *All STAs*. You should now see the mobile station addresses displayed. The signal strength is shown in the right column for the APs and the portable wireless computers (STAs). The channel number (or numbers) you recorded earlier should now be displayed along with the MAC address for each wireless station.

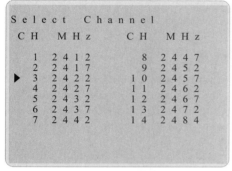

Figure 6-43: Channel Selection

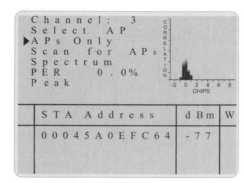

Figure 6-44: AP Only Mode

TIP: Pressing the ENT key will toggle between "APs only" and "All STAs".

7. At the previous step, you should have identified all APs and portable computer stations in the network. In the remaining steps, you will make further measurements to make sure that the units are operating with good signal quality and received signal strength.

8. On the Main Screen, scroll to the *Select AP* option. Press **ENT** until the arrow on the lower screen indicates the AP you want to evaluate in the following steps. Press **ENT** and note that the asterisk appears next to the selected AP address. Scroll down to the *PEAK* option and press **ENT** again. You will see the screen shown in Figure 6-45.

9. The Peak display shows the MAC address for the selected AP as well as signal strength and Packet Error Rate (PER). The bottom line displays the Service Set Identification (SSID). Walk around the perimeter of the lab while maintaining the Peak

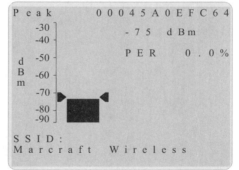

Figure 6-45: Peak Screen

display on the tester. Note the variation in signal strength at various distances from the AP. Proceed to other areas of the facility where you expect the portable workstations to be deployed in typical use in the network. Make a plot of any areas where the signal strength drops below –80 dBm. Wherever coverage appears to be inadequate, move the AP to a new location to experiment with optimizing the coverage area. Survey the area again to check for proper coverage.

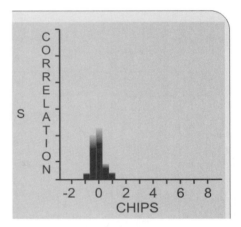

```
              0 0 0 4 5 A 0 E F C 6 4

P a c k e t    E r r o r    R a t e

R a t e          P E R          U s a g e

      1 M b      0 . 0 %       8 9 . 8 %
      2 M b      0 . 0 %        0 . 0 %
5     5 M b      0 . 0 %        0 . 0 %
    1 1 M b      0 . 0 %       1 0 . 1 %
```

Figure 6-46: Packet Error Rate

10. With the same *AP* selected, press the **ESC** key and scroll to the *Packet Error Rate (PER)* option and press **ENT**. You will see the display shown in Figure 6-46.

For the APs selected in previous steps, this screen shows transmission speed, packet error rate, and usage percentage for each rate. Note that the address for the selected AP appears at the top of the screen.

TIP: The Peak display has an audio sound option that provides a signal for listening to a counter as you move to various locations. The audio can be toggled on and off by pressing the **8** key on the keypad.

11. Press the **ESC** key to return to the main screen display. Toggle to select **All STAs**. Select one wireless workstation and stand with the Grasshopper tester near the AP. Have a partner move the portable station to various areas in the facility. Make note of the received signal strength and packet error rate for each portable. This exercise is a way to determine the coverage area and check for dead spots in the coverage area.

12. Multipath measurements are provided on the Main Screen, as shown in Figure 6-47. Multipath signals are displayed in a power spectrum versus CHIPs on the bottom scale. Multipath energy can be analyzed as you survey various areas in the building. Radio signals that are less than one CDMA chip rate apart may cause problems. Try to document areas where the multipath energy may increase packet error rate with the use of the multipath display. Where multipath problems are severe, try looking for direct paths in the next step.

TIP: You may want to review chipping rates and direct sequence spread spectrum modulation in Chapter 5. Review on this subject is located in the *"CDMA and Spread Spectrum Basics"* section.

Figure 6-47: Multipath Signals

13. Where multiple paths may exist for radio signals in a facility, it may be necessary to use a directional antenna with the Grasshopper tester. A directional antenna is useful when looking for the source of direct paths versus walls or structures that are causing the multipath signals. As an option to the omnidirectional antenna, a directional antenna can be installed on the Grasshopper. Using the cable assembly and antenna assembly shown in Figure 6-48, install the omnidirectional antenna.

14. The last step in the survey is to look for sources of radio frequency interference that could cause problems in a wireless LAN environment. Press **ESC** and then press the **SETUP** key. Select the channel you are using for the AP in the lab.

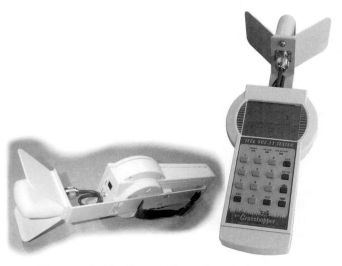

Figure 6-48: Omnidirectional Antenna

15. Press the **ENT** key to return to the Main Screen display. Scroll to the *Spectrum* option and press **ENT**. The screen shown in Figure 6-49 will appear. This figure is an example of interference that may come from various sources that operate in the 2.4 GHz ISM band. Move throughout the lab and network coverage area with the display set to the spectrum screen. Note that the channel being scanned is displayed at the top of the screen. You can toggle to the hold mode to see low duty cycle interference (and the AP transmissions) by using the **1** key on the keypad

TIP: You may note a source of radio frequency energy in the ISM band by moving near an operating microwave oven.

TIP: The Grasshopper user manual refers to the spectrum display as the Radio Signal Strength Indicator (RSSI).

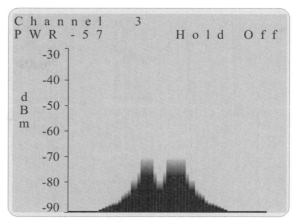

Figure 6-49: Interference Spectrum

16. Complete the lab by removing the battery pack and placing it in the charger. If you have a spare battery pack, make sure that both packs are stored in the packing case fully charged.

Lab 6 Questions

1. What type of information is provided on the Peak display screen?

2. Explain how the AP MAC addresses are found in the network.

3. What procedure would you use to measure the signal strength of a wireless workstation at a specified location in a room?

4. What is the meaning of the value of signal strength labeled "dBm"?

Lab 7 – Setup and Installation of a Wireless Data Terminal

<u>Lab Objectives</u>

Identify the components of a Palm VIIx handheld organizer and wireless data terminal.

Understand the applications and features of the Palm VIIx terminal.

Practice entering data.

Install the Palm VIIx software.

Use the Palm VIIx with a computer.

<u>Materials Needed</u>

- Palm VIIx wireless data terminal

- Palm VIIx cradle

- 2 AA batteries

- Handbook for VII handheld series

- Installation software CD

<u>Instructions</u>

1. Review the materials needed before you start. Identify the accessories including the cradle and software CD.

2. Take a moment to identify the front panel and rear panel components and controls shown in Figure 6-50.

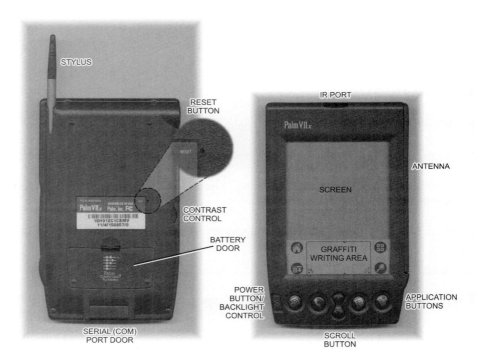

Figure 6-50: Front and Rear Panels

3. Locate and remember the names for the front panel controls shown in Figure 6-51. The application icons are permanently displayed as shown and include the applications launcher, menus, calculator, and the icon to find data in the text of your basic applications. The five control buttons below the display as indicated in Figure 6-51 are, from left to right, the date book, address book, scroll button, to do list, and memo pad.

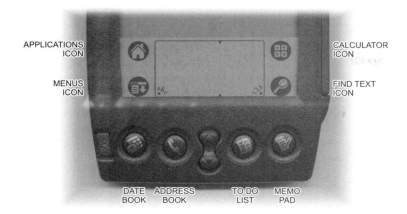

Figure 6-51: Icons and Controls

4. Locate the two AA batteries. On the rear panel of the Palm VII, open the battery compartment door and install the batteries. Observe the polarity shown on the inside of the battery compartment. This starts the charging sequence for the wireless transmitter

5. After the batteries are installed, the setup sequence will begin. Remove the stylus located on the right rear of the Palm VII. Follow the prompts for setup by tapping on the screen when directed by the screen messages. Tap on the target during the calibration step, as shown in Figure 6-52. Do not use objects other than stylus to tap or write on the screen.

6. Step 3 of the setup will ask you to set the country, clock, and date. After reading the information in step 4 concerning web clipping, tap the next button. The next screen shows that the setup is complete.

7. Proceed with the next step about learning to enter text. Tap the next tab on the screen. The information on the next screen explains the four options for entering text.

8. To open an application, tap the application launcher icon. Note that the time, battery level, and application category are displayed. Use the scroll bar to see all the application icons.

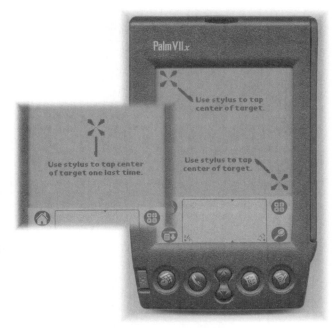

Figure 6-52: Calibrating the Screen

TIP: You can also open an application by writing freehand in the square area at the bottom of the screen with the stylus. This is called a Graffiti character. Practice writing a capital "W" with the stylus. The screen will scroll automatically to all applications starting with "W". Practice writing Graffiti characters to scroll to other applications.

Practice using menus and launching applications in the following steps.

TIP: The pick list arrow at the top right of the applications menu screen shows the categories of applications. Tapping this arrow will display categories such as All, Games, Main, System, Utilities, Unfiled, PalmNet, and Edit categories. The categories are shown in Figure 6-53. Note that some of the applications are for the wireless Internet access. They will be covered later.

Figure 6-53: Selecting Applications

9. With the **All** category selected in the pick list, tap the **Memo Pad** icon. This will show the title area at the top of the display highlighted. Tap any one of the memos. Tap the highlighted area. You should now see three menus available: Record, Edit, and Options. The *Record* option menu is highlighted and contains the commands New Menu, Delete Memo, and Beam Memo shown in Figure 6-54. This is an example of how the menus and commands within the menus are used for most of the main applications.

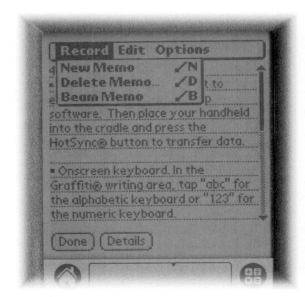

Figure 6-54: Selecting Menus

Figure 6-55: Calculator

10. The Calculator applications can be launched as follows. Tap the **Calculator** icon on the applications menu or press the **Calculator** button. Press the **Calculator** button and practice using the calculator, shown in Figure 6-55.

11. Earlier, you learned about the four steps for entering text. In the next step, you will practice entering text from the keyboard. Tap on the **Address** application on the main screen. Tap on the lower left portion of the Graffiti area containing the **abc** icon. The keyboard will appear, as shown in Figure 6-56. Practice writing a few words and numbers using the keyboard.

12. Online tips are available with the Palm handheld. Tap the **i** icon in the upper right corner. After you review the tip, tap **Done**.

13. Text can also be entered using the Graffiti area. Review the four concepts for Graffiti writing on pages 31 to 40 in the Palm Handbook.

14. After reviewing the Graffiti writing tips, tap the **application launcher** icon. The display showing the memo pad will appear, as shown in Figure 6-57. Tap the **Memo Pad** application icon. Practice entering text and numbers in the Graffiti area.

Figure 6-56: Entering Text with the Keyboard

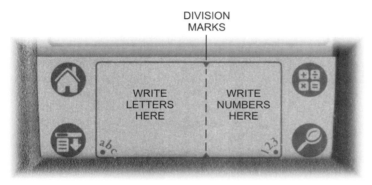

Figure 6-57: Graffiti Area

The next steps will install the Palm software on a computer. The Palm software enables you to work with the handheld applications on your computer, perform backups using the HotSync application, and export and import data. You can also print your Date Book, Address Book, To Do List, and Memo Pad information on any printer.

15. Turn off the computer and connect the cradle to a serial port on the computer. Do not place the handheld on the cradle at this time. Turn the computer on.

16. Insert the Palm Desktop Software CD in the drive. The menu shown in Figure 6-58 will appear. Click on **Install**.

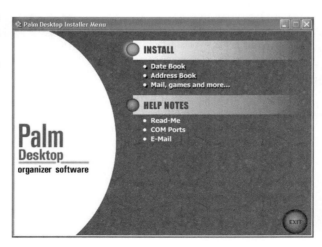

Figure 6-58: Install Software Menu

17. After reading the welcome message shown in Figure 6-59, click **Next**.

18. Read and comply with the three Pre-Install steps shown in Figure 6-60 and continue by clicking on **Next**.

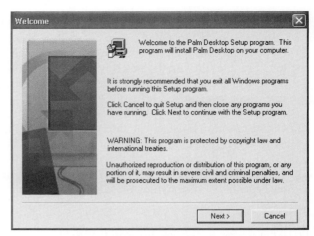

Figure 6-59: Welcome to Install

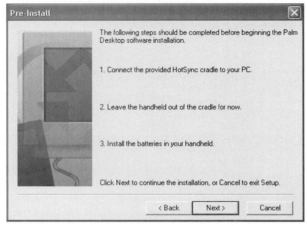

Figure 6-60: Pre-Install Setup

19. When the *Setup Type* screen appears, shown in Figure 6-61, select **Typical**. Enter the drive and directory where you want to install the software. The default is the C:\ drive.

20. The next screen, shown in Figure 6-62, will be displayed if you have Microsoft Outlook installed. If you have Microsoft Outlook installed on your computer, you will be prompted to choose between linking to Outlook or the Palm Desktop. Select **Synchronize with Palm Desktop** and click on **Next**.

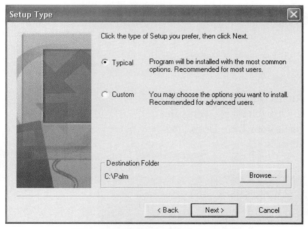

Figure 6-61: Setup Type

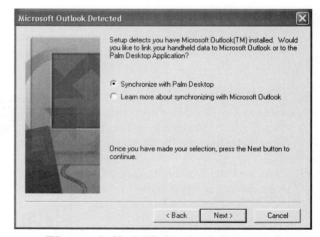

Figure 6-62: MS Outlook Detected

21. The next screen will help you select a serial port configuration. Follow the instructions shown in Figure 6-63. Place the Palm handheld on the cradle and click on **OK**. The software will proceed with the installation.

22. The next screen, shown in Figure 6-64, will appear with information about the web clipping applications. Click on **Next** to continue.

Figure 6-63: Serial Port Setup

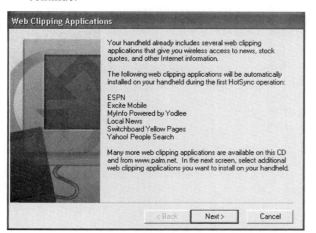

Figure 6-64: Web Clipping Applications

23. The next screen, shown in Figure 6-65, contains a list of Juno web clipping applications. Note that these applications will be downloaded later during a HotSync operation. Leave all items checked and click on **Next** to continue.

24. The next screen, shown in Figure 6-66, describes the *iMessenger* application that allows you to send text messages to any Internet e-mail address and *Mail*, which allows you to manage your e-mail with your desktop program. Click on **Next** to continue.

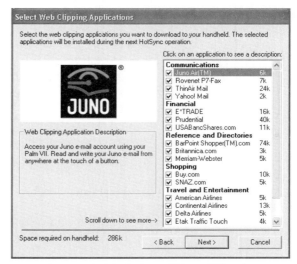

Figure 6-65: Select Web Clipping Applications

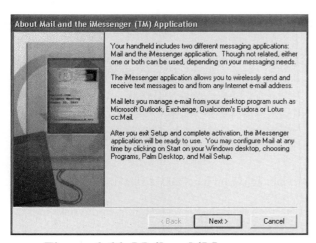

Figure 6-66: Mail and iMessenger

25. The screen shown in Figure 6-67 indicates that the Palm software installation is complete. Click on **Finish** to exit Palm Desktop Setup.

Figure 6-67: Setup Complete

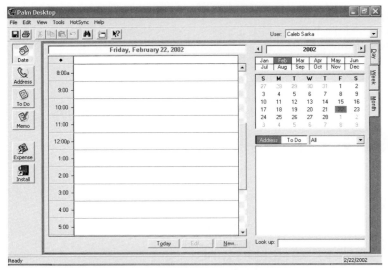

Figure 6-68: Palm Desktop

26. Push the **HotSync** button on the cradle. The HotSync program you have installed will now synchronize the desktop and Palm files. This procedure may be followed to maintain synchronization with all of your Palm files in the future.

27. On the computer desktop, click on **Palm Desktop** and the submenu **Palm Desktop**. The display shown in Figure 6-68 will appear. This is the duplicate application launcher for your desktop. (If you have Microsoft Excel installed on your computer, you can now use the integrated spreadsheet application by clicking on Expense.) Practice opening and using the other Palm desktop applications.

Lab 7 Questions

1. Explain the procedure for synchronizing the information on a desktop application with Palm data.

2. Name four methods for entering text on the Palm VIIx handheld.

3. What feature enables you to select applications according to defined categories?

4. What is the procedure for opening the on-screen keyboard display to enter information?

Lab 8 – Activating the Palm.Net Service

Lab Objectives

Set up and activate the Palm.Net service on a Palm handheld terminal.

Check for adequate received signal strength from the nearest Palm.Net base station.

Understand the procedure for sending and receiving e-mail.

Understand how to access a "web clipping" web site on the Palm.Net.

Conduct a test for the receipt of an iMessenger text message.

Send a message using the iMessenger application.

Materials Needed

- Palm VIIx wireless data terminal

- Palm VIIx cradle

- 2 AA batteries

- Handbook for Palm VII handheld series

- *Palm.Net Service Plan* booklet

- *Read This First* booklet

Instructions

1. This lab will take you through the steps for activating wireless Internet service for the Palm VIIx handheld. These instructions assume that you have completed all the steps in the previous instructions. Activating the service will require a credit card account.

2. Read the *Palm.Net Service Plan* booklet. Use the service area map to determine whether your location is covered by the areas shown in the map. With Palm.Net service you pay for the amount of information transmitted and received. Select the type of service you wish to activate.

3. To activate the service, raise the antenna as indicated in Figure 6-69. Follow the on-screen instructions and tap the **Next** button after reading the information on each page.

4. To activate wireless service, you will need to select a user name and password, and have a credit card available. The on-screen menu will suggest user names. After completing this step, record your user name and password in the *Read This First* booklet.

5. When the screen menu indicates that the wireless service has been activated, you may use the Palm.Net applications. Keep the antenna in the upright position.

TIP: Power is removed to conserve battery life after a few seconds of inactivity. Press the green **power on** button at the bottom left of the screen when you are ready to resume.

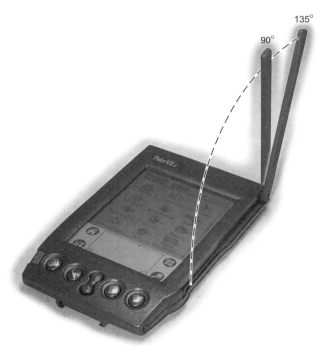

**Figure 6-69:
Raise Antenna**

6. Tap the **launch applications** button. With the applications screen display shown in Figure 6-70, tap the **pick list arrow** in the upper right corner of the screen. Tap the **Palm.Net** option. Tap the onscreen **Diagnostics** icon. This will display the signal strength of the wireless network base station closest to you, as shown in Figure 6-71. Change locations to obtain the best reading in the room. Notice how the position of the antenna can affect received signal strength. Notice the change in signal strength when you move to an outdoor location.

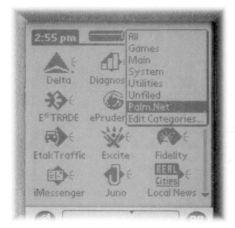

Figure 6-70: Palm.Net Application

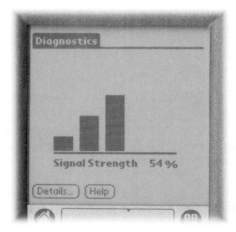

Figure 6-71: Signal Strength

7. Using the *Palm.Net* applications screen, tap on the **weather** icon. The Weather Channel web clipping page will appear as shown in Figure 6-72. Tap on **Weather News** and read the national weather summary for today. Tap on the back arrow at the top of the screen. Tap on **Select a City**. Then tap on **your state**. When the list of cities appears, select your **present location**. The local weather will appear, as shown in Figure 6-73. Scroll to see the future week weather forecast.

Figure 6-72: Weather Channel Web Clipping

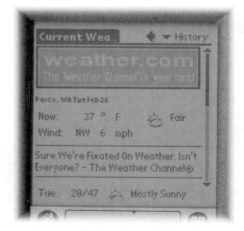

Figure 6-73: Local Weather

8. Use the back arrow to back out of the Weather Channel to the *Palm.Net applications* screen. Tap the **iMessenger** icon. Tap on **New**. The display shown in Figure 6-74 will appear. Use the Graffiti screen or the keyboard for typing. Type or write an **e-mail address** in the *To:* field for the person you wish to receive your e-mail message. Type a **subject** in the *Subj:* field. Enter a **short message** in the *Body:* field. (You can send a test message to yourself if you wish. Use your ID mail address selected during the activation.)

TIP: Your e-mail address is (*your ID*)@palm.net. You selected your ID when you activated the service. Also, if you wish to use the keypad instead of the Graffiti pad, tap on the **abc** icon on the bottom left of the display. Be sure to type the correct text in the three fields on the *iMessenger* screen.

9. When you have completed the message, tap on **Check** and **Send**.

Figure 6-74: New Message

10. Return to the main *Palm.Net* applications screen. Tap on the **ABC news** icon. When the World Front Page web clipping display appears, as shown in Figure 6-75, tap on **World** to read news headlines.

TIP: The Palm.Net is an "always on" network protocol. You do not log on and off. Access is initiated by raising the antenna. You are billed within your service plan for the number of kilobytes you transmit and receive.

11. Return to the *iMessenger* application by tapping on the **iMessenger** icon. The display shown in Figure 6-76 will appear. Tap on the **pick list** to show the *Inbox*. This allows you to check for incoming messages.

Figure 6-75: ABC News Application

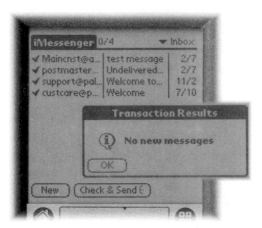

Figure 6-76: iMessenger

Lab 8 Questions

1. How many kilobytes (kB) are allowed for sending and receiving with the basic service plan?

2. What procedure would you use to check for the signal strength from the nearest Palm.Net base station?

3. How is information entered on the Palm handheld for composing a message?

4. What is the procedure for checking for new e-mail messages?

Key Points Review

In this chapter, the following key points were introduced concerning wireless networks. Review the following key points before moving into the Review Questions section to make sure you are comfortable with each point and to verify your knowledge of the information.

- Without a set of standards, the capability for cellular carriers to exchange information concerning roaming subscribers would be severely limited.

- Standards also facilitate the ability to confirm the identity of a cellular mobile station as a legal subscriber, thereby protecting carriers from fraud and cloning of cell phones. This feature adopted in the network standards is called authentication.

- The ANSI-41-C has been replaced with the current version called Cellular Radiotelecommunications Intersystem Operations (ANSI/TIA/EIA-41-D-97).

- The Mobile Switching Center (MSC) is the main control entity in the network that provides the interface between the PSTN and the base station.

- The Home Location Register (HLR) database contains all specific subscriber service data including the profile of services that each subscriber is authorized to use on the network.

- The Visitor Location Register (VLR) is a database associated with an HLR. It contains dynamic information about subscribers who are currently active in the area served by the associated MSCs.

- The term "visitor" used in the TIA/EIA-41 standard is sometimes confusing because the VLR contains information about active home subscribers who are active as well as visitors from another service area.

- The Authentication Center (AC) is a database that manages authentication and encryption information of subscribers. The AC may be collocated with the MSC or HLR or it may be located at another location on the network. An AC may serve several MSCs in the network.

- The Equipment Identity Register (EIR) is an optional database that contains the list of all valid mobile equipment on the network. The equipment in this database is listed by the International Mobile Equipment Identity (IMEI).

- The Base Station Controller (BSC) handles all mobile management tasks for one or more Base Transceiver Systems (BTS).

- The Base Transceiver System (BTS) contains the radio transmitter, receiver, antenna, and tower that define a cell site.

- The Mobile Station (MS) is the radio communications device that is designed to handle the protocols on the air interface.

- Registration is the process that occurs when a cell phone or any cellular wireless device classified as a mobile station informs the network of its presence in the cellular system.

- Power-off registration is performed when the power off switch is pressed. This notifies the network that the phone is no longer available for receiving calls.

- Location-based registration occurs when the mobile station has moved beyond a specified threshold of distance since the previous registration.

- Zone or area based registration occurs when the mobile station travels to a new area within the same network. This reduces the number of cells that need to be paged to notify the subscriber of an inbound call.

- Time-elapsed registration is performed by the mobile station when a time interval since the last registration is exceeded.

- Authentication consists of the processing executed between a mobile station and the network to confirm that the mobile station is who it claims to be prior to honoring a request for service.

- The procedures for authentication involve similar processing routines in the mobile station and the authentication center. The results are then compared in the authentication center to confirm that the answer to the algorithm is the same in both cases.

- Voice call processing and call delivery proceed after the authentication of the originating mobile station.

- The Internet provides access to a rapidly growing number of information sources worldwide.

- The Internet uses the TCP/IP suite of protocols. These protocols do not support complete mobility.

- The standard IP addressing protocol does not support full mobility when the attachment point changes during a session. Enhancements to TCP/IP are needed to support full mobility. These enhancements are referred to as Mobile IP.

- Portable computing is typically defined as connecting to an office LAN from a remote location when away from home using an Internet Service Provider.

- Full mobility requires the network to maintain the connection when the computer changes the point of attachment by entering another area served by a different MSC and IWF.

- Mobile IP is published as RFC 2002 and the associated documents RFC 2003 through 2006 by the IETF.

- Mobile IP routers with the enhanced software are called Home Agents (HA) and Foreign Agents (FA).

- Wireless Local Area Networks (WLAN) offer additional capabilities for mobile networking using radio technology and portability for a wide variety of applications.

- Cellular radio networks use licensed radio spectrum. Wireless LANs use special unlicensed radio bands.

- In the U.S. most wireless LAN products are designed for use in one of the ISM (Industrial, Scientific, and Medical) license-free bands. The FCC provides no protection from interference among the users of the ISM band.

- The High Performance Radio Local Area Network (HIPERLAN) was developed in Europe by the European Telecommunications Standards Institute (ETSI).

- HIPERLAN is not likely to take market share from IEEE 802.11 products except perhaps in Europe.

- HomeRF is a wireless LAN standard supported by the HomeRF Working Group (often referred to as the HomeRF WG).

- The Bluetooth wireless technology provides the capability for the replacement of cables that connect one device to another with a short-range radio link.

- The Bluetooth wireless technology supports both point-to-point and point-to-multipoint connections. The radio interface operates in the 2.4 GHz ISM band and employs Frequency-Hopped Spread Spectrum (FHSS) modulation.

- In June 1997, the IEEE completed work on the initial standard for wireless LANs, defined as 802.11. This standard specified a 2.4 GHz operating frequency with data transmission rates of 1 and 2 Mbps.

- The 802.11b standard was the result of a study by the IEEE that looked at ways to achieve a higher speed that would make the technology more competitive with other emerging standards.

- The ability to achieve higher data rates in the 802.11b specification was made possible in part by using a higher level modulation.

- When planning the layout for or troubleshooting a wireless LAN, you must know the FCC rules governing the limitations on radiated transmitter power and the use of directional antennas.

- An 802.11b standard supports the following two topologies: Independent Basic Service Set (IBSS) networks and Extended Service Set (ESS) networks.

- Mobile stations in a IBSS network can communicate with each other as long as they remain within the BSS. Stations in the BSS are said to be operating in the "ad hoc" mode.

- An access point acts as a wireless bridge to the wired LAN.

- It is possible for transmissions in the 2.4 GHz band to radiate outside the immediate area where an office or campus network would be located.

- The IEEE 802.11b standard provides a security feature called Wired Equivalent Privacy (WEP). WEP provides a means for encrypting data transmission and authenticating the nodes.

REVIEW QUESTIONS

The following questions test your knowledge of the material presented in this chapter:

1. What feature included in the cellular network standards is used to confirm the identity of a cellular mobile station as a legal subscriber?

2. What algorithm is used to generate the 128-bit SSD_A and SSD_B codes?

3. What database, normally associated with the MSC, contains all specific subscriber service data including the profile of services that each subscriber is authorized to use?

4. What is the initial process that occurs in a cellular network when a mobile station is first powered up?

5. Briefly describe the limitations of portability in a wireless IP environment.

6. Briefly describe the purpose and function of the "care of" address as applied to mobile IP networks.

7. What protocol, specified in RFC 2003, enables both the permanent IP address and the "care of" address to be contained within the same packet?

8. Describe the purpose of the IWF.

9. Explain the definition of the term "tunneling" as applied to mobile IP networks.

10. What is the FCC name for the radio frequency bands used for WLANs and other unlicensed wireless services?

11. What WLAN standard was developed in Europe under the sponsorship of the European Telecommunications Standards Institute (ETSI)?

12. What data transmission rates are supported by the IEEE 802.11b WLAN standard?

13. Describe an ad hoc wireless LAN network.

14. What feature of the 802.11b standard provides for security of transmitted data packets?

15. What type of WLAN specified in the 802.11b standard provides access to other network elements via a distribution system?

WIRELESS BROADBAND NETWORKS

LEARNING OBJECTIVES

U pon completion of this chapter, you should be able to perform the
following tasks:

1. State the general definition of a broadband network.

2. Describe the general characteristics of an optical network.

3. List the characteristics and advantages of satellite networks.

4. Provide a brief description of DSL technology.

5. Identify the uses of ISDN.

6. Describe the features of fixed wireless broadband networks.

7. Describe the transmission speeds available with the MMDS.

8. Draw a diagram of a 12-sector MMDS cell plan.

9. Describe the types of modulation used by MMDS networks.

10. Evaluate the conditions that influence Internet download speeds.

11. Create a chart of the MMDS spectrum.

12. List the primary services available with the LMDS.

13. Draw a diagram of a basic LMDS network.

14. Describe the frequency band plan for the LMDS.

15. Show a method for calculating system capacity for the LMDS band.

16. Draw a chart showing the 39 GHz frequency band plan.

17. List the advantages and problems associated with the 39 GHz band.

18. Describe the functional purpose of the local loop in telephone networks.

19. List five advantages of a WLL solution over copper wire local loop service.

20. Describe two advantages for the use of optical networks.

21. Draw a diagram showing the optimum optical spectrum available for free-space
optical hardware.

22. Describe the bandwidth values offered by free-space optical networks.

Wireless Broadband Networks

INTRODUCTION

This chapter introduces another segment of the wireless technologies known as broadband wireless access. The FCC has allocated the largest amount of radio spectrum to this service. The bands are identified by numerous names and classifications, all administered by the Wireless Telecommunications Bureau of the FCC. Both wired and wireless broadband services are covered in this chapter with comparisons between the main systems in each category.

Wireless broadband access is a very large segment of the wireless industry. In this chapter, we will concentrate on the two primary services in the broadband technology areas. They are the Multichannel Multipoint Distribution Service (MMDS) and the Local Multipoint Distribution Service (LMDS), which are the dominant fixed wireless broadband access technologies with the largest share of the radio spectrum.

Broadband Networks

> **Broadband networks** provide transmission of data at high speeds.

Broadband networks

This is the brief and rather unclear definition of broadband found in most technical sources. The term has been widely used for many years to describe wired network transmission systems, typically at T1 rates of 1.544 Mbps and above. In this chapter, you will learn more about wireless broadband networks that carry information at rates of 30 Mbps and above.

Cellular mobile voice networks covered in the previous chapter are designed to carry voice conversations. They are designed to restrict the bandwidth associated with each call to maximize spectrum efficiency.

> Broadband networks operate in fixed locations. They are structured to transmit information between two points and often between multiple locations at very high transmission rates of 200 kbps to 30 Mbps using radio frequencies in the range of 2 GHz to 40 GHz.

In the next section we will compare some well-known wired and wireless broadband technologies.

Optical Networks

Fiber-optic cable transmission is the leading non-wireless broadband technology.

Fiber-optic cable uses infrared waves to transmit data, voice, and video information. Optical networks are expensive because of the high cost of the cable and the trenching necessary to lay the cable. The total cost of laying fiber is generally between $90 and $250 per foot. This makes fiber too expensive except for long haul circuits. Broadband wireless, however, performs well in the short haul network class.

Optical systems transmit data in the gigabit range.

An optical technology called Dense Wavelength Division Multiplexing (DWDM) has been developed to send multiple data streams over the same cable. Each data stream is transmitted on a distinct optical wavelength on the same cable. The DWDM feature allows a 2.4 Gbps signal to be multiplied up to 16 times. A sixteen-channel system supports up to 40 Gbps in each direction. A 40-channel system under development will support 100 Gbps, the equivalent of ten STM-64/OC-192 transmitters (10×9.6 Gbps).

Satellite Networks

Satellites offer long haul broadband communications capability.

Satellites have a high up-front cost of launching. They have a general bandwidth capacity of 30 Mbps. Satellites are distance-insensitive and can provide intercontinental communications. Satellites are also used for television cable channel and broadcast network distribution to TV cable companies. Broadband networks using satellite technology are covered in more detail in Chapter 8.

Digital Subscriber Link

Digital Subscriber Link (DSL) refers to a special wireline technology that provides more bandwidth over existing copper telephone lines serving a customer's premise.

Digital Subscriber
Link (DSL)

DSL creates more bandwidth on the existing copper phone lines through special hardware attached at both the user's premises and in the phone network. Speeds vary depending on the type of service and generally range from 384 kbps to 1.5 Mbps for both uplink and downlink. Speeds are typically provided that are higher on the downlink side, where the higher rates are more beneficial for Internet access. Similar to ISDN, DSL provides shared data service and voice service over the same copper wires.

Although DSL is spreading, it is not deployed in many areas and its availability depends on a host of technical and economic factors, including the distance between the user's location and the carrier's central office. The user's fixed-line service must be located no more than 18,000 feet (approximately 3.4 miles) from a central office or remote terminal because the farther the user is from the hardware in the phone network, the less bandwidth the user will receive. For example, 7 Mbps might be available at 12,000 feet, but only 1 Mbps at 18,000 feet.

Integrated Services Digital Network

Integrated Services Digital Network (ISDN) is another technology that uses existing copper telephone lines to obtain high-speed digital service.

Integrated Services Digital Network (ISDN)

ISDN offers the capability to have shared voice and data circuits over the same twisted pair of copper wires from the customer location to the telephone company local switching office. Basic Rate Interface (BRI) service provides up to 128 kbps. This is a major improvement over analog modems at 28.8 and 56 kbps service. ISDN service is not available in all areas. It is also, similar to DSL, limited to customers located within approximately 18,000 feet (3.4 miles) from the local office switching center.

Leased High-Capacity Circuits

Leased line service is a high data rate service provided by the regional Bell operating companies and competitive local exchange carriers. A T1 standard circuit provides 1.544 Mbps.

A T3 circuit provides a capacity of 45 Mbps. Leased lines are expensive and used primarily by businesses for high-speed access to the Internet and corporate intranets. These dedicated lines, similar to ISDN and DSL, offer the "always connected" feature, which is a major advantage over dial-up service.

TV Cable

Cable TV's coaxial connections to the home can deliver data at speeds up to 100 times faster than standard dial-up modems.

Cable is another alternative that bypasses the local telephone service for high-speed Internet access. Cable service speeds can range up to 10 Mbps, but most service is in the 256 kbps to 3 Mbps range. Cable service is a shared bandwidth technology. All users on a single circuit split the available bandwidth; therefore, the service is sporadic depending on the number of users active at one time.

Fixed Wireless Broadband Access Networks

Fixed wireless

Fixed wireless is a licensing term used by the FCC to define radio service where the transmitting stations are permanently installed (fixed) at specific geographical locations.

Fixed wireless networks provide access to the PSTN and the Internet as an alternative technology to DSL, cable, and leased dedicated lines. Fixed wireless networks are therefore referred to as "broadband wireless access" or BWA technologies. Multichannel Multipoint Distribution Service (MMDS) and Local Multipoint Distribution Service (LMDS) are the primary technologies used in the fixed service as defined by the FCC. These two wireless broadband services are the major topics we will focus on in this chapter.

Networks using microwave radio transmission equipment were designed by AT&T and later MCI as an alternative to copper wire media. Microwave transmission towers provided high-speed wireless circuit paths across bodies of water or mountainous terrain where copper or fiber optic cable installation was impractical. Other market forces opened up new uses for microwave broadband services in the 1970s.

MMDS historically has been used for one-way "wireless cable" video and instructional closed circuit television service.

WorldCom and Sprint have acquired MMDS licenses in most major markets in the U.S.

Local Multipoint Distribution Service (LMDS) is similar to MMDS, but uses higher frequencies with limited range between customer locations and transmitter locations.

Microwave Propagation Characteristics

In Chapter 4, the problems associated with propagation at high frequencies were explained. Rain, atmospheric attenuation, path loss, blockage, and multipath are all problems associated with frequency bands above 1 GHz. As the frequency increases, the problems get worse. Engineers must consider all these issues when designing microwave systems. The radio link design and architecture selected typically targets an availability number in the 99.99 percent region or higher.

The frequency bands assigned for the fixed wireless service occupy various portions of the spectrum from 2.1 to 39 GHz. The propagation characteristics are quite different across such a wide range of frequencies.

For example, a radio transmission over a one-mile path using the 39 GHz band will encounter approximately 235 times more power loss than a transmission in the 2.1 GHz band. Other environmental features such as rain, foliage, terrain, and clear line of site between transmitter and receiver are important issues the design engineer must consider.

At the higher frequency bands, propagation is affected by the weather.

The 24, 28, and 38 GHz bands are attenuated severely when the rain rate (amount per unit of time) is heavy. As a result, a shorter path is required between the transmitter site and the receiver location at these frequencies. Adaptive power control is needed to maintain the required quality of service with respect to bit error rates. The MMDS band, with its longer wavelength, is not affected by rain.

The multipath problem with the upper bands has been the subject of much research. A modulation technique known as Orthogonal Frequency Division Multiplexing (OFDM) has been used with success in other WLAN systems such as HIPERLAN 2. OFDM has been used with the Digital Video Broadcast (DVB) and Digital Audio Broadcasting (DAB) standards. OFDM spreads the data to be transmitted over a large number of carriers, each modulated at a low bit rate. OFDM shows promise of major improvements in performance for broadband wireless access technologies. Additional detail on OFDM will be given in Chapter 9.

Multichannel Multipoint Distribution Service

Multichannel Multipoint Distribution Service (MMDS) is a point-to-point and point-to-multipoint broadband microwave radio service. It utilizes frequencies in the 2.1 and 2.6 GHz bands.

Multichannel
Multipoint
Distribution Service
(MMDS)

Within the wireless industry, the term "microwave" has become the common name for radio services using frequency bands above 1 GHz (wavelengths less than 30 cm). There is, however, no established standard for where the microwave spectrum begins or ends.

The business objective of this type of wireless service is to provide high-speed data access to the Internet using radio channels. It also provides so-called "last mile" access to the PSTN for long distance carriers. It is the wireless competition for Digital Subscriber Line (DSL), cable modem access, fiber-optic cable, and T1 broadband service.

MMDS was originally conceived as a microwave radio distribution service for closed circuit television and business data services using analog transmission in two 6 MHz channels. This service was identified in the early 1970s as the Multichannel Distribution Service (MDS). In 1975, some firms used the MDS bands to distribute HBO and other subscriber-paid TV channel programming. The MDS service providers received program content from the cable channel service providers, as illustrated in Figure 7-1. The MDS facility housed the satellite receiving equipment and the frequency converters. Each video channel was converted to the MDS transmitter frequencies and was broadcast with a scrambled signal. The subscriber's set top converter box unscrambled the signal and provided the tuner/decoder functions for direct connection to the TV antenna input connector. Other TV distribution services also used licensed spectrum in the 2 GHz band in the 1970s. The Instructional Television Fixed Service (ITFS) was a service used for closed circuit broadcasting for educational purposes by universities and other nonprofit organizations. The wireless TV delivery companies were unable to compete successfully with the improvements that were to come to the cable industry. Analog channels utilized 6 MHz of spectrum per channel. As cable companies expanded with 100+ channel offerings using digital technologies and fiber-optic links, insufficient spectrum was available for the MDS license holders to compete with the coaxial cable delivery systems. Since MMDS was originally licensed as a one-way video transmission service, it was not able to offer other full-duplex voice and data services.

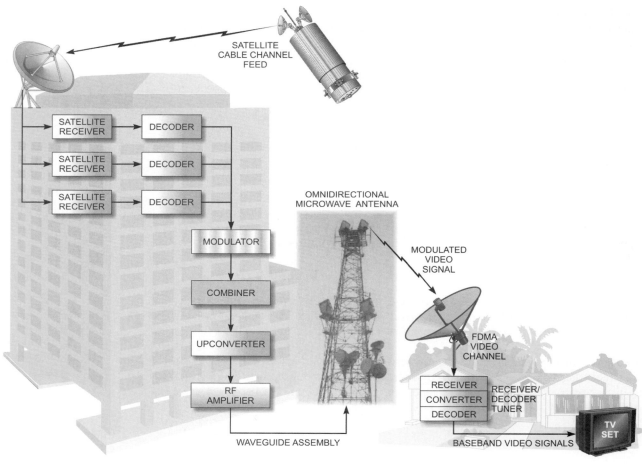

Figure 7-1: MMDS Video Distribution Service

This changed in 1998 when the FCC opened up the MMDS spectrum to bi-directional use. This changed the interest in MMDS. Internet service providers have started to use the MMDS spectrum for high-speed Internet access as well as video teleconferencing, voice, and data services.

MMDS has had limited success as a competitive high-speed data service as well as a TV cable channel distribution service. Some of the pioneers in the MMDS industry went bankrupt in the late 1990s and 2000 business years. The FCC has also provided some restructuring to encourage the use of the MMDS spectrum for high-speed digital data services and Internet access using the expanded spectrum in the 2.1 and 2.5 MHz bands. Major telecommunications companies acquired licenses in 2000 and 2001 to operate point-to-multipoint service using advanced digital technologies and non–line-of-sight modulation techniques.

MMDS Network Architecture

> MMDS networks are structured to carry high-capacity data traffic between customer locations and a centrally located MMDS microwave gateway station.

The frequency band used for MMDS supports paths approximately 35 miles in length. In order to maximize the use of available spectrum, most MMDS carriers use cells to provide coverage for metropolitan areas or business and university campus locations. Current MMDS systems utilize cable modem technology to deliver 10 to 30 megabits per second (Mbps) downstream and 32 kbps to 10 Mbps upstream. MMDS provides line-of-sight connectivity between customer locations and MMDS base stations or central hub stations.

MMDS Cells

MMDS cells are not designed for mobile subscribers. There is no requirement for MMDS license holders to use cellular architecture. Cells are practical solutions used by the carriers to make maximum use of available spectrum. MMDS cells serve customer locations that are permanently located in business offices, college campuses, or private residences. Cells are not necessarily contiguous. MMDS cells can be located as necessary to provide coverage of the area served by the individual MMDS licenses and customer plant sites or other high-density metropolitan business locations.

> With appropriate terrain characteristics, a single MMDS cell can cover a 35-mile radius, or 3,850 square miles.

> MMDS requires line-of-sight for reliable quality of service.

This means that if the receiver cannot "see" the transmitter hub station or relay station, the path will not work with any degree of reliability. Some new modulation formats, such as orthogonal frequency division multiplexing (OFDM), may support non–line-of-sight paths.

Frequency reuse is an important consideration with MMDS. Each MMDS service provider may deploy cells according to specific market objectives. One solution used by MMDS service providers is the use of sectors, as illustrated in Figure 7-2. The cell design uses directional antennas. In the example, the MMDS hub connects various customer sites with a high-speed backbone using fiber-optic or broadband wireless link to an Internet service provider. The sectors in the cell use directional antennas to form 30-degree sectors. This allows six sectors for channel A and six sectors for channel B to cover an entire city or metropolitan area. Twelve channels are available, each carrying separate traffic with only two channels of spectrum utilized. The channel reuse factor in the example is therefore 6.

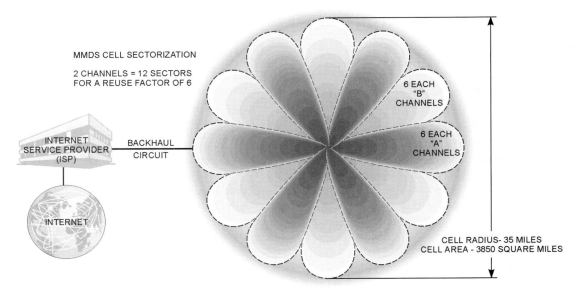

Figure 7-2: MMDS Cells

> Several factors determine the cell size including terrain, type of modulation, access method (FDMA or TDMA), and quality of service goals.

Each sector may utilize spectrum for forward links, reverse links, or both.

MMDS Internet Access

High-speed wireless Internet access is one of the major applications offered to private homes and business centers by MMDS service organizations. MMDS provides the connection between customer computer networks and the Internet. Figure 7-3 illustrates the connectivity between network elements. Hub stations or relay base stations provide the routing and radio access for uplink and downlink radio channels using both FDMA and TDMA access protocols. Customer sites are offered "always connected" service for local area corporate networks. Routers at the customer location handle IP address routing of requests to the hub station using the customer premise equipment (CPE) radio transceiver. The hub station provides the routing to an Internet service provider and forwards responses to the customer network over the downlink channel. The MMDS carrier may provide the ISP service as well as the wireless access service. Hub stations may use cellular and sectored network architectures depending on subscriber locations and density.

Uplink and downlink speeds are capable of handling constant bit rate (CBR) applications; however, the range of data rates, shown in the example in Figure 7-3, may vary due to factors unrelated to the radio link characteristics. MMDS service providers claim that the Internet alone can affect the download speeds shown in the example due to one or more of the following conditions:

- Backhaul traffic usage between the MMDS cell hub and the ISP

- Time of day and Internet traffic dynamics

- Bandwidth limitations of the web site server and the Internet

- Type of file transfer (FTP files download faster than HTML web pages)

- Speed and latency of routers

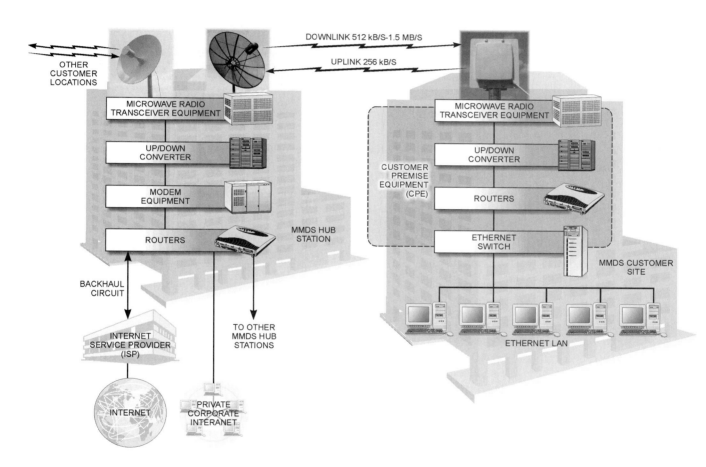

Figure 7-3: MMDS Wireless Internet Access

Multichannel Multipoint Distribution Service Band Plan

The MMDS frequency band plan is shown in Figure 7-4. The MMDS spectrum is divided into 6 MHz channels. During the 1970s, the FCC allocated frequencies in the 2.1 and 2.6 GHz bands for distribution of commercial TV broadcasting as a competitive alternative to cable TV service. Each 6 MHz channel served a single analog TV channel.

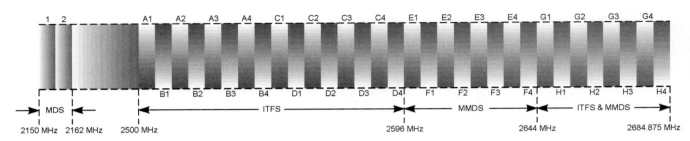

MDS - MULTIPOINT DISTRIBUTION SERVICE
ITFS - INSTRUCTIONAL TELEVISION FIXED SERVICE
MMDS - MULTICHANNEL MULTIPOINT DISTRIBUTION SERVICE

NOTE: ALL CHANNELS ARE 6 MHz WIDE EXCEPT H4.
H4 CONTAINS 32 SUBCHANNELS AT 125 kHz, EACH = 3.875 MHz.

Figure 7-4:
MMDS Band
Allocation - 2.1 and
2.6 GHz Band Plan

The Instructional Television Fixed Service (ITFS) served colleges, universities, and other noncommercial organizations for educational closed-circuit television applications. The Multipoint Distribution Service (MDS) contains two 6 MHz channels and was the first allocated spectrum for cable channel TV service. Technology advances in coaxial cable service soon advanced to allow 100+ channels on cable. The MDS service was too limited to compete with cable service.

> The MMDS, MDS, and ITFS spectrum was restructured and expanded after the Telecommunications Act of 1996 to accommodate the growing need for point-to-multipoint digital data service.

The FCC reallocated eight of the original ITFS channels (group E and F) to what is now the MMDS band. Digital modulation techniques with forward error correction (FEC) and Motion Picture Experts Group (MPEG) digital video compression technology enabled reliable high-speed wireless service to compete with the traditional high-speed wired technologies.

Other microwave radio services not shown have been less successful commercially. The Digital Electronic Messaging Service (DEMS) originally was allocated 100 MHz of spectrum in the 18 MHz band. The Teligent Corporation was the only operator in this band. They convinced the FCC to reallocate the DEMS band to a 400 MHz segment in the 24 GHz band. The Wireless Communications Service (WCS) was also inaugurated in a narrow segment (only 20 MHz) of the 2.3 MHz band.

Local Multipoint Distribution Service

The **Local Multipoint Distribution Service (LMDS)** is another broadband technology. It operates in a much higher frequency band than the MMDS. It operates in the 28 and 31 GHz bands in the U.S. and the 26 GHz band in Europe. These bands are called millimeter wave bands because the wavelength at 28 GHz is only .0107 meters (approximately 10 millimeters).

Local Multipoint Distribution Service (LMDS)

LMDS uses cells that are smaller than the cell sizes described for MMDS because of the higher atmospheric attenuation at 28 GHz compared to the 2.5 GHz band used by MMDS.

LMDS networks can provide two-way broadband wireless access services including:

- Video

- Internet access

- Voice and IP traffic over Asynchronous Transfer Mode (ATM).

Local Multipoint Distribution Service Network Architecture

A LMDS network can be composed of a series of cells that each deliver point-to-multipoint services to subscribers. Because of the higher frequencies used by this band, the attenuation of radio signal is much higher than the MMDS band. Each transmitter in a cell serves a relatively small area, about two to three miles in diameter.

This small cell size means that the LMDS network requires a large number of antennas. As cellular and PCS industry experience has shown, this can be troublesome, since there are a limited number of places where antennas and hub equipment can be installed.

Local Multipoint Distribution Service Band Plan

LMDS is allocated 1300 MHz of radio spectrum. Prior to the opening of the 39 GHz band, this was the largest amount of spectrum ever assigned by the FCC to any single type of wireless service.

This is more than seven times the total combined amount of the cellular band (50 MHz) and PCS (120 MHz) band.

As illustrated in Figure 7-5, the LMDS band is allocated in two blocks in the 28 GHz and 31 GHz band. The FCC began auctions for the LMDS bands in 1998. Additional auctions continue to be announced from time to time. The spectrum auctions are broken into Block A Basic Trading Areas (BTAs) and Block B Basic Trading Areas (BTAs) following the same pattern as the PCS auctions discussed in Chapter 5.

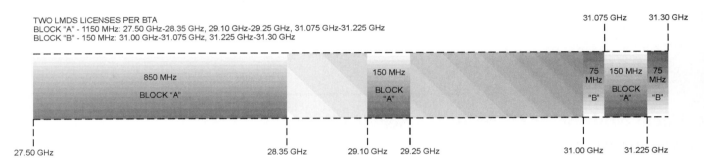

Figure 7-5: LMDS Band Allocation - 28 and 31 GHz Band Plan

Local Multipoint Distribution Service Capacity

The LMDS system capacity is defined as the total speed of transmission that is possible with a given amount of spectrum.

In reality, many variables can affect system capacity. Environmental features such as rain rate, line of sight, foliage, buildings, and terrain can impact system capacity. Physical features such as antenna height, available bandwidth, modulation scheme, and frequency reuse patterns also influence the total capacity.

The modulation choices are 4-QAM, 16-QAM, and 64-QAM. The available data rate is very dependent on modulation. For example, with 850 MHz of contiguous spectrum in the 28 GHz LMDS band, the system capacity can be estimated as follows:

Example 1:

> 64-QAM has a theoretical spectral efficiency of 5 bits/second/hertz, therefore:
>
> 850 MHz × 5 bits/second/hertz = 4250 Mbps of capacity

Example 2:

16-QAM has a theoretical spectral efficiency of 3.5 bits/second/hertz

850 MHz × 3.5 bit/second/hertz = 2975 Mbps of capacity

In actual performance, the total capacity is much lower. Higher level modulation requires a much higher signal-to-noise ratio (SNR). 64-QAM, for example, must have excellent link conditions capable of supporting a minimum of 10^{-12} to 10^{-14} bit error rates. This can be accomplished at the expense of much smaller cell sizes and the resulting higher system deployment cost.

LMDS total available capacity can be broken down between customers in nearly arbitrary fashion. If the uplink and downlink capacity is divided evenly, the service is classified as symmetrical (uplink and downlink bit rates are the same). The capacity is more often arranged asymmetrically (unequal up and down rates) with the higher capacity available on the downlink, which follows the model of most broadband wireless and wired Internet access technologies.

The preceding values assume a raw data rate capacity without any overhead. In reality, protocols are used to permit multiple users to access the channels in a time division duplex (TDD) mode or frequency division duplex (FDD) mode that reduces the net effective rate to some value less than the raw bit rate.

The 39 GHz Band Plan

The 39 GHz band is the highest frequency band currently licensed for use in wireless broadband applications.

With a wavelength of approximately 7 millimeters, several high-performance, highly directional antenna designs that support good frequency reuse are possible in this band. The 39 MHz band has great potential for broadband service with 1400 MHz of new spectrum available. It has been largely pioneered by Winstar, a microwave communications service corporation.

The technology for radio design at 39 GHz and above was still in the research stage in the late 1990s. Auctions for this band were completed in the year 2000 with the award of 2,173 licenses in various metropolitan business areas. The FCC allocation provides 1400 MHz, as shown in Figure 7-6. The band is broken into 14 paired segments of 50 MHz each. 700 MHz separates the forward and reverse links. The 39 GHz band is used for point-to-point and multipoint broadband service similar to LMDS. Due to the severe atmospheric attenuation in this band, cells are limited to 1 mile in radius or 3.14 square miles in area.

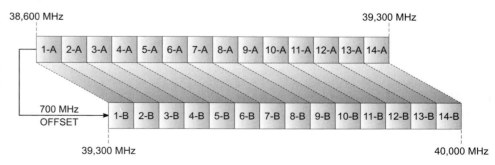

Figure 7-6: 39 GHz Band Plan

Wireless Local Loop

Local loop is the term used since the birth of the telephone system to describe the connection between the customer and the local office that provides switching center access to the PSTN.

The circuit is a closed loop of copper wire that connects the telephone to the local switch and provides dial tone and path for analog speech or modem connections from the home or business to the telephone switching center. Copper wires have been the primary media used to provide what is called the local loop "last mile" in the telephone network for over a century. Figure 7-7 depicts the local loop wired path between the telephone subscriber and the local office.

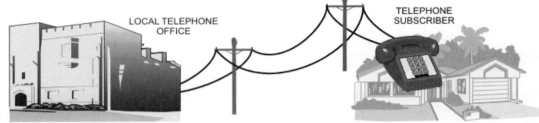

Figure 7-7: The Local Loop

Figure 7-8 illustrates a **Wireless Local Loop (WLL)** configuration, which provides businesses and private homes access to the PSTN using a radio link. WLL cells can range between 2 and 20 miles depending on range and individual product lines. WLL offers several advantages. In the U.S., we have a very high degree of penetration among the population for access to a PSTN. There are, however, many remote areas in the U.S. and other countries of the world where copper has not been installed in the ground for access to public telephone systems.

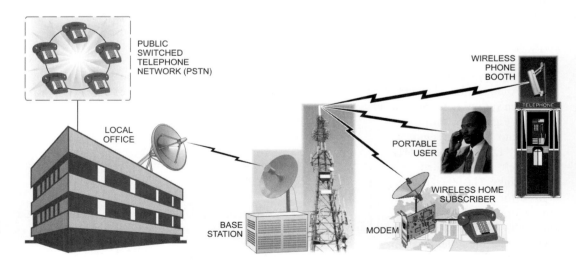

Figure 7-8: Wireless Local Loop Services

A WLL approach for voice services is an economical approach over copper wire for a number of reasons:

- Wireless local loop service can be installed much faster and at lower cost in remote areas.

- Wires installed in the ground or above-ground utility poles require access to terrestrial easements, private and public lands. The cost is often prohibitive based on population density and return on investment for the local exchange carriers (LECs) in developing countries and remote areas of the world.

- Broadband radio access technologies such as MMDS and LMDS can offer high data rate services for wireless Internet access at competitive rates for wireline access technologies such as DSL, cable, and ISDN.

- WLL technology and hardware has evolved from existing cellular designs and mature digital voice standards; therefore, there are no research and development startup costs for WLL product lines.

- WLL is not a mobile wireless technology. It is not designed for intercellular handoffs. This reduces the cost and complexity of the radio access interface and base station equipment.

Wireless Local Loop Radio Technologies

> WLL technology uses many of the radio access methods developed for cellular and broadband wireless access.

Both CDMA (ANSI-95) and TDMA (ANSI-136) standards are widely used by WLL vendors. Several companies also offer WLL-specific designs that are proprietary.

There are no standards developed specifically for WLL systems. Several WLL products are built upon the same technology used in cellular voice networks.

Broadband Free-Space Optical Systems

Free-space optical (FSO) systems are another class of broadband technologies. FSO uses free space as a transmission media for pulses of light waves. Free-space optical systems are so named to differentiate between fiber-optic broadband systems, where the light waves are enclosed in a cable, and systems that transmit light waves using "free space" as the medium.

Free-space optical
(FSO)

> Free-space optical systems use laser beams focused in a very narrow beam width between a transmitter source and a receiver.

Since a beam transmits in only one direction at a time, two links are usually provided for full-duplex service. They are configured for point-to-point communications. The components used are similar to in-the-ground fiber optic systems. The main advantages for free-space optical systems are as follows:

- They are capable of operating at extremely high data rates.

- Optical links can be set up on very short notice at minimal cost. They are ideal for emergency restoration of communications links during a disaster.

The Optical Spectrum

The optical spectrum and related bands used in optical free-space communications are illustrated in Figure 7-9. Most optical systems use the $780 - 900$ nanometer (nm = 1×10^{-9} meters) band and the $1500 - 1600$ nm band. The wavelength of the light is measured either in angstroms (1×10^{-10} meters) or nanometers. So 1 nanometer equals 10 angstroms. The 1300 nm band is avoided because it has poor propagation characteristics through the atmosphere.

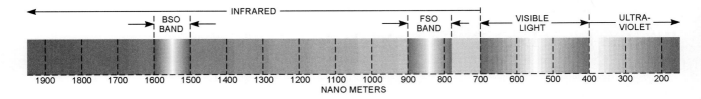

**Figure 7-9:
The Optical
Spectrum**

The FCC does not require a license for any type of electromagnetic wave for frequencies higher than 300 MHz (wavelengths less than 1 mm).

Optical System Capacity

All commercial optical systems use simple On-Off Keying (OOK) modulation, therefore the modulation scheme is not a factor in determining spectral efficiency as was the case with LMDS. Typical commercial systems on the market have a capacity of 1 Gbps. Most of the components used in free-space optical designs are similar to fiber-optic components.

Service providers are free to allocate optical bandwidth as the market and customer needs require. Most providers have backbone links running as high as 622 Mbps (OC-12), which they then feed off lower speed links to customers in the range of 20 Mbps to 155 Mbps.

Free-space optical links are suitable for relatively short distance communications caused by atmospheric absorption, rain, and fog.

Link separation is governed by the rate of divergence of the laser beam.

Key Points Review

In this chapter, the following key points were introduced concerning wireless broadband networks. Review the following key points before moving into the Review Questions section to make sure you are comfortable with each point and to verify your knowledge of the information.

- Broadband networks provide transmission of data at high speeds.

- Broadband networks operate in fixed locations. They are structured to transmit information between two points and often between multiple locations at very high transmission rates of 200 kbps to 30 Mbps using radio frequencies in the range of 2 GHz to 40 GHz.

- Fiber-optic cable transmission is the leading non-wireless broadband technology.

- Optical systems transmit data in the gigabit range.

- Satellites offer long haul broadband communications capability.

- Digital Subscriber Link (DSL) is a special wireline technology used to obtain more bandwidth over existing copper telephone lines serving a customer's premise.

- Integrated Services Digital Network (ISDN) is another technology that uses existing copper telephone lines to obtain high-speed digital service.

- Leased line service is a high data rate service provided by the regional Bell operating companies and competitive local exchange carriers. A T1 standard circuit provides 1.544 Mbps.

- Cable TV's coaxial connections to the home can deliver data at speeds up to 100 times faster than standard dial-up modems.

- "Fixed wireless" is a licensing term used by the FCC to define the radio service where transmitting stations are permanently installed (fixed) at specific geographical locations.

- MMDS historically has been used for one-way "wireless cable" video and instruction closed-circuit television service.

- The frequency bands assigned for the fixed wireless service occupy various portions of the spectrum from 2.1 to 39 GHz. Propagation characteristics are quite different across such a wide range of frequencies.

- At the higher frequency bands, propagation is affected by the weather.

- Multichannel Multipoint Distribution Service (MMDS) is a point-to-point and point-to-multipoint broadband microwave radio service. It utilizes frequencies in the 2.1 and 2.6 GHz bands.

- MMDS networks are structured to carry high-capacity data traffic between customer locations and a centrally located MMDS microwave gateway station.

- With appropriate terrain characteristics, a single MMDS cell can cover a 35-mile radius, or 3,850 square miles.

- MMDS requires line-of-sight for reliable quality of service.

- Several factors determine the cell size including terrain, type of modulation, access method (FDMA or TDMA), and quality of service goals.

- The MMDS, MDS, and IFTS spectrum was restructured and expanded after the Telecommunications Act of 1996.

- LMDS uses cells that are smaller than the cell sizes described for MMDS because of the higher atmospheric attenuation at 28 MHz compared to the 2.5 GHz band used by MMDS.

- A LMDS network can be composed of a series of cells that each deliver point-to-multipoint services to subscribers. Each transmitter in a cell serves a relatively small area, about two to three miles in diameter.

- LMDS is allocated 1300 MHz of radio spectrum. Prior to the opening of the 39 GHz band, this was the largest amount of spectrum ever assigned by the FCC to any single type of wireless service.

- The LMDS system capacity is defined as the total speed of transmission that is possible with a given amount of spectrum.

- The 39 GHz band is the highest frequency band currently licensed for use in wireless broadband applications.

- Local loop is the term used since the birth of the telephone system to describe the connection between the customer and the local office that provides switching center access to the PSTN.

- WLL technology uses many of the radio access methods developed for cellular and broadband wireless access.

- Free-space optical systems use laser beams focused in a very narrow beamwidth between a transmitter source and a receiver.

- The FCC does not require a license for any type of electromagnetic wave of frequencies higher than 300 MHz (wavelengths less than 1 mm).

- Service providers are free to allocate optical bandwidth as the market and customer needs require. Most providers have backbone links running as high as 622 Mbps. (OC-12), which they then feed off lower speed links to customers in the range of 20 Mbps to 155 Mbps.

- Free-space optical links are suitable for relatively short distance communications caused by atmospheric absorption, rain, and fog.

REVIEW QUESTIONS

The following questions test your knowledge of the material presented in this chapter:

1. Describe the fundamental differences between cellular networks and fixed service broadband wireless networks.

2. What technology allows multiple streams of data to be transmitted over the same fiber-optic cable?

3. What are the two most popular broadband wireless access technologies?

4. Explain how the weather can affect transmission in the MMDS and LMDS frequency bands.

5. What technology is being introduced in microwave transmission systems to improve performance and reduce the effects of multipath?

6. What process is used to grant licenses for operation of radio transmitting stations in the MMDS and LMDS frequency bands?

7. List five conditions that may affect download speeds from the Internet when using MMDS wireless access?

8. Explain why cell size is smaller for LMDS networks when compared to MMDS cell size.

9. Explain the relationship between modulation and cell size when calculating network capacity.

10. Calculate the total bandwidth capacity in 850 MHz of LMDS spectrum when using 16-QAM modulation.

11. What is the main operating constraint associated with the 39 GHz frequency band?

12. List five advantages for the use of wireless local loops compared to copper wire local loop technology.

13. Name two advantages of the use of free-space optical broadband systems.

14. What type of license is required for operation of a free-space optical transmission system?

15. What is the typical capacity of a backbone circuit using free-space optical technology?

CHAPTER
8

SATELLITE COMMUNICATIONS SYSTEMS

LEARNING OBJECTIVES

LEARNING
OBJECTIVES

Upon completion of this chapter and its related lab procedures, you should be able to perform the following tasks:

1. Describe the objectives of a satellite communications system.

2. List the three basic elements of a satellite system and the functions that each performs.

3. Describe the terms used for orbital element sets.

4. List the features of a geostationary satellite orbit.

5. Discuss the use of medium earth orbiting satellites for navigational applications.

6. List the features and uses for low earth orbits.

7. Recognize the difference between "big LEOs" and "little LEOs."

8. Identify the frequency bands used for satellite communications.

9. Identify the FCC documentation source for rules and regulations governing satellite communications.

10. Describe the services and applications for the three categories of satellite services listed in the FCC rules and regulations.

11. Describe how orthogonal polarization is used for frequency reuse in a satellite communications system. List the two main types of polarization.

12. Describe the main components of satellite bus architecture.

13. Compare the payload components and the bus components of a satellite.

14. Describe the two types of attitude control systems used by satellites.

15. Describe the licensing rules and regulations for satellite earth stations.

16. List the advantages of the use of parabolic dishes for satellite earth stations.

17. Calculate the antenna gain for a satellite earth station and describe the parameters that are used in the formula for antenna gain.

18. List the features that determine the beamwidth of a parabolic antenna. Show why beamwidth is an important factor in earth station design and installation.

19. Calculate the beamwidth of a satellite earth station antenna.

20. Identify the main services and business areas served by domestic satellites.

21. List the frequency bands used for the Mobile Satellite Service.

22. List the main reasons why LEO satellites are preferred over GEO satellites for satellite voice communications systems.

23. Describe why satellite constellations are used for low earth orbit systems.

24. List two examples of satellite systems using constellations of satellites.

25. Discuss the operational objectives of the radiodetermination satellite service.

26. List the operational differences between the U.S. Global Positioning System and the Russian Global Navigation Satellite System.

27. Describe the basic features of a satellite position and location system and why very accurate clocks are needed for the position determination solution.

28. Calculate the distance from the Earth to a satellite using the time-elapsed value and the speed of light.

CHAPTER 8

Satellite Communications Systems

INTRODUCTION

In 1945, Arthur C. Clarke, a RAF electronics officer and member of the British Interplanetary Society wrote an article in a publication called *Wireless World*. Clarke described a then futuristic idea of placing artificial earth satellites in orbit around the Earth. His vision predicted that satellites could be placed in orbit at an altitude of 22,300 miles where the satellites would circle the Earth in synchronism with the Earth's rotation, thereby appearing to remain stationary with respect to an earth station. Clarke envisioned three satellites equally spaced in orbit to provide a communications relay system around the world. Today, Clarke's prediction is a reality. Satellites are circling the Earth in geosynchronous orbit and carry video and high-speed data to virtually every corner of the Earth.

Satellites now serve in such roles as maritime safety and warning for ships at sea, voice communications for climbers on Mt. Everest, earth resources monitoring, global weather forecasting, and precise navigation position location anywhere on the Earth. World news is now relayed in real time to all countries of the world. Television viewers were able to view the Olympic games from Australia in progress via international satellite relay stations in space. Television programming with 200-channel digital service is now available from satellites 22,300 miles in space direct to home subscribers using small 18-inch receiving dish-type antennas.

In this chapter you will explore how satellites are launched, constructed, and sustained as a wireless communications systems in space. Different orbital configurations and their commercial applications are discussed. Earth station features are covered in detail with emphasis on antenna design and the requirements for antenna performance with geostationary satellites. Sample problems and exercises are included for calculating antenna gain and beamwidth.

The concept and purpose of launching satellites in low earth orbit are discussed. The types of satellite-based voice and messaging systems used for low earth orbit systems are also covered.

The 1970s saw the development and launch of a constellation of satellites by the U.S. Department of Defense and the former Soviet Union for providing precision location and navigation for military units. These systems now serve civilian, terrestrial, maritime, and aviation users in a variety of applications. Cellular telephone systems now obtain precision time for synchronizing the base stations from the U.S. Global Positioning System (GPS) satellites' on-board atomic clocks. Navigation satellite systems are being integrated with third generation digital cellular phone systems for aiding 911 calls for mobile users. In this chapter we will explore the basics of how satellite navigation systems determine position location for terrestrial users.

Elements of a Satellite Communications System

A satellite system consists of a radio repeater station in space that is revolving around the Earth in an earth orbit that links with a number of earth stations on the ground. The objective of a satellite system is to provide a communications link between two or more locations on the Earth that can cover longer distances than terrestrial microwave relay stations. As indicated in Figure 8-1, a satellite acts as a radio relay system with sufficient altitude to overcome the horizon limitations of a terrestrial microwave network.

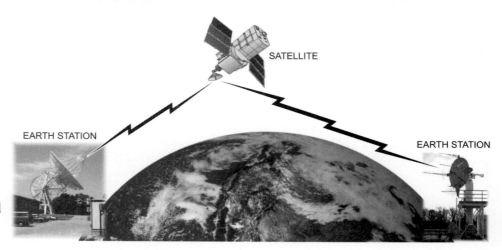

SATELLITE

EARTH STATION

EARTH STATION

Figure 8-1: Satellite and Earth Stations

Depending upon the altitude, a satellite can provide coverage over a very wide geographical area, avoiding the cost of several ground-based transmission facilities.

A satellite system consists of three basic system subsystems. They are the space segment, the earth segment, and the ground control segment.

space segment

- The **space segment** contains the satellite and all on-board systems necessary to complete a transmit and receive radio link with the ground stations. The satellite maintains communications with earth stations that may be transmitting voice, high-speed data, or video information. The satellite also uses other radio links to earth stations that monitor the status of equipment on board.

earth station segment

- The **earth station segment** contains all of the earth terminals that access the radio link on the satellite. Earth stations also provide gateways to terrestrial data and voice networks.

ground control segment

- The **ground control segment** maintains and monitors the health and status of the satellite. The control system may control one or more satellites depending on the type of orbit and network design features. Control stations may also be distributed in different geographical locations as necessary to have proper radio access to the satellites in a specific network.

Satellite Orbits

Satellites are placed in various orbits by rockets called **launch vehicles**. The amount of energy required to place a satellite in orbit depends on several factors including satellite weight, altitude of the required orbit, and the geographical location of the launch site. Satellites may be launched into earth orbits or, with sufficient energy, they may be launched to escape the Earth's gravity for outer space exploration. Several nations now have satellite launch facilities and provide commercial launch services for both commercial and military customers. Various launch vehicles used by different countries are shown in Figure 8-2.

launch vehicles

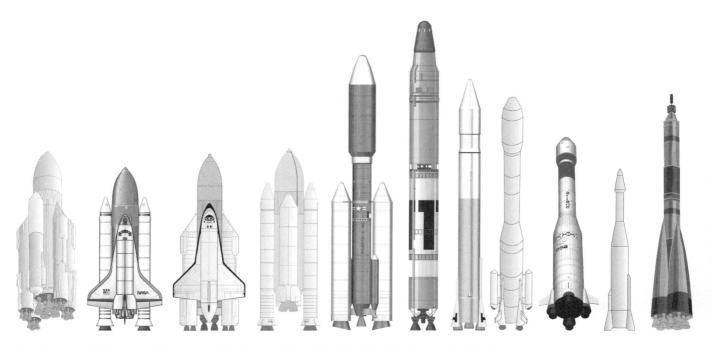

Satellites may be launched into a nearly infinite variety of earth orbits. Each type of orbit is planned in advance and selected to serve a specific application. Satellite motion can be described and predicted using a set of known mathematical laws of motion. Satellite orbital parameters are described using a set of mathematical values called the orbital element sets. Element sets are used by computers to catalog satellites and determine their exact three-dimensional location in space at any specific point in time. Element sets for all earth orbiting satellites are maintained by such agencies as NASA and the United States Space Command. Software programs designed to predict, update, and track the location of all satellites in earth orbit use the element sets associated with each satellite. The following section summarizes some of the terms used by satellite engineers to describe satellite orbits.

Figure 8-2: International Launch Vehicles

Satellite Orbit Terminology and Definitions

As shown in Figure 8-3, satellites may be launched into various orbits. They may circle the Earth following either a circular or elliptical path. Orbits may be tilted or inclined with respect to the equator. If the inclination angle is zero, the orbit follows the equator. Some orbits may be perpendicular to the equator. This type of orbit is called a polar orbit.

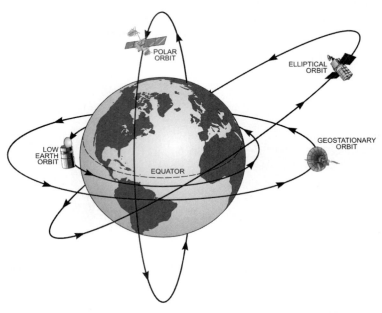

Figure 8-3: Types of Earth Orbits

Figure 8-4 shows the closest approach of the orbit to the Earth's surface, called the **perigee**, and the farthest point of the orbit, called the **apogee**. If the apogee and perigee points are equal, the orbit is circular. If they are different, the orbit is an ellipse. When an inclined satellite orbit crosses the equator going from south to north, the point is called the **ascending node**. When the satellite orbit crosses the equator going from north to south, the point is called the **descending node**.

perigee

apogee

ascending node

descending node

Figure 8-4: Inclination Angle, Apogee, and Perigee showing Ascending and Descending Nodes

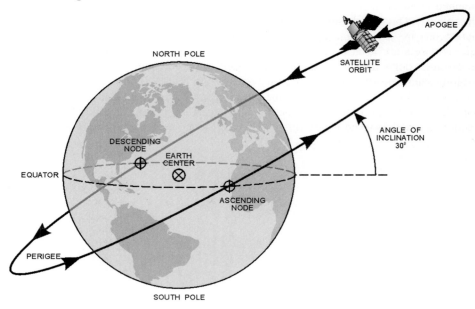

After the main features of the orbit are defined, a value is needed for the "size" of the orbit. The larger the orbit, the higher the altitude or distance from the Earth and the longer the orbital period. Since the third law of motion, established by Johann Kepler (1571–1630), proved a precise relationship between the satellite speed and its distance from the Earth we can define the size of the orbit if we know the speed. This parameter in satellite language is called **Mean Motion**. High-altitude satellites go slower than low-altitude satellites with a known inverse mathematical relationship between speed and altitude.

Mean Motion

The geometry is now defined by the preceding definitions for the satellite orbit in three-dimensional space. The only parameter left to define is the location of the satellite in the orbit at a point in time. The Epoch is like a snapshot in time, which is the "tick" mark for when to determine the satellite location. The **Mean Anomaly** is the term then used to specify where the satellite is located on the orbital path at Epoch time.

Mean Anomaly

According to Kepler's laws of motion, a satellite's orbital plane, which is an imaginary flat surface connecting all points of the orbit, always passes through the center of the Earth.

In the absence of any external disturbing forces, a satellite will maintain the same orbit forever. In reality, several small forces act on an orbiting satellite to change its orbit. If the satellite is in low orbit, a very small atmospheric drag may eventually cause a satellite to slow and fall to earth. The sun and moon also cause the orbit to be altered slightly. These forces are collectively called orbital perturbations.

Geostationary and Geosynchronous Earth Orbits

Our earthbound clocks consider the Earth's rotation to be measured relative to the sun's (mean) position. 24-hour days are based upon a mean solar day. However, the mean solar day is not the rotational period of the Earth that we're interested in. A **geosynchronous** satellite completes one orbit around the Earth in the same time that it takes the Earth to make one rotation in inertial (or fixed) space relative to the stars. This time period is known as one sidereal day and is equivalent to $23^h56^m04^s$ of mean solar time.

geosynchronous

For a satellite's orbit period to be geosynchronous, it must be approximately 35,786 kilometers (19,323 nautical miles or 22,236 statute miles) above the Earth's surface. To stay over the same spot on Earth, a geostationary satellite also has to be directly above the equator at an altitude of 35,786 km, as illustrated in Figure 8-5; otherwise, from the Earth the satellite would appear to move in a lazy figure 8, slowly drifting in a north-south path over the equator once per day. Figure 8-5 illustrates a geosynchronous satellite whose orbit is directly over the equator. This type of orbit in addition to being geosynchronous is also **geostationary**. It appears to remain in a fixed position when viewed from a fixed earth station. A geostationary satellite can be utilized by ground stations within the coverage area between 76 degrees north and south latitudes. A geostationary satellite can cover approximately 38% of the Earth's surface. From the geostationary position, the Earth's disk subtends an angle of 17.4 degrees.

geostationary

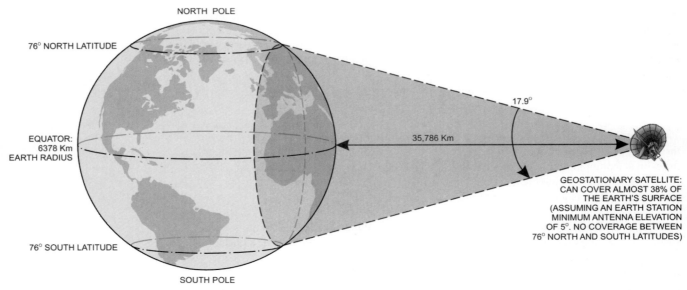

NORTH POLE

76° NORTH LATITUDE

EQUATOR:
6378 Km
EARTH RADIUS

76° SOUTH LATITUDE

SOUTH POLE

17.9°

35,786 Km

GEOSTATIONARY SATELLITE:
CAN COVER ALMOST 38% OF
THE EARTH'S SURFACE
(ASSUMING AN EARTH STATION
MINIMUM ANTENNA ELEVATION
OF 5°. NO COVERAGE BETWEEN
76° NORTH AND SOUTH LATITUDES)

**Figure 8-5:
Geostationary
Satellite Earth
Coverage**

Uses of Geostationary Orbits

Geostationary satellites are widely used throughout the world for a variety of applications. Earth stations can be installed with fixed mounts pointed to the satellite's position. The FCC issues licenses for the Fixed Satellite Service (FSS) for both space systems and earth stations. The term "fixed" is used by the FCC to distinguish a class of terminals that are licensed for use at a permanent (fixed) geographical location. Earth stations that are receive only (RO) type terminals do not require a FCC license. Typical applications for geostationary satellites include:

- Long distance video and digital data communications

- Weather and earth resource monitoring

- Direct to home television subscription service

- High-speed Internet access

- Network and cable channel broadcast television programming distribution to ground network television broadcasting affiliates and cable head ends

- Military secure communications networks

- Marine safety

- Intercontinental communications

Because of their higher altitude, geostationary satellites are relatively expensive to launch. The long distance round trip delay between transmissions from earth stations, even at the speed of light, adds up to 1/4 of a second. Satellite and ground station processing delays can add another 1/2- to 3/4-second delay. This delay is referred to as latency and can cause problems for voice communications and some types of terrestrial network communications protocols.

Geostationary Orbital Slots

Geostationary orbital slots are assigned on a global basis according to the longitudinal position above the equator. In the past, geostationary satellites have been positioned at 3-degree intervals along the orbital ring around the earth. The FCC now permits satellites to be placed into slots separated by only 2 degrees over the U.S. As an example, Figure 8-6 shows a portion of the geostationary ring with 2-degree longitudinal spacing. Licensing and coordination of orbital slot allocations among nations is handled jointly by the International Telecommunications Union (ITU) and the U.S. Federal Communications Commission. In response to the huge demand for orbital slots, the FCC and ITU have progressively reduced the required spacing down to only 2 degrees for C-band and Ku-band satellites. This places additional demand on the design of earth stations for precise pointing accuracy and mechanical stability.

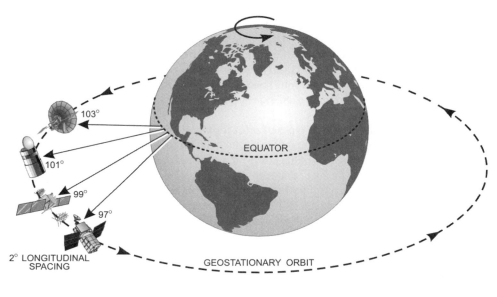

Figure 8-6: Geostationary Orbital Slots

Medium Earth Orbits

Medium Earth Orbits (MEO) range from any orbit less than synchronous to a few hundred miles in altitude.

MEO satellites are used in groups called constellations to provide specific services. The United States Global Positioning System (GPS) and the Russian Global Navigation Satellite System (GLONASS) navigation satellite systems orbit the Earth at altitudes of 12,548 miles and 11,865 miles, respectively. Both systems employ 24 satellites in various orbital planes to provide uniform global coverage. MEO orbits provide coverage with a fixed minimum number of satellites always visible at any time, anywhere on the Earth. This feature is favored for satellite navigation systems where global uniform coverage is essential for proper position determination accuracy.

Low Earth Orbits

Low earth orbit (LEO)

Low Earth Orbit (LEO) satellites circle the Earth at altitudes typically between 300 miles and 800 miles above the earth. LEO orbits are used for mobile satellite phone technology as well as paging and message delivery systems.

The FCC calls this class of service the Mobile Satellite Service (MSS) because it is licensed for mobile users. The earth stations other than gateway stations are intended for mobile users using portable handsets. MSS provides cellular type service on a global basis without the limitations of regional terrestrial cellular networks.

LEO systems require several satellites to provide continuous coverage because each satellite is able to "see" a smaller area of the Earth than higher altitude satellites. The coverage area of a satellite is called the "footprint" although much smaller than the GEO satellite footprint. MSS low earth orbit satellites are like cells moving over the users. A constellation of satellites between 45 and 66 satellites is typical for complete earth coverage. Satellites are distributed in several planes with inclinations spaced properly to offer the best coverage. Each satellite and its cell of coverage is used to relay calls to gateway stations.

Low earth orbits are the preferred types of orbit for mobile satellite phone and paging systems. A primary advantage of LEO satellites is that transmitting terminals on the Earth don't have to be very powerful because of the low orbit. LEOs also are smaller than the large GEO satellites, less costly, and cheaper to launch. Although more LEOs are needed to support telecommunications, the propagation delay is considerably less. The major advantages claimed by the LEO service providers is that telephone calls will be possible from handheld mobile units to any place on the Earth, just like a global cellular network.

Benefits of the Use of the Low Earth Orbit

LEO satellites enjoy several advantages over other types of orbits for satellite telephony and non-voice messaging systems:

- LEO satellites can offer mobile telephone service to areas where there is insufficient population to justify a terrestrial based cellular network. This includes not only many developing countries but 80% of the U.S. as well. This explains why most of the initiatives for LEOs have come from the United States.

- Many developing countries are interested in LEO systems as an alternative to investing in expensive terrestrial telephone systems.

- Communication via LEOs does not suffer from the objectionably long transmission delays for voice communications associated with geosynchronous systems.

- Satellite phones do not require high-power transmitters or highly directional antennas that need to be continually pointed toward the satellite. In practice, transmit powers can be much lower than 1 watt.

- Satellites in low earth orbit are technically less complex and more robust than geosynchronous satellites.

Polar Orbits

Polar orbits

Polar orbits circle the Earth with an inclination of near 90 degrees. As the name implies, polar orbits circle the Earth in a north-south direction at right angles to the Earth's rotation passing over the poles.

> Satellites in polar orbit are used for search and rescue, mapping, military surveillance, and environmental monitoring of the Earth's surface. A polar orbiting satellite can eventually view the entire surface of the Earth.

Polar orbits can be adjusted to scan successive areas of the Earth during each orbital pass. The satellite circles the poles while the Earth turns below, as shown in Figure 8-7. Polar orbits may be used at various LEO or MEO altitudes depending on the specific mission.

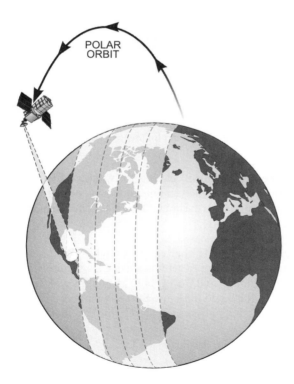

Figure 8-7: Polar Orbit

Satellite Frequency Bands

Frequency bands for satellite communications are defined in the Federal Communications Commission (FCC) Code of Federal Regulations, Title 47, Part 25. Satellite communications reach beyond the borders of nations, therefore international cooperation is necessary to avoid harmful interference. This is accomplished by international bodies such as the ITU Radio Regulation Board (RRB) and the International Telecommunications Union (ITU).

Part 25 of the FCC rules and regulations for satellite communications is partitioned into three types of service, as illustrated in Figure 8-8. The FCC rules call for spectrum allocations for the Fixed Satellite Service (FSS), the Mobile Satellite Service (MSS), and the Radiodetermination Satellite Service (RDSS). Newer rules and regulations for the Direct Broadcast Service (DBS) in the U.S. are contained in Part 100 of the FCC rules.

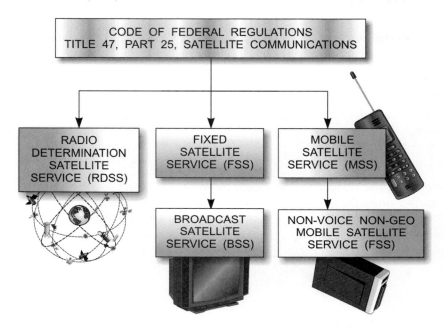

**Figure 8-8:
Satellite Communication
Services**

The **Fixed Satellite Service (FSS)** is allocated a large share of the radio spectrum because this type of service is the largest business segment of the satellite industry.

The Broadcast Satellite Service (BSS) is used for what is called direct broadcast service (DBS). The BSS delivers "cable type" commercial television programming to subscribers with small receive only terminals in the home. The BSS shares frequencies with the fixed satellite service.

The fixed satellite service uses different frequency bands for the uplink and downlink radio channels. This is similar in concept to cellular systems, which require separate frequencies for Frequency Division Duplex (FDD) operation. The uplink and downlink band plan for the fixed satellite service is illustrated in Figure 8-9. Some of the frequency bands are used in international service and other portions are used in the U.S. domestic satellite service.

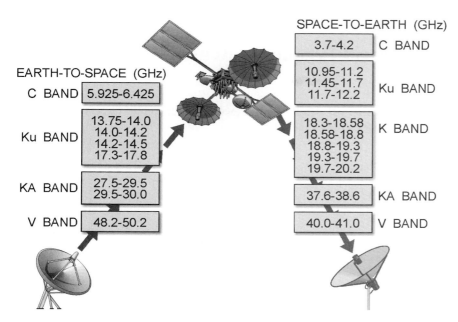

EARTH-TO-SPACE (GHz)

C BAND	5.925-6.425
Ku BAND	13.75-14.0 14.0-14.2 14.2-14.5 17.3-17.8
KA BAND	27.5-29.5 29.5-30.0
V BAND	48.2-50.2

SPACE-TO-EARTH (GHz)

3.7-4.2	C BAND
10.95-11.2 11.45-11.7 11.7-12.2	Ku BAND
18.3-18.58 18.58-18.8 18.8-19.3 19.3-19.7 19.7-20.2	K BAND
37.6-38.6	KA BAND
40.0-41.0	V BAND

Figure 8-9: Fixed Satellite Service Band Plan

The **Mobile Satellite Service (MSS)** has two main categories of spectrum allocations. The first is for satellite mobile phone services and the second class of service is for non-voice systems such as paging and messaging services. Each of these frequency band allocations is intended for use with low earth orbit (LEO) satellite systems. The satellite-based voice systems are often labeled "big LEOs" and the narrow-band non-voice satellite systems are referred to as "little LEOs." Table 8-1 shows the FCC-allocated uplink and downlink spectrum plan for mobile satellite services.

Table 8-1: Mobile Satellite Band Plan

Non-Voice Non-Geostationary (Little LEO's)	
Earth-to-Space	148 - 149.9 MHz 149.9 - 50.05 MHz 399.9 - 400.95 MHz
Space-to-Earth	137 - 138 MHz 400.15 - 401 MHz
Voice Non-Geostationary (Big LEO's)	
User-to-Satellite	1610 - 1626.5 MHz
Satellite-to-User	1613.8 - 1626.5 MHz (secondary) 2483.5 - 2500 MHz

The **Radio Determination Satellite Service (RDSS)** uses radio spectrum in the L and S bands. This service is used for position location and navigation using a constellation of satellites and small, portable earth stations. Examples of RDSS systems are explained later in this chapter.

Satellite Band Frequency Reuse

C-Band and Ku-band satellites use several transponders (receiver-transmitter chains) each with a bandwidth of 54 MHz or more.

With 500 MHz of satellite bandwidth in the 6/4 GHz or 14/12 GHz band, 8 transponders can be provided each using 54 MHz of bandwidth with a center-to-center spacing of 61 MHz. Satellites employ frequency reuse schemes to increase the number of transponders to $2 \times 8 = 16$ transponders using a frequency reuse factor of 2. Some satellites carry both C band and Ku band transponders. Frequency reuse can be accomplished with polarized feed assemblies, where one set of transponders uses one polarization, and the other set reuses the same frequency band using a cross (orthogonal) polarization. Isolation of the two polarizations can be maintained at 30 dB or more by staggering the center frequencies of each of the cross-polarized transponders. Two types of orthogonal polarization are used. Linear polarization uses vertical and horizontal polarized feed assemblies. Circular polarization is transmitted in a helical rotating pattern, with right-hand circular polarization (RHCP) rotating in a clockwise direction as seen from the satellite, and left-hand circular polarization (LHCP) signals rotating in a counterclockwise direction. A graphical representation of each orthogonal polarization method is given in Figure 8-10.

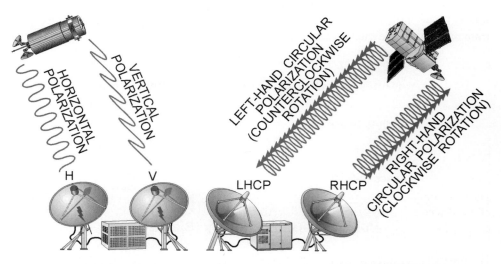

Figure 8-10: Frequency Reuse Using Orthogonal Polarization

Satellite System Elements and Design Features

Satellite spacecraft designs vary depending on the tasks to be performed. All satellites use a basic component architecture with the subsystem components similar to those outlined in Figure 8-11.

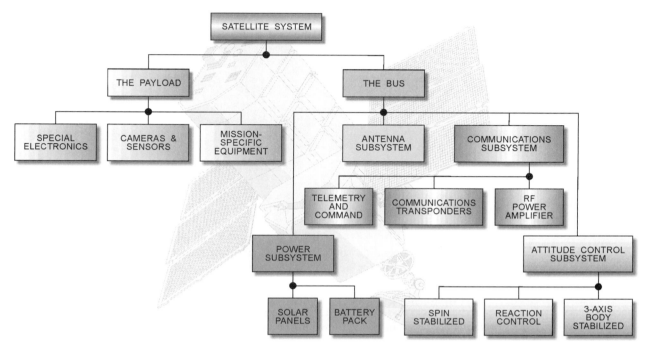

Figure 8-11: Satellite System Architecture

The Bus

The bus is designed to carry the payload. It contains all the components necessary for control of the spacecraft. These subsystems include a minimum of four important functions. They are **power**, **attitude control**, **antenna**, and **communications**.

The **power subsystem** provides all electrical power for the onboard components such as the communications transponders, computers, payload electronics, cameras, attitude control, reaction control thrusters, and the telemetry and command radio transmitters and receivers. Solar panels collect power from the sun and charge the onboard batteries. The batteries may be used when the satellite is in the eclipse phase of an orbit.

The solar panels may be arranged on large extended panels, as shown in Figure 8-12, or wrapped around the cylinder of the satellite, as illustrated in Figure 8-13.

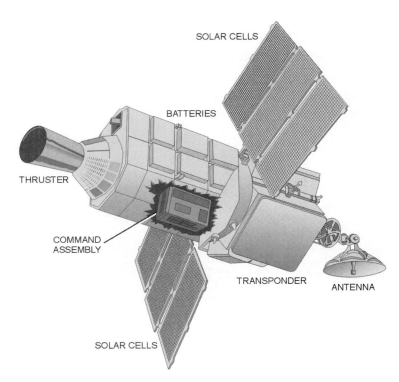

Figure 8-12:
Three-Axis Body
Stabilized Satellite

on-orbit attitude
control subsystem

All satellites depend on some type of **on-orbit attitude control subsystem** to keep the antenna pointed towards the Earth. There are two types of attitude-controlled satellites. They are spin-stabilized and three-axis body-stabilized satellites. Spin-stabilized satellites act as a large gyroscope, maintaining stability by rotating about their axis of maximum moment of inertia. This cylinder-shaped satellite has a spinning section and a de-spun section that is kept stationary by counter rotation so that the antennas mounted on it are kept pointing earthward.

A typical spin-stabilized satellite is shown in Figure 8-13. The spin-axis orientation is determined by on-board sensors that can determine the location of the Earth using an infrared device. The dual sensors each determine the edge of the earth horizon against the colder temperature of space. The spin axis is then focused on equal sides of the horizon, which keeps the spin axis centered on the Earth.

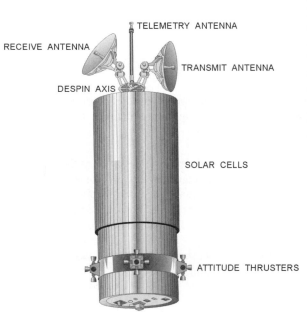

Figure 8-13: Spin-Stabilized Satellite

Three-axis body-stabilized satellites are controlled about their three axes. They are the yaw, pitch, and roll axes, illustrated in Figure 8-14. The satellite is aligned with the local vertical and the normal to orbit plane.

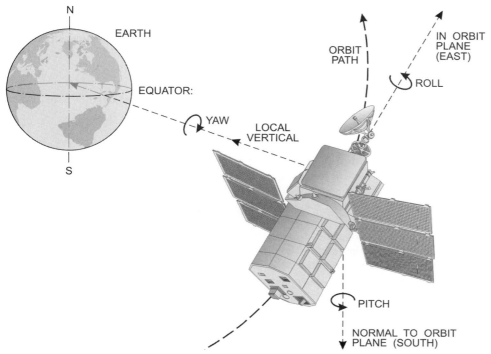

Figure 8-14: Satellite, Yaw, Pitch, and Roll Axes

Antennas are mounted facing the Earth, as illustrated in Figure 8-15. A typical three-axis body stabilized satellite is shown in Figure 8-12.

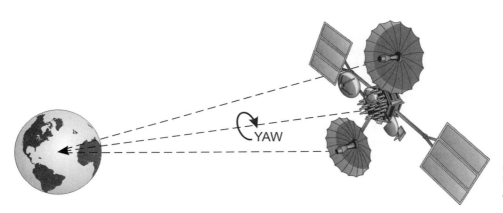

Figure 8-15: Antenna Axis Facing the Earth

The **reaction control** components are the thrusters and propulsion fuels used to correct the satellite orientation when commanded by the ground satellite control facility. The thrusters are needed occasionally to correct the orbit when orbital perturbations occur. Geostationary satellites are also equipped with an apogee kick motor that is used to propel the satellite from the lower intermediate altitude to the final higher geostationary orbit during the final phase of the launch sequence.

The **antenna subsystem** for a communications satellite includes a receiving antenna for capturing the uplink signal and a transmitting antenna for transmitting the amplified uplink signal back to the earth. Some satellites may use several antennas for dual band operation. Communications satellites may employ steerable spot-beam antennas for placing the uplink and downlink footprints upon selected areas of the Earth. Non-communications satellites, such as weather, environmental research, military surveillance, and navigation satellites, may receive commands from the earth station, but do not function as repeater stations. All satellites are configured with command and telemetry antennas for maintaining full-duplex communications links with ground control stations.

Antennas are designed to provide cross polarization reception and transmission as explained earlier under the section on satellite frequency bands.

The **communications subsystem** consists of a repeater system that receives on one band of frequencies and transmits back (transponds) to the Earth on another. As an example, a Ku band repeater would receive signals in the uplink frequency band of 14 – 14.5 GHz. After amplification and conversion steps, the signal would be transmitted on the downlink in the frequency band of 11.7 – 12.2 GHz.

The downlink signal is amplified by a power amplifier prior to being passed to the downlink antenna feed system. The satellite amplifies all channels by approximately 60 dB prior to forwarding the information on the downlink. The telemetry and command system continually transmits to the earth all information needed by the ground controllers, such as the amount of thruster fuel remaining, satellite attitude, battery condition, temperature, power consumption, and payload status. The command system controls the satellite at all times by receiving and decoding the commands received from the ground station.

The Payload

The **payload** consists of all equipment that has been installed on the satellite for the performance of a specific application. If the satellite were designed as a communications repeater, the payload would consist of a series of transponders. Payloads for weather satellites could contain cameras for taking pictures of cloud formations, infrared sensors, and various instruments for observing weather conditions on the Earth. Other on-board systems may include instrumentation for earth resources monitoring or high-resolution mapping. Direct broadcast satellites are equipped with special spot-beam antennas and high-power amplifiers for delivery of digital television programming for cable head ends or consumer direct broadcast satellite (DBS) reception.

Satellite Earth Stations

In addition to licensing the space station segment of the satellite system, the FCC is responsible for licensing earth stations.

Satellite earth stations provide the radio link between the Earth and the satellite.

Earth stations are divided into two broad categories. One type provides both radio transmitter and receive modules for communicating in full-duplex mode between the ground station and the satellite. The other types are receive only (RO) type terminals. RO terminals do not require a FCC license.

> The FCC defines an **earth station** as a complex of transmitters, receivers, and antennas used to relay and/or receive communications traffic (voice, data, and video) through space to and from satellites in both geostationary orbits and non-geostationary orbits.

earth station

The predominant frequency bands shown earlier in this chapter for earth station transmissions are the C-band, Ku-band, K-band, and Ka-band for fixed satellite services, and the 1.6/2.4 GHz (L and S) bands and the 137-138/148-149.9 MHz (UHF) bands for mobile satellite service.

The FCC rules for earth station licensing are contained in part 25 of Title 47 of the Code of Federal Regulations. Several classes of earth stations are specified in the FCC rules. They are:

- Fixed earth station (transmit/receive)

- Temporary-fixed earth station (non-permanent, transportable, transmit/receive)

- Fixed earth station (receive-only)

- Fixed earth station (VSAT network, 12/14 GHz)

- Developmental earth station (fixed or temporary-fixed)

- Mobile earth station (handheld units and vehicle-mounted units)

The FCC also issues one blanket license for a large number of technically identical earth stations. This is easier to administer for terminal systems such as Very Small Aperture Terminals (VSATs) and Satellite News Gathering (SNG) vans used for remote site programming by broadcast television stations.

Earth Station Antennas and Feed Assemblies

Satellite ground stations use the "dish" type of antenna shown in Figure 8-16. Dish antennas are so named because of the bowl-shaped appearance of the reflector. The size of the antenna is represented by the diameter (d). The reflector surface is designed in the shape of a parabolic curve. The parabolic curve has the property of reflecting all incident rays arriving along the reflector's axis of symmetry to a common focus located to the front and center of the dish. This is the location of the antenna feed point. The feed point is where the incoming waves are focused into a device commonly called a feed horn or feed assembly. The parabolic dish is also used for transmitting radio signals up to the satellite. The feed horn illuminates the parabolic reflector from the feed assembly and radiates the energy into a beam-shaped pattern along the axis of symmetry, as shown in Figure 8-16. A flashlight reflector is a common type of parabolic surface.

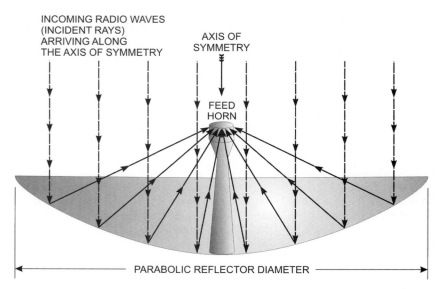

INCOMING RADIO WAVES
(INCIDENT RAYS)
ARRIVING ALONG
THE AXIS OF SYMMETRY

AXIS OF
SYMMETRY

FEED
HORN

PARABOLIC REFLECTOR DIAMETER

Figure 8-16:
Parabolic Antenna
Reflector and Feed Point

The parabolic antenna's ability to amplify signals is primarily governed by the accuracy of this parabolic curve. Poor antenna performance can result from imperfections or distortions in the reflector surfaces during manufacture. It is also important that the surface of the reflector is not warped or damaged in any way during the installation. Antenna efficiency is represented in design formulas as the eta factor. A perfect antenna would have $\eta = 1$. This as rarely achieved in practice. One or more of the following characteristics can affect the antenna efficiency.

- Main reflector illumination efficiency

- Spillover efficiency

- Phase efficiency

- All structures that support the feed assembly and cause blockage of the aperture

- Feed system dissipation loss

- Impedance mismatch (VSWR) loss due to reflections at the feed port

- Tolerance efficiency of the reflector surface

Each of the above features will vary according to the type of feed assembly used, size of the reflector, and cost. Typical efficiencies range from $\eta = .5$ to .6 with .55 being typical. Efficiency affects the total gain but not the beamwidth. The beamwidth of the antenna is governed by antenna dimensions and frequency. Antenna gain is also governed by antenna dimension, frequency, and the efficiency factor (η). The formulas and design criteria for prime focus parabolic antennas and aperture efficiencies are complex. The formulas used in the sample problems in this section assume an efficiency (η) of .55.

Parabolic antennas are ideal for satellite communications for the following reasons:

- They have very high gain at the higher microwave frequencies used by satellites. High gain performance is necessary to provide the required uplink and downlink power for the long haul to and from a satellite.

- Parabolic antennas can be steered using the proper mounts to accurately follow a satellite if necessary or to position the antenna along the geostationary arc.

Calculating Antenna Gain

Antenna gain can be expressed as the relative increase in power achieved by focusing the radio energy in a particular direction.

Gain is determined by using an isotropic antenna as an ideal theoretical lossless antenna that radiates power equally well in all directions. An antenna that focuses the radiated energy in a narrow beam is then compared to the isotropic antenna. The difference is called antenna gain.

As previously stated, parabolic antenna gain is determined by the operating frequency, efficiency, and antenna size. The following formula can be used to calculate antenna gain with an efficiency of .55:

$$G = 60.7 f^2 d^2$$

where:

f = frequency in GHz
d = antenna diameter in meters

Example problem:

Calculate the gain of a parabolic dish antenna with an efficiency of .55, an operating frequency of 11.2 GHz, and an antenna diameter of 3 meters.

$$G = 60.7 f^2 d^2$$

$$G = 60.7 \times (11.2)^2 \times (3)^2$$

$$G = 68,527.872$$

An antenna with the above operating parameters would provide over 68,000 times as much power in a narrow beam than an isotropic antenna with a gain of 1. Antenna gain is expressed in dB rather than the actual power ratio; therefore, the gain in dB becomes:

$$G \text{ (dB)} = 10 \log \frac{P_1}{P_2}$$

$$G \text{ (dB)} = 10 \log \frac{68,527.872}{1} = 48.35 \text{ dB}$$

Calculating Antenna Beamwidth

Antenna beamwidth is another parameter that is related to antenna gain, antenna diameter, and operating frequency. Antenna beamwidth is measured at the half-power points of the main beam, as illustrated in Figure 8-17.

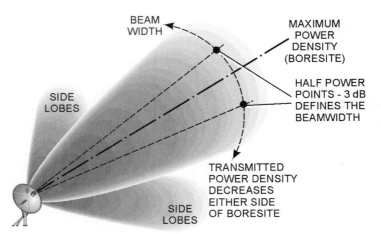

**Figure 8-17:
Antenna Beamwidth**

A line drawn from the center of the reflector into space is called the boresite of the antenna. The antenna radiation pattern can be viewed as a cone with the vertex at the antenna surface. The angle of the arc formed by the cone is called the antenna beamwidth. The radiated power drops off on either side of the boresite. Beamwidth is measured at the points on either side of the boresite where the power drops off to one-half the power density (−3 dB) at the center. Other energy fields are called side lobes. Beamwidth can be calculated using the formula:

$$BW = \frac{70\lambda}{d}$$

where:

λ = wavelength in meters
d = antenna diameter in meters

Example problem:

Calculate the beamwidth of a 3-meter parabolic dish satellite antenna operating with an uplink C-band frequency of 3.95 GHz.

Converting f to λ:

$$\lambda = \frac{300}{f(\text{MHz})} = \frac{300}{3950} = .0759 \text{ meters}$$

$$BW \text{ (degrees)} = \frac{70\lambda}{d}$$

$$BW = \frac{70 \times .0759}{3} = 1.77 \text{ degrees}$$

Example problem:

Determine the beamwidth for a 3-meter parabolic dish antenna when the frequency is changed to a Ku-band uplink frequency of 11.2 GHz.

Converting f to λ:

$$\lambda = \frac{300}{f(\text{MHz})} = \frac{300}{11,200} = .0268 \text{ meters}$$

$$\text{BW (degrees)} = \frac{70\lambda}{d}$$

$$\text{BW} = \frac{70 \times .0268}{3} = .625 \text{ degrees}$$

Antenna beamwidth is important because, as mentioned earlier, geostationary satellites in the United States are separated by only 2 degrees in the orbital arc. If the beamwidth were wider than 2 degrees, the antenna would "see" more than one satellite.

Figure 8-18 shows the dramatic reduction in beamwidth that occurs when using the higher Ku-band frequency with a 3-meter antenna aperture. Also shown is the degradation in performance that would occur using a 1-meter antenna for a C-band uplink. The 1-meter parabolic antenna would see three satellites at the same time (with 2-degree spacing) due to the wider beam width of a 1-meter versus a 3-meter antenna. This would be undesirable since adjacent satellites could encounter interference from transmissions in the same frequency band intended for the satellite located in the boresite orbital slot. Satellite earth stations must therefore be carefully designed to meet the required gain, aiming accuracy, platform stability, and beamwidth for the selected frequency band in use.

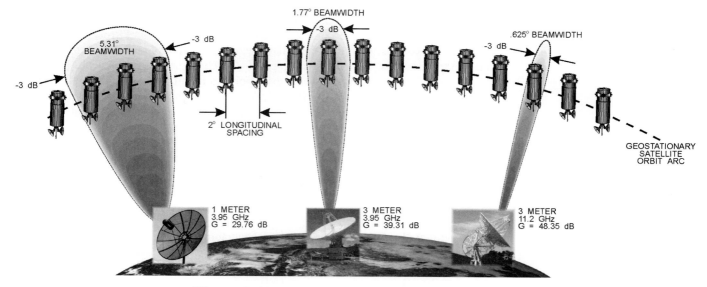

Figure 8-18: Antenna Beamwidth Comparisons

Antenna Feed Polarizers

As mentioned earlier, satellites often use dual polarization to maximize frequency reuse. This allows two independent radio carriers to use the same frequency band. One carrier uses horizontal polarization, while the other carrier using the same frequency uses vertical polarization. This permits frequency reuse and doubles the satellite capacity.

C-band satellites used in International service, such as the INTELSAT fleet of satellites, typically use circular polarization. For the best possible reception of circularly polarized satellite transmissions, a feedhorn is required that is designed to receive this type of signal from the satellite. Although standard linear vertical and horizontal polarized feedhorns can still pick up any circular polarized signal, the signal will be severely degraded. Several manufacturers offer special feedhorns that can receive both the linear and circular polarization formats. Many linearly polarized feedhorns also can be modified to receive circularly polarized signals.

Other Satellite Services

Most satellite communications are maintained throughout the world using the geostationary fixed satellite service. There are also other services that are important and are discussed in the following paragraphs.

Domestic Fixed Satellite Service

Domestic C-band 4/6 GHz satellites fall into three categories based on the markets they serve. "Cable" satellites distribute television programming to cable head ends to homes equipped with backyard Television Receive Only (TVRO) "dishes." "Broadcast" satellites distribute network programming to affiliates and syndicated programming to affiliates and independent stations. The third category of 4/6 GHz satellites is generally used for point-to-point transmission (as opposed to the point-to-multipoint operation of the other categories) of video and data signals. Most new domestic satellites use the higher frequency Ku Band for VSAT networks, broadcast television, and digital audio entertainment.

Mobile Satellite Service

Mobile Satellite Service (MSS) is a service for mobile subscribers using cellular-type handsets communicating with Low Earth Orbit (LEO) satellites.

The FCC has allocated spectrum for two classes of MSS described as follows:

- **Non-voice non-geostationary service** operating in the VHF and UHF frequency bands. This service is often referred to as "little LEOs" or small, lightweight, low-cost satellites. Little LEOs are designed to provide messaging, paging, store and forward packet networks, vehicle tracking, and location and point. Systems have been proposed using LEO constellations consisting of 36 to 48 small satellites.

- **Non-geostationary voice service** operating in the L and S bands. These systems are referred to as "big LEOs." They are designed to support cellular-type voice service in locations where regular terrestrial cellular service is not available. Users can communicate directly to a LEO satellite that relays the call to a terrestrial gateway station for access to the PSTN. They provide cellular-type voice communications on a global scale covering all areas of the world. Two big LEO systems have been declared operational and contain constellations of 48 and 66 satellites. User calls are handled by satellites passing over the coverage area in a "moving cell" type pattern, as shown in Figure 8-19.

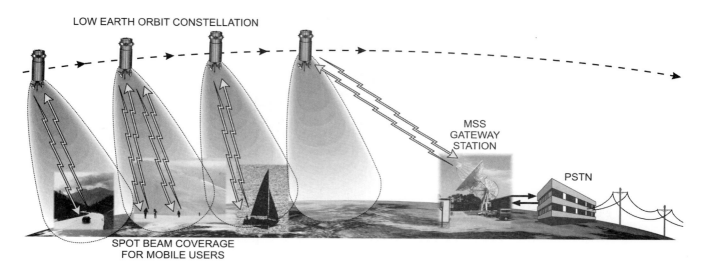

LOW EARTH ORBIT CONSTELLATION

MSS GATEWAY STATION

PSTN

SPOT BEAM COVERAGE FOR MOBILE USERS

Figure 8-19: Mobile Satellite Service Voice Network

Radiodetermination Satellite Service

Radiodetermination satellite service (RDSS) is a term used in the radio regulations to describe a service that uses satellites for navigational purposes. The FCC has allocated radio spectrum in the L and S bands for the RDSS.

The United States and Russia each operate satellite systems that provide position location information. The United States Department of Defense (DOD) operates and maintains the Global Positioning System (GPS). Russia operates a similar system called the Global Navigation Satellite System (GLONASS). Both systems operate on a similar principle of measuring the time difference of arrival of radio transmissions from several satellites with precisely known orbital element sets and accurate clocks.

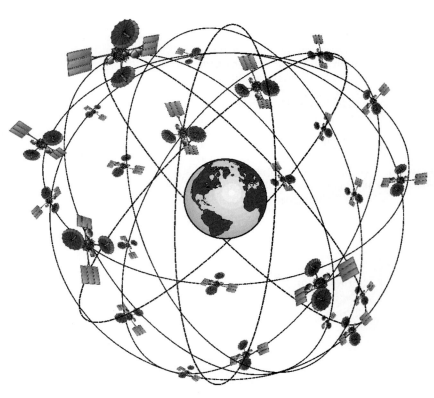

The concept of a satellite navigation system is built on the premise that a navigation service should provide global coverage with a minimum number of satellites always in view anywhere on the Earth. In order to meet this requirement and provide a high quality of service, several non-geostationary satellites are arranged in several orbital planes called constellations. The GPS and GLONASS constellations each consist of 24 satellites. The distribution of satellites in the constellations vary slightly between the two systems. The GPS constellation of 24 satellites in 8 planes is illustrated in Figure 8-20. This provides a minimum of 6 satellites in view at all times, anywhere on the Earth.

Table 8-2 highlights the features of the two systems.

Figure 8-20: Global Positioning System Constellation of 24 Satellites

Table 8-2: GPS and GLONASS Specifications

	GLONASS	GPS
Number of Satellites	24	24
Orbital Planes	3	6
Satellites Per Plane	8	4
Altitude	19,100 km	20,180 km
Inclination Angle	64.8°	55°
Revolution Period	11 hr 15 min 40 sec	11 hr 58 min 00 sec
Access	FDMA	CDMA
Frequency Band	1620 - 1614.94 MHz	L1 - 1575.42 MHz L2 - 1227.6 MHz

Position Location Determination Using Satellites

Although designed as a military support system, GPS is a system that will be integrated into many common civilian telecommunications and wireless monitoring functions in the future. Cell phone networks are now required to provide position location information to emergency response (E-911) centers. Auto tracking systems now provide auto owners with position location and security systems using the GPS constellation. GPS provides highly accurate time, velocity, and positional data as well as meeting the primary positioning service. Depending on the mode of use and the equipment, high-precision measurements accurate to less than one meter can be made for geodetic applications.

> Determining the location of any GPS receiver on the Earth requires a solution in three dimensions: latitude, longitude, and altitude.

A minimum of four satellites is required to determine latitude, longitude, and altitude. GPS also provides a solution for velocity information. Referring to Figure 8-21, if the distance (x) is known, we can only determine that the receiver is located anywhere on the sphere circumscribed around satellite A. In other words, every point on sphere A is the same distance from the satellite. If we can measure the distance (y) to the second satellite B, the location of the receiver can be narrowed to an area of the universe anywhere on the circumference (a) formed by the intersection of spheres A and B. The third satellite C, is needed to reduce the uncertainty to two points on circumference (a). This is sufficient for determining the receiver location because one of the points lies on the Earth's surface and the other point lies in outer space or deep in the Earth. The computers used in the GPS receiver have various methods for distinguishing the correct points from the false points. This three-satellite solution would be sufficient if we had perfect synchronization of time between the receiver and the satellite. To understand the importance of time, it is necessary to examine how the distances (range) to satellites A, B, and C can be measured.

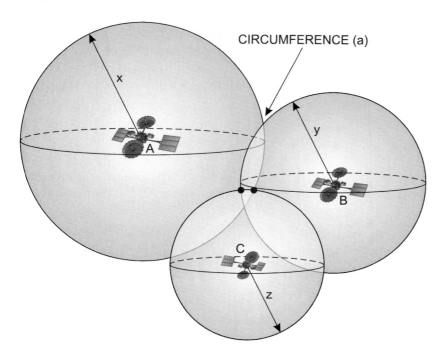

CIRCUMFERENCE (a)

Figure 8-21: GPS Three-Satellite Solution

GPS satellites use Code Division Multiple Access (CDMA) modulation using separate PN codes for each satellite in the constellation. The same PN codes can also be generated in the GPS receiver. Figure 8-22 illustrates how the two codes can be used to determine the propagation delay from the satellite to the receiver. This assumes that the clocks are extremely accurate and synchronized in the satellite and the receiver to know the start of the code pattern. Any errors between the two clocks would result in large errors in the range measurement. GPS design engineers solved the problem in an interesting way. Atomic clocks are used in each satellite for accurate time synchronization. The receiver would be very expensive if it also contained an atomic clock. Measuring the distance to the fourth satellite solves the problem. The GPS receiver starts a solution for a point in space where the spheres form the unique point of intersection, as illustrated in Figure 8-21, but the clock in the receiver is assumed to be incorrect to some degree because it is not perfectly synchronized with the satellite clock. The solution will be imperfect because the intersection point of the spheres does not form a single point due to the clock offset; therefore, the computer in the GPS receiver starts adjusting its clock (forward or backward) until a perfect single-point intersection is solved. This is the point where the receiver clock and the satellite atomic clock are perfectly synchronized. The ranging problem is now easily solved with the two clocks perfectly synchronized.

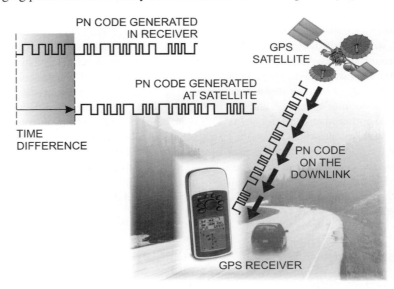

Figure 8-22:
Measuring Distance to a
GPS Satellite

The following example illustrates how the distance to a GPS satellite is measured when the clocks are synchronized:

Example 1

Assume that the GPS receiver measures a time delay of .06 seconds for the PN code received on the downlink from a GPS satellite. What is the calculated distance (d) to the satellite in miles?

$d = $ (speed) $\times$ (time)
Speed of light = 186,000 miles per second
Time = .06 seconds then;
$d = 186,000 \times .06 = 11,160$ miles

The calculation for position determination also requires the known instantaneous precise position of each of the satellites in space. The GPS receiver maintains a table called an **almanac** that tells where a satellite should be in space at any given time. Forces such as the pull of the moon, the sun, and the planets change the GPS satellite's orbit very slightly, but the Department of Defense ground stations monitors its exact position and transmits the information to the satellite. Each satellite then includes the adjustments in the downlink message. The GPS receiver is therefore able to update its almanac in memory when powered on. This orbital position information of the entire constellation is referred to **ephemeris** data. Each GPS satellite has its own atomic clock; however, slight errors, although extremely small, must be accounted for. Ground control stations monitor each satellite and uplink information relative to clock errors in the constellation. Each satellite broadcasts the clock error offset information along with the ephemeris data on the downlink signal.

GPS Errors

The GPS system has been designed to provide a high degree of accuracy. However, there are still unavoidable errors. Added together, these errors may cause a deviation of +/−50-100 meters from the actual GPS receiver position. Experience shows that the errors are often much less. There are several sources for these variable errors, the most significant of which are:

- *Ephemeris Errors/Clock Drift/Measurement Noise* – As mentioned earlier, GPS downlink signals contain information about ephemeris (orbital position) errors, and about the rate of clock drift for the broadcasting satellite. The data concerning ephemeris errors may not exactly model the true satellite motion or the exact rate of clock drift. Distortion of the signal by measurement noise can further increase positional error. The disparity in ephemeris data can introduce 1-5 meters of positional error, clock drift disparity can introduce 0-1.5 meters of positional error, and measurement noise can introduce 0-10 meters of positional error.

- *Selective Availability* – Ephemeris errors should not be confused with Selective Availability (SA), which is the intentional alteration of the time and ephemeris signal by the Department of Defense. SA can introduce 0-70 meters of intentional positional error. On May 1, 2000, the U.S. Government ordered SA to be turned off. Civilian and military users now obtain the same precision of measurement of approximately 10 meters or better. The Department of Defense reserves the right to turn SA on at a later date if it is considered to be in the best interests of national defense.

- *Atmospheric Conditions* – The ionosphere and troposphere refract the GPS signals. This causes the speed of the GPS signal in the ionosphere and troposphere to be different from the speed of the GPS signal in space. Therefore, the distance calculated from "Signal Speed x Time" shown in an earlier example will be different for the portion of the GPS signal path that passes through the ionosphere and troposphere and for the portion that passes through space.

- *Multipath* – A GPS signal bouncing off a reflective surface prior to reaching the GPS receiver antenna is referred to as multipath. Because it is difficult to completely correct multipath error, even in high-precision GPS units, multipath error is a serious concern to the GPS user.

Table 8-3 is commonly known as the GPS Error Budget, which lists the most common sources of error in GPS positions.

**Table 8-3:
GPS Error Budget**

Source	Uncorrected Error Level
Ionosphere	0-30 meters
Troposphere	0-30 meters
Measurement Noise	0-10 meters
Ephemeris Data	1-5 meters
Clock Drift	0-1.5 meters
Multipath	0-1 meter
Selective Availability	0-70 meters (when activated by the Defense Department)

As discussed above, several external sources introduce errors into a GPS position. While the errors discussed above always affect accuracy, another major factor in determining positional accuracy is the alignment, or geometry, of the group of satellites (constellation) from which signals are being received. The geometry of the constellation is evaluated for several factors, all of which fall into the category of Dilution Of Precision, or DOP.

- *DOP* – DOP is an indicator of the quality of the geometry of the satellite constellation. Your computed position can vary depending on which satellites you use for the measurement. Different satellite geometries can magnify or lessen the errors in the error budget described above. For example, Figure 8-23 shows how the satellite geometry may produce a worst-case DOP because of the smaller angles between satellites at some instant in time.

**Figure 8-23:
Poor Dilution of
Precision**

Figure 8-24 shows a lesser DOP with better satellite geometry. A greater angle between the satellites lowers the DOP, and provides a better measurement. A higher DOP indicates poor satellite geometry, and an inferior measurement configuration. This error changes with time for any position on Earth due to the relative motion of the satellites.

**Figure 8-24: Good
Dilution of Precision**

Lab 9 – Comparing the Features of GPS Receivers

<u>Lab Objective</u>

This exercise is designed to provide you with a guide for purchasing a GPS receiver. You will become familiar with the terms, nomenclature, physical and electrical specifications, and performance features of common handheld GPS receivers used primarily for hobby and recreational purposes.

<u>Instructions</u>

1. Review the GPS Receiver Buyers Comparison Guide shown in Table 8-4.

> ┌─ NOTE ──────────────────────┐
> │ Look at the features that are most sensitive to │
> │ the cost of each GPS receiver. │
> └────────────────────────────┘

Table 8-4: GPS Receiver Buyer's Comparison Guide

Receiver Features	Vendor 1	Vendor 2	Vendor 3	Vendor 4	Vendor 5
Channel Tracking	Parallel 12	Parallel 12	Multiplex 8	Parallel 12	Parallel 12
Accuracy	15 meters	15 meters	30 meters	15 meters	15 meters
Internal Memory	1.44 MB	20 MB	None	15 MB	1.44 MB
Additional Maps	CD-ROM	CD-ROM	None	CD-ROM	CD-ROM
Display (W x H inches)	2.2 x 1.5	2.2 x 1.5	2.2 x 1.5	2.4 x 3.5	1.8 x 3.5
Display Type	Gray LCD	Gray LCD	Gray LCD	Gray LED	16 Color LCD
Pixels	160 x 100	256 x 160	160 x 100	160 x 240	128 x 240
Screen Orientation	Vertical	V and H	Vertical	V and H	V and H
Alarms	No	No	No	Yes	Yes
GPS Antenna	Built-in	Detachable	Built-in	Detachable	Detachable
Weight	9.5 oz	10 oz	9 oz	1.2 lb	1.3 lb
Temperature Range	5–150 F	5–150 F	10–100 F	0–160 F	5 – 150 F
Waterproof	1PX7	1PX7	1PX2	1PX7	1PX7
Voltage Range	10-30 V	10-30 V	10-30 V	10-40 V	10-40 V
Battery Life	36 HR	26 HR	36 HR	16 HR	2.5 HR
Number of Waypoints	500	500	100	600	600
Electronic Compass	No	Yes	No	Yes	Yes
Barometric Altimeter	No	Yes	No	No	No
WAAS Capable	No	Yes	No	No	No
Price	$350	$550	$220	$600	$850

TIP: Table 8-4 shows many of the most important features found in several GPS receiver specifications. The table does not include some terms or features that are proprietary or unique among leading vendors. The purpose of this table is to help you become familiar with general specifications and state of the art for receiver technology and to serve as an aid when reviewing published specifications prior to purchasing a GPS receiver.

2. Use the glossary of terms listed in Table 8-5 to analyze the receiver features listed in the GPS Receiver Buyers Comparison Guide.

3. Table 8-6 lists the various applications and preferences a typical buyer may use when selecting a GPS receiver. Study the evaluation criteria. Select and rank the preferred vendor's products as first choice, acceptable, or unacceptable for each of the listed criteria.

TIP: Some products may have identical features that may be equally acceptable for a given application. For example, most receivers have 12 parallel channel tracking capability, therefore all may rank as acceptable for this characteristic.

4. Evaluate the five products for position determination accuracy and number of parallel or serial channels tracked.

5. Examine the receivers that provide external map and software update capability. Determine where this feature may be important for a specific application.

6. Determine the differences in display type, resolution, and size. Determine where this feature may be worth the extra cost.

7. Screen orientation may be important when mounting the unit in an auto or airplane cockpit.

8. Pick the importance of alarms for any specific application. Remember that alarms are used as an auxiliary function.

9. Determine the application where detachable antennas are important.

10. Analyze how environmental specifications are important for certain applications.

11. Most units use the same voltage from an external power source. If it is not practical to use a cigarette lighter adapter, pay attention to battery life.

12. Number of waypoints usually exceeds the number needed for most applications.

13. WAAS may be important to some applications. Determine whether the cost is consistent with the application.

TIP: Be sure to review the glossary for the explanation of the WAAS.

14. Rank the costs that you believe are consistent with the features offered.

TIP: This ranking may vary with the budget of the individual buyer. List the receivers according to how you personally would evaluate performance against cost.

Table 8-5: GPS Glossary of Terms

Term	Meaning
Accuracy	The nominal position location accuracy that can be expected under ideal receiving conditions by the GPS receiver. The accuracy shown in the Buyers Comparison Guide indicates the accuracy with Selective Availability (SA) turned off.
Additional Maps	A feature that provides for downloading different map databases from a CD-ROM using a computer serial or USB port connection. Some GPS receivers do not provide map upgrade options other than the internally stored map. All receivers will give you your latitude, longitude, and altitude, but they do not all show your location on a detailed map. Decide what kind of map you will need and make sure that the receiver offers that type of map for downloading.
Alarms	A feature that sounds an audible alarm when certain conditions exist, such as when a turn or change in course from a planned route is required.
Barometric Altimeter	Device that measures altitude by sensing a change in atmospheric pressure. This is an alternate method of indicating altitude other than the GPS method of altitude determination.
Battery Life	Duration of time for use of rechargeable batteries or the life expectancy of alkaline batteries. Some handheld receivers can accept external power, which is handy if you plan to be driving all day with your GPS on and do not want to drain the batteries. Car, boat, or airplane in-dash GPS receivers run on an external power source provided by the larger unit it's hooked up to. These devices are not mobile.
Channel Tracking	Indicates the number of channels that can be tracked at the same time. Multiplex receivers have only one channel. They pick up one satellite signal at a time, cycling through a few satellites. They work much better in open environments, as their connection can easily be disrupted by buildings or other obstacles. The most affordable models use multiplex receivers. Parallel-channel receivers have several channels, and lock onto many satellites at the same time. They do not lose satellite connections very easily and they can pinpoint the location more exactly. These receivers were once fairly expensive, but there are several affordable models now on the market. If you plan to use your receiver in a big city or mountainous areas, you should probably get one with at least 12 parallel channels.
Display Size	Screen size dimensions. Larger screens provide more detail but usually cost more.
Display Type	Describes screen characteristics such as color or gray scale. Also describes physical type of display used, such as Light Emitting Diode (LED) or Liquid Crystal Display (LCD).
Electronic Compass	Compass that maintains heading information when you are standing still. Without this feature, GPS receivers normally only display heading information while you are in motion.
GPS Antenna	Denotes internal or external mounted antenna. Detachable antennas provide a means for location of the antenna away from the receiver in a more favorable location for best reception of satellite signals.

Table 8-5: GPS Glossary of Terms (continued)

Term	Meaning
Internal Memory	Amount of memory that is available for downloading map programs. If you plan to use route-mapping and track logging extensively, you will want a receiver that has enough memory. Consider how many waypoints you would want to store and find out what a receiver's maximum storage capability is. Also, look for a receiver with a backup system that will hold onto your information while you change the receiver's batteries.
Number of Waypoints	Lists the maximum number of waypoints that can be stored. This number usually exceeds what the average person would use under normal situations. This relates to the amount of memory contained in the receiver.
Pixels	Number of pixels in the horizontal and vertical dimensions. This is a measure of screen resolution similar to computer display screens.
Price	Usually manufacturers list price. GPS receivers are often discounted by dealers. It pays to shop around.
Screen Orientation	Indicates whether the screen orientation uses the short and long dimension of the screen in a vertical or horizontal display. Many receivers allow the orientation to be optionally changed to suit the type of mounting used.
Temperature Range	Design parameters for maximum and minimum operating temperature range.
Voltage Range	Voltage required for proper operation. Values are given for the range of voltages from external sources such as an auto, aircraft, or boat electrical system.
WAAS Capable	WAAS is an acronym for the Wide Area Augmentation System. WAAS uses a series of ground stations to monitor the accuracy and integrity of GPS signals. The information is transmitted up to GEO satellites where it is broadcast to WAAS-capable GPS receivers. WAAS is primarily used for integrity verification by the aviation industry. This is a great feature if you plan to use the receiver in a critical application such as air navigation.
Waterproof	If you will be using GPS on a boat or while hiking, you should look for a receiver with good waterproofing. Some receivers are sealed so that they are completely waterproof, while others are merely constructed so that they resist water. Many GPS manufacturers use the European International Electrotechnical Commission (IEC) 529 specification for water immersion testing. iPX7 indicates protection against falling water when tilted up to 15 degrees in each direction from the normal operating position.
Weight	Total weight of the receiver in ounces. This is important if the receiver is to be used for camping or hiking.

Table 8-6: GPS Receiver Evaluation Criteria

Vendor 1	First Choice	Acceptable	Unacceptable
General Performance and Tracking Accuracy			
Downloadable Map Storage Capacity			
Display Quality			
Automobile Travel Applications			
Marine Applications			
Aviation Applications			
Hiking Applications			
External Antenna			
Battery Endurance			
Cost			

Vendor 2	First Choice	Acceptable	Unacceptable
General Performance and Tracking Accuracy			
Downloadable Map Storage Capacity			
Display Quality			
Automobile Travel Applications			
Marine Applications			
Aviation Applications			
Hiking Applications			
External Antenna			
Battery Endurance			
Cost			

Vendor 3	First Choice	Acceptable	Unacceptable
General Performance and Tracking Accuracy			
Downloadable Map Storage Capacity			
Display Quality			
Automobile Travel Applications			
Marine Applications			
Aviation Applications			
Hiking Applications			
External Antenna			
Battery Endurance			
Cost			

Table 8-6: GPS Receiver Evaluation Criteria (continued)

Vendor 4	First Choice	Acceptable	Unacceptable
General Performance and Tracking Accuracy			
Downloadable Map Storage Capacity			
Display Quality			
Automobile Travel Applications			
Marine Applications			
Aviation Applications			
Hiking Applications			
External Antenna			
Battery Endurance			
Cost			

Vendor 5	First Choice	Acceptable	Unacceptable
General Performance and Tracking Accuracy			
Downloadable Map Storage Capacity			
Display Quality			
Automobile Travel Applications			
Marine Applications			
Aviation Applications			
Hiking Applications			
External Antenna			
Battery Endurance			
Cost			

Lab 9 Questions

1. Which features did you determine had minimal impact on cost?

2. Which three features did you believe were the most important when making a selection?

3. Which application was the most important for your specific needs and preferences?

Lab 10 – Using a GPS Receiver

<u>Lab Objective</u>

This exercise is designed to provide you with a "hands-on" experience using a GPS receiver. You will be able to enter data, select various options for the display, use various menus, select navigation routes, and monitor the status of the satellites used by the receiver during the progress of the lab procedure.

<u>Materials Needed</u>

- Garmin eMap model GPS receiver
- Two AA batteries
- Automobile/licensed driver
- Garmin eMap Owners Manual
- Garmin eMap Quick Start

<u>Instructions</u>

This lab exercise will take you through the steps for navigating with a Garmin eMap model GPS receiver. You must find a location outside with a clear view of the sky to conduct most of the steps in this lab. The GPS receiver depends upon receiving signals from as many satellites as possible. At least four satellites will always be in view, but the angles from your position to the satellites will vary with time. For some of the steps, you will need to place the GPS receiver on the dash of an auto as close as possible to the windshield for best results.

1. Review the function and location of each control located on the front panel and left side of the receiver, as illustrated in Figure 8-25. These controls will be referenced in this lab. This is the same image as page 1 of the user manual for the Garmin eMap.

2. Install two AA batteries in the receiver.

3. Move to an outside area. Press the power on switch. The antenna is enclosed in the top of the receiver. Hold the receiver in a comfortable position where you can see the screen with the top of the receiver pointing upward. The Garmin navigation warning message will appear. Press **ENTER** after reading the message. The first time you use the receiver, the unit will update the satellite almanac information in its database. This may take a several minutes. This delay will decrease the next time you use the receiver. If you receive a *Poor Satellite Reception* message, read the options. Move to a better location.

Figure 8-25: eMap Receiver

After a short delay, the map screen will appear, as shown in Figure 8-26, showing your present location. The top of the screen shows the time, distance traveled, and speed. This display will provide information about your destination later when waypoints are used.

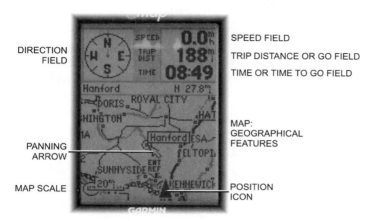

DIRECTION FIELD
SPEED FIELD
TRIP DISTANCE OR GO FIELD
TIME OR TIME TO GO FIELD
MAP: GEOGRAPHICAL FEATURES
PANNING ARROW
MAP SCALE
POSITION ICON

Figure 8-26: Map Display

TIP: The position icon is a small pointer that shows your present location and direction when you are moving. The map scale can be changed by pressing the **in** and **out** buttons. The map scale is shown on the bottom left of the display for each zoom position. It is possible to zoom in so far that the resolution of the map data and the appearance of the map are no longer accurate. When this occurs the word "overzoom" will appear below the map scale. The error is still minimal. Pressing the **menu** button one time shows the map options. Pressing the **menu** button twice displays the Main Menu.

4. You can set the screen contrast and backlighting function by pressing the backlight/contrast key on the left side of the case. When the display appears, as shown in Figure 8-27, use the rocker arm left and right side to change the screen contrast. Press the **ENTER** key to save. Backlighting can be turned on and off by toggling the backlight switch on and off. You can also turn the backlight function on and off when the display is on by moving the rocker keypad up for on and down for off.

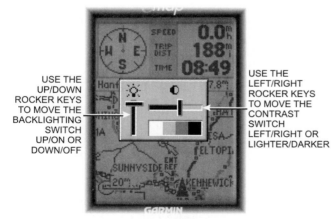

USE THE UP/DOWN ROCKER KEYS TO MOVE THE BACKLIGHTING SWITCH UP/ON OR DOWN/OFF

USE THE LEFT/RIGHT ROCKER KEYS TO MOVE THE CONTRAST SWITCH LEFT/RIGHT OR LIGHTER/DARKER

Figure 8-27: Backlight/Contrast Setting

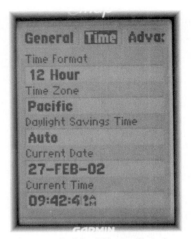

Figure 8-28: Time Setting

TIP: The backlight will turn off automatically to preserve battery life. You can also change the settings by pressing **menu** two times to select the main menu and selecting **setup** and **backlight timeout**.

5. GPS updates the internal receiver clock during satellite acquisition; however, you will need to set the clock to your time zone. Press the **menu** button two times. Highlight *Setup* and press **ENTER**. Use the rocker keypad to select **Time**. Scroll up and down using the rocker key. Press the **ENTER** key for submenus. Highlight the information for your time zone, as shown in Figure 8-28. Use the **ESC** (escape) key to return to the map display.

You are now ready for an exercise to track your position while moving in a vehicle and create a track log. Select a route covering a distance of at least 5 miles. Select a partner to conduct the navigation for safety purposes. Do not attempt to perform this exercise while driving alone.

6. Before starting this step, press the power on switch. Observe if the map display is showing your current location. Hold or place the GPS receiver in your vehicle near the windshield. You can check the quality of the reception by pressing the **menu** button twice to select the main menu. Scroll down to *GPS Info:* and press the **ENTER** key. The sky view display, shown in Figure 8-29, shows the number of satellites being tracked, signal strength, and approximate overhead azimuth and elevation of each satellite in view. Your current latitude, longitude, speed, accuracy, and elevation will also be displayed along with receiver status.

Figure 8-29: Sky View Display

TIP: The *use indoors map* option stops the satellite reception and conserves power usage. When you want to resume tracking, press the **menu** key, highlight *Use Outdoors*, and press the **ENTER** key. When the use indoor mode is on, this does not mean that the receiver is receiving valid data. It is only used to stop processing and updating your position to conserve battery life when satellites are out of range.

7. Drive to the destination that you have planned for this part of the lab exercise. Follow your path on the moving map display. When you are ready to return to the original point, stop and store the track with the following steps.

Figure 8-30: Save Back Through Menu

8. Press the **menu** button twice to select the Main Menu. Scroll down and highlight *Tracks*. Press the **ENTER** key. Highlight the **save** button and press the **ENTER** key. This will display the *Save Back Through* menu, shown in Figure 8-30. Scroll down to set the time frame you want to save (midnight is OK) and press **ENTER**. This will display a track identification number with a data and time. Scroll and highlight *Show on Map Page* and press **ENTER**. This will place a check mark to have the track shown on the Main Map Page. Highlight the "OK" button and press **ENTER**. Press **ESC** twice to return to the Main Map.

9. To return the starting point, highlight the saved track you just entered and press the **ENTER** key. Highlight the *Trackback* and press the **ENTER** key. On this menu for Direction to Navigate select **Reverse**. This will return you to the Main Map Page.

10. Follow the navigation path on the Main Map Page back to the starting point. During your trackback navigation, you will see the display shown in Figure 8-31.

TRACKBACK
PATH

**Figure 8-31:
Trackback Path**

11. In addition to the direction arrow, you will hear two audible beeps as you approach a point where you will be making a turn to follow the trackback route. An onscreen message will also alert you concerning an approaching turn. The upper display will show an arrow directing you for your return route, the distance to go, time to go, and current speed. Return to the starting point.

The following steps will help you create a route. In addition to the trackback navigation option, it is possible to create waypoints to guide you to a preferred location. The waypoints in this exercise will vary according to your specific location.

12. Press the **menu** button twice to display the Main Menu. Highlight *Tracks* and press **ENTER**. Highlight the *New* field, as shown in Figure 8-32. Press the **ENTER** key. This will show the new route page as empty.

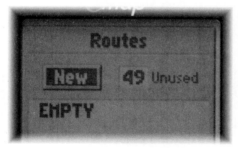

Figure 8-32: Routes Menu

13. Press the menu key to show the pop-up options. Highlight the *Add Waypoint* and press the **ENTER** key to show the Find menu, shown in Figure 8-33. Any category can now be used to create a waypoint on the route list. Highlight *Cities* and press **ENTER**. Scroll down the list of cites. Note that the distance and direction to each city from your present location is displayed near the bottom of the screen as you scroll, as shown in Figure 8-34.

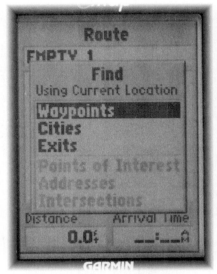

Figure 8-33: Find Menu　　　　Figure 8-34: Cities Menu

TIP: Only Waypoints, Cities, and Exits will be displayed unless additional map data cards have been installed. Points of Interest, Addresses, and Intersections will be grayed out.

14. Highlight a city as your destination and press **ENTER**. This display will show the information about your selected city. To add this city to your route, highlight the *OK* button, as shown in Figure 8-35, and press the **ENTER** key to add the city to your route.

TIP: You may continue to add as many waypoints as you wish by repeating the previous steps.

**Figure 8-35:
Selecting a City**

You are now ready to navigate again by using the new route you have created. When you are finished adding the last waypoint, press the menu key. This will display the route options menu shown in Figure 8-36.

15. Select **Start Navigation** and press the **ENTER** key. Return to the map page when you are ready to travel. When navigating with routes, the same information will be displayed as you observed in the trackback exercise. The arrow will point to the waypoint with speed, distance, and time to go displayed with the Main Map page.

16. Navigating from your present location to a city near you can be simplified in the next step by using the *Go To* feature.

17. Select the **Main** Menu, highlight **Routes**, and press the **ENTER** key. User the rocker key to highlight **New** and press the **ENTER** key. This will display the route page as you have seen before.

18. Press the menu key, highlight **Add Waypoint**, and press the **ENTER** key.

19. Highlight **Cities** and press the **ENTER** key. Scroll down the list of cities in your area. Highlight a City and press the **ENTER** key. The City Information display, shown in Figure 8-37, will appear.

20. Use the rocker keypad to select **Go To** and press the **ENTER** key. The Main Map display will appear. When you start moving, the map will show your progress in the direction of your city. The top of the map will show the direction to the destination, the distance, time to go, and speed.

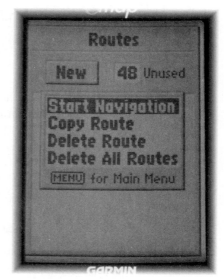

Figure 8-36: Route Options Menu

Figure 8-37: City Information Menu

Lab 10 Questions

1. How do you display the current number of satellites being tracked?

2. How is the remaining battery capacity monitored?

3. What controls are available for changing the scale of the Main Map display?

4. What is the purpose of the *Use Indoors* option on the map page menu?

5. How do you check for your current latitude and longitude position?

Key Points Review

In this chapter, the following key points were introduced concerning satellite communications systems. Review the following key points before moving into the Review Questions section to make sure you are comfortable with each point and to verify your knowledge of the information.

- The objective of a satellite system is to provide a communications link between two or more locations on the Earth that can cover longer distances than terrestrial microwave relay stations.

- Depending upon the altitude, a satellite can provide coverage over a very wide geographical area, avoiding the cost of several ground-based transmission facilities.

- A satellite system consists of three basic subsystems. They are the space segment, the earth segment, and the ground control segment.

- When the period of rotation of the Earth exactly matches the orbital period of a satellite, the satellite is said to be in geosynchronous (geo) orbit.

- Medium Earth Orbits (MEO) range from any orbit less than synchronous to a few hundred miles in altitude.

- Low Earth Orbit (LEO) satellites circle the Earth at altitudes typically between 300 miles and 800 miles above the earth. LEO orbits are used for mobile satellite phone technology as well as for paging and message delivery systems.

- Satellites in polar orbit are used for search and rescue, mapping, military surveillance, and environmental monitoring of the earth's surface. A polar orbiting satellite can eventually view the entire surface of the earth.

- The Fixed Satellite Service (FSS) is allocated a large share of the radio spectrum because this type of service is the largest business segment of the satellite industry.

- C-Band and Ku-band satellites use several transponders (receiver-transmitter chains), each with a bandwidth of 54 MHz or more of bandwidth.

- Satellite earth stations provide the radio link between the Earth and the satellite.

- The FCC defines an earth station as a complex of transmitters, receivers, and antennas used to relay and/or receive communications traffic (voice, data, and video) through space to and from satellites in both geostationary satellite orbits and non-geostationary satellite orbits.

- Parabolic antennas have very high gain at the higher microwave frequencies used by satellites. High gain performance is necessary to provide the required uplink and downlink power for the long haul to and from a satellite.

- Parabolic antennas can be steered using the proper mounts to accurately follow a satellite if necessary or to position the antenna along the geostationary arc.

- Antenna gain can be expressed as the relative increase in power achieved by focusing the radio energy in a particular direction.

- Determining the location of any GPS receiver on the Earth requires a solution in three dimensions: latitude, longitude, and altitude.

- Antenna beamwidth is important because, as mentioned earlier, geostationary satellites in the United States are separated by only 2 degrees in the orbital arc. If the beamwidth were wider than 2 degrees the antenna would "see" more than one satellite.

- Mobile Satellite Service (MSS) is a service for mobile subscribers using cellular-type handsets communicating with Low Earth Orbit (LEO) satellites.

- Other errors due to refraction and propagation contribute to the errors in the position determination; however, the GPS still provides position accuracies of 10 meters or better under ideal conditions.

REVIEW QUESTIONS

The following questions test your knowledge of the material presented in this chapter:

1. Name the three basic system elements of a satellite system.

2. Name five types of satellite services that are provided using geostationary satellites.

3. Explain how geostationary satellites are assigned, licensed, and identified as to their authorized orbital location.

4. List two reasons why LEO satellites are favored over GEO satellites for satellite mobile telephone service.

5. Explain why some satellites are placed in polar orbits.

6. Under title 47 of the Federal Regulations for telecommunications, where would you look for the rules and regulations governing the satellite Direct Broadcast Service (DBS)?

7. Briefly explain the basis and concept for the terms "big LEOs" and "little LEOs."

8. What are the two main assemblies that form the basic architecture communications satellites?

9. Explain the main function of a satellite attitude control system.

10. List the three characteristics of a parabolic "dish" type satellite earth station antenna that determine its gain.

11. Why is the beamwidth of an earth station antenna important when receiving or transmitting a signal on a geosynchronous satellite?

12. Explain the main business goal for the big LEO satellite voice systems.

13. Describe the basic principle used by the GPS and GLONASS satellite navigation systems for determining position location on the Earth.

14. How many satellites are used by the GPS?

15. What is the minimum number of GPS satellites required to obtain a three-dimensional position determination by a GPS receiver?

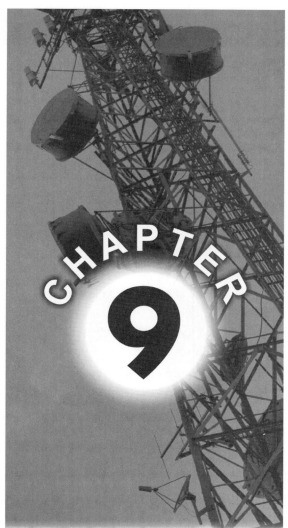

CHAPTER

9

ADVANCED
WIRELESS SYSTEMS

Upon completion of this chapter, you should be able to perform the following tasks:

1. Describe the objectives of the International Mobile Telecommunications (IMT) 2000 initiative.

2. Discuss the advantages of 3G radio technologies over 2G radio technologies.

3. Discuss the differences between the W-CDMA and CDMA2000 specifications.

4. Identify the difference between the TDMA and CDMA third generation proposals.

5. Discuss the features of 2.5G systems.

6. Describe how users share the bandwidth of a W-CDMA channel.

7. Describe the objective of Orthogonal Frequency Division Multiplexing (OFDM).

8. Describe how multipath can disrupt wireless communications links.

9. Define how intersymbol interference limits the data transmission rate.

10. Discuss the main objective of the Ultra-Wideband (UWB) technology.

11. Describe the meaning of duty cycle.

12. Calculate the bandwidth of a UWB modulated signal.

13. Describe the design objectives of the Teledesic satellite communications system.

14. Show how the satellites are arranged in orbit for the Teledesic system.

15. Identify the types of terminals used with the Teledesic network.

16. Explain the data transmission rates of the Teledesic satellite network.

Advanced Wireless Systems

INTRODUCTION

This chapter provides an overview of the emerging technologies and standards for the wireless industry. Each topic in this chapter will affect the way we communicate in the next decade and beyond.

The third generation, or simply "3G", is the next phase of the mobile wireless industry. Practically all of the industrial nations of the world have been involved for the first time ever in developing a telecommunications architecture for global users. In this chapter, you will discover the details of each of the third generation mobile wireless technologies and review the radio spectrum allocation issues associated with 3G.

More efficient use of the radio spectrum is a continuing effort for research organizations and the telecommunications industry. In this chapter, you will discover how two technologies are approaching this problem. They are called Orthogonal Frequency Division Multiplexing (OFDM) and Ultra-Wideband (UWB).

In the near future satellite communications will play a major role in the expansion of high-speed third generation global telecommunications networks. One such system, called Teledesic, plans to cover the earth with a constellation of 288 satellites. This chapter covers the technical and operational issues associated with the Teledesic system.

Third Generation Wireless Standards

Third generation, or simply "3G", is a term used to describe the wireless standards and technology that will provide data transmission speeds up to 2 Mbps and a host of broadband services on existing and upgraded mobile cellular networks.

Third generation

The first and second generation of mobile wireless communications systems were covered in Chapter 5 and 6. The evolution of these systems to a future third generation of mobile wireless communications systems has focused on higher data rates, improved voice quality, and a wide range of data services.

The transition from the second generation to third generation includes some enhancements that are called "2.5G".

These interim systems include higher data rates on the evolutionary path to systems, which qualify as "third generation compatible." The evolution of mobile wireless standards through three generations and a list of the many acronyms used in this chapter are illustrated in Figure 9-1. (Since the publication of the 3G standards, some of the U.S. carriers using TDMA as the second generation radio access technology have opted to follow the GSM evolution to W-CDMA instead of the TDMA UWC-136 standard.)

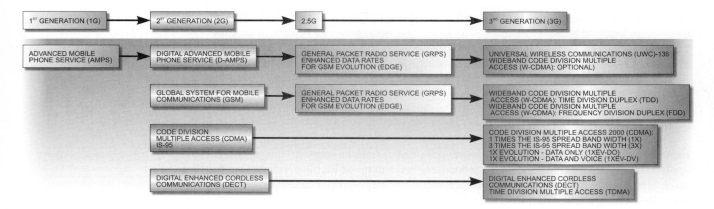

Figure 9-1: The Evolution of Mobile Cellular Standards

International Mobile Telecommunications 2000

The International Telecommunications Union (ITU), a charter organization of the United Nations in Geneva, Switzerland, is the leading organization for global standards development. The ITU initiated an effort under the banner name of **International Mobile Telecommunications 2000**. They set a goal for users of mobile telecommunications systems on a worldwide basis to be able to use the same phone number using global roaming across two major core networks (GSM and ANSI-41C) in North America, Europe, and Asia. To meet this goal, the initial standards development effort needed to evolve from second generation radio interfaces and networks. The hope was that "seamless" roaming could be accomplished for mobile wireless users.

> The call for proposals for third generation mobile Radio Transmission Technologies (RTTs) was initiated by the IMT-2000 study group in 1998 and resulted in several submissions from Standards Development Organizations (SDOs) from various countries.

The IMT2000 group received many different submissions based on TDMA (Time Division Multiple Access) and CDMA (Code Division Multiple Access) technology. In December 1999, the IMT-2000 team recommended five mobile/terrestrial radio interface standards, shown in Figure 9-2. The illustration highlights the IMT-2000 term for each third generation standard.

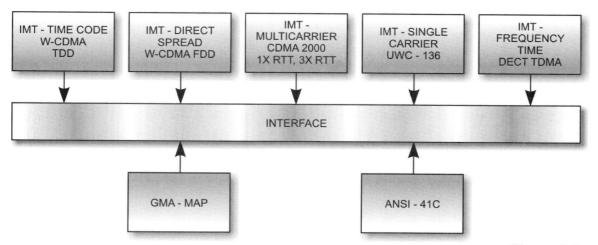

Figure 9-2:
IMT2000 Specifications

The ETSI/GSM (European Telecommunications Standards Institute/Global System for Mobile Communications) companies, such as Nokia and Ericsson, are backing the W-CDMA (Wideband Code Division Multiple Access).

U.S. vendors, including Qualcomm and Lucent Technologies, are backing CDMA2000.

The W-CDMA FDD (Frequency Division Duplex – FDD) standard uses paired radio channels for forward and reverse links. The W-CDMA TDD standard uses the same channel for both transmit and receive (Time Division Duplex – TDD).

Work was continued after the acceptance of the five radio technology standards by two principal standards groups. The **3G Partnership Project** (**3GPP**) is concentrating on Evolved GSM networks and W-CDMA radio access technologies.

3G Partnership Project (3GPP)

The **3G Partnership Project 2** (**3GPP2**) was organized to concentrate on the development and evolution of the American National Standard (ANSI-41C core network) and the CDMA2000 radio access technologies.

3G Partnership Project 2 (3GPP2)

While "voice only" wireless cellular networks have become very popular, the emergence and rapid growth of the Internet has now started the trend for mobile users to expect the same capabilities for high-speed access to the Internet and corporate intranets when away from home or the office.

Third generation handsets offer high-speed access to video-on-demand, short message service, music, entertainment, location services, mobile commerce, news, and e-mail. All of these expanded third generation services are potential applications for future mobile subscribers.

Examples of 3G handsets that are compatible with the third generation CDMA2000 1X standard at 144 kbps are shown in Figure 9-3.

Figure 9-3:
CDMA 1X Handsets

CDMA2000

> CDMA2000 is intended to provide 3G services over mobile radio networks that conform to the ANSI document "TIA/EIA-41," which includes existing (2G) ANSI-95 and TDMA (ANSI-136) nets.

In the future, it is expected that the multicarrier mode of CDMA2000 will be extended to allow connection to core networks based on GSM's Mobile Application Part (MAP).

The multicarrier mode is very similar to the frequency-division multiplexing form of W-CDMA. The dissimilarities stem from the need to allow the ANSI-95 mode to work in the enhanced 3G network just as GSM handsets can be accommodated in the W-CDMA extension.

ANSI-2000 Release A is the name given to CDMA2000 (the latter is the marketing name) by the Telecommunications Industry Association (TIA) standards group. The Release A standard was completed in March 2000. Since the publication of Release A, other evolutions of CDMA2000 have been published and are covered in the following paragraphs.

CDMA2000 supports a range of chip rates; all can be expressed by:

$$N \times 1.2288 \text{ Mcps}$$

where:

$$N = 1, 3, 6, 9, 12$$

When $N > 1$, there are two ways by which CDMA2000 can spread the signal. The first one, Multi-Carrier, basically demultiplexes the message signal into N information signals and spreads each of those on a different carrier, at a chip rate of 1.2288 Mcps. The second one, **Direct-Spread**, simply spreads the message signal directly with a chip rate of $N \times 1.2288$ Mcps. In the Multi-Carrier mode, each carrier has an ANSI-95 (second generation standard) signal format. Figure 9-4 illustrates the channelization of both direct spread and multicarrier features of CDMA2000. CDMA2000 also allocates guard band regions of 625 kHz on each side of the allocated bandwidth to prevent interference with neighboring bands. The figure for the total bandwidth must include these two guard regions.

> **NOTE**
>
> Mcps stands for MegaChips Per Second. It is a term used in CDMA technology. CDMA uses a separate code sequence to spread the radio carrier over a wide band of frequencies. These bits carry no information and are called Chips. This term is used to distinguish the spreading code bits (Chips) from the information bits added to the signal which carry the data and are transmitted at a much lower rate. Remember, the information rate is specified in bits per second (bps) or kilobits per second (kbps) or megabits per second (Mbps).

Direct-Spread

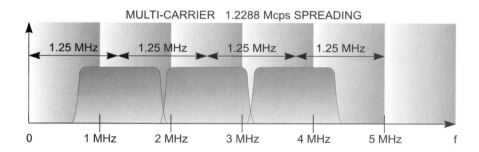

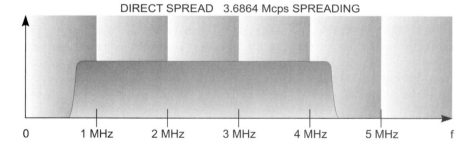

Figure 9-4: Multicarrier and Direct Spread Comparisons

> **TEST TIP**
>
> Know the chip rates and occupied bandwidth for the CDMA2000 standard.

This is a summary of the important features of CDMA2000:

- Modulation chip rate: $N \times 1.2288$ megachips/second
  - Where $N = 1, 3, 6, 9,$ and 12
- Data rates: Both packet-switched and circuit-switched data rates up to 144 kbps
  - Later phases up to 4 Mbps
  - 20 ms frame structure
  - 2x voice capacity
- Pulse shaping: same as ANSI-95
- Modulation:
- Uplink:
  - BPSK data modulation
  - QPSK spreading modulation
- Downlink:
  - QPSK data modulation
  - QPSK spreading modulation
- Detection: coherent for both uplink and downlink
- Channel spacing: $(N + 1) \times 1.25$ MHz

1x-Evolution – Data Only

The 1x-Evolution – Data Only (1x EV-DO) specification is another evolution of CDMA2000. The 1x EV-DO standard, promoted and developed by Qualcomm, was also known earlier as High Data Rate (HDR). It was first published as ANSI-856 in November 2000. The 1x in the specification name refers to the chip rate of 1.2288 Mcps used in the second generation ANSI-95 specification.

The features of 1x EV-DO include:

- Proposed by Qualcomm in 1998 as High Data Rate (HDR)
- Standardized as TIA/EIA-856 in November, 2000
- Dedicated data channel of 1.25 MHz
- Asymmetric Packet Data Channels
  - Up to 2.4 Mbps on the downlink
  - Average downlink throughput during peak sector use – 600 Mbps
  - Up to 307 kbps on the uplink
  - Average uplink throughput during peak sector use – 220 Mbps

- Compatible with existing ANSI-95 2G architecture as well as CDMA2000

- Data and voice on separate dedicated channels

- Adaptive data rates adjust according to the channel signal-to-interference ratio (SIR) as often as 1.67 ms

- Uses QPSK modulation as well as higher level 8-PSK and 16-QAM modulation when the channel SIR permits

1x-Evolution – Data and Voice

A second evolution of the core CDMA2000 specification is called 1xEvolution – Data and Voice (1x EV-DV). It provides the capability to mix data and voice.

1x EV-DV also supports higher data rates than 1x EV-DO.

The features of 1x EV-DV include:

- Agreement reached on baseline framework by the 3GPP2 on October 2001

- Targeted for new standard release May 2002 as ANSI-2000 Revision C

- Supports both data and voice

- Published standard proposed target date for March 2002

- Asymmetric adaptive data rates proposed

    - Up to 3.072 Mbps packet data on the downlink when the signal-to-interference ratio (SIR) permits

    - 1.024 Mbps packet data on the uplink when the SIR permits

- Supports adaptive higher level modulation up to 64-QAM

- Other CDMA2000 evolving proposals by industry teams under consideration by the 3GPP2 in 2002

W-CDMA

Wideband Code Division Multiple Access (W-CDMA) is the standard developed in Europe by the European Telecommunications Standards Institute (ETSI) and member industry participants. W-CDMA is also known under the ETSI name of **Universal Mobile Telecommunications System (UMTS)**.

Universal Mobile Telecommunications System (UMTS)

The W-CDMA standard has been integrated (*harmonized* is the typical term used in 3G) with an earlier proposal submitted by China. This standard employs a time division duplex use of the same channel, thereby conserving spectrum through a "ping-pong" technique where the forward and reverse channels are used alternately on the same radio channel. When the spectrum is shared in this manner it is called W-CDMA time division duplex (TDD). The other access mode used by W-CDMA uses paired radio channels. This mode uses separate channels for the forward and reverse links and is appropriately named W-CDMA frequency division duplex (FDD).

The maximum W-CDMA chip rate is 3.84 Mcps and yields a modulated carrier about 5 MHz wide. System operators can deploy multiple carriers, each of which occupies 5 MHz. Moreover, in a W-CDMA system, multiple end-users can share each 5 MHz channel. In Figure 9-5, for example, five users share a single channel. Three operate at fixed data rates, while the data rates of the remaining two are variable. The W-CDMA system deals with this situation by continually changing the way it distributes the channel's bandwidth among the five users, adjusting the spreading factors of each user every 10 ms. System cellular operators can allocate more channels as the need for new subscribers and market conditions warrant.

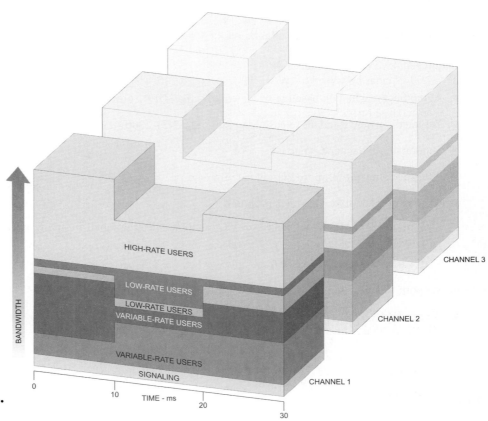

**Figure 9-5:
W-CDMA Dynamic
Bandwidth Allocation vs.
Time**

Summaries of the main features of W-CDMA are as follows:

- Bandwidth: 1.25 MHz, 5 MHz, 10 MHz, and 20 MHz
  - Bandwidth of carriers is chosen to match integral numbers of GSM 200 kHz channels for high data rate users

- Chip rates: 3.84 Mcps per channel. Each channel occupies 5 MHz of bandwidth
  - Several users share the bandwidth of a channel at the same time
- Frame rate: 10 ms
- Modulation: QPSK for data and BPSK for spreading the carrier
- Data rates: Variable

UWC-136

The UWC-136 is a third generation standard that has evolved from the North American TDMA-based ANSI-136 second generation standard.

As indicated earlier, the UWC Consortium has added W-CDMA as one of the options for third generation evolution.

The UWC-136 standard has two modes:

- 136HS (High Speed) Indoor
- UWC-136HS Outdoor

136HS Outdoor (also called EGPRS-136), is almost identical to the GSM packet radio scheme called **Enhanced Data Rate for GSM Evolution** (**EDGE**). Packet radio techniques are coupled with adaptive modulation—Gaussian minimum-shift keying (GMSK) and eight-phase phase-shift keying (8-PSK)—to give EDGE all of the 3G features except for its 2-Mb/s indoor data rate. The TDMA and GSM communities enjoy a common radio interface in EDGE for outdoor and typical mobile applications. Indoor office applications up to 2 Mb/s are accommodated by two radio techniques: the TDD mode in the W-CDMA interface on the GSM side, and 136HS Indoor on the TDMA side. The bandwidth and frame structure are illustrated in Figure 9-6.

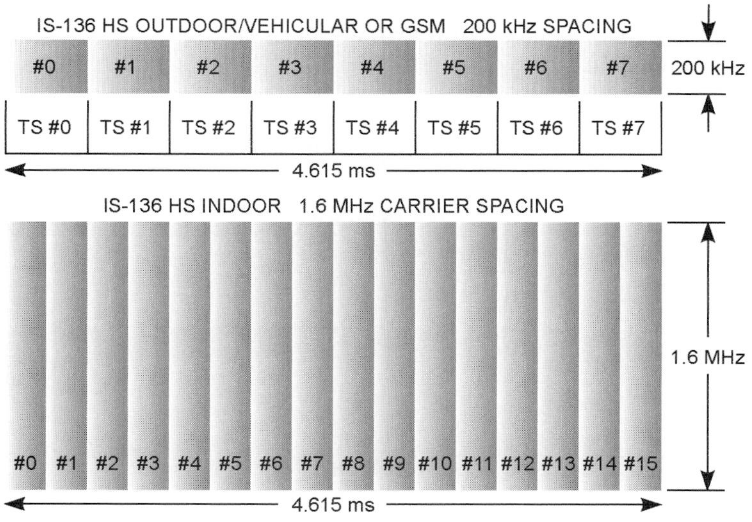

136HS Outdoor

Enhanced Data Rate for GSM Evolution (EDGE)

Figure 9-6: ANSI-136HS 200 kHz and 1.6 MHz Bandwidths

136HS Indoor uses 1.6 MHz carriers with two types of bursts: a sixteenth and a sixty-fourth of a 4.615 ms TDMA frame. The breakout of the time slots is illustrated in Figure 9-7. The high user data rates are accommodated in the 1/16-length burst, while the intermediate user rates can be accommodated in the 1/64-length burst. Adaptive modulation in both kinds of bursts lets 136HS Indoor adapt itself to a wide variety of user applications. The 1.6 MHz carriers are 8 x the 200 kHz bandwidth of the GSM carriers in order to accommodate UWC-136 and GSM multi-mode handsets in the future.

TEST TIP

Be able to describe the frame size and channel bandwidth of the HS Indoor and HS Outdoor modes or UWC-136.

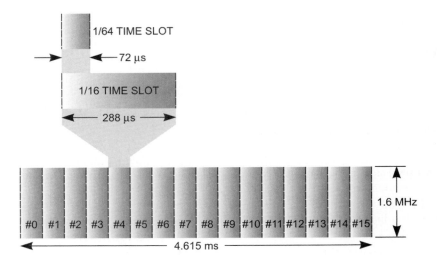

**Figure 9-7:
ANSI-136 Indoor - Two
Types of Time Slots**

New Radio Spectrum for 3G Systems

> New radio spectrum will be required for third generation wireless services.

In 1992, the International Telecommunications Union (ITU) World Administrative Radio Conference (WARC-92) identified 230 MHz of spectrum for terrestrial use for IMT-2000 on a worldwide basis. An additional 519 MHz was proposed during the World Radiocommunications Conference (WRC-2000) in Istanbul in 2000. These bands are illustrated in Figure 9-8.

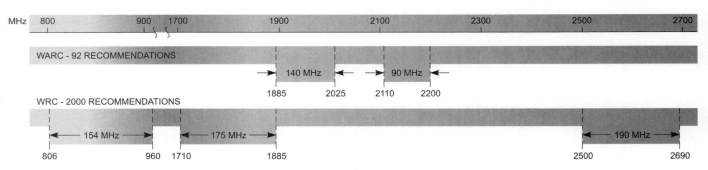

Figure 9-8: ITU Proposed 3G Spectrum

There currently is no global consensus (often referred to as "harmonization") as to how the frequency bands identified at the WARC-92 and WRC-2000 meetings will be used to implement 3G. The FCC continued in 2001 with several spectrum studies and reports on how to allocate radio spectrum in the United States for the third generation wireless services. The majority of international study groups show a preference for establishing 3G spectrum in the 2 GHz band.

> Finding the radio frequencies that are appropriate for 3G services is a complex problem.

The frequency bands proposed in Figure 9-8 for international use are already widely used for the fixed wireless services in the United States, such as the ITFS, MDS, and MMDS discussed in Chapter 7. The military also is a major user of various systems that occupy the 1710 to 1850 MHz spectrum region. The FCC has considered options such as incumbent relocation, segmenting the bands, or engineering some type of sharing of these bands. The FCC findings indicate that implementation of either the segmentation or relocation options would significantly affect deployment of, and impose considerable costs on the ITFS/MDS/MMDS services. One study suggests, for example, that the cost to ITFS/MDS/MMDS operations over a ten-year period could be up to $19 billion.

In March 2001, the FCC announced in a published final report that any of the options would require considerable time to implement and significant costs to re-engineer and deploy systems; and delivery of fixed wireless broadband services to the public and educational users would be delayed or, in rural areas or smaller markets, never be realized. The relocation option also would require other services to relocate, and the time and costs to move those additional services would be significant, ranging from approximately $10.2 to 30.4 billion. These costs would need to be balanced with the broad-based benefits to prospective users and the national economy of deploying both 3G and fixed wireless broadband systems.

WRC-2000 identified additional spectrum for possible use by terrestrial IMT-2000 systems, including the 806-960, 1710-1885, and 2500-2690 MHz bands. The WRC-2000 results allow countries flexibility in deciding how to implement IMT-2000 systems. The conference recognized that in many countries the frequency bands identified for 3G use are likely to be heavily encumbered by equally vital services that for either strategic or economic reasons cannot be readily displaced or relocated. Furthermore, not all countries in the world require equal amounts of spectrum to support future wireless services. The availability of spectrum to be used for future wireless services depends on current spectrum usage, ease of deployment of future radio-based systems, and possible transition of incumbents to different frequency bands. The various bands now allocated to mobile services in the 2 GHz band for the U.S., Europe, and Japan are shown in Figure 9-9.

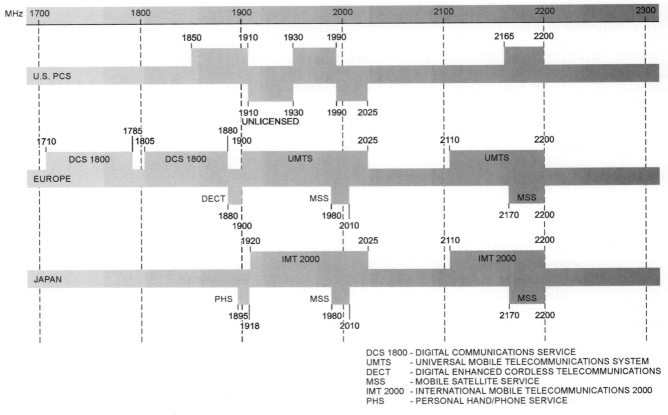

DCS 1800 - DIGITAL COMMUNICATIONS SERVICE
UMTS - UNIVERSAL MOBILE TELECOMMUNICATIONS SYSTEM
DECT - DIGITAL ENHANCED CORDLESS TELECOMMUNICATIONS
MSS - MOBILE SATELLITE SERVICE
IMT 2000 - INTERNATIONAL MOBILE TELECOMMUNICATIONS 2000
PHS - PERSONAL HAND/PHONE SERVICE

**Figure 9-9:
Mobile
Communications -
Spectrum Allocations
in the 2 GHz Band**

In the United States, the 698-746, 746-794, 806-960 (includes present cellular band), 1710-1850, 1850-1990 (present PCS bands), 2110-2150, 2160-2165, and 2500-2690 MHz bands could be considered for use by future 3G systems. The FCC has initiated the *Advanced Wireless Services* proceeding to identify spectrum that could be made available for use by 3G systems.

New Wireless Technologies

In order to maintain the growth of wireless communications, either new spectrum must be allocated or more efficient use of existing spectrum must be developed.

Two promising technologies offer more efficient modulation techniques. The first is called Orthogonal Frequency Division Multiplexing. The second new technology is called Ultra-Wideband.

A satellite system called Teledesic is in development in the United States. It may have the potential for revolutionizing the way high-speed wireless communications are offered to all countries.

The next section covers the details of these emerging global communications systems.

Orthogonal Frequency Division Multiplexing

Orthogonal Frequency Division Multiplexing (OFDM) is a special type of transmission technology gaining wide acceptance in wireless communications. It has several design objectives that are summarized as follows.

Frequencies and modulation techniques used for fixed and mobile wireless systems at 800 MHz and higher are sensitive to multipath signals, which degrade transmission quality. OFDM uses special digital signal processing methods and burst communications to offset the effects of multipath.

Standard Frequency Division Multiplexing (FDM) channels must be separated with guard channels to avoid adjacent channel interference. OFDM is able to use FDM channels that are spaced much closer and overlap each other without using guard bands. OFDM technology squeezes FDM channels closer together so that they overlap, as depicted in Figure 9-10, and do not require guard bands.

┌ **TEST TIP** ┐
Be able to describe the objectives of OFDM.

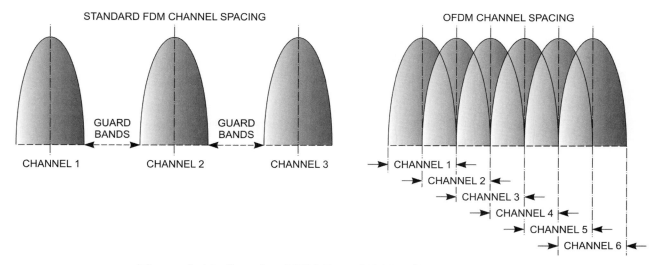

Figure 9-10: Standard FDM vs. OFDM Channel Spacing

OFDM uses many subchannels to offset the impact of frequency-selective fading on a single frequency.

All of these improvements may sound too good to be true; however, as you may expect, they do not come free.

In order to examine the advantages of OFDM in greater depth, we must first look into the causes of channel impairment called **intersymbol interference (ISI)**.

A symbol is the change in the state of a radio carrier wave as a result of modulation. For example, a symbol may be the result of a change in the phase state of a QPSK-modulated radio carrier. Since a QPSK modulation scheme can have one of four states, one state change of the phase can represent 2 bits. Possible combinations are 11, 10, 01, and 00. We therefore can conclude that the bit rate of a QPSK is 2 times the symbol rate. Symbol rate and baud rate are the same; however, symbol rate is the preferred term used in modern communications systems. (For a review of symbols and modulation basics, please refer to Chapter 4 and the section on modulation.) Since the symbol rate determines channel occupied bandwidth and the symbol carries the digital information, care must be taken to maintain the integrity of the symbol in the radio channel between the transmitter and the receiver.

┌─ TEST TIP ─┐

Be able to describe the cause of ISI.

ISI is caused by the delayed multipath components in a transmitted symbol. A multipath environment is shown in Figure 9-11. The echoes of the primary signal appear at the receiver at a slightly later time. The receiver samples the received symbol at a specific time. The received signal is a combination of a primary and several multipath signals. The delayed multipath components add or subtract at sample time in the receiver. When the phase displacements of a multipath signal, such as paths 1 and 2, are 180 degrees out of phase with the primary, the signal could be lost. The echoed signals may trail off into other subsequent symbols depending on the symbol rate. The total amount of delay interference for a specific environment is called the **delay spread**, which could typically be 4 to 6 µs. Delay spread is the residual effect of all the reflected signals arriving at a delay interval following the primary signal.

delay spread

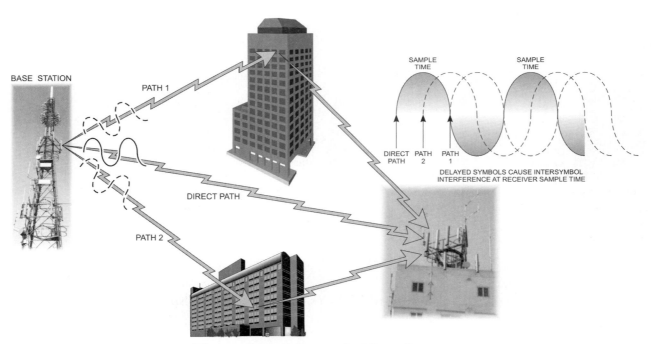

Figure 9-11: Intersymbol Interference

OFDM solves the multipath problem by expanding the distance between symbols to a value longer than the delay spread.

A high-speed bit stream that would normally be modulated on a single carrier is multiplexed into several closely spaced subcarriers, as indicated in Figure 9-10. For example, a high-speed data stream of 5 Mbps has a time interval between symbols of 1/T or .2 μs. In a multipath environment with a delay spread of 6 μs, the delayed symbols would cause ISI with the following 30 symbols. With an OFDM multicarrier system of 800 subchannels, the 5 Mbps bit stream would be multiplexed into 800 subchannels. In this example, the data rate on each subcarrier would be 5 Mbps/800 = 6.25 kbps. The interval between each symbol on each subcarrier now becomes 1/T which is 160 μs. Any delayed symbols arriving in delay spread of 6 μs would not cause any ISI.

OFDM gets the term "orthogonal" in its name because the subcarriers are shaped using a function called the **Fast Fourier Transform (FFT)**.

Fast Fourier Transform (FFT)

The transmitter and receiver perform the FFT and inverse FFT between the time domain to the frequency domain, as shown in Figure 9-12. For a relatively long time, the complexity of a real-time Fourier transform appeared prohibitive, not to mention the problem of stability of oscillators in the transmitter and receiver, the linearity required in RF power amplifiers, and the power back off required. With advances in digital signal processing, it is now a practical as well as a useful improvement for wireless transmission systems.

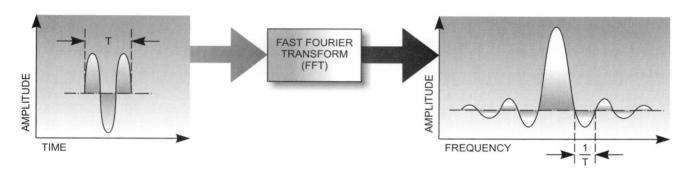

The frequency components of a frequency domain plot of a time varying signal involve an infinite summation of the harmonic waves in the pulse, which starts with the fundamental frequency and increases with integer multiples of the fundamental base frequency of 1/T. A time varying signal may therefore be expressed as a composite of sine waves and their respective amplitudes. Figure 9-12 shows that the frequency domain pulse has zero crossings spaced at intervals equal to 1/T, where T is the symbol interval. The OFDM signal processing fixes the spacing of the subcarriers such that the zero crossings are spaced 1/T apart, with T being the symbol interval.

Figure 9-12: FFT Function of a Modulated Subcarrier

Figure 9-13 shows the subcarriers spaced at 1/T. The peaks of each subcarrier at sample time are located on the zero crossings of all the other subcarriers' spectra crossings. Although they overlap with each other, they will not interfere with each other as long as the exact spacing is maintained. This is referred to in OFDM as **spectral orthogonality**.

spectral orthogonality

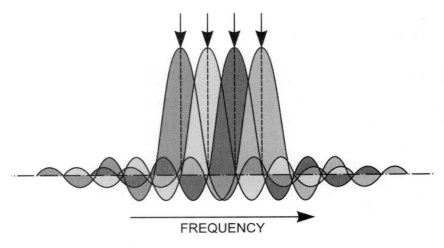

**Figure 9-13:
Multichannel Frequency
Domain Signals**

FREQUENCY

Summing up the advantages of OFDM:

- OFDM is able to overcome some of the transmission problems where there is non–line-of-sight alignment between the transmitter and receiver.

- It is possible to dramatically decrease the ISI in microwave transmission systems with OFDM.

- Even with the uses of several subchannels, the total spectrum used is less than the bandwidth that would be required in an equivalent non-OFDM multicarrier FDM system with the required guard bands.

- By spreading the bandwidth over several subchannels, the OFDM system is less sensitive to frequency-selective fading.

- Higher data rates are possible without the greater impact of ISI.

OFDM is now gaining widespread use in applications such as:

- Digital audio broadcasting

- Wireless LANs (802.11a)

- HYPERLAN Phase 2

- Digital Terrestrial Television Broadcasting (DTTB). In the DTTB OFDM transmission standard, about 2,000 to 8,000 subcarriers are used.

Ultra-Wideband

Ultra-Wideband (UWB) is a wireless digital transmission technology that uses the radio spectrum in a special way.

In May of 2000, the FCC posted a notice of proposed rulemaking on the UWB technology. The rules will determine how UWB radio technology could be used in the general public interest without causing interference to other radio services. The technology offers the potential for an array of new wireless communications applications. It also could revolutionize the radio air interface access methods for existing mobile cellular systems. New applications include covert communications, short range personal communications networks, short range precision radar systems, sensing systems, ground penetration radar systems, and locating objects or people behind walls or enclosures. Many of the radar-based applications have already been developed using UWB technology and are used commercially under several product names.

There are several names adopted for UWB technology. They include time domain, carrier free, impulse transmission, carrierless, and others.

The specific technique under consideration by the FCC was developed by Time Domain Inc. This approach to UWB along with a chipset is patented under the name "PulsON^{tm}".

UWB technology is not new. It uses a short pulse method for spreading the spectrum of a modulated radio carrier similar to the more popular spread spectrum modulation techniques covered earlier in Chapter 5.

UWB employs one of the theoretical spread spectrum techniques, overlooked to some extent except for military applications called **Time-Hopped Spread Spectrum (THSS)** modulation.

Time-Hopped Spread Spectrum (THSS)

The primary difference between time-hopped spread spectrum modulation and other more familiar types of spread spectrum modulation, such as Direct-Sequence Spread Spectrum (DSSS) and Frequency-Hopped Spread Spectrum (FHSS), is the duty cycle. Let me define what is meant by the duty cycle of a radio transmission.

TEST TIP

Know the definition of duty cycle.

UWB uses a series of very short pulses. This is similar to a conventional radar system where the transmitter is on for a very short period of time followed by a relatively long interval for the receiver to "listen" for returned radar pulses from the target. The ratio of the **time on** to the **time off** interval is referred to as the duty cycle. DSSS and FHSS systems use a continuous carrier with a 100% duty cycle. Time-hopped spread spectrum (THSS) may have a duty cycle of 1% or less where the transmitter is off most of the time.

time on

time off

The spreading of the radio carrier spectral energy over a given bandwidth of radio spectrum for any radio transmitted wave is related to a mathematical function called the Fourier transform discussed earlier, which establishes a mathematical relationship between the frequency domain and the time domain. This tells us that a pulse duration (T) of a radio burst of transmitted power in seconds (in the time domain) has an occupied bandwidth 2/T hertz of a radio spectrum (in the frequency domain).

Example problem:

The occupied bandwidth of a radio transmission in hertz (BW) is equal to 2 divided by the time duration of the transmitted pulse in seconds. Therefore, as shown in the following equation, a transmitted pulse of one nanosecond (1×10^{-9} seconds) duration would result in an occupied bandwidth of 2 GHz.

$$BW = \frac{2}{T} = \frac{2}{1 \times 10^{-9}} = 2 \times 10^{9} = 2\,GHz$$

So far, this indicates that we can create a wideband radio signal by turning the radio carrier off except for short periods of time. But we need to have some type of digital modulation scheme other than on/off keying to make some use of the transmitted pulse. Pulse Position Modulation (PPM) is used as illustrated in Figure 9-14. A transmission of a single sine wave called a **monopulse** is varied in the time domain about the nominal position. In the example shown, 20 million pulses would be transmitted each second. The monopulse interval would be 50 nanoseconds. The **pulse interval** shown in the example is different from the **pulse width** of each monopulse. With pulse position modulation, the deviation of the pulse from the nominal position can be used to represent a logical 1 or a logical 0. In this example, a logical "0" may be represented by transmitting the monopulse 50 picoseconds (1×10^{-12} seconds) early. A logical "1" may be represented by transmitting a monopulse 50 picoseconds late. The low duty cycle would provide a dramatic decrease in the power consumption for transmitters in a model UWB system.

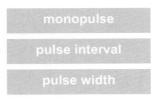

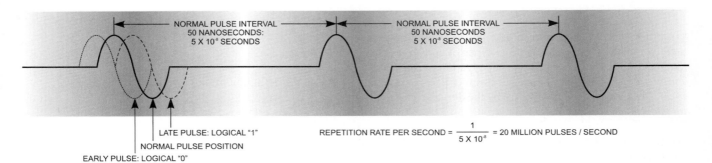

Figure 9-14: Pulse Position Modulation

The nominal position of each pulse in Figure 9-14 assumed an equal interval between each pulse. If we have a system of two transmitters and two receivers, there would be no way to separate the two channels since they each use the same periodic monopulse interval. DSSS (CDMA) uses different PN codes to segregate users sharing the same radio spectrum. With time-hopping UWB, different PN codes are used to randomize the time of each pulse. Each transmit/receive pair in a time-hopped system would therefore use an assigned unique PN code, thereby establishing a separate channel for each user. Figure 9-15 shows an example of how a time-hopped multiple access system could be developed.

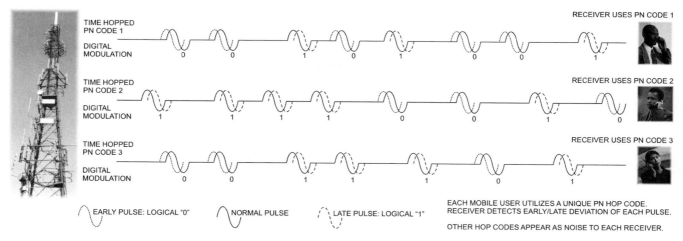

TIME HOPPED PN CODE 1
DIGITAL MODULATION
0 0 1 0 1 0 0 1
RECEIVER USES PN CODE 1

TIME HOPPED PN CODE 2
DIGITAL MODULATION
1 1 1 1 0 0 1 0
RECEIVER USES PN CODE 2

TIME HOPPED PN CODE 3
DIGITAL MODULATION
0 0 1 1 1 0 1
RECEIVER USES PN CODE 3

EARLY PULSE: LOGICAL "0"   NORMAL PULSE   LATE PULSE: LOGICAL "1"

EACH MOBILE USER UTILIZES A UNIQUE PN HOP CODE. RECEIVER DETECTS EARLY/LATE DEVIATION OF EACH PULSE.

OTHER HOP CODES APPEAR AS NOISE TO EACH RECEIVER.

There currently are no standards developed for future UWB communications systems. Figure 9-15 shows a theoretical approach for a time-hopped UWB system.

Figure 9-15: Time-Hopped UWB System

Because of their low duty cycle, low spectral power density, and high processing gain, time-hopped UWB systems provide the possibility of a wireless radio interface access methodology that could accommodate a large number of users sharing the same radio spectrum in a mobile radio system of the future.

The Teledesic Satellite Network

An advanced concept in broadband satellite communications called Teledesic (pronounced tel-e-DEH-sic) is planned for operational status in 2005. Teledesic is designed to carry broadband digital transmission telecommunications.

Teledesic will be functionally equivalent to a high-speed packet switched network in the sky.

The network will consist of 288 satellites split into 12 planes each with 24 satellites.

TEST TIP
Know the planned number of satellites for Teledesic.

The satellite illustrated in Figure 9-16 uses a phased-array antenna. Each of these satellites is a node in a fast-packet-switch network and has optical communication links with other satellites in the same and adjacent orbital planes. The satellites create a "mesh" network of routers in orbit similar to the terrestrial Internet packet-switched topology.

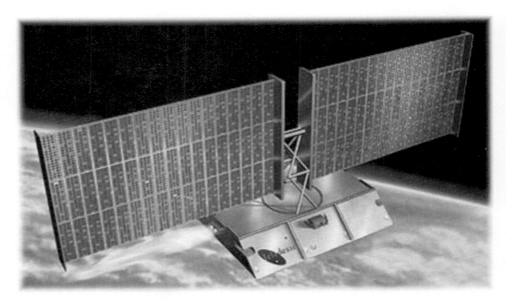

Figure 9-16:
Teledesic Satellite

The Teledesic network will offer broadband multimedia services such as Internet access to individual home clients. Gateway stations will also serve corporate clients and Internet service providers. Figure 9-17 shows the overall network structure for space and terrestrial components. The satellites use both inter-satellite links and satellite-earth station links. Users will have downlink capacities of 64 Mbps and 2 Mbps or more on the uplink. The capacity of the Teledesic network is advertised to support aggregate rates of 500 Mbps within any circular area of 100 km radius. The network will provide bandwidth-on-demand packet service, which is similar to modern terrestrial packet data services. The constellation uses the uplink Ka-band segment of 28.6 to 29.1 GHz. The downlink operates at K-band between 18.8 to 19.3 GHz.

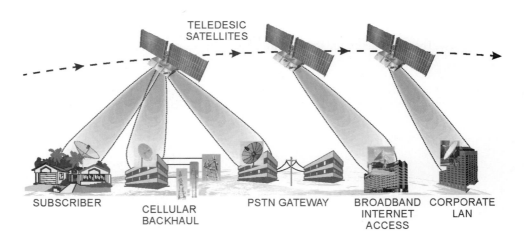

Figure 9-17:
Teledesic Network

The satellite orbital parameters are illustrated in Figure 9-18. The near-polar orbits are sun-synchronous and are inclined at 8.5 degrees. The 12 planes are separated by 15 degrees. The satellite orbital altitude is 1350 km. The offset between the earth polar axis and the orbital axis is shown as 8.5 degrees. This makes the convergence of the 12 orbital planes offset from the polar axis, as shown in the illustration. The constellation design has 24 satellites planned for each of the 12 planes. This design will provide uniform coverage for all of the world's population.

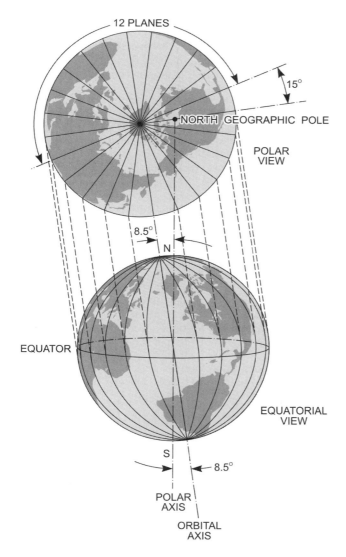

Figure 9-18:
Teledesic Satellite Orbits

Teledesic proposes a wide variety of terminals, with bit rates ranging from a minimum of 16 kbps, plus 2 kbps for signaling, up to 2.048 Mbps (128 basic channels of 16 kbps) for the mobile subscriber terminals. The mobile terminal antennas have a diameter ranging from 8 cm to 1.8 meters, and an average output power ranging from 0.01 W to 4.7 W. The antenna diameter is determined by maximum output power, maximum channel rate, climatic region, and availability requirements.

A small number of fixed-site terminals operating between OC-3 (155 Mbps) and OC-24 (1.2 Gbps) can also be supported in the design.

Key Points Review

In this chapter, the following key points were introduced concerning the emerging technologies and standards for the wireless industry. Review the following key points before moving into the Review Questions section to make sure you are comfortable with each point and to verify your knowledge of the information.

- "Third generation" or simply "3G", is a term used to describe the wireless standards and technology that will provide high-speed data transmission speeds up to 2 Mbps and a host of broadband services on existing and upgraded mobile cellular networks.

- The transition from the second generation to the third generation includes some enhancements that are called "2.5G".

- The call for proposals for third generation mobile Radio Transmission Technologies (RTTs) was initiated by the IMT-2000 study group in 1998 and resulted in several submissions from Standards Development Organizations (SDOs) from various countries.

- The ETSI/GSM (European Telecommunications Standards Institute/Global System for Mobile Communications) companies, such as Nokia and Ericsson, are backing the W-CDMA (Wideband Code Division Multiple Access).

- U.S. vendors, including Qualcomm and Lucent Technologies, are backing CDMA2000.

- While "voice only" wireless cellular networks have become very popular, the emergence and rapid growth of the Internet has now started the trend for mobile users to expect the same capabilities for high-speed access to the Internet and corporate intranets when away from home or the office.

- CDMA2000 is intended to provide 3G services over mobile radio networks that conform to the ANSI document "TIA/EIA-41," which includes existing (2G) ANSI-95 and TDMA (ANSI-136) nets.

- The 1x-Evolution – Data Only (1x EV-DO) specification is another evolution of CDMA2000. The 1x EV-DO standard, promoted and developed by Qualcomm, was also known earlier as High Data Rate (HDR).

- A second evolution of the core CDMA2000 specification is called 1x-Evolution – Data and Voice (1x EV-DV). It provides the capability to mix data and voice.

- Wideband Code Division Multiple Access (W-CDMA) is the standard developed in Europe by the European Telecommunications Standards Institute (ETSI) and member industry participants. W-CDMA is also known under the ETSI name of Universal Mobile Telecommunications System (UMTS).

- The UWC-136 is a third generation standard that has evolved from the North American TDMA-based ANSI-136 second generation standard.

- New radio spectrum will be required for third generation wireless services.

- Finding the radio frequencies that are appropriate for 3G services is a complex problem.

- In order to maintain the growth of wireless communications, either new spectrum must be allocated or more efficient use of existing spectrum must be developed.

- Two promising technologies offer more efficient modulation techniques. The first is called Orthogonal Frequency Division Multiplexing. The second new technology is called Ultra-Wideband.

- A satellite system called Teledesic is in development in the United States. It may have the potential for revolutionizing the way high-speed wireless communications are offered to all countries.

- OFDM solves the multipath problem by expanding the distance between symbols to a value longer than the delay spread.

- OFDM gets the term "orthogonal" in its name because the subcarriers are shaped using a function called the Fast Fourier Transform (FFT).

- OFDM is able to overcome some of the transmission problems where there is non–line-of-sight alignment between the transmitter and the receiver.

- It is possible to dramatically decrease the ISI in microwave transmission systems with OFDM.

- Even with the uses of several subchannels, the total spectrum used is less than the bandwidth that would be required in an equivalent non-OFDM multicarrier FDM system with the required guard bands.

- By spreading the bandwidth over several subchannels, the OFDM system is less sensitive to frequency-selective fading.

- Higher data rates are possible without the greater impact of ISI.

- Ultra-Wideband (UWB) is a term used for a wireless digital transmission technology that uses the radio spectrum in a special way.

- UWB employs one of the theoretical spread spectrum techniques, overlooked to some extent except for military applications called time-hopped spread spectrum (THSS) modulation.

- Because of their low duty cycle, low spectral power density, and high processing gain, time-hopped UWB systems provide the possibility of a wireless radio interface access methodology that could accommodate a large number of users sharing the same radio spectrum in a mobile radio system of the future.

- An advanced concept in broadband satellite communications called Teledesic (pronounced tel-e-DEH-sic) is planned for operational status in 2005. Teledesic is designed to carry broadband digital transmission telecommunications.

- The network will consist of 288 satellites split into 12 planes each with 24 satellites.

- Teledesic proposes a wide variety of terminals, with bit rates ranging from a minimum of 16 kbps, plus 2 kbps for signaling, up to 2.048 Mbps (128 basic channels of 16 kbps) for the mobile subscriber terminals. The mobile terminal antennas have a diameter ranging from 8 cm to 1.8 meters, and an average output power ranging from 0.01 W to 4.7 W. The antenna diameter is determined by maximum output power, maximum channel rate, climatic region, and availability requirements.

The following questions test your knowledge of the material presented in this chapter:

1. What are the two core networks that are used for handling mobile communications in North America, Europe, and Asia?

2. What are the two main CDMA technologies that are competing for a role in the global 3G market?

3. What radio transmission technology is used in the North American UWC-136HS standard?

4. Describe the evolutionary path for the CDMA2000 standard.

5. Describe the 3G technology standard that is referred to as "UMTS."

6. Briefly describe the functional features of EDGE.

7. Explain how the features of UWC-136 Indoor will allow dual-mode phones to be used with the UWC-136HS Indoor mode as well as be backward compatible with GSM networks.

8. Describe the problems faced by the United States in using radio spectrum recommended by the WARC 92 and WRC 2000 conferences for international 3G harmonization (sharing of the same spectrum by all countries).

9. Describe what is meant by the term "delay spread."

10. Describe the purpose of OFDM. What type of problem is it designed to solve?

11. What are some advantages offered by Ultra-Wideband technology?

12. What is pulse position modulation?

13. What type of communications is the Teledesic satellite system designed to handle?

14. How many satellites are planned for the Teledesic Network?

15. What are the data rates possible with Teledesic for higher performance fixed earth stations?

Glossary

10Base-T An Ethernet specification defined by the IEEE 802.3 committee officially as specification 802.3i. The specification describes a 10 Mb/s wire speed using baseband information. The specification requires a star topology using Category 3, 4, or 5 unshielded twisted pair (UTP)cable.

100Base-T An Ethernet IEEE 802.3u specification utilizing a wire speed of 100 Mb/s using Category 5 unshielded twisted pair (UTP) cable.

1x-Evolution – Data Only (1x EV-DO) A specification that is an evolution of CDMA2000. The 1x EV-DO standard, promoted and developed by Qualcomm Inc., was also known earlier as High Data Rate (HDR). It was first published as ANSI-856 in November 2000.

1x-Evolution – Data and Voice A second evolution of the core CDMA2000 specification developed by Qualcomm Inc. It provides the capability to mix data and voice. It also supports higher data rates than 1x EV-DO.

A

Access channel A channel used by the mobile station only in the ANSI-95 CDMA cellular standard for originating calls and responding to a page.

Adjacent channel interference A type of interference caused by spillover of some of the transmitted energy between two or more radio frequency carrier waves operating on adjacent channels.

Advanced Mobile Phone Service (AMPS) The first generation analog cellular standard TIA/EIA-553 using FM modulation.

Air interface The multiple access standard of a wireless network; technologies include AMPS, TDMA, CDMA, and GSM.

A-key A secret number issued to a cellular phone that is used in conjunction with a subscriber's shared secret data (SSD) information for authentication in the network.

Alexanderson alternator An electromechanical device developed in the early 20th century by Ernst Alexanderson for the generation of radio carrier waves. This invention led to the development of commercial broadcasting in the U.S.

Almanac A database maintained by a satellite positioning system that tells where each satellite should be in space at any given instant in time.

Algorithm A computational process or structured procedure for producing a desired result

ANSI (American National Standards Institute) A U.S. organization chartered to accredit standards developed by a variety of professional standards bodies and industry groups.

ANSI-41 A standard for intersystem network operation of the U.S. cellular system published by the Telecommunications Industry Association (TIA) subcommittee TR45.2. ANSI-41 has been revised several times since its original publication in January 1991. The ANSI-41-C has been replaced with the current version called Cellular Radiotelecommunications Intersystem Operations (ANSI/TIA/EIA-41-D-97).

ANSI-54 The D-AMPS standard first published in 1990 and revised as ANSI-54B in 1991. This standard offered several improvements over the AMPS. ANSI-54B was later finalized as TIA/EIA-54B, which was subsequently rescinded in 1996 and replaced by TIA/EIA-627.

ANSI-95 Initially the official air interface specification for CDMA operating in the 800 MHz cellular band. Several improvements and iterations resulted in the release of ANSI-95A. Later features not included in ANSI-95A were subsequently published as TSB-74 and, finally, the ANSI-J-STD-008 brought forth CDMA standards for operation in the new 1900 MHz PCS band. In 1998, the revised standard was released as TIA/EIA-95 (no longer an "interim standard") and incorporated all the features of ANSI-95, J-STD-008, and TSB-74, and additional features.

ANSI-136 A second-generation digital cellular standard using Time Division Multiple Access (TDMA) as the air interface. The ANSI-136 (now TIA/EIA-136B) specification is a newer TDMA specification incorporating a digital control channel in place of the AMPS analog control channel. The 136B standard is a migration to new services from the TIA/EIA 627/ANSI-54 specification.

Antenna gain The ratio of the power radiated by an antenna in a favored direction compared with a theoretical reference antenna called an isotropic antenna that radiates equal energy in all directions.

ATM (Asynchronous Transfer Mode) A high-speed, high-bandwidth transmission technology used for multimedia, video, data, and voice transmission using fixed-length data transmission units called cells.

Atomic clock A very precise clock used in GPS satellites that operates using the elements cesium or rubidium. A cesium clock has an error of one second per million years.

Attenuation Loss of transmitted power as it propagates through space. Attenuation is also used to describe the loss or reduction of voltage, current, or power as it travels through a medium.

Audion The name of an early experimental vacuum tube invented by Lee DeForest.

Authentication The process used in security systems to verify the identity of a source of a transmission or to verify the eligibility of an individual to receive service or transmissions.

Authentication Center A protected database that stores and processes secret keys required to authenticate and provide encryption for wireless telephones.

Azimuth The horizontal direction from one point on the Earth to another measured clockwise in degrees (0-360) from a north or south reference line. An azimuth is also called a bearing.

B

Bandwidth The transmission capacity of a medium in terms of a range of frequencies. A greater bandwidth indicates the ability to transmit a greater amount of data over a given period of time.

Basic Trading Area (BTA) A regional area based on economics and used as a basis for allocating licenses for wireless common carriers. The total amount of broadband spectrum available is evenly divided between Metropolitan Trading Areas (MTAs) and BTAs. BTAs represent rural and suburban areas, whereas MTAs represent urban areas. These divisions were drawn from work performed by the Rand McNally Corporation.

Beamwidth Antenna beamwidth is measured at the half-power points of the main beam that projects from the antenna aperture. The antenna half-power point radiation pattern can be viewed as a cone with the vertex at the antenna surface. The angle of the arc formed by the cone is called the antenna beamwidth.

Bit (Binary digit) A unit of digital information consisting of a binary one or a zero. A bit is the fundamental unit of storage or transmission in digital systems.

Bit Error Rate (BER) Numerically equal to the number of bits received that are in error divided by the total number of bits transmitted. Since the ratios typically involve large numbers, the BER is typically specified as a power of ten. A system that is operating with a BER of 3 bits in error per million bits of transmitted digital information is said to have a BER of $3/1,000,000$ or 3×10^{-6}.

Bit rate The number of binary digits transmitted over a media. Bit rate is specified as the number of bits transmitted in one second (b/s).

Blank and burst encoding A signaling format used on the Advanced Mobile Phone Service (AMPS) forward voice channel when a call is in progress. It is implemented by turning the FM voice transmission off for approximately 100 ms and inserting a "burst" of digital data commanding the mobile station to perform specific tasks.

Broadband A type of communications medium capable of transmitting a relatively large amount of data over a given period of time.

Broadcast Satellite Service (BSS) An FCC category of wireless service also known as Broadcast Satellite Service (BSS) and Direct Broadcast Service (DBS) in the industry. The BSS delivers "cable type" commercial television programming to subscribers with small receive only terminals in the home. The BSS shares frequencies with the fixed satellite service.

Bus A transmission path where a number of devices are all attached to a common media. Also used to describe the support system on a satellite.

C

Carrier wave A radio wave used to convey or "carry" baseband information between a radio transmitter and receiver. A radio wave is also referred to as the carrier wave or the signal. As the name implies, some type of information is carried by the medium between a radio transmitter and receiver.

CDMA (Code Division Multiple Access) A multiple access technique for sharing a radio spectrum resource among several users. CDMA uses spread spectrum modulation where the resulting occupied bandwidth is in excess of the normal bandwidth used in standard modulation techniques. With CDMA, subscribers in a single cellular coverage area share the same carrier frequency at the same time with each station separated by orthogonal code sequences contained in the composite modulated carrier. Second generation CDMA networks use one radio carrier for all mobiles and another carrier for all base stations. It therefore operates in the FDD mode.

CDMA Frequency Assignment A 1.2288 MHz segment of spectrum centered at a discrete frequency identified by the CDMA channel number. The allowable channels are centered on one of the 30 kHz channels of the AMPS system.

cdmaOne A promotional logo used by the CDMA Development Group for the ANSI-95 CDMA standard.

CDMA2000 A third generation brand name for a CDMA specification evolved from the ANSI-95 second generation cellular standard.

Cell sectoring A technique used by cellular service providers to increase the capacity of a cellular network by reducing co-channel interference. With sectoring, the base station uses a set of directional antennas to establish the transmit and receive patterns that favor a specific direction. Three-sector or 6-sector cells are often used in high-traffic areas.

Cellular radio A technology consisting of a series of cells with dedicated radio channels that provide mobile telephone service and access to the Public Switched Telephone Network (PSTN). The cellular concept provides for reusable frequencies with structured spacing between cells using the same radio channels.

Channel coding A processing function used in wireless communications systems that adds extra bits to the information bit stream to provide forward error correction.

Chip Informal term used in CDMA systems to refer to either a binary element of a spreading sequence, or to the time interval that it occupies, such as $1/1.2288$ MHz $= 813.8$ nanoseconds.

Chipping code A pseudorandom binary code used in direct sequence spread spectrum radio systems to modulate and spread the radio frequency radio carrier energy over a wide bandwidth. Chipping code is a term used to differentiate the spreading code bits from the information bits.

Chirped FM Pulsed Spread Spectrum A pulsed spread spectrum technique where the radio carrier is turned on and off in a fixed period and fixed on-off time interval. When the transmitter is pulsed on, the frequency is varied (swept) over a fixed range, creating an additional spreading of the carrier. Chirped FM systems are used primarily in radar systems to obtain better resolution in range measurements.

Clock bias The difference between the indicated clock time in a GPS receiver and true universal time.

Cochannel interference A type of interference that occurs between two channels operating on the same frequency. Cochannel interference may occur in cellular systems where two cells are improperly isolated by distance or terrain. Wireless systems are often designed to tolerate a minimal amount of cochannel interference.

Continuous waves A term that describes a radio carrier wave with constant frequency over time. The term was first used to describe early radio transmissions using alternators, which produced "continuous" waves as opposed to keyed "on and off" transmissions produced by wireless telegraphy transmitters.

Control channel A radio channel used in cellular systems for call management, setup, and status monitoring of a voice channel.

COSPAS – SARSAT An international satellite system that provides space communications supporting the GMDSS. The system is a joint effort supported by the U.S., Canada, Russia and France.

D

D-AMPS A North America second generation digital wireless cellular standard that evolved from the AMPS analog standard for cellular communications. D-AMPS was the result of a study and subsequent design by the Telecommunications Industry Association (TIA) for an improved service for the AMPS analog standard ANSI-553. The TIA activity led to the publication of the D-AMPS standard ANSI-54 in 1990 and ANSI-54B in 1991. This standard offered several improvements over the AMPS. ANSI-54B was later finalized as TIA/EIA-54B, which was subsequently rescinded in 1996 and replaced by TIA/EIA-627.

dBm Power measured in dB relative to one milliwatt.

dBW Power measured in dB relative to one watt.

Decibel The decibel (dB) is a method of expressing the ratio of two quantities. The expression is in terms of the logarithm to base 10 of the ratio instead of the raw ratio. This is done for convenience in expressing the ratio of numbers many magnitudes apart with decibel numbers that are not as large.

Department of Defense (DOD) A U.S. Government executive agency responsible directly to the President for the management of the military services.

Direct Sequence Spread Spectrum (DSSS) A type of spread spectrum technology used with the cellular ANSI-95 standard air interface. In this case the carrier wave is modulated with a binary bit stream. The repetitive rate of the bit stream is called the chipping rate and is much larger than the information bit rate. This code is called the spreading code. As an example, ANSI-95 uses a spreading code of 1.2288 megachips per second.

Discriminator A circuit used in a frequency modulation (FM) receiver to recover the audio information from the carrier wave.

Doppler shift The apparent increase or decrease in frequency observed by a receiving station due to the relative difference in the velocity vector between the radio transmitter and receiver. Doppler shift is also observed in the audio spectrum when a sound appears to change frequency when the velocity vectors between the sound source and the listener are changed.

Dual-band phone In the U.S. cellular industry, a cell phone that can operate in the 800 MHz band and the 1900 MHz (PCS) band.

Dual-mode phone A wireless handset capable of operating in the analog mode and a digital mode.

E

Edison effect A phenomenon first observed and named by Sir John Fleming in 1901. He observed the dark coating on the inside of an early light bulb, which later led to experiments with vacuum tubes and the development of radio receivers.

Effective Isotropic Radiated Power (EIRP) The transmitted power multiplied by the antenna gain referenced to an ideal isotropic radiator. EIRP is larger than ERP in the same direction by the gain of an ideal dipole relative to an isotropic radiator, which is 2.1 dB.

Effective Radiated Power (ERP) The transmitted power multiplied by the antenna gain referenced to a half-wave dipole.

Electronic serial number (ESN) The electronic serial number is burned into a cell phone programmable read-only memory at the factory by the manufacturer. It is used as a permanent 32-bit identification number. Any attempt to alter the ESN makes the phone unusable.

Encapsulation A process where a packet or frame is enclosed within another packet for the purpose of hiding the header of the encapsulated packet during transit through a network. At the destination, the receiving node recognizes the encapsulation header and recovers the original enclosed packet.

Equipment identity register An optional database used within a cellular network that contains the list of all valid mobile equipment on the network. The equipment in this database is listed by the International Mobile Equipment Identity (IMEI). If a mobile station shows up on the network that is reported stolen or is not approved for use, it can be marked as invalid and will be denied service.

Ethernet A local area network set of protocols initially developed in the early 1970s by Bob Metcalfe and David Boggs of the Xerox Palo Alto Research Center (PARC) in California. The IEEE later developed a set of standards known as the IEEE 802.3 Ethernet standards, based on the original Xerox Ethernet architecture.

Extremely High Frequency (EHF) The radio frequency band that extends from 30 GHz to 300 GHz.

F

Fast Ethernet A term used to describe the IEEE 802.3u 100baseT Ethernet specification with a wire speed of 100 Mb/s.

Federal Communications Commission (FCC) A U.S. Government regulatory agency that manages administration of telecommunications policy for the commercial industry in the U.S. The FCC conducts the issuing of wireless operator licenses and public spectrum auctions for qualified corporations and common carriers.

Fixed earth station A station operating in the satellite service intended to be used at a specified fixed point specified in an FCC license. The definition is usually reserved for stations that transmit and receive information from a satellite and excludes unlicensed receive only (RO) terminals

Frequency Division Duplex (FDD) A type of simultaneous two-way communications between two stations where each station has a dedicated channel. With FDD, both stations can transmit and receive at the same time.

Frequency-Hopping Spread Spectrum (FHSS) A method where the radio carrier frequency is shifted in selected increments across a given frequency band. The pseudorandom code is used to provide a pattern for the frequency shifts.

Fixed Satellite Service (FSS) A service defined in the FCC rules that includes satellites operating in geosynchronous orbit with a fixed orbital location and earth stations operating at a fixed geographical location.

Footprint The area of the Earth's surface covered by a satellite transmitter antenna pattern.

Forward control channels (FOCC) The wireless channel used from the base station to mobile station by the Advanced Mobile Phone Service (AMPS) for call setup and supervisory control.

Frame A term often used interchangeably with the term "packet." The most accepted definition for the term "frame" denotes an information unit that contains control and address information corresponding to the OSI layer 2 or the Data Link layer.

Free-space optical A term used to describe a wireless technology that uses light waves to transmit information through the atmosphere. The words "free space" are used to differentiate between atmospheric media and "closed" fiber-optic cables that also use light waves to transmit digital information.

Frequency Within the context of wireless communications terminology, a measure of the number of cycles or reversals of an electromagnetic field that occur in one second. Also used to express the quantity of units of information transmitted in a given time period expressed in hertz.

G

Geostationary satellite A geosynchronous satellite that orbits the Earth directly over the equator. To stay over the same spot on Earth, a geostationary satellite also has to be directly above the equator at an altitude of 35, 786 km; otherwise, from the Earth the satellite would appear to move in a lazy figure 8, slowly drifting in a north-south path over the equator once per day.

Geosynchronous satellite A satellite that completes one orbit around the Earth in the same time that it takes the Earth to make one rotation in inertial (or fixed) space relative to the stars. This time period is known as one sidereal day and is equivalent to $23^h56^m04^s$ of mean solar time. A geosynchronous satellite may also be geostationary if the orbital plane passes through the equator.

Gigabit Ethernet A term used to describe the IEEE 802.3z 1000baseX Ethernet specification operating at a wire speed of 1 Gigabit/s.

Global Maritime Distress and Safety System (GMDSS) A worldwide terrestrial radio and satellite communications system designed for maritime safety. The GMDSS employs several cooperative terrestrial and satellite radio communications systems designed to provide reliable distress alerts, weather warnings, search and rescue operations, and preservation of life on the high seas.

Global Positioning System (GPS) A global navigation system based on a constellation of 24 satellites plus spares orbiting the earth at an altitude of 11,865 miles and providing very precise, worldwide positioning and navigation information 24 hours a day. Also referred to as the NAVSTAR system.

GLONASS: The Russian **Glo**bal **Na**vigation **S**atellite **S**ystem (GLONASS) similar to the U.S. Global Positioning System. GLONASS consists of a constellation of 24 satellites at an altitude of 12,548 miles.

Ground waves Electromagnetic waves below 2 MHz that travel generally parallel to the surface of the Earth.

GSM An acronym for Groupe Spécial Mobile (commonly referred to as Global System for Mobile Communications in English). The GSM standard uses TDMA as the air interface and is maintained by the European Telecommunications Standards Institute (ETSI). The standard is widely used in Europe and to a lesser extent in the U.S.

H

Handoff The dynamic transfer of the control of a wireless voice or data session from one cellular base station to another base station without a break in the transmission.

Handset Any type of handheld device used to transmit and receive calls within a wireless system.

Home Location Register A database containing all specific subscriber service data, including the profile of services that each subscriber is authorized to use on a mobile cellular network. Every mobile station has a "home" location determined by the service area of the service provider.

I

Initial Voice Channel Designation Message (IVCDM) A message transmitted during call setup on the Advanced Mobile Phone Service (AMPS) forward control channel to the mobile station. The IVCDM contains the voice channel number, the associated Supervisory Audio Tone (SAT) for the mobile to repeat back, and the Voice Channel Mobile Attenuation Code.

Instructional Television Fixed Service (ITFS) A service provided by one or more fixed microwave stations operated by an educational organization and used to transmit instructional information to fixed receiving locations. The FCC uses the term "fixed" to describe radio services that operate licensed radio transmitter stations at permanently installed transmitting and receiving locations, as opposed to "mobile" services that do not operate from stationary locations.

International Maritime Organization (IMO) An international treaty organization, sponsored by the United Nations, that administers maritime policy affecting passenger ships and other types of ships of weight 300 tons and over. IMO rules affecting radio, which have treaty status in the U.S., are included in the Safety of Life at Sea (SOLAS) Convention. The most current system adopted by the IMO and SOLAS is the Global Maritime Distress and Safety System (GMDSS)

International Mobile Satellite Organization A corporation that provides mobile communications for land, sea, and air terminals on a national scale. The company was formed from the privatization of the former International Maritime Satellite Organization (INMARSAT) in 1994.

International Organization for Standardization (ISO) A worldwide group of national standards bodies representing 130 countries. "ISO" is a word, derived from the Greek "isos," meaning "equal," which is the root of the prefix "iso-" that occurs in a host of terms, such as "isometric" (of equal measure or dimensions) and "isonomy" (equality of laws, or of people before the law). From "equal" to "standard," the line of thinking that led to the choice of "ISO" as the name of the organization is easy to follow. In addition, the name ISO is used around the world to denote the organization, thus avoiding the plethora of acronyms resulting from the translation of "International Organization for Standardization" into the different national languages of members, e.g., IOS in English, OIN in French (from Organization Internationale de Normalisation). Whatever the country, the short form of the Organization's name is always ISO. ISO's work results in international agreements, which are published as International Standards. One of the popular standards developed by ISO is the Open Standards Interconnect (OSI) model.

International Telecommunications Union (ITU) An international organization founded in Paris in 1865 as the International Telegraph Union. It took its present name in 1934 and became a specialized agency of the United Nations in 1947. The ITU is an intergovernmental organization, within which the public and private sectors cooperate for the development of telecommunications. The ITU adopts international regulations and treaties governing all terrestrial and space uses of the frequency spectrum as well as the use of the geostationary-satellite orbit, within which countries adopt their national legislation. The ITU has taken the initiative in the promotion and implementation of third generation international wireless standards called IMT-2000

Internet A worldwide collection of networks that make up an international Internet where almost any client can access any server. The Internet was initially a U.S. Government network called the ARPANET, which was named from the developing organization, the Advanced Research Projects Agency.

Inter-working function (IWF) A processing system used in a cellular mobile switching center that provides the interface between two dissimilar networks. An IWF is normally used to provide an interface between cellular networks and the Internet.

Intranet A network that provides connections between different facilities within a single corporation, government organization, or university campus.

Inverse square law A law of physics applied in wireless network design that states that as the distance from a transmitting source to the receiver doubles (x2), the received power density decreases by the inverse of the square of 2.

J

K

L

Local Area Network (LAN) Local area networks are composed of several personal computers connected together and limited in scope to a single building or facility. A LAN enables computer users to share files and peripheral equipment.

Local Multipoint Distribution Service (LMDS) An FCC term for a fixed, broadband wireless service authorized for use in the 28 and 31 GHz bands, providing full-duplex voice, high-speed Internet access, data, and video transmission links. LMDS typically uses a series of cells that each deliver point-to-multipoint services to subscribers. (Often referred to incorrectly as local multipoint distribution system.)

Local oscillator A radio circuit used in superheterodyne receivers to provide a tunable local radio carrier wave. The local oscillator output is mixed with the incoming desired radio carrier frequency to produce an intermediate frequency signal.

Low Earth Orbit (LEO) A satellite orbit between 300 and 800 miles above the earth.

M

Mcps Megachips per second (10^6 chips per second). Chips is the term used to describe the spreading code bits used in direct sequence spread spectrum systems.

Mean time between failures (MTBF) A statistical measure of the reliability of a system or set of components based upon the known failure rates of the individual components within the system. MTBF is usually expressed in hours.

Medium earth orbit A satellite orbit in the range of 10,000 to 15,000 miles above the Earth.

Metropolitan Area Network (MAN) A network that encompasses an area larger than a LAN. It serves larger areas within a city or, as the name implies, a major metropolitan area. An example is the fire, police, emergency response teams and 911 emergency answering locations that would each be configured as LANs, but are connected together in a single metropolitan region to form a MAN.

Metropolitan Statistical Area (MSA) A division of one of 306 urban areas designated by the FCC. These divisions were used in establishing the areas where two competing carriers were granted licenses to operate the first generation AMPS cellular systems. The FCC also designated 428 rural areas called Rural Service Areas (RSAs) for dual AMPS licenses.

Metropolitan Trading Area (MTA) A regional area based on economics and used as a basis for allocating spectrum licenses for wireless common carriers. The total amount of broadband spectrum available is evenly divided between MTAs and Basic Trading Areas (BTAs). BTAs represent suburban and rural areas, whereas MTAs represent urban areas. These divisions were drawn from work performed by the Rand McNally Corporation.

Microwave radio A somewhat arbitrary term used to describe wireless communications systems operating at frequencies higher than 2 GHz. (In some parts of the FCC rules such as part 21, covering domestic public fixed radio services, microwave frequencies are defined as frequencies above 890 MHz.)

Mobile Identification Number (MIN) A number programmed from the keypad into the cell phone memory by the service provider when service is initiated. It is a 34-bit number that is governed by the handset 10-digit phone number.

Mobile Satellite Service (MSS) An FCC-defined class of wireless service that uses mobile hand sets and low earth orbiting satellites. The MSS is designed to provide global connectivity that delivers cellular type voice and data services beyond the location constraints imposed by terrestrial cellular base stations.

Mobile station A portable handset, data terminal, modem, or any other radio equipment authorized to communicate with a cellular base station.

Mobile Switching Center (MSC) The main control and mobile management facility in a cellular network that provides the interface between the PSTN and the cellular base stations.

Mobile telephone service A mobile radio system prior to cellular radio deployment that provided mobile stations with access to the Public Switched Telephone Network (PSTN).

Mobile Telephone Switching Office (MTSO) An out-of-fashion synonym for Mobile Switching Center.

Mobitex A wireless data network architecture used in several countries. Mobitex was developed by Ericsson. It is operated in the U.S. by Bell South Wireless Data (BSWD), now part of Cingular Wireless.

Multichannel Multipoint Distribution Service (MMDS) Often referred to incorrectly as the multichannel multipoint distribution *system*. MMDS is an FCC term used to describe a wireless service in accordance with rules contained in title 47, part 21 for Domestic Public Fixed Radio Services for the delivery of broadcast television service to homes in competition with cable delivery of video. MMDS rules were changed in 1999 to provide two-way multimedia service. MMDS channels are allocated for use in the frequency band 2596 MHz to 2644 MHz and associated 125 kHz channels. As a result of FCC rule changes in 1999, MMDS has found new applications markets in two-way multimedia, video, voice, and high-speed Internet access markets.

Multipath A condition often encountered in wireless networks where several paths exist for radio signals between the radio transmitter and receiver. This condition may cause severe fluctuations in received signal strength at the receiver location.

Multipoint Distribution Service (MDS) A domestic public radio service rendered on microwave frequencies from one or more fixed stations transmitting to multiple receiving facilities located at fixed points. MDS also may encompass transmissions from response stations to response station hubs or associated fixed stations. MDS is authorized for use in the 2.1 GHz band.

N

National Telecommunications and Information Administration (NTIA) A U.S. Government agency that regulates all use of radio onboard any federal government vessel, including military vessels, in U.S. ports and waters. NTIA rules do not apply outside the federal government. NTIA rules, applied in Title 47, Code of Federal Regulations (CFR), Part 300, are contained in the Manual of Regulations and Procedures for Federal Radio Frequency Management (NTIA manual).

NAVSTAR (NAVigation Satellite Timing and Ranging) The official U.S. Government name given to the GPS satellite system.

ns Nanosecond (10^{-9} second).

O

On-Off Keying (OOK) A type of modulation where the radio carrier or light beam is turned on and off by a modulating baseband binary bit stream.

Orthogonal Frequency Division Multiplexing (OFDM) A signal-processing technique that enables FDM channels to be spaced much closer and to overlap each other without using guard bands. OFDM therefore uses radio spectrum more efficiently than standard FDM channels. OFDM also solves the multipath problem by expanding the distance between symbols to a value longer than the delay spread. A high-speed bit stream that would normally be modulated on a single carrier is multiplexed into several closely spaced subcarriers, thereby permitting wider symbol spacing on each channel.

Overhead Message A type of message used in the CDMA ANSI-95 standard. An overhead message is sent by the base station on the Paging Channel to communicate base-station-specific and system-wide information to mobile stations.

P

Packet A unit of information formatted for transmission over a network that contains control and header information that corresponds to the OSI Network layer or layer 3. "Packet" is often used interchangeably (and incorrectly) with "frame."

Packet radio A system of equipment and protocols used to transmit packets over wireless networks

Paging The act of seeking a cellular mobile station in order to deliver an incoming call.

Paging channel A channel used in the CDMA ANSI-95 standard. It is used to communicate with the mobile station when the mobile is not using a traffic channel. The mobile station will begin monitoring the paging channel after it has set its timing to the system time provided by the sync channel.

PalmNet Brand name for a wireless data network and Internet access service offered by Palm Inc.

Parallel Channel Receiver A term used in GPS receiver performance specifications. It defines a receiver design that uses multiple receiving circuits for the purpose of tracking a number of GPS satellites simultaneously.

Path loss The amount of attenuation or signal strength loss that occurs between a transmitter and a receiver in a radio link. Path loss is determined by the operating frequency and the distance between the transmitter and the receiver.

Payload All equipment that has been installed on a satellite for the performance of a specific application excluding the bus. Typical payload missions include military reconnaissance, earth resource monitoring, communications, atomic clocks, deep space research, and search and rescue.

Peer network A peer network (or peer-to-peer network), as the name suggests, is where all computers on a LAN are peers and have equal status concerning sharing rights and capabilities. Any individual machine can share data or peripherals with any other machine on the same local area network.

Personal Communications Service (PCS) Any of several types of wireless voice and/or data communications systems, operating in the 1900 MHz radio band typically incorporating digital technology. PCS licenses are most often used to provide services similar to advanced cellular mobile or paging services. However, PCS can also be used to provide other wireless communications services, including services that allow people to place and receive communications while away from their home or office, as well as wireless communications in homes, office buildings, and other fixed locations. The FCC has issued PCS licenses for use in the 1900 MHz band through public auctions.

Phase Shift Keying (PSK) A digital modulation technique widely used in wireless communications systems. PSK modulation changes the phase of the carrier wave within a set of discrete values that represent one or more bits of input data. With PSK 2^n discrete values of phase represent "n" bits of data since "n" bits can take on any of 2^n different combinations.

Phased array antenna An antenna design that provides dynamic beam steering and beamwidth modification of the radiation pattern by dynamically altering the phase shift between successive elements in an array antenna.

Pilot channel A radio channel used in the ANSI-95 air interface transmitted by the base station continuously. Mobile stations scan for the strongest pilot channel when power is first applied.

PN sequence A pseudonoise sequence used in spread spectrum systems. A periodic binary sequence approximating, in some sense, a Bernoulli (coin tossing) process with equally probable outcomes.

Polar orbit A satellite orbit inclined near 90 degrees with respect to the equator.

Power Control Bit An ANSI-95 control bit sent in every 1.25 ms interval on the Forward Traffic Channel to signal the mobile station to increase or decrease its transmit power.

Processing gain In spread spectrum systems, the ratio expressed in dB between the spreading code bit rate (chipping rate) and the information bit rate. For example, a system using a chipping rate of 1 Mcps and an information bit rate of 10 kb/s would have a processing gain of 100 or 20 dB.

Propagation A term used in wireless communications describing the movement of electromagnetic energy from a transmission source to a receiving location.

Propagation loss The total reduction in radiant power density between the transmitting antenna and the receiving antenna. Propagation loss includes both spreading (free space) loss and attenuation loss. For non–line-of-sight situations it also includes diffraction loss around obstacles. It does not include antenna gain or feeder loss. Sometimes called *isotropic* propagation loss.

Protocols A formal set of rules that establishes the method of handling data transmissions in both wired and wireless networks.

Pseudorange A measure of the apparent propagation time from the GPS satellite to the receiver antenna, expressed as a distance. The apparent propagation time is determined from the time shift required to align a replica of the GPS code generated in the receiver with the received GPS code. The time shift is the difference between the time of signal reception (measured in the receiver time frame) and the time of emission (measured in the satellite time frame). Pseudorange is obtained by multiplying the apparent signal-propagation time by the speed of light. Pseudorange differs from the actual range by the amount that the satellite and receiver clocks are offset, by propagation delays, and other errors, including those introduced by selective availability.

Q

Quadrature Amplitude Modulation (QAM) A type of digital modulation that uses a combination of phase shift keyed (PSK) and amplitude modulation (AM).

R

Radio Determination Satellite Service An FCC wireless regulated service category used for position location and navigation. The RDSS employs a constellation of satellites in precisely known orbits and small handheld or portable earth stations to provide positioning services.

Radio Frequency Interference (RFI) Electromagnetic radiation usually caused by electrical devices or natural phenomena that emit radio wave energy in the same frequency band that is used for commercial purposes.

Radio receiver An electronic device that intercepts radio transmissions and recovers the transmitted baseband information.

Rake receiver A multiple set of receiver channels employed in ANSI-95 cellular systems to separately detect multipath components. The combining is performed in a correlator where the received CDMA carrier is added to the delay-compensated copies of the signal that result from reflections in an urban environment.

Registration The process by which a mobile cellular station makes its presence known to a base station to facilitate call delivery.

Reverse control channel (RECC) The wireless channel used from the mobile station to the base station by the Advanced Mobile Phone Service (AMPS) for call setup and supervisory control.

Rician fading The fading process characteristic of radio signals when there is a strong line-of-sight signal path and multiple non-direct signal paths.

Rural Service Areas (RSAs) A geographical area originally designated by the FCC for allocating AMPS cellular licenses to two competing companies called the wireline and non-wireline carriers. A similar plan encompassing urban areas was developed for 306 Metropolitan Statistical Areas (MSAs).

S

Satellite bus A bus contains all satellite system components other than the payload. It contains all of the components necessary for control of the spacecraft. These subsystems include a minimum of four important functions. They are the power, attitude control, antenna, and communications subsystems.

Search and Rescue Radar Transponder (SART) A device required by international maritime law to be carried aboard ocean-going ships to aid in the location of lifeboats or rafts.

Serving MSC The MSC that currently has the mobile station obtaining service at one of its cell sites within its coverage area.

Shared Secret Data (SSD) A bit pattern stored in the mobile station and known by the base station. SSD is used to support the authentication procedures and voice privacy. Shared Secret Data is maintained during power off.

Short waves A historical term used to define all of the electromagnetic spectrum above the AM broadcast band. In more modern terms, it describes the range of frequencies between 3 MHz and 30 MHz.

Signaling System 7 (SS7) This is an international, standard, Common Channel Signaling (CCS) system. It defines the architecture, network elements, interfaces, protocols, and management procedures for a network which transports control information between network switches and between switches and databases.

Sky waves Radio frequencies between approximately 2 MHz and 30 MHz. Sky waves travel in a straight line until they encounter a layer of the atmosphere called the ionosphere. At that point they are refracted by the ionosphere and eventually reach a point on the Earth at some distance from the transmitting station. Sky waves are used primarily for long distance radio communications links.

Soft handoff The process used in the ANSI-95 CDMA cellular standard where a mobile station is assigned to more than one cell using multiple PN codes but the same radio frequency during a handoff sequence. This provides a more reliable cell transition than the analog and TDMA air interface, where a frequency change is required during a "hard" handoff.

Source encoder A data processing function used in wireless digital transmission systems that accepts input digital data and analog voice information for processing prior to the modulator and channel coding functions.

Spread spectrum A radio transmission technology in which the bandwidth of the carrier is considerably greater than that normally required to convey the baseband information. Spread spectrum technology provides excellent resistance to interference from other radio signals and intentional jammers at the expense of wider spectrum utilization.

Superheterodyne A type of radio receiver design where the incoming radio signal is mixed with a variable tuned locally generated radio carrier to produce a fixed intermediate frequency (IF). The superheterodyne circuit was invented by Edwin Howard Armstrong in the 1920s. It led to the development of high-performance, low-cost radio broadcast receivers. It is the fundamental design used in most modern wireless communications receivers.

Supervisory Audio Tone (SAT) A radio link integrity control feature used with the Advanced Mobile Phone Service (AMPS) After a call has been established the base station transmits a selected continuous SAT on the Forward Voice Channel (FVC). The mobile station turns the signaling tone around and sends it back to the base station. This is the confirmation to the base station that the mobile station radio link is still connected and has not been lost due to a fade or other anomaly.

Supervisory Audio Tone (SAT) color code A special code used during a cell handoff by the Advanced Mobile Phone Service (AMPS). When a handoff occurs, the base station sends a special "SAT Color Code" notifying the mobile station about what SAT to expect in the new cell. This tone is picked up by the mobile after changing to a new assigned VC. This is the confirmation to the base station that the handoff was executed correctly.

Supervisory Tone (ST) A 10 kHz audio burst used on the Advanced Mobile Phone service (AMPS) reverse voice channel for link control and management tasks.

Sync channel One of the channels used in the ANSI-95 CDMA standard. It carries a repeating message that identifies the station, and the absolute phase of the pilot bit sequence. The data rate is always 1200 bits/second. The sync channel allows the mobile station to establish a bit timing reference between the base station and the mobile station. All CDMA base stations are time synchronized.

System Identification Number (SID) A 15-bit number programmed into a cell phone during service activation that identifies the service provider. It is used by the cellular network to determine whether a subscriber is operating in its "home" system or it is roaming in another service provider's network.

T

Telecommunications Industry Association (TIA) An organization accredited by the American National Standards Institute (ANSI) to develop voluntary industry standards. The TIA's Standards and Technology Department is composed of five divisions, which sponsor more than 70 standards-setting groups.

Time Division Duplex (TDD) A type of two-way communication where users share the same channel but not at the same time. The information frames are multiplexed or "ping-ponged" between the transmit and receive paired stations using the same radio channel.

Token Ring A local area network standard developed by IBM in the 1970s. It was later adopted as IEEE standard 802.5.

Traffic channel A communication path between a mobile station and a base station for user and signaling traffic in the CDMA ANSI-95 standard. The term Traffic Channel implies a Forward Traffic Channel and Reverse Traffic Channel pair.

Transponder A paired transmitter and receiver contained within a communications satellite to receive and amplify uplink transmissions. The transponder converts the uplink amplified signal to the downlink frequency and retransmits the signal to a ground station.

Triangulation A method of determining the location of an unknown point by using the laws of trigonometry. Triangulation is the basic process used by the Global Positioning System (GPS) for position location determination.

Trunking A process used in early mobile radio systems where mobile stations shared a number of available channels by scanning for an unused channel.

U

Ultra High Frequency (UHF) The radio frequency band that extends from 300 MHz to 3 GHz.

Ultra-Wideband (UWB) A technology loosely defined as any transmission scheme that occupies a bandwidth more than 25% of a center frequency or more than 1.5 GHz. UWB employs one of the theoretical spread spectrum techniques, overlooked to some extent except for military applications called time-hopped spread spectrum (THSS) modulation. The primary difference between time-hopped spread spectrum modulation and the other more familiar types of spread spectrum modulation, such as direct-sequence spread spectrum (DSSS) and frequency-hopped spread spectrum (FHSS), is the duty cycle.

Unique Challenge-Response Procedure An exchange of information between a mobile station and a base station in a cellular system for the purpose of authenticating the mobile station identity. The procedure is initiated by the base station and is characterized by the use of a challenge-specific random number.

Universal Time Coordinated (UTC) A universal time standard, referencing the time at Greenwich, England. Also referred to as Greenwich Mean Time (GMT) or Zulu time.

V

Very High Frequency (VHF) The radio frequency band that extends from 30 MHz to 300 MHz.

Very Small Aperture Terminals (VSATs) A satellite earth station terminal usually with an aperture size of one meter or less. VSATs are used for point-to-multipoint networks using geosynchronous satellites.

Visited system The cellular system that is providing service to a roaming mobile station.

Visitor Location Register A database used in cellular systems associated with an HLR. It contains dynamic information about subscribers who are currently active in the area served by the associated MSCs. The term "visitor" used in the TIA/EIA-41 standard is sometimes confusing because the VLR contains information about active home subscribers who are active as well as visitors from another service area. The VLR database content varies due to the dynamics of active mobile stations. In contrast, the HLR is a database with a relatively stable set of entries about current home subscribers, whether active or inactive.

Voice Channel Mobile Attenuation Code (VMAC) A code used by the Advanced Mobile Phone Service (AMPS) base station during call setup. The VMAC tells the mobile what power level to use after it moves from the control channel to the assigned voice channel.

W

Walsh code A set of 64 codes used to provide separation (orthogonality) for each channel in the CDMA ANSI-95 standard. Walsh codes are also used on the reverse link but for another purpose. It is the Walsh code that provides the separation or "code division" feature for the forward channels in CDMA.

Wavelength The distance in meters measured between wave crests of a transverse electromagnetic wave. Wavelength is inversely proportional to frequency and related to the speed of light. Given the speed of light as 300×10^6 meters per second, a radio frequency of 300 MHz would have a wavelength of 1 meter.

W-CDMA Wideband Code Division Multiple Access (W-CDMA) is a third generation spread spectrum wireless standard developed in Europe by the European Telecommunications Standards Institute (ETSI) and member industry participants. W-CDMA is also known under the ETSI name of Universal Mobile Telecommunications System (UMTS).

Wide Area Network (WAN) A large network covering long distances and typically operated by regional Bell operating companies or private carriers. WANs are used to provide high-capacity connections between corporate enterprises, telephone exchanges, packet switched networks, and metropolitan areas.

Wireless local loop A class of wireless service that bypasses the copper connection between a business or home and the local telephone switch with a radio link.

Wireless telegraphy An early radio transmission technique used primarily between ships and shore stations. Morse code characters were transmitted and received as dots and dashes by keying the radio transmitter on and off. It was the first primitive digital wireless transmission system.

Wireline carrier One of two carriers granted a license for providing AMPS for each service area during the initial cellular first generation in the 1980s. Local telephone companies were called the wireline carriers. The others were the non-wireline carriers. Each company in each area took half the spectrum available.

Wire speed The transmission bit rate of packets or frames within a wired network. Wire speed is used to differentiate between the bit rate within a packet from the throughput rate, which takes into account the idle time between packets. With varying network loads, throughput decreases, whereas wire speed is a constant unaffected by network loading. For example, a 100Base-T Ethernet LAN has a fixed wire speed of 100 Mbps. The throughput bit rate is a variable number considerably less than 100 Mbps.

X

Y

Z

Acronyms

A

AC	Authentication Center
AGC	Automatic Gain Control
AM	Amplitude Modulation
AMPS	Advanced Mobile Phone Service
ANSI	American National Standards Institute
ASK	Amplitude Shift Keying
ATM	Asynchronous Transfer Mode
AVC	Automatic Volume Control

B

BSC	Base Station Controller
BSS	Broadcast Satellite Service
BTA	Basic Trading Area
BTS	Base Transceiver System

C

CAVE	Cellular Authentication and Voice Encryption
CCIA	Computer Communications Industry Association
CDG	CDMA Development Group
CDMA	Code Division Multiple Access
CDR	Call Detail Record
CFR	Code of Federal Regulations
CNES	Centre National d'Etudes Spatiales (French)
COSPAS	Russian words "Cosmicheskaya Sistyema Poiska Avariynich Sudov," meaning Space System for the Search of Vessels in Distress
CPE	Customer Premise Equipment
CPFSK	Continuous Phase Frequency shift Keying
CSMA/CD	Carrier Sense Multiple Access with Collision Detection
CTIA	Cellular Telecommunications and Internet Association

D

DAB	Digital Audio Broadcast
DAMPS	Digital Advanced Mobile Phone Service
DBS	Direct Broadcast Service
DCCH	Digital Control Channel
DDTB	Digital Terrestrial Television Broadcasting
DECT	Digital European Cordless Telecommunications
DEMS	Digital Electronic Messaging Service
DND	Department of National Defense (Canada)
DOD	Department of Defense
DOP	Dilution of Precision
DPSK	Differential Phase Shift Keying
DSC	Digital Selective Calling
DSL	Digital Subscriber Link
DSSS	Direct Sequence Spread Spectrum
DVB	Digital Video Broadcast
DWDM	Dense Wavelength Division Multiplexing

E

EDGE	Enhanced Data rate for GSM Evolution
EHF	Extremely High Frequency
EIA	Electronics Industry Association
EIR	Equipment Identity Register
ELF	Extremely Low Frequency
ELT	Emergency Locator Transmitter
EPIRB	Emergency Position Indicating Radio Beacon
ESN	Electronic Serial Number
ESS	Electronic Switching System
ETSI	European Telecommunications Standards Institute

F

FA	Foreign Agent
FCC	Federal Communications Commission
FDD	Frequency Division Duplex
FDMA	Frequency Division Multiple Access
FEC	Forward Error Correction
FFT	Fast Fourier Transform
FHSS	Frequency-Hopped Spread Spectrum
FM	Frequency Modulation
FOCC	Forward Control Channel
FOIRL	Fiber-Optic Inter Repeater Link
FSK	Frequency Shift Keying
FSO	Free-Space Optical
FSS	Fixed Satellite Service

FTP	File Transfer Protocol		LMDS	Local Multipoint Distribution Service
FVC	Forward Voice Channel		LNP	Local Number Portability
			LOS	Line Of Sight
			LUT	Local User Terminal

G

M

GEO	Geostationary Earth Orbit		MAHO	Mobile Assisted Handoff
GLONASS	Global Navigation Satellite System		MAN	Metropolitan Area Network
GMDSS	Global Maritime Distress and Safety System		MCC	Mission Control Center
GOES	Geostationary Operational Environmental Satellites		MDS	Multichannel Distribution Service
			MEO	Medium Earth Orbit
GPS	Global Positioning System		MF	Medium Frequency
GSM	Groupe Spécial Mobile (Commonly referred to as Global Systems for Mobile Communications in English)		MIN	Mobile Identification Number
			MMDS	Multichannel Multipoint Distribution Service
			MODEM	Modulator Demodulator
			MPEG	Motion Picture Experts Group
H			MS	Mobile Station
			MSAU	Multistation Access Unit
HA	Home Agent		MSC	Mobile Switching Center
HBO	Home Box Office		MSS	Mobile Satellite Service
HIPERLAN	High Performance Local Area Network		MSK	Minimum Shift Keying
HF	High Frequency		MSS	Mobile Satellite Service
HLR	Home Location Register		MTA	Metropolitan Trading Area
HTML	HyperText Markup Language		MTS	Mobile Telephone Service

I

N

IEEE	Institute for Electrical and Electronics Engineers		NASA	National Aeronautics and Space Administration
IETF	Internet Engineering Task Force		NAVTEX	Navigational Telex
IMEI	International Mobile Equipment Identity		NIC	Network Interface Card
IMO	International Maritime Organization		NOAA	National Oceanic and Atmospheric Administration (NOAA)
IMT-2000	International Mobile Telecommunications – 2000		NTIA	National Telecommunications and Information Administration
IMTS	Improved Mobile Telephone Service			
IP	Internet Protocol			
IPCP	Internet Protocol Control Protocol		**O**	
IS	Interim Standard			
ISDN	Integrated Services Digital Network		OC	Optical Carrier
ISI	Intersymbol Interference		OFDM	Orthogonal Frequency Division Multiplexing
ISM	Industrial, Scientific, and Medical		1xEV-DO	One time Evolution – Data Only
ISO	International Organization for Standardization		1xEV-DV	One time Evolution – Data & Voice
ISP	Internet Service Provider		OOK	On-Off Keying
ITFS	Instructional Television Fixed Service			
ITU	International Telecommunications Union		**P**	
IVCDM	Initial Voice Channel Designation Message			
IWF	Inter-Working Function		PAM	Pulse Amplitude Modulation
			PC	Personal Computer
			PCIA	Personal Communication Industry Association
L			PCM	Pulse Code Modulation purchase
			PCS	Personal Communications Services
LAN	Local Area Network		PDA	Personal Digital Assistant
LED	Light Emitting Diode		PLB	Personal Locator Beacon
LEO	Low Earth Orbit			
LEOLUT	LEO Local User Terminal			
LF	Low Frequency			
LHCP	Left-Hand Circular Polarization			

PN	Pseudonoise	
PPM	Pulse Position Modulation	
PPP	Point-to-Point Protocol	
PSK	Phase Shift Keying	
PSTN	Public Switched Telephone Network	
PTM	Pulse Time Modulation	
PWM	Pulse Width Modulation	

Q

QAM	Quadrature Amplitude Modulation
QPSK	Quadrature Phase Shift Keying
OQPSK	Offset Quadrature Phase Shift Keying

R

RAF	Royal Air Force
RAS	Remote Access Servers
RCC	Radio Common Carrier
RDSS	Radiodetermination Satellite Service
RECC	Reverse Control Channel
RF	Radio Frequency
RHCP	Right-Hand Circular Polarization
RO	Receive Only
ROM	Read Only Memory
RSSI	Radio Signal Strength Indicator
RTT	Radio Transmission Technology
RVC	Reverse Voice Channel

S

SA	Selective Availability
SAR	Search and Rescue
SARSAT	Search and Rescue Satellite-Aided Tracking
SART	Search And Rescue Radar Transponder
SAT	Supervisory Audio Tone
SDO	Standards Development Organization
SDH	Synchronous Digital Hierarchy
SHF	Super High Frequency
SID	System Identification Number
SME	Signal Message Encryption
SNG	Satellite News Gathering
SOLAS	Safety Of Life At Sea
SONET	Synchronous Optical Network
SSD	Shared Secret Data
ST	Signaling Tone
STD	Standard
STM	Synchronous Transfer Module
STP	Shielded Twisted Pair

T

TIA	Telecommunications Industry Association
TCP	Transmission Control Protocol
TDD	Time Division Duplex
TDMA	Time Division Multiple Access
3GPP	Third Generation Partnership Project
3GPP2	Third Generation Partnership Project Two
THSS	Time-Hopped Spread Spectrum
TP	Twisted Pair
TRF	Tuned Radio Frequency

U

UHF	Ultra High Frequency
UMTS	Universal Mobile Telecommunications System
USCG	United States Coast Guard
UTC	Universal Time Coordinated
UTP	Unshielded Twisted Pair
UWB	Ultra-Wideband

V

VC	Voice Channel
VHF	Very High Frequency
VLAN	Virtual Local Area Network
VLF	Very Low Frequency
VLR	Visitor Location Register
VMAC	Voice Channel Mobile Attenuation Code
VOIP	Voice Over Internet Protocol
VSAT	Very Small Aperture Terminal
VSWR	Voltage Standing Wave Ratio
VTS	Vessel Traffic Service

W

WAAS	Wide Area Augmentation System
WAN	Wide Area Network
WAP	Wireless Application Protocol
WARC	World Administrative Radio Conference
W-CDMA	Wideband Code Division Multiple Access
WCS	Wireless Communications Service
WLAN	Wireless Local Area Network
WLL	Wireless Local Loop
WRC	World Radiocommunications Conference

Index

G

Geostationary (GEO), 34, 301
Geostationary orbital slots, 303
Geosynchronous, 301
Gigabit Ethernet, 89
Global cellular, 182
Global Maritime Distress and Safety System (GMDSS), 31
Global System for Mobile Communication (GSM), 186
GPS errors, 323
Graffiti area, 261
Graphics Interchange Format (GIF), 65
Grasshopper, 254
Ground control segment, 298
Ground wave, 14, 113
GSM 1800, 187
GSM 1900, 187
GSM 400, 186
GSM 900, 186
GSM system, 189

H

Half duplex, 65, 141
Handoff, 163
Handoff and control functions, 199
Handoff phase, 177
Handover, 163
Hardline, 78
High Performance Radio Local Area Network (HIPERLAN), 238
Home Location Register (HLR), 161, 189, 217
HomeRF, 239
Hybrid media networks, 82
Hybrid network, 60

I

IEEE 802.11b standard, 240
Infrared, 81
Infrastructure mode, 244
Infrastructure network, 93
Initialization phase, 176
Inmarsat earth station terminals, 36
Inmarsat satellites, 36
Institute for Electrical and Electronic Engineers (IEEE), 28
Integrated Services Digital Network (ISDN), 277
Interference, 117
International Maritime Organization (IMO), 30
International Maritime Satellite Organization (Inmarsat), 35
International Mobile Telecommunications (IMT) 2000, 23, 344
International Organization for Standardization (ISO), 22, 25
International Telecommunications Union (ITU), 3, 26
Internet, 53
Intersymbol interference (ISI), 356
Intersystem cooperation, 216

Intranets, 53
Inverse square law, 118
Ionosphere, 114
IP encapsulation, 233
IP in IP protocol, 233
Isotropic radiator, 135
ITU band designations, 109

J

Joint Photographic Experts Group (JPEG), 64

K

Keying, 143

L

Lambda (λ), 106
Launch vehicles, 299
Leased line service, 277
Licensed, 236
Line of sight (LOS), 115, 124
Link budgets, 124
Local Area Networks (LANs), 51
Local Multipoint Distribution Service (LMDS), 285
Local Multipoint Distribution Service band plan, 286
Local Multipoint Distribution Service capacity, 286
Local Multipoint Distribution Service network architecture, 285
Local oscillator frequency, 11
Local User Terminal (LUT), 35
Location-based registration, 221
Log10, 116
Low Earth Orbit (LEO), 33, 304

M

Maxwell, James Clark, 103
Mean Anomaly, 301
Mean Motion, 301
Medium Earth Orbit (MEO), 303
Megachips, 194
Mesh network topology, 57
Message, 68
Messaging networks, 235
Metropolitan Area Network (MAN), 52
Microwave, 111
Microwave antennas, 136
Microwave propagation characteristics, 278
Microwave radio, 80
Minimum Shift Keying (MSK), 144
Mission Control Center (MCC), 35
MMDS cells, 281

Y

Z

Wireless Lab Equipment
Hands-on Learning = 90% Retention*

Hands-on... the leading instructional method for retention of information.

*According to Certification Magazine, retention rates associated with hands-on learning are as high as 90%.

Grasshopper™ Handheld Wireless Receiver

A handheld, wireless receiver designed specifically for sweeping and optimizing LANs. The instrument measures coverage of direct sequence CDMA networks which allows the user to measure and determine the Access Point, Packet Error Rate and RSSI signal levels aiding in locating the hub and access points throughout a building. It detects and differentiates from narrow-band multipath interferences, such as microwave ovens and frequency hopping systems. It also features a display, keypad, and battery pack for true portability. Works with any IEEE 802.11b DSSS access point.

Catalyst™ 1900 Series Switch

The Cisco Catalyst® 1900 Series Ethernet Switch provides industry-leading performance and end-to-end network integration. The Catalyst 1900 switch has up to 25 10BaseT switched Ethernet ports (including the AUI switch port on the back panel), each port providing users or groups of users, dedicated 10-Mbps bandwidth to resources within the network. These ports connect to other 10BaseT-compatible devices, such as single workstations and 10BaseT hubs. The switch also has two 100BaseT switched ports, and is designed for plug-and-play operation, requiring only that you assign basic IP information to the switch and connect it to the other devices in your network.

Palm VII™ organizer

The Palm VII organizer can be easily connected to the Internet, without using a wire or an external modem. It accesses a wide spectrum of information available on the Internet and views that information in an easily read format on the organizer screen. You can use your Palm VII organizer to stay connected by wireless Internet messaging anytime, anywhere.

eMAP™ GPS Receiver

The eMAP GPS receiver contains a 12-parallel channel GPS receiver, weighs six ounces, and boasts an extra-large display for showing more map data, while operating for up to 12 hours on two AA batteries. The internal basemap, containing information on North and South America, includes state and country boundaries, lakes, rivers, streams, airports, cities, towns, coastlines, U.S., state, and interstate highways and exit information.

Wireless Access Point

The Wave2Net Wireless Access Point is used as the wireless equivalent of a "hub", splitting and distributing 11 Mbps connections to as many as 256 wireless clients through 2 adjustable dipole high-power antennas. It features network management software, USB console support, roaming, and supports advanced WEP security. Testing to WiFi compliance certification allows interoperability with all major networking 802.11b DSSS equipment.

Wireless PCI Adapter

The Wave2Net Wireless PCI Adapter includes all the features you've been looking for in a single card. Full Privacy 128-bit WEP guarantees the highest security available and its high-quality construction and strict compliance to industry leading specifications assures you will be able to use with any major wireless brand. You can expect fast connections with an improved antenna and processor design along with extended roaming for seamless connections throughout your network.

Laptop Computer

This Laptop Computer features Windows XP, 700+ MHz CPU, 10+ GB hard disk drive, FDD 3.5", 800 X 600 SVGA, CD-ROM, and v.90 Modem.

EtherFast® 10/100 PCMCIA Card

The EtherFast® 10/100 PC Card will connect the notebook computer to a 10BaseT or 100BaseTX network. Ready to run in both half and full duplex modes, the EtherFast® 10/100 PC Card supports speeds of 10 Mbps, 20 Mbps, 100 Mbps, or 200 Mbps. The card adjusts its speed and duplex to almost any 10BaseT or 100BaseTX network automatically. The EtherFast® 10/100 PC Card includes a 32 KB buffer for fast file transfers, low voltage operation, hot swap compatibility, advanced error correction, 16-bit architecture, and complete software suite.

MARCRAFT
Your IT Training Provider
(800) 441-6006   www.mic-inc.com

Complete and Affordable Classroom Management

Classroom management just got a heck of a lot easier. Thanks to TEAMS 32 you're relieved from many of the mundane and time-consuming tasks involved in managing a classroom. It can even eliminate a lot of the paper-work...maybe all of it! And you get back the time to do what you actually want to do: teach. TEAMS 32 is flexible enough to fit any classroom size or style, whether traditional or a more complex rotational system. Manage your classroom the way you want! Classroom records are kept and updated automatically, including individual student test performance, attendance, class rosters, and other student information.
For more details and a sample CD, simply call **(800) 441-6006**

A+ Certification
This book provides you with training necessary for the A+ Certification testing program that certifies the competency of entry-level (6 months experience) computer service technicians. The A+ test contains situational, traditional, and identification types of questions. All of the questions are multiple choice with only one correct answer for each question. The test covers a broad range of hardware and software technologies, but is not bound to any vendor-specific products. The program is backed by major computer hardware and software vendors, distributors, and resellers. A+ Certification signifies that the certified individual possesses the knowledge and skills essential for a successful entry-level (6 months experience) computer service technician, as defined by experts from companies across the industry.

CompTIA seal of approval

Network+ Certification
Network+ is a CompTIA vendor-neutral certification that measures the technical knowledge of networking professionals with 18-24 months of experience in the IT industry. The test is administered by NCS/VUE and Prometric™. Discount exam vouchers can be purchased from Marcraft. Earning the Network+ certification indicates that the candidate possesses the knowledge needed to configure and install the TCP/IP client. This exam covers a wide range of vendor and product neutral networking technologies that can also serve as a prerequisite for vendor-specific IT certifications. Network+ has been accepted by the leading networking vendors and included in many of their training curricula. The skills and knowledge measured by the certification examination are derived from industry-wide job task analyses and validated through an industry wide survey. The objectives for the certification examination are divided in two distinct groups, Knowledge of Networking Technology and Knowledge of Networking Practices.

CompTIA seal of approval

i-Net+ Certification
The i-Net+ certification program is designed specifically for any individual interested in demonstrating baseline technical knowledge that would allow him or her to pursue a variety of Internet-related careers. i-Net+ is a vendor-neutral, entry-level Internet certification program that tests baseline technical knowledge of Internet, Intranet and Extranet technologies, independent of specific Internet-related career roles. Learning objectives and domains examined include: Internet basics, Internet clients, development, networking, security, and business concepts. Certification not only helps individuals enter the Internet industry, but also helps managers determine a prospective employee's knowledge and skill level.

CompTIA seal of approval

Server+ Certification

The Server+ certification deals with advanced hardware issues, such as RAID, SCSI, multiple CPUs, SANs, and more. This certification is vendor-neutral with a broad range of support, including core support by 3Com, Adaptec, Compaq, Hewlett-Packard, IBM, Intel, EDS Innovations Canada, Innovative Productivity, and Marcraft. This book focuses on complex activities and solving complex problems to ensure servers are functional and applications are available. It provides an in-depth understanding of installing, configuring, and maintaining servers, including knowledge of server-level hardware implementations, data storage subsystems, data recovery, and I/O subsystems.

Fiber Cabling Installers Certification prepares technicians for the growing demand for qualified cable installers who understand and can implement fiber optic technologies. These technologies cover terminology, techniques, tools and other products in the fiber optic industry. This text/lab book covers the basics of fiber optic design, installations, pulling and prepping cables, terminations, testing, and safety considerations. Labs cover ST-compatible and SC connector types, both multi- and single-mode cables and connectors. Learn about insertion loss, optical time domain reflectometry, and reflectance. This text covers mechanical and fusion splices and troubleshooting cable systems. This text/lab covers the theory and hands-on skills needed to prepare you for fiber optic entry-level certification.

The Complete Data Cabling Installer Certification provides the IT industry with an introductory, vendor-neutral certification for skilled personnel that install Category 5 copper data cabling. The Marcraft Complete Data Cabling Installer Certification Training Guide provides students with the knowledge and skills required to pass the Data Cabling Installer Certification exam and become a certified cable installer. The DCIC is recognized nationwide and is the hiring criterion used by major communication companies. Therefore, becoming a certified data cable installer will enhance your job opportunities and career advancement potential.

Security Installers Certification is an expert introduction to the security alarm industry, for those who have limited or no previous knowledge of the industry. After successfully completing the Security Installers Certification program, you'll be prepared for employment in the security industry as a technician or a field installer.

This manual is also beneficial to sales reps for enhancing their technical knowledge. No matter what your background may be or what your educational intentions are, this manual offers you a wealth of information and will answer all of your security installation questions.

The Complete Introductory Computer Course is an entry-level course. It prepares students for the more challenging A+ Certification course. It also provides a careerport™ into the fast-growing IT industry. The MC-2300 is a 45 hour, easy-to-understand exploration of basic computer hardware, software, and troubleshooting. This course helps build students confidence and basic computer literacy. The fully illustrated 198-page Theory Text/Lab Guide provides an easy-to understand exploration of the basics of computers: basic computer architecture and operation, step-by-step computer hardware assembly, computer hardware and functions, common software packages, consumer maintenance practices, and troubleshooting a "sick" computer. The reusable MC-2300 Intro Computer Trainer comes with all the necessary hardware, software, and tools to perform over 30 hands-on Lab Explorations.

The Complete Introductory Networking Course is a superbly-illustrated theory text and lab guide all in one. It not only provides a great way for students to explore over 45 hours of easy-to-understand basic Networking topics, but also develops job skills for starting them on the path towards a new high-tech career!

This manual guides you through such activities as: installation and configuration of local, area network hardware, peer-to-peer networking functions, sharing computer resources, mapping to remote resources, and consumer level network troubleshooting. The Complete Introductory Networking Course provides an excellent starting point for IT Certification, including Microsoft's MCSE, Novell's CNA, and Cisco's CCNA.

The Complete Introductory Internet Course takes advantage of the growing demand for qualified Internet technicians. This 45 hour course explores easy-to-understand basic Internet topics and helps develop Internet skills.

This manual guides you through such activities as: configure e-mail accounts, design a basic HTML page, setup basic firewall for security, and establish Internet connection sharing. The Complete Introductory Internet Course provides an excellent starting point for IT Certification, including CompTIA's i-Net+ and Prosoft's Certified Internet Webmaster (CIW).

A GAME OF THRONES

THE GRAPHIC NOVEL

VOLUME 2

GEORGE R. R. MARTIN

A GAME OF THRONES

THE GRAPHIC NOVEL

VOLUME 2

ADAPTED BY DANIEL ABRAHAM

ART BY TOMMY PATTERSON

COLORS BY IVAN NUNES

LETTERING BY MARSHALL DILLON

ORIGINAL SERIES COVER ART BY

MIKE S. MILLER AND MICHAEL KOMARCK

BANTAM BOOKS • NEW YORK

Paintings on pages vi and 182 by Michael Komark.
Paintings on pages 2, 32, 62, 92, 122, and 152 by Mike S. Miller.

Published in the United States by Bantam Books, an imprint of the Random House Publishing Group, a division of Random House, Inc. New York.

Bantam Books is a registered trademark and the Bantam colophon is a trademark of Random House, Inc.

ISBN 978-0-440-42322-5
eBook ISBN 978-0-345-53560-3

Printed in the United States of America on acid-free paper.

www.bantamdell.com

9 8 7 6 5 4 3 2 1

Graphic novel interior design by Foltz Design.

Visit us online at www.DYNAMITE.com

Follow us on Twitter @dynamitecomics

Like us on Facebook /Dynamitecomics

Watch us on YouTube /Dynamitecomics

Nick Barrucci, CEO / Publisher
Juan Collado, President / COO
Joe Rybandt, Senior Editor
Josh Johnson, Art Director
Rich Young, Director Business Development
Jason Ullmeyer, Senior Graphic Designer
Keith Davidsen, Marketing Manager
Josh Green, Traffic Coordinator
Chris Caniano, Production Assistant

CONTENTS

A GAME OF THRONES

THE GRAPHIC NOVEL

VOLUME 2

ISSUE #7

THE CRABS HAD ARRIVED FROM EASTWATCH ONLY THAT MORNING, PACKED IN A BARREL OF SNOW, AND THEY WERE SUCCULENT.

ARE YOU CERTAIN YOU MUST LEAVE US SO SOON?

PAST CERTAIN. MY BROTHER JAIME MAY DECIDE THAT YOU HAVE CONVINCED ME TO TAKE THE BLACK.

YOU'RE A CUNNING MAN, TYRION. WE HAVE NEED OF MEN OF YOUR SORT ON THE WALL.

THEN I SHALL SCOUR THE SEVEN KINGDOMS FOR DWARFS AND SHIP THEM ALL TO YOU, LORD MORMONT.

LANNISTER MOCKS US.

ONLY YOU, SER ALLISER.

YOU HAVE A BOLD TONGUE FOR SOMEONE WHO IS LESS THAN HALF A MAN. PERHAPS YOU AND I SHOULD VISIT THE YARD. MAKE YOUR JAPES WITH STEEL IN YOUR HAND.

I *HAVE* STEEL IN MY HAND, SER ALLISER, ALTHOUGH IT APPEARS TO BE A CRAB FORK.

SHALL WE DUEL?

HA HA HA HA

HA HA HA HA HA

TO THE VICTOR THE SPOILS! I CLAIM THORNE'S SHARE OF THE CRABS.

YOU ARE A WICKED MAN TO PROVOKE OUR SER ALLISER SO.

CHIP THE ICE OFF YOUR EYES, MY LORD. SER ALLISER THORNE SHOULD BE MUCKING OUT YOUR STABLES, NOT DRILLING YOUR YOUNG WARRIORS.

THE WATCH HAS NO SHORTAGE OF STABLEBOYS. THAT SEEMS TO BE ALL THEY SEND US THESE DAYS. STABLEBOYS AND THIEVES AND RAPERS.

MORE WINE, TYRION?

YOU HAVE A GREAT THIRST FOR A SMALL MAN.

PATCHES OF SNOW CRUNCHED BENEATH HIS FEET AS HIS BOOTS BROKE THE NIGHT'S CRUST, AND HIS BREATH STEAMED BEFORE HIM LIKE A BANNER.

THE KING'S TOWER AWAITED WITH ITS PROMISE OF WARMTH AND A SOFT BED, YET TYRION FOUND HIMSELF WALKING PAST IT, TOWARD THE VAST PALISADE OF THE WALL.

HE ENTERED THE IRON CAGE AND PULLED ON THE BELL ROPE. THREE QUICK PULLS.

HE HAD TO WAIT AN ETERNITY. LONG ENOUGH TO BEGIN TO WONDER WHY HE WAS DOING THIS. HE HAD ALMOST DECIDED TO FORGET HIS WHIM WHEN THE CAGE GAVE A JERK AND BEGAN TO ASCEND.

HE MOVED UP SLOWLY BY FITS AND STARTS, AND THEN MORE SMOOTHLY. THE GROUND FELL AWAY AND THE CAGE SWUNG. HE COULD FEEL THE COLD OF THE METAL THROUGH HIS GLOVES.

SEVEN HELLS, IT'S THE DWARF.

AND WHAT WILL YOU BE WANTING AT THIS TIME OF NIGHT?

A LAST LOOK.

LOOK ALL YOU WANT. JUST HAVE A CARE YOU DON'T FALL OFF. THE OLD BEAR WOULD HAVE OUR HIDES.

WHO GOES THERE? HALT!

IF I HALT FOR TOO LONG, I'LL FREEZE IN PLACE, JON.

I GAVE YOU NOTHING BUT WORDS.

THEN GIVE YOUR WORDS TO BRAN, TOO.

YOU'RE ASKING A LAME MAN TO TEACH A CRIPPLE HOW TO DANCE. HOWEVER SINCERE THE LESSON, THE RESULT IS LIKELY TO BE GROTESQUE.

STILL, I KNOW WHAT IT IS TO LOVE A BROTHER. I WILL GIVE BRAN WHAT HELP IS IN MY POWER.

THANK YOU, MY LORD OF LANNISTER.

FRIEND.

MOST OF MY KIN ARE BASTARDS, BUT YOU'RE THE FIRST I'VE HAD TO FRIEND.

MY UNCLE IS OUT THERE. THE FIRST NIGHT THEY SENT ME UP HERE, I THOUGHT: UNCLE BENJEN WILL RIDE BACK TONIGHT. HE NEVER CAME, THOUGH.

IF HE DOESN'T COME BACK, GHOST AND I WILL GO FIND HIM.

I BELIEVE YOU.

BUT WHO WILL GO FIND YOU? HE WONDERED.

HER FATHER HAD BEEN FIGHTING WITH THE SMALL COUNCIL AGAIN. ARYA COULD SEE IT ON HIS FACE WHEN HE CAME TO THE TABLE. LATE AGAIN, AS HE HAD BEEN SO OFTEN.

MY LORD.

BE SEATED.

THE TALK IN THE YARD IS THAT WE SHALL HAVE A TOURNEY, MY LORD. KNIGHTS FROM ALL OVER THE REALM ARE COMING IN HONOR OF YOUR APPOINTMENT AS HAND.

DO THEY ALSO SAY IT'S THE LAST THING IN THE WORLD I WOULD HAVE WANTED?

A TOURNEY? WILL WE BE PERMITTED TO GO, FATHER?

I MUST ARRANGE ROBERT'S GAMES AND PRETEND TO BE HONORED FOR HIS SAKE. THAT DOES NOT MEAN I MUST SUBJECT MY DAUGHTERS TO THIS FOLLY.

OH PLEASE. I WANT TO SEE.

PRINCESS MYRCELLA WILL BE THERE, MY LORD, AND SHE IS EVEN YOUNGER THAN SANSA.

IT WOULD LOOK QUEER IF YOUR FAMILY DID NOT ATTEND.

I SUPPOSE SO. I SHALL ARRANGE A PLACE FOR YOU, SANSA.

FOR BOTH OF YOU.

I DON'T CARE ABOUT THEIR STUPID TOURNEY.

IT WILL BE A **SPLENDID** EVENT. YOU SHAN'T BE WANTED.

ENOUGH! I AM WEARY UNTO DEATH OF THIS ENDLESS WAR. YOU ARE SISTERS. I EXPECT YOU TO BEHAVE LIKE SISTERS!

PRAY EXCUSE ME. I FIND I HAVE A SMALL APPETITE TONIGHT.

BACK AT WINTERFELL, ARYA HAD LOVED NOTHING BETTER THAN TO SIT AT HER FATHER'S TABLE AND LISTEN TO HIM TALK. EVERY DAY, A DIFFERENT MAN WOULD BE ASKED TO JOIN THEM.

NO ONE TALKED TO HER HERE. SHE DIDN'T CARE. SHE LIKED IT THAT WAY. SHE HATED THE SOUNDS OF THEIR VOICES, THE WAY THEY LAUGHED, THE STORIES THE TOLD.

THEY'D LET THE QUEEN KILL LADY. THEY'D LET THE HOUND KILL MYCAH. NO ONE HAD RAISED A VOICE OR DRAWN A BLADE.

THIS IS NO TOY FOR CHILDREN, LEAST OF ALL FOR A GIRL. WHAT WOULD SEPTA MORDANE SAY IF SHE KNEW YOU WERE PLAYING WITH SWORDS?

I WASN'T *PLAYING!* AND I HATE SEPTA MORDANE!

ENOUGH! I OUGHT TO SNAP THIS TOY ACROSS MY KNEE AND PUT AN END TO THIS NONSENSE.

DO YOU KNOW THE FIRST THING ABOUT SWORD FIGHTING?

STICK THEM WITH THE POINTY END?

THAT...IS THE ESSENCE OF IT, I SUPPOSE.

I WAS TRYING TO LEARN. I ASKED MYCAH TO PRACTICE WITH ME.

I *ASKED* HIM. I WAS MY FAULT THAT...

YOU ARE TOO YOUNG TO BE BURDENED WITH ALL MY CARES. BUT YOU ARE ALSO A STARK OF WINTERFELL. YOU KNOW OUR WORDS.

WINTER IS COMING.

WHEN THE SNOW FALLS AND THE WHITE WIND BLOWS, THE LONE WOLF DIES. BUT THE PACK SURVIVES.

IF YOU MUST HATE, HATE THOSE WHO TRULY DO US HARM. SEPTA MORDANE IS A GOOD WOMAN AND SANSA IS YOUR SISTER. YOU NEED HER AS SHE NEEDS YOU.

I DON'T HATE SANSA. NOT REALLY.

WE HAVE COME TO A DARK AND DANGEROUS PLACE, CHILD. THE WILLFULLNESS, THE RUNNING OFF, THE ANGRY WORDS...IT'S TIME TO BEGIN GROWING UP.

HERE.

GO ON. IT'S YOURS.

I CAN KEEP IT? FOR TRUE?

FOR TRUE.

AND TRY NOT TO STAB YOUR SISTER, WHATEVER THE PROVOCATION.

THE NEXT MORNING, SHE APOLOGIZED TO SEPTA MORDANE AND ASKED FOR HER PARDON.

THREE DAYS AFTER, HER FATHER'S STEWARD SENT HER TO THE SMALL HALL.

YOU'RE LATE, BOY.

TOMORROW YOU WILL BE HERE AT MIDDAY.

HE HAD AN ACCENT. THE LILT OF THE FREE CITIES. BRAAVOS OR MYR.

WHO ARE YOU?

YOUR DANCING MASTER.

TOMORROW, YOU WILL CATCH IT.

THIS IS NOT A GREATSWORD THAT IS NEEDING TWO HANDS. YOU WILL TAKE THE BLADE IN ONE HAND ONLY.

LEFT IS GOOD. ALL IS REVERSED, IT WILL MAKE YOUR ENEMIES MORE AWKWARD. DO NOT SQUEEZE SO TIGHT.

WHAT IF I DROP IT?

CAN YOU DROP PART OF YOUR ARM?

NINE YEARS SYRIO FOREL WAS FIRST SWORD TO THE SEALORD OF BRAAVOS. LISTEN TO HIM, BOY.

"IN THAT DARKNESS, THE OTHERS CAME FOR THE FIRST TIME. THEY WERE COLD THINGS, DEAD THINGS THAT HATED IRON AND FIRE AND THE TOUCH OF THE SUN. AND EVERY CREATURE WITH HOT BLOOD IN ITS VEINS."

"THEY SWEPT OVER HOLDFASTS AND CITIES AND KINGDOMS LEADING HOSTS OF THE SLAIN. THEY HUNTED THE MAIDS THROUGH THE FROZEN FORESTS AND FED THEIR DEAD SERVANTS ON THE FLESH OF HUMAN CHILDREN."

"NOW THESE WERE THE DAYS BEFORE THE ANDALS CAME. THE KINGDOMS THEN WERE THE KINGDOMS OF THE FIRST MEN WHO HAD TAKEN THE LANDS FROM THE CHILDREN OF THE FOREST. YET HERE AND THERE, THE CHILDREN STILL LIVED IN THEIR WOODEN CITIES AND HOLLOW HILLS."

"AND THE FACES IN THE TREES KEPT WATCH."

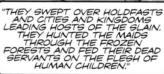

"AS COLD AND DEATH FILLED THE EARTH, THE LAST HERO SET OUT TO SEEK THE CHILDREN IN HOPES THAT ANCIENT MAGICS COULD WIN WHAT THE ARMIES OF MEN HAD LOST. HE SET OUT INTO THE DEAD LANDS WITH A SWORD, A HORSE, A DOG, AND A DOZEN COMPANIONS."

"FOR YEARS HE SEARCHED, UNTIL HE DESPAIRED OF FINDING THE CHILDREN OF THE FOREST IN THEIR SECRET CITIES."

"ONE BY ONE, HIS FRIENDS DIED. AND HIS HORSE. AND FINALLY EVEN HIS DOG."

"HIS SWORD FROZE SO HARD, THE BLADE SNAPPED WHEN HE TRIED TO USE IT."

"THE OTHERS SMELLED THE HOT BLOOD IN HIM."

"THEY CAME SILENT ON HIS TRAIL, STALKING HIM WITH PACKS OF PALE WHITE SPIDERS AS BIG AS HOUNDS--"

BANG

HODOR!

WE HAVE VISITORS AND YOUR PRESENCE IS REQUIRED, BRAN. TYRION LANNISTER AND SOME MEN OF THE NIGHT'S WATCH.

I'M LISTENING TO A STORY.

WAIT, MY LITTLE LORD. VISITORS ARE NOT SO PATIENT, AND OFTTIMES THEY BRING STORIES OF THEIR OWN.

YOU LANNISTERS HAD BEST REMEMBER THAT.

HODOR, BRING MY BROTHER HERE.

YOU SAID YOU HAD BUSINESS WITH BRAN. WELL, HERE HE IS, LANNISTER.

I'M TOLD YOU WERE QUITE THE CLIMBER.

TELL ME, HOW IS IT YOU HAPPENED TO FALL THAT DAY?

I NEVER...

THE CHILD DOES NOT REMEMBER ANYTHING OF THE FALL, OR THE CLIMB THAT CAME BEFORE IT.

INTERESTING.

MY BROTHER IS NOT HERE TO ANSWER YOUR QUESTIONS.

DO YOUR BUSINESS AND BE ON YOUR WAY.

I HAVE A GIFT FOR YOU.

DO YOU LIKE TO RIDE, BOY?

MY LORD, THE CHILD HAS LOST THE USE OF HIS LEGS. HE CANNOT SIT A HORSE.

NONSENSE. WITH THE RIGHT HORSE AND THE RIGHT SADDLE, EVEN A CRIPPLE CAN RIDE.

I'M *NOT* A CRIPPLE!

THEN I'M NOT A DWARF. MY FATHER WILL REJOICE TO HEAR IT.

START WITH AN UNBROKEN YEARLING WITH NO OLD TRAINING TO BE UNLEARNED. GIVE THIS TO YOUR SADDLER.

YOU DRAW NICELY, MY LORD. YES. THIS OUGHT TO WORK.

I SHOULD HAVE THOUGHT OF IT MYSELF.

WILL I TRULY BE ABLE TO RIDE?

YOU WILL. AND I SWEAR TO YOU, BOY, ON HORSEBACK YOU WILL BE AS TALL AS ANY OF THEM.

IS THIS SOME TRAP? WHAT IS BRAN TO YOU? WHY SHOULD YOU HELP HIM?

YOUR BROTHER JON ASKED IT OF ME, AND I HAVE A TENDER SPOT IN MY HEART FOR CRIPPLES, BASTARDS, AND BROKEN THINGS.

THE WOLVES DO NOT LIKE YOUR SMELL, LANNISTER.

PERHAPS IT'S TIME I TOOK MY LEAVE--

GRR GRR GRR

ISSUE #8

LORD ARRYN'S DEATH WAS A GREAT SADNESS FOR US ALL, MY LORD. I WOULD BE MORE THAN HAPPY TO TELL WHAT I CAN OF THE MANNER OF HIS PASSING.

WOULD YOU CARE FOR SOME REFRESHMENT? A CUP OF ICED MILK SWEETENED WITH HONEY? I FIND IT REFRESHING IN THE HEAT.

THAT WOULD BE MOST KIND.

ICED MILK FOR THE KING'S HAND AND MYSELF, CHILD.

THE SMALLFOLK SAY THAT THE LAST YEAR OF SUMMER IS ALWAYS THE HOTTEST. ON DAYS LIKE THIS, I ENVY YOU NORTHERNERS YOUR SUMMER SNOWS.

MAEKAR'S SUMMER WAS HOTTER THAN THIS AND NEAR AS LONG.

THERE WERE EVEN FOOLS WHO TOOK IT TO MEAN THAT THE GREAT SUMMER HAD COME AT LAST.

BUT THEN WE HAD A SHORT AUTUMN AND A TERRIBLE LONG WINTER.

SWEET CHILD. THANK YOU. YOU MAY GO.

NOW WHERE WERE WE? OH, YES. YOU ASKED ABOUT LORD ARRYN...

I DID.

IF TRUTH BE TOLD, HE HAD NOT SEEMED HIMSELF FOR SOME TIME. HIS SON WAS SICKLY AND HIS LADY WIFE SO ANXIOUS SHE WOULD SCARCE LET THE BOY OUT OF HER SIGHT. SMALL WONDER THAT HE SEEMED MELANCHOLY.

WHAT CAN YOU TELL ME OF HIS LAST ILLNESS?

IT IS POSSIBLE, MY LORD, BUT I DO NOT THINK IT LIKELY. EVERY HEDGE MAESTER KNOWS THE COMMON POISONS, AND LORD ARRYN DISPLAYED NONE OF THE SIGNS.

AND THE HAND WAS LOVED BY ALL. WHAT SORT OF MONSTER WOULD DARE TO MURDER SUCH A NOBLE LORD?

I HAVE HEARD IT SAID THAT POISON IS A WOMAN'S WEAPON.

WOMEN, CRAVENS...AND EUNUCHS. THE LORD VARYS WAS A BORN A SLAVE IN LYS, DID YOU KNOW?

THE KING WAS AT LORD ARRYN'S BEDSIDE. WAS THE QUEEN WITH HIM?

NO. SHE AND THE CHILDREN WERE MAKING THE JOURNEY TO CASTERLY ROCK WITH HER FATHER, LORD TYWIN.

I WOULD BE CURIOUS TO EXAMINE THE BOOK YOU LENT JON ARRYN THE DAY BEFORE HE FELL ILL.

I FEAR YOU WOULD FIND IT OF LITTLE INTEREST. A PONDEROUS TOME ON THE LINEAGES OF THE GREAT HOUSES. BUT IF YOU WISH, I SHALL HAVE IT SENT TO YOUR CHAMBERS.

I THANK YOU FOR YOUR HELP. I HAVE TAKEN ENOUGH OF YOUR TIME.

COME TO ME AS OFTEN AS YOU LIKE, LORD EDDARD. I AM HERE TO SERVE.

YES, NED THOUGHT, BUT WHOM?

THERE WAS NO AVOIDING THE HEAT. HE COULD FEEL THE SILK TUNIC CLINGING TO HIS CHEST. THICK, MOIST AIR COVERED THE CITY LIKE A DAMP WOOLEN BLANKET.

THE RIVERSIDE HAD GROWN UNRULY AS THE POOR FLED THEIR HOT, AIRLESS WARRENS TO JOSTLE FOR SLEEPING PLACES NEAR THE WATER, WHERE THE ONLY BREATH OF WIND COULD BE FOUND.

THERE WAS NO RELIEF IN THE TOWER OF THE HAND.

ARYA? WHAT ARE YOU DOING?

SYRIO SAYS A WATER DANCER CAN STAND ON ONE TOE FOR HOURS.

WHICH TOE?

ANY TOE.

MUST YOU DO YOUR STANDING HERE? IT'S A LONG, HARD FALL DOWN THESE STEPS.

SYRIO SAYS A WATER DANCER NEVER FALLS.

FATHER, WILL BRAN COME LIVE WITH US NOW?

NOT FOR A LONG TIME, SWEET ONE. HE NEEDS TO GET HIS STRENGTH BACK. FOR NOW, IT IS ENOUGH THAT HE IS ALIVE.

HE WAS GOING TO BE A KNIGHT OF THE KINGSGUARD. CAN HE STILL BE A KNIGHT?

NO. BUT HE MAY BE LORD OF A GREAT HOLDFAST AND SIT ON THE KING'S COUNCIL. OR RAISE CASTLES. OR SAIL A SHIP ACROSS THE SUNSET SEA. OR BECOME HIGH SEPTON.

BUT HE WILL NEVER RUN BESIDE HIS WOLF AGAIN, HE THOUGHT WITH A SADNESS TOO DEEP FOR WORDS.

NOR WILL HE EVER LIE WITH A WOMAN, OR HOLD HIS OWN SON IN HIS ARMS.

CAN I BE A KING'S COUNCILOR AND BUILD CASTLES AND BECOME HIGH SEPTON?

YOU WILL MARRY A KING AND RULE HIS CASTLE. YOUR SONS WILL BE KNIGHTS AND LORDS. AND MAYBE HIGH SEPTON.

NO. THAT'S SANSA.

RAT. PIMPLE. HELP OUR STONE HEAD HERE. THE THREE OF YOU OUGHT TO BE ABLE TO MAKE LADY PIGGY SQUEAL.

SER ALLISTER HAD OFTEN SET TWO FOES AGAINST HIM, BUT NEVER THREE. HE WOULD GO TO SLEEP BRUISED AND BLOODIED TONIGHT.

STAY BEHIND ME.

THREE TO TWO WILL MAKE FOR BETTER SPORT.

THREE.

WHY ARE YOU WAITING?

KNOW YOUR FOE, SER RODRIK HAD TAUGHT JON ONCE. HALDER WAS BRUTALLY STRONG, BUT SHORT OF PATIENCE. HE HAD NO TASTE FOR DEFENSE.

THIS MUMMER'S FARCE HAS GONE ON LONG ENOUGH FOR TODAY.

FOR AN INSTANT, I THOUGHT I FINALLY HAD YOU, SNOW.

FOR AN INSTANT, YOU DID.

DID HE HURT YOU?

I'VE BEEN BRUISED BEFORE.

MY NAME IS SAMWELL TARLY, OF HORN...I MEAN, I **WAS** OF HORN HILL. MY FATHER IS LORD RANDYLL.

IF YOU WANT, YOU CAN CALL ME SAM.

WHY DIDN'T YOU GET UP AND FIGHT?

I WANTED TO, TRULY. I JUST... COULDN'T.

I FEAR I'M A COWARD. MY LORD FATHER ALWAYS SAID SO.

YOU WERE HURT. TOMORROW, YOU'LL DO BETTER.

NO, I WON'T. I NEVER DO BETTER.

LIFE AT CASTLE BLACK FOLLOWED CERTAIN PATTERNS. THE MORNINGS WERE FOR SWORDPLAY, THE AFTERNOONS FOR WORK. THE BLACK BROTHERS SET NEW RECRUITS TO MANY DIFFERENT TASKS, TO LEARN WHERE THEIR SKILLS LAY.

THAT AFTERNOON, THE WATCH COMMANDER SENT JON TO THE WINCH CAGE WITH FOUR BARRELS OF FRESH-CRUSHED STONE TO SCATTER OVER THE FOOTPATHS ATOP THE WALL.

JON FOUND HE DID NOT MIND. HE COULD THINK HERE—AND HE FOUND HIMSELF THINKING OF SAMWELL TARLY AND, ODDLY, TYRION LANNISTER.

MOST MEN WOULD RATHER DENY A HARD TRUTH THAN FACE IT, THE DWARF HAD TOLD HIM.

THE WORLD WAS FULL OF CRAVENS WHO PRETENDED TO BE HEROES. IT TOOK A QUEER SORT OF COURAGE TO ADMIT COWARDICE AS SAM HAD.

HIS SORE SHOULDER MADE THE WORK SLOW. DUSK WAS SETTLING OVER THE NORTH AS JON SIGNALED THE WINCH MEN TO LOWER HIM DOWN.

IS THAT A WOLF?

DIREWOLF. IT'S THE SIGIL OF MY FATHER'S HOUSE.

OURS IS A STRIDING HUNTSMAN. I HATE TO HUNT.

LET'S GO OUTSIDE. HAVE YOU SEEN THE WALL?

I'M FAT, NOT BLIND. OF COURSE I SAW IT. IT'S SEVEN HUNDRED FEET HIGH.

I NEVER THOUGHT IT WOULD BE LIKE THIS, WITH ALL THE BUILDINGS FALLING DOWN. AND I NEVER SAW SNOW UNTIL LAST MONTH.

THEY WON'T MAKE ME GO UP THERE, WILL THEY? I DON'T LIKE HIGH PLACES.

I DON'T UNDERSTAND. IF YOU'RE TRULY SO CRAVEN, WHY ARE YOU *HERE*?

SAM'S ROUND FACE SEEMED TO CAVE IN ON ITSELF AND HE BEGAN TO CRY-HUGE CHOKING SOBS THAT MADE HIS WHOLE BODY SHAKE.

JON COULD ONLY STAND AND WATCH. IT SEEMED THE TEARS WOULD NEVER END.

IT WAS GHOST WHO KNEW WHAT TO DO. THE FAT BOY CRIED OUT, STARTLED...THEN HIS SOBS TURNED TO LAUGHTER.

JON LET THE SILENCE BREATHE. IN TIME, SAMWELL TARLY BEGAN TO SPEAK.

THE TARLYS WERE A FAMILY OLD IN HONOR, AND SAMWELL WAS BORN HEIR TO RICH LANDS, A STRONG KEEP, AND THE GREATSWORD HEARTSBANE, FORGED OF VALYRIAN STEEL AND PASSED FROM FATHER TO SON FOR FIVE HUNDRED YEARS.

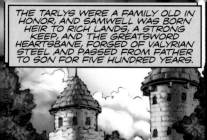

BUT WHATEVER PRIDE RANDYLL TARLY MIGHT HAVE FELT AT SAM'S BIRTH VANISHED AS THE BOY GREW PLUMP, SOFT, AND AWKWARD.

SAM LOVED TO LISTEN TO MUSIC, TO WEAR SOFT VELVETS, AND TO PLAY IN THE CASTLE KITCHENS WITH THE COOKS. HE GREW ILL AT THE SIGHT OF BLOOD.

A DOZEN MASTERS-AT-ARMS CAME AND WENT FROM HORN HILL, TRYING TO TURN HIM INTO THE KNIGHT HIS FATHER WANTED. HE WAS CURSED AND CANED, SLAPPED AND STARVED.

WARLOCKS CAME FROM QARTH, PROMISING THEIR RITES WOULD MAKE HIM BRAVE. WHEN SAM GOT SICK AND RETCHED, LORD RANDYLL HAD THEM SCOURGED.

AFTER THREE GIRLS IN AS MANY YEARS, LADY TARLY GAVE LORD RANDYLL A SECOND SON. DICKON WAS A FIERCE, ROBUST CHILD, AND SAMWELL HAD SEVERAL YEARS OF PEACE WITH HIS MUSIC AND HIS BOOKS.

WHERE HAVE YOU BEEN?

TALKING WITH SAM.

HE TRULY IS CRAVEN. AT SUPPER, HE WAS TOO SCARED TO SIT WITH US.

LORD OF HAM THINKS HE'S TOO GOOD TO SIT WITH US.

I SAW HIM EAT A PORK PIE. DO YOU THINK IT WAS A BROTHER?

STOP IT!

LISTEN TO ME...

HE TOLD THEM HOW IT WAS GOING TO BE. HE PERSUADED SOME, CAJOLED SOME, SHAMED OTHERS, AND MADE THREATS WHERE THREATS WERE REQUIRED.

AT THE END, ALL AGREED. EXCEPT RAST.

YOU GIRLS DO AS YOU PLEASE, BUT IF THORNE SENDS ME AGAINST LADY PIGGY, I'M SLICING OFF A RASHER OF BACON.

IT WAS A FORTNIGHT BEFORE HE FOUND THE COURAGE TO TO JOIN IN THEIR TALK, BUT IN TIME HE WAS LAUGHING AT PYP'S FACES AND TEASING GRENN WITH THE BEST OF THEM.

I DON'T KNOW WHAT YOU DID, JON, BUT I KNOW YOU DID IT.

I'VE... NEVER HAD A FRIEND BEFORE.

WE'RE NOT FRIENDS.

WE'RE BROTHERS.

WE'RE FORTUNATE MY BROTHER STANNIS IS NOT WITH US. YOU REMEMBER THE TIME HE PROPOSED TO OUTLAW BROTHELS?

THE KING ASKED IF HE'D LIKE TO OUTLAW EATING, SHITTING, AND BREATHING TOO.

I HAVE HEARD ENOUGH ABOUT WHORES FOR ONE DAY. UNTIL THE MORROW.

THE RED KEEP AND THE HAND'S TOURNEY WERE CHAFING HIM RAW. NOR WERE THEY THE ONLY THINGS.

THE LINEAGES AND HISTORIES OF THE GREAT HOUSES OF THE SEVEN KINGDOMS, WITH DESCRIPTIONS OF MANY HIGH LORDS AND NOBLE LADIES AND THEIR CHILDREN.

PYCELLE HAD SPOKEN TRULY. IT MADE PONDEROUS READING. YET JON ARRYN HAD ASKED FOR IT.

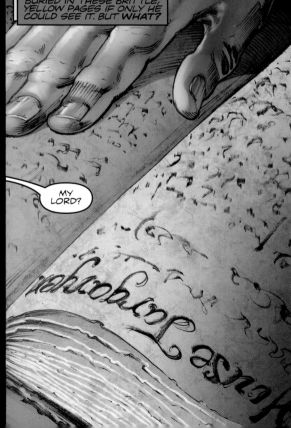

THERE WAS SOME TRUTH BURIED IN THESE BRITTLE, YELLOW PAGES IF ONLY HE COULD SEE IT. BUT WHAT?

MY LORD?

JORY. I'VE PROMISED THE CITY WATCH TWENTY OF MY GUARD UNTIL THE TOURNEY IS DONE. I RELY ON YOU TO MAKE THE CHOICE.

DID YOU FIND THE STABLEBOY?

THE WATCHMAN NOW, MY LORD. HE SWEARS HE'LL NEVER TOUCH ANOTHER HORSE. HE CLAIMS TO HAVE KNOWN LORD ARRYN WELL. THE HAND USED TO BRING HIS MOUNTS CARROTS AND APPLES.

THE BOY WAS THE LAST OF LITTLEFINGER'S FOUR. SER HUGH HAD BEEN BRUSQUE AND UNINFORMATIVE. THE SERVING GIRL HAD BEEN PLEASANT AND SAID LORD JON HAD BEEN READING MORE THAN WAS GOOD FOR HIM.

THE POTBOY HAD BEEN FULL OF KITCHEN GOSSIP ABOUT THE NEW SET OF PLATE LORD ARRYN HAD COMMISSIONED. THE KING'S OWN BROTHER, STANNIS BARATHEON, HAD HELPED TO DESIGN IT.

CARROT'S AND APPLES. DID OUR WATCHMAN RECALL ANYTHING OF NOTE?

HE SAYS THAT LORD ARRYN OFTEN WENT RIDING WITH STANNIS BARATHEON. ONCE TO A BROTHEL.

THE HAND OF THE KING VISITED A BROTHEL WITH STANNIS BARATHEON?

WHICH BROTHEL?

THE BOY DIDN'T KNOW. THE GUARDS WOULD.

A PITY THAT LYSA CARRIED THEM OFF TO THE VALE. EVERYONE WHO MIGHT KNOW WHAT HAPPENED TO JON ARRYN IS A THOUSAND LEAGUES AWAY.

I SUPPOSE YOU'D BEST BEGIN VISITING WHOREHOUSES.

HARD DUTY, MY LORD.

PERHAPS LORD STANNIS WILL RETURN FOR ROBERT'S TOURNEY.

THAT WOULD BE A STROKE OF FORTUNE, MY LORD.

IN OTHER WORDS, NOT BLOODY LIKELY.

WINE FOR THE KING'S HAND!

I AM TOBHO MOTT, MY LORD. PLEASE COME IN. PUT YOURSELF AT EASE.

IF YOU ARE IN NEED OF NEW ARMS FOR THE HAND'S TOURNEY, YOU HAVE COME TO THE RIGHT SHOP. PERHAPS A BLADE? I WORKED IN QOHOR AS A BOY AND KNOW THE SPELLS TO TAKE VALYRIAN STEEL AND WORK IT ANEW.

DID YOU MAKE A SUIT OF PLATE FOR LORD ARRYN?

THE HAND DID CALL UPON ME WITH LORD STANNIS. I REGRET TO SAY THEY DID NOT HONOR ME WITH THEIR PATRONAGE.

THEY ONLY ASKED TO SEE THE BOY.

HE HAD NO NOTION OF WHO THE BOY MIGHT BE. BUT IF LORD ARRYN AND STANNIS HAD COME FOR THAT...

I SHOULD LIKE TO SEE THE BOY AS WELL.

YOU KNOW WHO THE BOY IS.

HE'S MY APPRENTICE. WHO HE WAS BEFORE HE CAME TO ME, THAT'S NONE OF MY CONCERN.

IF THE DAY COMES WHEN HE'D RATHER WIELD A SWORD THAN FORGE ONE, SEND HIM TO ME. UNTIL THEN, YOU HAVE MY THANKS.

MY LORD.

DID YOU FIND ANYTHING, MY LORD?

I DID.

AND IT STILL LEFT HIM WONDERING WHAT JON ARRYN HAD WANTED WITH A KING'S BASTARD.

AND WHY WAS IT WORTH HIS LIFE?

ISSUE #9

TWO ROOMS, THAT'S ALL THERE IS. THEY'RE UNDER THE BELL TOWER, BUT WE'RE FULL UP. IT'S THOSE OR THE ROAD.

LEAVE YOUR BOOTS DOWNSTAIRS. THE BOY WILL CLEAN THEM.

WE HAD BEST MAKE HASTE IF WE HOPE TO EAT TONIGHT, MY LADY. THOSE WHO COME LATE TO THE TABLE DON'T EAT.

IT MIGHT BE BETTER IF WE WERE NOT KNIGHT AND LADY, BUT COMMON TRAVELERS. FATHER AND DAUGHTER ON SOME FAMILY BUSINESS?

AS YOU SAY, MY LADY...

MY DAUGHTER.

SEVEN BLESSINGS TO YOU, GOODFOLK. ARE YOU BOUND TO THE TOURNEY AT KING'S LANDING?

MY NAME'S MARILLION. DOUBTLESS YOU'VE HEARD ME PLAY SOMEWHERE? I WAS MADE TO SING FOR KINGS AND HIGH LORDS.

SANSA HAD ATTENDED THE HAND'S TOURNEY WITH SEPTA MORDANE AND JEYNE POOLE, AND IT HAD BEEN BETTER THAN THE SONGS.

THEY WATCHED THE HEROES OF A HUNDRED SONGS RIDE FORTH, EACH MORE FABULOUS THAN THE LAST.

THE KINGSLAYER RODE BRILLIANTLY. HE OVERTHREW SER ANDAR ROYCE AND MARCHER LORD BRYCE CARON AS EASILY AS IF HE WERE RIDING AT RINGS, THEN TOOK A HARD-FOUGHT MATCH FROM BARRISTAN SELMY.

SER RENLY FELL TO THE HOUND WITH SUCH VIOLENCE HE SEEMED TO FLY OFF HIS HORSE. HIS HEAD HIT THE GROUND WITH AN AUDIBLE CRACK THAT MADE THE CROWD GASP, BUT IT WAS ONLY ONE GOLDEN ANTLER ON HIS HELM SNAPPING OFF.

LATER, A HEDGE KNIGHT IN A CHEQUERED CLOAK DISGRACED HIMSELF BY KILLING BERIC DONDARRION'S HORSE AND WAS DECLARED FORFEIT. LORD BERIC PUT HIS SADDLE TO A NEW MOUNT AND WAS PROMPTLY KNOCKED OFF IT BY THE WARRIOR PRIEST THOROS OF MYR.

SER ARON SANTAGAR AND LOTHOR BRUME TILTED THRICE WITHOUT RESULT. SER ARON FELL AFTERWARD TO LORD JASON MALLISTER, AND BRUNE TO YOHN ROYCE'S YOUNGER SON ROBAR.

THE MOST TERRIFYING MOMENT OF THE DAY CAME DURING SER GREGOR CLEGANE'S SECOND JOUST WHEN THE POINT OF HIS LANCE RODE UP AND STRUCK A YOUNG KNIGHT FROM THE VALE UNDER THE GORGET.

SANSA HAD NEVER SEEN A MAN DIE. SHE OUGHT TO HAVE BEEN CRYING, BUT THE TEARS WOULD NOT COME.

IT WOULD HAVE BEEN DIFFERENT IF IT HAD BEEN JORY OR SER RODRIK OR FATHER, SHE TOLD HERSELF. THIS YOUNG STRANGER FROM THE VALE OF ARRYN WAS NOTHING TO HER.

THE WORLD WOULD FORGET HIS NAME NOW. THERE WOULD BE NO SONGS SUNG FOR HIM.

IN THE END IT CAME TO FOUR: THE HOUND AND HIS MONSTROUS BROTHER GREGOR, THE KINGSLAYER...

...AND LORAS TYRELL, THE KNIGHT OF FLOWERS.

AFTER EACH VICTORY, SER LORAS WOULD REMOVE HIS HELM, RIDE SLOWLY AROUND THE FENCE, AND FINALLY PLUCK A WHITE ROSE AND THROW IT TO SOME FAIR MAIDEN IN THE CROWD.

WHEN HIS WHITE MARE STOPPED IN FRONT OF HER, SHE THOUGHT HER HEART WOULD BURST.

SWEET LADY, NO VICTORY IS HALF SO BEAUTIFUL AS YOU.

HIS LAST MATCH OF THE DAY WAS AGAINST THE YOUNGER SER ROYCE, BUT SANSA'S EYES WERE ONLY FOR SER LORAS.

TO THE OTHER MAIDENS, HE HAD GIVEN WHITE ROSES.

SHE INHALED ITS SWEET FRAGRANCE AND SAT CLUTCHING IT LONG AFTER SER LORAS HAD RIDDEN OFF.

YOU MUST BE ONE OF HER DAUGHTERS. YOU HAVE THE TULLY LOOK.

I'M SANSA STARK. I HAVE NOT HAD THE HONOR, MY LORD.

PRINCE JOFFREY HAD NOT SPOKEN A WORD TO HER SINCE THE AWFUL THING HAD HAPPENED, AND SHE DARED NOT SPEAK TO HIM.

AT FIRST, SHE'D THOUGHT SHE HATED HIM FOR WHAT THEY'D DONE TO LADY. BUT AFTER SHE'D WEPT HER EYES DRY, SHE'D TOLD HERSELF THAT IT HAD NOT BEEN JOFFREY'S DOING. NOT TRULY.

THE QUEEN HAD DONE IT. SHE WAS THE ONE TO HATE. HER AND ARYA.

NOTHING BAD WOULD HAVE HAPPENED EXCEPT FOR ARYA.

SER LORAS HAS A KEEN EYE FOR BEAUTY, SWEET LADY.

HE WAS TOO KIND. SER LORAS IS A TRUE KNIGHT.

DO YOU THINK HE WILL WIN TOMORROW, MY LORD?

NO.

MY DOG WILL DO FOR HIM. OR PERHAPS MY UNCLE JAIME.

AND IN A FEW YEARS, WHEN I AM OLD ENOUGH TO ENTER THE LISTS, I SHALL DO FOR THEM ALL.

THE SERVANTS KEPT THE CUPS FILLED ALL NIGHT, BUT SHE NEEDED NO WINE. SHE WAS DRUNK ON THE MAGIC OF THE NIGHT, GIDDY WITH GLAMOUR.

COURSES CAME AND WENT—A SOUP OF BARLEY AND VENISON, SALADS OF SWEETGRASS AND PLUMS, SNAILS IN HONEY AND GARLIC—AND JOFFREY WAS THE SOUL OF COURTESY.

NO!

DO NOT TELL ME WHAT TO DO, WOMAN! I AM KING HERE, DO YOU UNDERSTAND?

I RULE HERE, AND IF I SAY THAT I WILL FIGHT TOMORROW, *I WILL FIGHT!*

HA! THE GREAT KNIGHT.

I CAN STILL KNOCK YOU IN THE *DIRT.* REMEMBER THAT, KINGSLAYER!

AS YOU SAY, MY LORD.

IT GROWS LATE. DO YOU NEED AN ESCORT BACK TO THE CASTLE?

NO. I MEAN TO SAY...YES, THANK YOU. I SHOULD BE GLAD OF SOME PROTECTION.

YOU TOO? YOU ARE A SOUR MAN, STARK.

YOUR GRACE, IT IS NOT SEEMLY THAT THE KING SHOULD RIDE INTO THE MELEE. IT WOULD NOT BE A FAIR CONTEST. WHO WOULD DARE STRIKE YOU?

NED SAW AT ONCE THAT SELMY HAD HIT THE MARK. THE DANGERS OF THE MELEE WERE ONLY A SAVOR TO ROBERT, BUT THIS TOUCHED HIS PRIDE.

WHY, ALL OF THEM, DAMN IT. IF THEY CAN. AND THE LAST MAN LEFT STANDING...

...WILL BE YOU.

THERE'S NOT A MAN IN THE SEVEN KINGDOMS WHO WOULD DARE RISK HURTING YOU.

ARE YOU TELLING ME THOSE PRANCING CRAVENS WILL *LET ME WIN?*

FOR A CERTAINTY.

GET OUT! GET OUT BEFORE I KILL YOU.

NOT YOU, NED.

DAMN YOU, NED STARK. YOU AND JON ARRYN. I LOVED YOU BOTH, AND YOU PUT ME ON A THRONE.

LOOK AT WHAT KINGING HAS DONE TO ME. GODS, TOO FAT FOR MY ARMOR. HOW DID IT COME TO THAT?

I SWEAR TO YOU, I WAS NEVER SO ALIVE AS WHEN I WAS WINNING THE THRONE, AND NEVER SO DEAD AS NOW THAT I'VE WON IT.

AND CERSEI. SHE'S LOVELY TO LOOK AT, BUT SHE'S *COLD*.

I'M SORRY FOR YOUR GIRL, NED. ABOUT THE WOLF. MY SON WAS LYING, I'D STAKE MY SOUL ON IT...

MORE THAN ONCE, I'VE DREAMED OF GIVING UP THE CROWN. TAKE SHIP FOR THE FREE CITIES WITH MY HORSE AND MY HAMMER. YOU KNOW WHAT STOPS ME? THE THOUGHT OF JOFFREY ON THE THRONE WITH CERSEI STANDING BEHIND HIM.

HOW COULD I HAVE MADE A SON LIKE THAT?

HE'S ONLY A BOY.

PERHAPS YOU'RE RIGHT. JON DESPAIRED OF ME OFTEN ENOUGH, YET I GREW INTO A GOOD KING.

AH, NED, SAY I'M A BETTER KING THAN AERYS AND BE DONE WITH IT. YOU NEVER COULD LIE FOR LOVE NOR HONOR.

SO WHO DO YOU THINK OUR CHAMPION WILL BE TODAY? HAVE YOU SEEN MACE TYRELL'S BOY? THE KNIGHT OF FLOWERS, THEY CALL HIM.

NOW, THERE'S A SON ANY MAN WOULD BE PROUD TO OWN TO.

THEY BROKE THEIR FAST ON BLACK BREAD AND GOOSE EGGS AND BACON. ALL TALK OF THE MELEE WAS FORGOTTEN, AND THAT BREAKFAST TASTED BETTER THAN ANYTHING EDDARD STARK HAD EATEN IN A LONG TIME.

AFTERWARD IT WAS TIME FOR THE TOURNAMENT TO RESUME.

A HUNDRED DRAGONS ON THE KINGSLAYER!

DONE! THE HOUND HAS A HUNGRY LOOK ABOUT HIM THIS MORNING.

EDDARD WOULD HAVE LIKED NOTHING BETTER THAN TO SEE *BOTH* OF THEM LOSE, BUT SANSA WAS WATCHING ALL MOIST-EYED AND EAGER.

HE HAD PROMISED TO WATCH THE FINAL LISTS WITH HER, AS SEPTA MORDANE WAS ILL.

BAM

I KNEW THE HOUND WOULD WIN!

IF YOU KNOW WHO'S GOING TO WIN THE SECOND MATCH, SPEAK UP NOW BEFORE LORD RENLY PLUCKS ME CLEAN.

SER GREGOR CLEGANE WAS CALLED THE MOUNTAIN THAT RIDES. SOME SAID IT HAD BEEN GREGOR WHO'D DASHED THE SKULL OF THE INFANT AEGON TARGARYEN. IT WAS WHISPERED THAT HE HAD RAPED THE MOTHER BEFORE PUTTING HER TO THE SWORD.

OH, HE'S SO BEAUTIFUL. DON'T LET SER GREGOR HURT HIM, FATHER.

THESE THINGS WERE NOT SAID IN HIS HEARING.

NOW, HOWEVER, SER GREGOR WAS HAVING TROUBLE CONTROLLING HIS STALLION.

AND IT BEGAN.

THE MOUNTAIN'S STALLION BROKE INTO A HARD GALLOP, PLUNGING FORWARD WILDLY. LORAS TYRELL'S MARE CHARGED FORWARD AS SMOOTH AS SILK.

CRASH

RAH!

LEAVE HIM BE.

NED SHOUTED "STOP HIM!" BUT HIS WORDS WERE LOST IN THE ROAR. EVERYONE ELSE WAS SHOUTING AS WELL.

IN THE NAME OF THE KING, STOP THIS MADNESS!

IS THE HOUND THE CHAMPION NOW?

I OWE YOU MY LIFE. THE DAY IS YOURS, SER.

I...AM NO SER.

RAH

HE TOOK THE VICTORY AND THE CHAMPION'S PURSE, AND FOR THE FIRST TIME IN HIS LIFE, THE LOVE OF THE COMMONS.

MUCH LATER, AFTER HE'D TAKEN THE GIRLS BACK TO THE CITY AND SEEN THEM BOTH SAFE IN BED, HE ASCENDED TO HIS ROOMS IN THE TOWER OF THE HAND.

THE HOUR WAS WELL PAST MIDNIGHT. DOWN BY THE RIVER, THE REVELS WERE ONLY BEGINNING TO DWINDLE.

TYRION LANNISTER'S DAGGER. BRAN'S FALL. THE DEATH OF JON ARRYN. ALL OF IT WAS LINKED, BUT THE TRUTH WAS AS CLOUDED NOW AS WHEN HE'D STARTED.

THE ARMORER'S APPRENTICE WAS THE KING'S SON, BUT NO BASEBORN CHILD COULD THREATEN ROBERT'S TRUEBORN CHILDREN...

A MAN TOSEE YOU, MY LORD. HE WON'T GIVE HIS NAME.

SEND HIM IN.

WHO ARE YOU?

A FRIEND. WE MUST SPEAK ALONE.

LEAVE US, JORY.

LORD VARYS?

I WILL NOT KEEP YOU LONG, MY LORD. BUT THERE ARE THINGS YOU MUST KNOW.

ISSUE #10

WE ARE TAKING HIM BACK TO WINTERFELL.

MY FATHER WILL WONDER WHAT'S BECOME OF ME. HE'LL PAY A HANDSOME REWARD TO ANY MAN WHO BRINGS HIM NEWS OF WHAT HAPPENED HERE TODAY.

THE IMP'S MEN COME WITH HIM. WE'LL THANK THE REST OF YOU TO STAY QUIET ABOUT WHAT YOU'VE SEEN HERE.

WORD WOULD BEGIN TO SPREAD THE INSTANT THEY WERE GONE. THE FREERIDER WITH THE GOLD COIN IN HIS POCKET WOULD FLY TO CASTERLY ROCK LIKE AN ARROW. YOREN WOULD TAKE WORD SOUTH. THAT FOOL SINGER MIGHT MAKE A LAY OF IT.

WALDER FREY'S MEN WOULD TAKE WORD TO HIM. FREY MIGHT BE SWORN TO RIVERRUN, BUT HE WAS A CAUTIOUS MAN. AT THE LEAST, HE WOULD SEND A BIRD WINGING SOUTH TO KING'S LANDING.

WE MUST RIDE AT ONCE. IF ANY OF YOU CHOOSE TO HELP US GUARD OUR CAPTIVE AND GET HIM TO WINTERFELL, I PROMISE YOU SHALL BE WELL REWARDED.

QUIET? IT WAS ALL TYRION COULD DO NOT TO LAUGH.

HE WAS NOT TRULY AFRAID. THEY WOULD NEVER GET TO WINTERFELL. RIDERS WOULD BE AFTER THEM WITHIN A DAY.

STILL, IT WAS A MISERABLE, POUNDING JOURNEY OVER ROUGH GROUND. THE HOOD MUFFLED ALL SOUND, AND THE RAIN SOAKED IT UNTIL IT WAS HARD TO BREATHE.

THE WRETCHED SINGER HAD EVEN COME ALONG, CONVINCED THERE WAS A GREAT SONG TO BE MADE FROM THIS.

TYRION WONDERED WHETHER THE BOY WOULD THINK THE ADVENTURE QUITE SO SPLENDID ONCE THE LANNISTER RIDERS CAUGHT UP WITH THEM.

THE RAIN HAD STOPPED AND DAWN LIGHT WAS SEEPING THROUGH THE WET CLOTH OVER HIS EYES WHEN CATELYN STARK CALLED THE DISMOUNT.

THIS... THIS IS THE *EASTERN* ROAD! YOU SAID WE WERE RIDING FOR *WINTERFELL*.

I DID. OFTEN, AND LOUDLY. NO DOUBT YOUR FRIENDS WILL RIDE THAT WAY WHEN THEY COME AFTER US.

EVEN NOW, DAYS LATER, THE MEMORY FILLED HIM WITH BITTER RAGE. ALL HIS LIFE, TYRION HAD PRIDED HIMSELF ON HIS CUNNING--THE ONLY GIFT THE GODS HAD SEEN FIT TO GIVE HIM. AND YET CATELYN STARK HAD OUTWITTED HIM.

NOW, AS TYRION WATCHED THE SELLSWORD BUTCHER HIS HORSE, HE CHALKED UP ONE MORE DEBT HE OWED THE STARKS.

NONE OF US WILL GO HUNGRY TONIGHT.

WANT A TASTE, DWARF?

MY BROTHER GAVE ME THAT MARE FOR MY TWENTY-THIRD NAME DAY.

THANK HIM FOR US, THEN. IF YOU EVER SEE HIM AGAIN.

TASTES WELL BRED.

PERHAPS THE DEAD MARE WAS THE LUCKY ONE. HE HAD HOURS OF RIDING AHEAD OF HIM, THEN A FEW MOUTHFULS OF FOOD AND A SHORT SLEEP ON COLD HARD GROUND, THEN ANOTHER NIGHT OF THE SAME. AND ANOTHER. AND THE GODS ONLY KNEW HOW IT WOULD END.

TO HORSE! BRONN, GUARD THE PRISONERS.

YES, MY LADY.

NO! ARM US. YOU NEED EVERY SWORD YOU CAN GET.

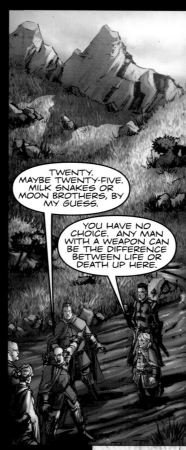

MAYBE TWENTY-FIVE. MILK SNAKES OR MOON BROTHERS, BY MY GUESS.

YOU HAVE NO CHOICE. ANY MAN WITH A WEAPON CAN BE THE DIFFERENCE BETWEEN LIFE OR DEATH UP HERE.

BUDABUMBUDABUM

GIVE ME YOUR WORD THAT YOU WILL PUT DOWN YOUR SWORDS AFTER THE FIGHT IS DONE.

ON MY HONOR AS A LANNISTER.

BUDABUMBUDABUM

ARM THEM.

I'VE NEVER FOUGHT WITH AN AXE.

PRETEND YOU'RE SPLITTING LOGS.

BUDABUM BUDABUM

BUDABUM BUDABUM

LOGS DON'T BLEED.

HE COULD HAVE SWORN THEY'D BEEN FIGHTING HALF A DAY, BUT THE SUN SEEMED SCARCELY TO HAVE MOVED AT ALL.

YOUR FIRST BATTLE? YOU NEED A WOMAN NOW. NOTHING LIKE A WOMAN AFTER A MAN'S BEEN BLOODED.

I'M WILLING IF SHE IS.

THE FREERIDERS BROKE INTO LAUGHTER. THAT WAS A START.

WE MUST PRESS ON WITH ALL HASTE, MY LADY. THEY WILL ATTACK AGAIN, AND WE MAY NOT SURVIVE A SECOND ATTACK.

WE WILL RIDE AT ONCE.

I'LL HAVE THAT BLADE BACK NOW, DWARF.

LET HIM KEEP IT. WE MAY HAVE NEED OF IT IF WE'RE ATTACKED AGAIN.

YOU HAVE MY THANKS, MY LADY.

AND AS I WAS SAYING BEFORE WE WERE INTERRUPTED, THERE IS A FLAW IN LITTLEFINGER'S FABLE. WHATEVER YOU MAY BELIEVE OF ME, I PROMISE YOU THIS. I NEVER BET AGAINST MY FAMILY.

FATHER SAID THE RED KEEP WAS SMALLER THAN WINTERFELL, BUT IN HER DREAMS IT HAD BEEN IMMENSE.

AN ENDLESS STONE MAZE WITH WALLS THAT SEEMED TO SHIFT AND CHANGE BEHIND HER.

SOMETIMES SHE WOULD HEAR HER FATHER'S VOICE, BUT ALWAYS FROM A LONG WAY OFF.

SHE LISTENED FOR THE SOUNDS OF PURSUIT, AND HEARD NOTHING. SHE WAS IN FOR IT IF THEY'D RECOGNIZED HER, BUT SHE DIDN'T THINK THEY HAD. SHE'D BEEN TOO FAST.

SWIFT AS A DEER.

SHE WONDERED WHERE SHE WAS.

SHE'D COUNT TO TEN THOUSAND. BY THEN IT WOULD BE SAFE TO COME CREEPING OUT AND FIND HER WAY HOME.

COME ALONG, M'LADY. YOU AND YOUR FATHER CAN FINISH YOUR TALK ON THE MORROW.

HOW MANY GUARDS DOES MY FATHER HAVE?

HERE AT KING'S LANDING? FIFTY.

YOU WOULDN'T LET ANYONE KILL HIM, WOULD YOU?

NO FEAR ON THAT COUNT, LITTLE LADY. LORD EDDARD'S GUARDED NIGHT AND DAY. HE'LL COME TO NO HARM.

WHAT IF A WIZARD WAS SENT TO KILL HIM?

WELL, AS TO THAT... WIZARDS DIE THE SAME AS OTHER MEN, ONCE YOU CUT THEIR HEADS OFF.

YOUR GRACE, I NEVER KNEW YOU TO FEAR RHAEGAR. HAVE THE YEARS SO UNMANNED YOU THAT YOU TREMBLE AT THE SHADOW OF AN UNBORN CHILD?

NO MORE, NED! NOT ANOTHER WORD. HAVE YOU FORGOTTEN WHO IS KING HERE?

NO, YOUR GRACE. HAVE YOU?

ENOUGH!

I'M SICK OF TALK. I'LL BE DONE WITH THIS OR BE DAMNED.

WHAT SAY YOU ALL?

SHE MUST BE KILLED.

WE HAVE NO CHOICE. SADLY, SADLY...

YOUR GRACE, THERE IS HONOR IN FACING AN ENEMY ON THE BATTLEFIELD, BUT NONE IN KILLING HIM IN HIS MOTHER'S WOMB.

FORGIVE ME, BUT I MUST STAND WITH LORD EDDARD.

MY ORDER SERVES THE REALM, NOT THE RULER. I ONCE COUNSELED KING AERYS, AND I BEAR THIS GIRL NO ILL WILL. BUT SHOULD WAR COME...

IS IT NOT WISER, EVEN KINDER, THAT DAENERYS TARGARYEN DIE NOW THAT TENS OF THOUSANDS MIGHT LIVE?

WHEN YOU FIND YOURSELF IN BED WITH AN UGLY WOMAN, THE BEST THING IS TO CLOSE YOUR EYES AND GET ON WITH IT. KISS HER AND BE DONE.

KISS HER?

A STEEL KISS.

WELL, THERE IT IS, NED.

THE ONLY QUESTION IS WHO CAN WE FIND TO KILL HER?

MORMONT CRAVES A PARDON.

HE CRAVES LIFE EVEN MORE. NOW POISON... THE TEARS OF LYS, LET US SAY...

POISON IS A COWARD'S WEAPON.

YOU SEND HIRED KNIVES TO KILL A GIRL AND QUIBBLE ABOUT HONOR?

I WILL NOT BE PART OF MURDER, ROBERT. DO AS YOU WILL, BUT DO NOT ASK ME TO FIX MY SEAL TO IT.

ISSUE #11

"WHEN WE REACH YOUR KEEP, SER DONNEL, WE MUST SEND FOR MAESTER COLEMON AT ONCE. SER RODRIK IS FERVERISH FROM HIS WOUNDS."

THE LADY LYSA HAS COMMANDED THE MAESTER TO REMAIN AT THE EYRIE AT ALL TIMES TO CARE FOR LORD ROBERT.

WE HAVE A SEPTON AT THE GATE WHO TENDS TO OUR WOUNDED.

CATELYN HAD MORE FAITH IN A MAESTER'S LEARNING THAN A SEPTON'S PRAYERS, AND WAS ABOUT TO SAY AS MUCH WHEN SHE SAW THE BATTLEMENTS AHEAD.

WHO WOULD PASS THE BLOODY GATE?

SER DONNEL WAYNWOOD, WITH THE LADY CATELYN STARK AND HER COMPANIONS.

I THOUGHT THE LADY LOOKED FAMILIAR. YOU ARE FAR FROM HOME, LITTLE CAT.

HOW CAN I BE WHEN YOU ARE WITH ME, UNCLE?

MAY WE ENTER THE VALE?

IN THE NAME OF ROBERT ARRYN, LORD OF THE EYRIE, DEFENDER OF THE VALE, TRUE WARDEN OF THE EAST, I BID YOU ENTER FREELY AND CHARGE YOU TO KEEP THE PEACE.

COME.

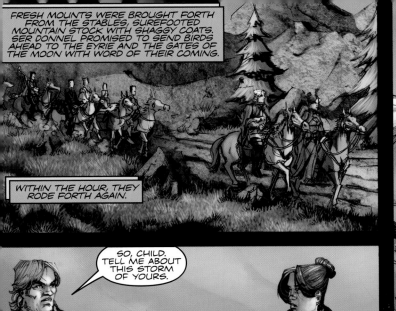

FRESH MOUNTS WERE BROUGHT FORTH FROM THE STABLES, SUREFOOTED MOUNTAIN STOCK WITH SHAGGY COATS. SER DONNEL PROMISED TO SEND BIRDS AHEAD TO THE EYRIE AND THE GATES OF THE MOON WITH WORD OF THEIR COMING.

WITHIN THE HOUR, THEY RODE FORTH AGAIN.

YOUR FATHER MUST BE TOLD.

IF THE LANNISTERS MARCH, WINTERFELL IS REMOTE AND THE VALE WALLED BEHIND MOUNTAINS, BUT RIVERRUN LIES RIGHT IN THEIR PATH.

SO, CHILD. TELL ME ABOUT THIS STORM OF YOURS.

IT TOOK LONGER THAN SHE WOULD HAVE BELIEVED TO TELL IT ALL.

LYSA'S LETTER AND BRAN'S FALL. THE ASSASSIN'S DAGGER AND LITTLEFINGER AND HER CHANCE MEETING WITH TYRION LANNISTER AT THE CROSSROADS INN.

I HAD THE SAME FEAR. WHAT IS THE MOOD IN THE VALE?

AND THERE IS THE BOY.

ANGRY. THE INSULT WAS KEENLY FELT WHEN THE KING NAMED JAIME LANNISTER TO AN OFFICE THE ARRYNS HAD HELD FOR THREE HUNDRED YEARS.

NOR IS YOUR SISTER ALONE IN WONDERING AT THE MANNER OF THE HAND'S DEATH. NONE DARE SAY JON WAS MURDERED, BUT SUSPICION CASTS A LONG SHADOW.

"LORD ROBERT. SIX YEARS OLD, SICKLY, AND PRONE TO WEEP IF YOU TAKE HIS DOLLS AWAY. HE IS JON ARRYN'S TRUEBORN HEIR, YET THERE ARE SOME WHO SAY HE IS TOO WEAK TO SIT HIS FATHER'S SEAT."

"SOME SAY NESTOR ROYCE-HIGH STEWARD THESE PAST FOURTEEN YEARS—SHOULD RULE UNTIL THE BOY COMES OF AGE. OTHERS THAT LYSA MUST MARRY AGAIN, AND SOON."

HUNGRY SHE WAS UNTIL THE PORTLY KNIGHT WHO COMMANDED THE WAYCASTLE OFFERED HER A SKEWER OF MEAT AND ONIONS. SHE ATE STANDING IN THE YARD WHILE THE STABLEHANDS PREPARED NEW MULES.

THEN IT WAS OUT AGAIN INTO THE STARLIGHT.

THE SECOND PART OF THE ASCENT SEEMED MORE TREACHEROUS. SHE COULD FEEL THE ALTITUDE NOW.

A HALF-DOZEN TIMES, MYA STONE HAD TO DISMOUNT AND CLEAR THE PATH OF FALLEN ROCK.

SNOW.

WE OUGHT TO KEEP GOING, MY LADY. IF IT PLEASE YOU.

YOU DON'T WANT YOUR MULE TO BREAK A LEG UP HERE, THE GIRL SAID, AND CATELYN WAS FORCED TO AGREE.

ABOVE SNOW, THE WIND WAS A LIVING THING.

THE STAIRS WERE CRACKED AND BROKEN FROM CENTURIES OF FREEZE AND THAW AND THE TREAD OF COUNTLESS MULES.

WHITEY'S A GOOD MULE, M'LADY. SURE OF FOOT EVEN ON ICE, BUT YOU NEED TO BE CAREFUL. HE'LL KICK IF HE DOESN'T LIKE YOU.

THE MULE SEEMED TO LIKE CATELYN, AND THERE WAS NO KICKING.

THERE WAS NO ICE, EITHER, AND SHE WAS GRATEFUL FOR THAT, AS WELL.

MY MOTHER SAYS THAT HUNDREDS OF YEARS AGO, THIS WAS WHERE THE SNOW BEGAN.

I CAN'T REMEMBER EVER SEEING SNOW THIS FAR DOWN THE MOUNTAIN.

WINTER IS COMING, CHILD, CATELYN WANTED TO TELL HER. PERHAPS SHE WAS BECOMING A STARK AT LAST.

BEST TO DISMOUNT FOR A BIT AND LEAD THE MULES.

THE WINDS CAN GET A BIT SCARY HERE.

SHE COULD **FEEL** THE EMPTINESS. THE VAST, BLACK GULF OF AIR.

THE WIND SCREAMED AT HER, TRYING TO PULL HER OVER THE EDGE. SHE COULD NOT MOVE FORWARD, AND THE MULE BEHIND BLOCKED HER RETREAT.

I'M GOING TO **DIE HERE,** SHE THOUGHT.

LADY STARK? ARE YOU WELL?

I... CANNOT DO THIS.

YES YOU CAN, MY LADY. KEEP YOUR EYES CLOSED IF YOU LIKE. TAKE MY HAND.

LET GO OF THE ROPE. WHITEY WILL TAKE CARE OF HIMSELF. JUST SLIDE YOUR FOOT FORWARD...

NOW ANOTHER.

EASY.

AND SO, FOOT BY FOOT, STEP BY STEP, THE BASTARD GIRL LED CATELYN ACROSS, BLIND AND TREMBLING, WHILE THE WHITE MULE FOLLOWED PLACIDLY BEHIND.

EVEN THE TOPLESS TOWERS OF VALYRIA COULD NOT HAVE BEEN MORE BEAUTIFUL THAN THE UNMORTARTED STONE WALL THAT WAS THE WAYCASTLE OF SKY.

THE STABLES AND BARRACKS ARE IN THERE.

THE LAST PART IS INSIDE THE MOUNTAIN.

IT'S SORT OF A CHIMNEY, LIKE A STONE LADDER MORE THAN PROPER STEPS.

IT WON'T BE MORE THAN AN HOUR.

THE LANNISTERS HAVE THEIR PRIDE, BUT THE TULLYS ARE BORN WITH BETTER SENSE.

I HAVE RIDDEN ALL DAY AND THE BEST PART OF A NIGHT.

HOW LONG MUST WE LINGER IN THESE RUINS BEFORE DROGO GIVES ME MY ARMY?

THE PRINCESS MUST BE PRESENTED TO THE *DOSH KALEEN*, AND—

THE CRONES, YES. AND THE MUMMERS SHOW OF A PROPHECY FOR THE WHELP IN HER BELLY. WHAT IS IT TO ME? I WAS PROMISED A CROWN, AND I MEAN TO HAVE IT.

THE DRAGON IS NOT MOCKED.

I PRAY THAT MY SUN-AND-STARS WILL NOT KEEP HIM WAITING TOO LONG.

YOUR BROTHER SHOULD HAVE WAITED IN PENTOS. ILLYRIO TRIED TO WARN HIM THAT HE HAD NO PLACE IN A KHALASAR.

HE WILL GO ONCE MY HUSBAND GIVES HIM THE TEN THOUSAND.

YES, KHALESSI, BUT...THE DOTHRAKI LOOK ON THESE THINGS DIFFERENTLY. KHAL DROGO WOULD SAY YOU WERE A GIFT, AND HE WILL MAKE A GIFT TO VISERYS IN HIS OWN TIME. YOU DO NOT *DEMAND* A GIFT.

IT'S NOT RIGHT TO MAKE HIM WAIT TO RECLAIM HIS THRONE.

VISERYS SAYS HE COULD SWEEP THE SEVEN KINGDOMS WITH TEN THOUSAND DOTHRAKI SCREAMERS.

"VAES DOTHRAK, THE CITY OF THE HORSELORDS."

NONE OF THE BUILDINGS ARE THE SAME?

THE DOTHRAKI DON'T BUILD. THESE WERE MADE BY SLAVES BROUGHT FROM THE LANDS THEY'VE PLUNDERED.

WHERE ARE THE PEOPLE WHO LIVE HERE?

ONLY THE CRONES OF THE DOSH KHALEEN DWELL PERMANENTLY IN THE SACRED CITY.

VAES DOTHRAK IS LARGE ENOUGH TO HOUSE EVERY MAN OF EVERY KHALASAR.

THE CRONES HAVE PROPHESIED THAT ONE DAY ALL THE KHALS WILL RETURN TO THE MOTHER OF MOUNTAINS AT ONCE, SO VAES DOTHRAK MUST BE READY.

KHALEESI. DROGO, WHO IS BLOOD OF MY BLOOD, COMMANDS ME TO TELL YOU THAT HE MUST ASCEND THE MOTHER OF MOUNTAINS THIS NIGHT TO SACRIFICE TO THE GODS FOR HIS SAFE RETURN.

ONLY MEN WERE ALLOWED TO SET FOOT IN THE MOTHER, AND IN TRUTH A NIGHT OF REST WOULD BE MOST WELCOME.

JHIQUI, A BATH, PLEASE.

TELL MY SUN-AND-STARS THAT I DREAM OF HIM, AND WAIT ANXIOUSLY FOR HIS RETURN.

DOREAH? FIND VISERYS AND ASK HIM TO SUP WITH ME. I WILL GIVE MY BROTHER HIS GIFTS TONIGHT. HE SHOULD LOOK LIKE A KING IN THE SACRED CITY.

IRRI, GO TO THE BAZAAR AND BUY FRUIT AND MEAT. ANYTHING BUT HORSE FLESH.

HORSE IS BEST. HORSE MAKES A MAN STRONG.

VISERYS HATES HORSE MEAT.

AS YOU SAY.

THE CLOTHING WAS MADE TO HER BROTHER'S MEASURE. TUNIC AND LEGGINGS OF WHITE LINEN. LEATHER SANDALS THAT LACED TO THE KNEE. A LEATHER VEST PAINTED WITH DRAGONS.

THE DOTHRAKI WOULD RESPECT HIM MORE, SHE HOPED, IF HE LOOKED LESS A BEGGAR.

HOW DARE YOU?

HOW DARE YOU SEND THIS WHORE TO GIVE ME COMMANDS?

I DIDN'T. I ONLY...DOREAH, WHAT DID YOU SAY?

KHALEESI, FORGIVE ME. I WENT TO HIM AS YOU BID, AND TOLD HIM YOU COMMANDED HIM TO JOIN YOU FOR SUPPER.

NO ONE COMMANDS THE DRAGON!

I AM YOUR KING!

SWEET BROTHER, PLEASE, THE GIRL MISSPOKE. I TOLD HER TO ASK YOU TO SUP WITH ME. IF IT PLEASES YOUR GRACE.

LOOK, THESE ARE FOR YOU.

WHAT IS THIS?

NEW RAIMENT. I HAD IT MADE FOR YOU.

DOTHRAKI RAGS. DO YOU PRESUME TO DRESS ME NOW? NEXT YOU'LL WANT TO BRAID MY HAIR!

YOU HAVE NO RIGHT TO A BRAID. YOU HAVE WON NO VICTORIES YET.

BUT THESE ARE GARMENTS FIT FOR A KHAL.

I AM THE LORD OF THE SEVEN KINGDOMS, SLUT! DO YOU THINK THAT BIG BELLY WILL PROTECT YOU IF YOU WAKE THE DRAGON?

ISSUE #12

WITH HIS LEGS UNABLE TO GRIP, THE SWAYING MOTION OF THE HORSE MADE BRAN FEEL UNSTEADY AT FIRST. BUT AFTER A TIME, THE RHYTHM BEGAN TO FEEL ALMOST NATURAL.

THERE WERE FEW PEOPLE IN THE VILLAGE OUTSIDE WINTERFELL'S WALLS.

OLD NAN SAID THAT WHEN THE SNOW FELL AND THE ICE WINDS HOWLED DOWN FROM THE NORTH, FARMERS LEFT THEIR FIELDS AND HOLDFASTS, AND THE WINTER TOWN CAME ALIVE.

BRAN HAD NEVER SEEN IT HAPPEN, BUT MAESTER LUWIN SAID THE END OF THE LONG SUMMER WAS NEAR AT HAND.

WINTER IS COMING.

SWEET KYRA! SHE SQUIRMS LIKE A WEASEL IN BED, BUT SAY A WORD TO HER IN THE STREET AND SHE BLUSHES LIKE A MAID.

DID I EVER TELL YOU ABOUT THE NIGHT THAT SHE AND BESSA—

NOT WHERE MY BROTHER CAN HEAR, THEON.

YOU ARE DOING WELL, BRAN.

THE BOY'S A STARK, TRUE ENOUGH. ONLY A STARK WOULD BE FOOL ENOUGH TO THREATEN WHEN SMARTER MEN WOULD BEG.

BRAN REALIZED WITH A START THAT THE MAN WORE BLACK RAGS. A DESERTER FROM THE NIGHT'S WATCH. HE REMEMBERED HIS FATHER SAYING THAT NO MAN WAS MORE DANGEROUS.

HIS LIFE IS FORFEIT IF HE IS TAKEN. HE WILL NOT FLINCH FROM ANY CRIME.

CUT OFF HIS COCK AND STUFF IT IN HIS MOUTH. THAT'LL SHUT HIM UP.

YOU'RE STUPID AS YOU ARE UGLY, HALI. BOY'S WORTH NOTHING DEAD. THINK WHAT MANCE WOULD GIVE FOR BENJEN STARK'S OWN BLOOD TO HOSTAGE!

YOU WANT TO GO BACK THERE, OSHA? MORE FOOL YOU.

THINK THE WHITE WALKERS WILL CARE THAT YOU HAVE A HOSTAGE?

THE CUT WAS QUICK AND CARELESS. BLOOD FLOWED, BUT THERE WAS NO PAIN. NOT EVEN A HINT OF FEELING.

STAND AWAY FROM MY BROTHER. PUT DOWN YOUR STEEL NOW AND I PROMISE YOU A PAINLESS DEATH.

HE'S A FIERCE ONE, HE IS. YOU MEAN TO FIGHT US, BOY?

DON'T BE A FOOL, LAD. YOU'RE ONE AGAINST FOUR. WE'LL THANK YOU FOR YOUR HORSE AND YOUR VENISON, AND YOU AND YOUR BROTHER CAN BE ON YOUR WAY.

DIREWOLVES...

DOGS. THERE'S NOTHING LIKE A WOLFSKIN CLOAK TO WARM A MAN.

TAKE THEM.

WINTERFELL!

MERCY, M'LORD.

ARE YOU HURT?

HE CUT MY LEG, BUT I COULDN'T FEEL IT.

A DEAD ENEMY IS A THING OF BEAUTY.

JON ALWAYS SAID YOU WERE AN ASS. I OUGHT TO CHAIN YOU IN THE YARD AND LET BRAN TAKE PRACTICE SHOTS AT *YOU*.

YOU SHOULD BE THANKING ME FOR SAVING YOUR BROTHER.

WHAT IF YOU'D MISSED THE SHOT? OR ONLY WOUNDED HIM? WHAT IF HIS HAND HAD JUMPED? YOU ONLY SAW HIS BACK. WHAT IF HE'D HAD A BREASTPLATE?

IT WOULD HAVE BEEN A VERY GOOD TIME TO KEEP HIS MOUTH SHUT AND HIS HEAD BOWED, BUT HIS MOOD WAS TOO FOUL FOR SENSE.

IS THAT THE BAD MAN, MOTHER? HE'S SO *SMALL.*

HE HAD FALTERED DURING THE LAST LEG OF THEIR CLIMB, AND BRONN HAD CARRIED HIM THE REST OF THE WAY. THE HUMILIATION POURED OIL ON THE FLAMES OF HIS ANGER.

THIS IS TYRION THE IMP, WHO MURDERED YOUR FATHER.

HE *KILLED* THE HAND OF THE KING!

OH, DID I KILL HIM TOO? IT WOULD SEEM I HAVE BEEN A BUSY LITTLE FELLOW. I WONDER WHEN I FOUND TIME TO DO ALL THIS SLAYING AND MURDERING.

YOU WILL SPEAK POLITELY TO MY SON OR HAVE CAUSE TO REGRET IT. THESE ARE TRUE KNIGHTS OF THE VALE AROUND YOU. EVERY ONE OF THEM WOULD DIE FOR ME!

AND SHOULD HARM COME TO ME, MY BROTHER WILL SEE THAT THEY *DO.*

CAN YOU FLY? DOES A DWARF HAVE WINGS? IF NOT, YOU WOULD BE WISE TO SWALLOW YOUR THREATS.

I MADE NO THREATS. THAT WAS A PROMISE.

YOU CAN'T HURT US. NO ONE CAN HURT US HERE!

TELL HIM, MOTHER!

NO ONE WILL HURT US, SWEET BOY.

THE EYRIE IS IMPREGNABLE.

NOT IMPREGNABLE.

MERELY INCONVENIENT.

YOU'RE A LIAR!

MOTHER, I WANT TO SEE HIM FLY!

SISTER, I BEG YOU TO REMEMBER THAT THIS IS MY PRISONER. I WILL NOT HAVE HIM HARMED.

MY SISTER'S LITTLE GUEST IS WEARY. SER VARDIS, TAKE HIM DOWN TO THE DUNGEON. A REST IN ONE OF OUR SKY CELLS WILL DO HIM GOOD.

I WILL REMEMBER THIS.

AND SO HE DID, FOR ALL THE GOOD IT DID HIM.

I TAKE IT BACK. NO FLUX FOR YOU, MORD.

I'LL KILL YOU MYSELF!

AT FIRST, HE HAD CONSOLED HIMSELF THAT THEY WOULDN'T DARE KILL HIM OUT OF HAND. NOW HE WAS NO LONGER CERTAIN. WITH EVERY DAY, HE GREW WEAKER.

HIS FATHER, HIS SISTER, HIS BROTHER. HE WONDERED WHICH HAD SENT THE FOOTPAD TO KILL THE STARK BOY, AND IF THEY'D ARRANGED THE DEATH OF JON ARRYN.

IF ARRYN HAD BEEN MURDERED, IT WAS DEFTLY DONE. SENDING AN OAF WITH A STOLEN KNIFE WAS CLUMSY.

AND WASN'T THAT PECULIAR...

PERHAPS THE DIREWOLF AND THE LION WERE NOT THE ONLY BEASTS IN THE WOOD. IF SO, SOMEONE WAS USING HIM AS CATSPAW, AND TYRION HATED BEING USED.

WELL. HIS MOUTH HAD GOTTEN HIM INTO THIS CELL. IT COULD DAMN WELL GET HIM OUT.

MORD! I WANT YOU!

MORD!

IT TOOK SOME TIME BEFORE HE HEARD THE FOOTSTEPS.

MAKING NOISE.

HOW WOULD YOU LIKE TO BE RICH, MORD?

CRACK

THAT WAS A STIFF ONE. I COULD USE A STRONG MAN LIKE YOU.

RICH AS THE LANNISTERS. THAT'S WHAT THEY SAY, MORD. MORE GOLD THAN YOU'LL SEE IN A LIFETIME.

IS NO GOLD.

THEY TOOK MY PURSE WHEN THEY CAPTURED ME, BUT THE GOLD IS STILL MINE. DELIVER A MESSAGE FOR ME, AND IT'S YOURS.

MESSAGE?

ONLY CARRY MY WORD TO YOUR LADY. TELL HER... ...TELL HER I WISH TO CONFESS MY *CRIMES.*

HE WAS LISTENING.

NO. YOU WILL FACE SER VARDIS ON THE MORROW.

SINGER! WHEN YOU MAKE A BALLAD OF THIS, BE CERTAIN YOU TELL THEM HOW LADY ARRYN DENIED THE DWARF A CHAMPION, AND SENT HIM FORTH BRUISED AND HOBBLING TO FACE HER FINEST KNIGHT.

I DENY YOU NOTHING! NAME YOUR CHAMPION, IMP. IF YOU THINK YOU CAN FIND A MAN TO DIE FOR YOU.

I'D SOONER FIND ONE TO KILL FOR ME.

NO ONE MOVED OR SPOKE OR MET HIS GAZE. FOR A LONG MOMENT, TYRION WAS SURE HE'D MADE A COLOSSAL BLUNDER.

AH, HELL. FINE...

I'LL STAND FOR THE DWARF.

TO BE
CONTINUED

Be sure not to miss
A GAME OF THRONES: THE GRAPHIC NOVEL, Volume 3
collecting issues 13–18, and with more special bonus content!
Coming soon.

AND NOW...

HERE IS A SPECIAL, INSIDER'S LOOK AT

THE MAKING OF

A GAME OF THRONES

THE GRAPHIC NOVEL

VOLUME 2

WITH COMMENTARY BY:

ANNE GROELL (SERIES EDITOR)

TOMMY PATTERSON (ARTIST)

DANIEL ABRAHAM (ADAPTER)

JASON ULLMEYER (DYNAMITE)

THE BIRTH OF A SCENE—
The Hand's Tourney

What we are going to do this time around is to examine the process of going from text to final graphic pages for one small snippet of the book—and with commentary by various members of the creative team—so you can get a sense of how the step-by-step process works behind the scenes.

For this exercise, we will be using the first five pages of the Hand's Tourney because it is one of the things we are proudest of. It is a scene of action, pageantry, and glorious detail—and also a scene that was trickily hard to adapt given the sheer scope of what George has packed into his pages.

Here, so you can see what Daniel was up against, is the relevant passage from the novel:

SANSA

Sansa rode to the Hand's tourney with Septa Mordane and Jeyne Poole, in a litter with curtains of yellow silk so fine she could see right through them. They turned the whole world gold. Beyond the city walls, a hundred pavilions had been raised beside the river, and the common folk came out in the thousands to watch the games. The splendor of it all took Sansa's breath away; the shining armor, the great chargers caparisoned in silver and gold, the shouts of the crowd, the banners snapping in the wind...and the knights themselves, the knights most of all.

"It is better than the songs," she whispered when they found the places that her father had promised her, among the high lords and ladies. Sansa was dressed beautifully that day, in a green gown that brought out the auburn of her hair, and she knew they were looking at her and smiling.

They watched the heroes of a hundred songs ride forth, each more fabulous than the last. The seven knights of the Kingsguard took the field, all but Jaime Lannister in scaled armor the color of milk, their cloaks as white as fresh-fallen snow. Ser Jaime wore the white cloak as well, but beneath it he was shining gold from head to foot, with a lion's-head helm and a golden sword. Ser Gregor Clegane, the Mountain That Rides, thundered past them like an avalanche. Sansa remembered Lord Yohn Royce, who had guested at Winterfell two years before. "His armor is bronze, thousands and thousands of years old, engraved with magic runes that ward him against harm," she whispered to Jeyne. Septa Mordane pointed out Lord Jason Mallister, in indigo chased with silver, the wings of an eagle on his helm. He had cut down three of Rhaegar's bannermen on the Trident. The girls giggled over the warrior priest Thoros of Myr, with his flapping red robes and shaven head, until the septa told them that he had once scaled the walls of Pyke with a flaming sword in hand.

Other riders Sansa did not know; hedge knights from the Fingers and

Highgarden and the mountains of Dorne, unsung freeriders and new-made squires, the younger sons of high lords and the heirs of lesser houses. Younger men, most had done no great deeds as yet, but Sansa and Jeyne agreed that one day the Seven Kingdoms would resound to the sound of their names. Ser Balon Swann. Lord Bryce Caron of the Marches. Bronze Yohn's heir, Ser Andar Royce, and his younger brother Ser Robar, their silvered steel plate filigreed in bronze with the same ancient runes that warded their father. The twins Ser Horas and Ser Hobber, whose shields displayed the grape cluster sigil of the Redwynes, burgundy on blue. Patrek Mallister, Lord Jason's son. Six Freys of the Crossing: Ser Jared, Ser Hosteen, Ser Danwell, Ser Emmon, Ser Theo, Ser Perwyn, sons and grandsons of old Lord Walder Frey, and his bastard son Martyn Rivers as well.

Jeyne Poole confessed herself frightened by the look of Jalabhar Xho, an exile prince from the Summer Isles who wore a cape of green and scarlet feathers over skin as dark as night, but when she saw young Lord Beric Dondarrion, with his hair like red gold and his black shield slashed by lightning, she pronounced herself willing to marry him on the instant.

The Hound entered the lists as well, and so too the king's brother, handsome Lord Renly of Storm's End. Jory, Alyn, and Harwin rode for Winterfell and the north. "Jory looks a beggar among these others," Septa Mordane sniffed when he appeared. Sansa could only agree. Jory's armor was blue-grey plate without device or ornament, and a thin grey cloak hung from his shoulders like a soiled rag. Yet he acquitted himself well, unhorsing Horas Redwyne in his first joust and one of the Freys in his second. In his third match, he rode three passes at a freerider named Lothor Brune whose armor was as drab as his own. Neither man lost his seat, but Brune's lance was steadier and his blows better placed, and the king gave him the victory. Alyn and Harwin fared less well; Harwin was unhorsed in his first tilt by Ser Meryn of the Kingsguard, while Alyn fell to Ser Balon Swann.

The jousting went all day and into the dusk, the hooves of the great warhorses pounding down the lists until the field was a ragged wasteland of torn earth. A dozen times Jeyne and Sansa cried out in unison as riders crashed together, lances exploding into splinters while the commons screamed for their favorites. Jeyne covered her eyes whenever a man fell, like a frightened little girl, but Sansa was made of sterner stuff. A great lady knew how to behave at tournaments. Even Septa Mordane noted her composure and nodded in approval.

The Kingslayer rode brilliantly. He overthrew Ser Andar Royce and the Marcher Lord Bryce Caron as easily as if he were riding at rings, and then took a hard-fought match from white-haired Barristan Selmy, who had won his first two tilts against men thirty and forty years his junior.

Sandor Clegane and his immense brother, Ser Gregor the Mountain, seemed unstoppable as well, riding down one foe after the next in ferocious style. The most terrifying moment of the day came during Ser Gregor's second joust, when his lance rode up and struck a young knight from the Vale under the gorget with such force that it drove through his throat, killing him instantly. The youth fell not ten feet from where Sansa was seated. The point of Ser Gregor's lance had snapped off in his neck, and his life's blood flowed out in slow pulses, each weaker than the one before. His armor was shiny new; a bright streak of fire ran down his outstretched arm, as the steel caught the light. Then the sun went behind a cloud, and it was gone. His cloak was blue, the color of the sky on a clear summer's day, trimmed with a border of crescent moons, but as his blood seeped into it, the cloth darkened and the moons turned red, one by one.

Jeyne Poole wept so hysterically that Septa Mordane finally took her off to regain her composure, but Sansa sat with her hands folded in her lap, watching with a strange fascination. She had never seen a man die before. She ought to be crying too, she thought, but the tears would not come. Perhaps she had used up all her tears for Lady and Bran. It would be different if it had been Jory or Ser Rodrik or Father, she told herself. The young knight in the blue cloak was nothing to her, some stranger from the Vale of Arryn whose name she had forgotten as soon as she heard it. And now the world would forget his name too, Sansa realized; there would be no songs sung for him. That was sad.

After they carried off the body, a boy with a spade ran onto the field and shoveled dirt over the spot where he had fallen, to cover up the blood. Then the jousts resumed.

Ser Balon Swann also fell to Gregor, and Lord Renly to the Hound. Renly was unhorsed so violently that he seemed to fly backward off his charger, legs in the air. His head hit the ground with an audible crack that made the crowd gasp, but it was just the golden antler on his helm. One of the tines had snapped off beneath him. When Lord Renly climbed to his feet, the commons cheered wildly, for King Robert's handsome young brother was a great favorite. He handed the broken tine to his conqueror with a gracious bow. The Hound snorted and tossed the broken antler into the crowd, where the commons began to punch and claw over the little bit of gold, until Lord Renly walked out among them and restored the peace. By then Septa Mordane had returned, alone. Jeyne had been feeling ill, she explained; she had helped her back to the castle. Sansa had almost forgotten about Jeyne.

Later a hedge knight in a checkered cloak disgraced himself by killing Beric Dondarrion's horse, and was declared forfeit. Lord Beric shifted his saddle to a new mount, only to be knocked right off it by Thoros of Myr. Ser Aron Santagar and Lothor Brune tilted thrice without result; Ser Aron fell afterward to Lord Jason Mallister, and Brune to Yohn Royce's younger son, Robar.

In the end it came down to four; the Hound and his monstrous brother Gregor, Jaime Lannister the Kingslayer, and Ser Loras Tyrell, the youth they called the Knight of Flowers.

Ser Loras was the youngest son of Mace Tyrell, the Lord of Highgarden and Warden of the South. At sixteen, he was the youngest rider on the field, yet he had unhorsed three knights of the Kingsguard that morning in his first three jousts. Sansa had never seen anyone so beautiful. His plate was intricately fashioned and enameled as a bouquet of a thousand different flowers, and his snow-white stallion was draped in a blanket of red and white roses. After each victory, Ser Loras would remove his helm and ride slowly round the fence, and finally pluck a single white rose from the blanket and toss it to some fair maiden in the crowd.

His last match of the day was against the younger Royce. Ser Robar's ancestral runes proved small protection as Ser Loras split his shield and drove him from his saddle to crash with an awful clangor in the dirt. Robar lay moaning as the victor made his circuit of the field. Finally they called for a litter and carried him off to his tent, dazed and unmoving. Sansa never saw it. Her eyes were only for Ser Loras. When the white horse stopped in front of her, she thought her heart would burst.

To the other maidens he had given white roses, but the one he plucked for her was red. "Sweet lady," he said, "no victory is half so beautiful as you." Sansa took the flower timidly, struck dumb by his gallantry. His hair was a mass of lazy brown curls, his eyes like liquid gold. She inhaled the sweet fragrance of the rose and sat clutching it long after Ser Loras had ridden off.

When Sansa finally looked up, a man was standing over her, staring.

He was short, with a pointed beard and a silver streak in his hair, almost as old as her father. "You must be one of her daughters," he said to her. He had grey-green eyes that did not smile when his mouth did. "You have the Tully look."

"I'm Sansa Stark," she said, ill at ease. The man wore a heavy cloak with a fur collar, fastened with a silver mockingbird, and he had the effortless manner of a high lord, but she did not know him. "I have not had the honor, my lord."

Septa Mordane quickly took a hand. "Sweet child, this is Lord Petyr Baelish, of the king's small council."

"Your mother was *my* queen of beauty once," the man said quietly. His breath smelled of mint. "You have her hair." His fingers brushed against her cheek as he stroked one auburn lock. Quite abruptly he turned and walked away.

By then, the moon was well up and the crowd was tired, so the king decreed that the last three matches would be fought the next morning, before the melee. While the commons began their walk home, talking of the day's jousts and the matches to come on the morrow, the court moved to the riverside to begin the feast. Six monstrous huge aurochs had been roasting for hours, turning slowly on wooden spits while kitchen boys basted them with butter and herbs until the meat crackled and spit. Tables and benches had been raised outside the pavilions, piled high with sweetgrass and strawberries and fresh-baked bread.

Here is how Daniel describes the process:

The tourney scene had a couple of aspects that were particularly challenging. First off, it's this amazing set piece. It's the Olympics of Westeros, which means all the pageantry and celebration and showmanship, and also blood and death and fear. Finding how to give the full physical scale to it meant finding places to pull back from the story and really let the art blow us away. But at the same time, there's a lot going on here. The death of Ser Hugh of the Vale is an important point in Eddard's story, and there are important character moments with Sansa and Ser Loras and Littlefinger. The narrative density of the scene is amazing.

In addition, the outline we had for the issue called for a reversal from the order in the book. Where the novel has Catelyn abducting Tyrion before the tourney, we were looking at reversing that and doing the tourney first, so as to end the issue with the dramatic moment of Tyrion ringed with swords. And there was one other lovely advantage to this order that wouldn't come clear until the graphic novel came together, which is that the last scene of the previous issue (Eddard wondering why knowledge of Robert's bastard son would have been important enough to die for) would have led directly into watching Ser Hugh of the Vale dying for it. It was a lovely transition, and so it was easy to overlook all the reasons that it actually wouldn't work.

Here is Daniel's initial draft of the adaptation:

GAME OF THRONES SCRIPT
ISSUE NINE
SCRIPT BY DANIEL ABRAHAM
BASED ON ORIGINAL WORK BY GEORGE R. R. MARTIN

PAGE ONE
NOTE TO THE COLORIST: All captions from page one to twelve should be in the
color set to indicate Sansa. Please check previous issues and match.

Panel One:
A small panel. We're close in on Ser Hugh of the Vale, recently knighted former
squire of Jon Arryn. He's lying on his back, looking up at us. He's wearing full plate
and a helmet (both of them bright and new) that lets us see his face. His cloak is
sky blue, with a border of crescent moons where the blood hasn't soaked it. Where
his gorget should protect his neck, there's an open space with a splintered length of
lance in it. We're very close in on him, so we can watch him bloody and dying.

        HUGH:
    nk ahhg ahk

Panel Two:
We're pulled back a little. Ser Hugh is on the ground of a jousting field, choking
to death on his own blood. We can't see the full tourney, but we see two or three
people standing over Ser Hugh. One of them is Gregor Clegane—the Mountain That
Rides—whom we've introduced before. He is also wearing full armor and has the
remains of a shattered lance in his massive hand. Ser Hugh's lance is on the green,
abandoned. One of the horses in full armor and barding is in the background.

        HUGH:
    ah...ak...
        HUGH:...

Panel Three:
Massive panel. We've pulled back even farther, and we can see the whole tourney
laid out before us. It's gigantic. At least a dozen jousting runs, a wide stretch of
cleared ground for the melee, tents, and pavilions. Banners are snapping in the breeze.
In the background, we can see two other knights jousting. There are commoners in
massive swarms, watching by the thousands. In the foreground, Sansa is sitting with
Jeyne Poole and Septa Mordane in among the high lords and ladies watching the
games. Sansa is wearing a green gown that complements her (auburn) hair. Jeyne is
looking away, her hand to her mouth in distress. Sansa is looking on calmly.
Let's pull out the stops here.

CAP: Sansa had ridden to the Hand's tourney with Septa Mordane and Jeyne Poole,
and it had been better than the songs. They watched the heroes of a hundred songs
ride forth, each more fabulous than the last.
CAP: The most terrifying moment of the day came during Ser Gregor Clegane's
second joust, when the point of his lance rode up and struck a young knight from
the Vale under the gorget. Sansa had never seen a man die. She ought to have been
crying, but the tears would not come.
CAP: It would have been different if it had been Jory or Ser Rodrik or Father, she
told herself. This young stranger from the Vale of Arryn was nothing to her. The
world would forget his name now. There would be no songs sung for him.

PAGE TWO

NOTE ON PAGES 2 & 3: This is the highlights from the full tourney. We're seeing action shots, but in every one of them, we should also see action going on around them.

Panel One:
Jaime Lannister on a huge white warhorse. He's wearing golden armor and the while cloak of the Kingsguard and has a fresh lance in his hand. He is beautiful, but he isn't playing to the crowd.

CAP: The Kingslayer rode brilliantly. He overthrew Ser Andar Royce and Marcher Lord Bryce Caron as easily as if he were riding at rings, then took a hard-fought match from Barristan Selmy.

Panel Two:
Sandor Clegane—the Hound—jousting against Renly Barratheon. Renly is wearing a helmet with golden horns and a full suit of armor, and the Hound is knocking him off the horse.

CAP: Ser Renly fell to the Hound with such violence he seemed to fly off his horse. His head hit the ground with an audible crack that made the crowd gasp, but it was only the golden antler on his helm snapping.

Panel Three:
An image of the whole tourney grounds as seen from above, like a crane shot. The details are less important than the impression of size and extent. There should be at least a couple of jousts happening in different lanes. The crowds are huge, and all around, commoners kept apart from the royalty.

CAP: Later, a hedge knight in a checkered cloak disgraced himself by killing Beric Dondarrion's horse and was declared forfeit. Lord Beric put his saddle to a new mount and was promptly knocked off it by the warrior priest Thoros of Mount.
CAP: Ser Aron Santagar and Lothor Brume tilted thrice without result. Ser Aron fell afterward to Lord Jason Mallister, and Brune to Yohn Royce's younger son Robar.

Panel Four:
This is a panel of Ser Loras Tyrell, the Knight of Flowers. We're seeing him out of the context of the tourney, almost like a portrait. He should be looking directly at the viewer, somewhat coyly. His plate mail is elaborately enameled with flowers. His hair is long, brown, and flowing. He's beautiful, with an almost anime-like androgyny.
CAP: In the end it came to four: The Hound and his monstrous brother Gregor, the Kingslayer...
CAP: ...And Loras Tyrell, the Knight of Flowers.

PAGE THREE

Panel One:
An evening scene. The moon is rising on the horizon, and the sky is twilight purple. Sansa is sitting where we saw her before, and over her shoulder we can see the jousting lane. Ser Loras Tyrell is on his horse, his hand lifted to the crowd in victory. His horse is covered in a blanket of red and white roses. Another knight, wearing bronze armor inscribed with eldrich runes (Robar Royce) is on the ground, with his squire coming to help him sit up.

CAP: After each victory, Ser Loras would remove his helm, ride slowly around the fence, and finally pluck a white rose and throw it to some fair maiden in the crowd.
CAP: His last match of the day was against the younger Ser Royce, but Sansa's eyes were only for Ser Loras.

Panel Two:
Ser Loras on his horse has stopped before Sansa. He's looking down at her gently. She's staring up at him, awestruck. It's still visibly an evening shot.

CAP: When his white mare stopped in front of her, she thought her heart would burst.
                LORAS:
      Sweet lady, no victory is half
         so beautiful as you.

Panel Three:
Close on Sansa, looking down at her cupped hands. She's holding a red rose from Ser Loras.

CAP: To the other maidens, he had given white roses.
CAP: She inhaled its sweet fragrance and sat clutching it long after he had ridden off.

Panel Four:
Sansa, in the foreground, looking out over the fields. Littlefinger is behind her, looking at her.

           LITTLEFINGER:
     You must be one of her daughters.
       You have the Tully look.

Panel Five:
Close on Sansa, turning to look over her shoulder. She's looking confused.

          SANSA:
       I'm Sansa Stark.
  I have not had the honor, my lord.

Panel One:
Septa Mordane talking to Sansa, Littlefinger is behind them. He's smiling, but the expression doesn't reach his eyes.

> MORDANE:
> Sweet child!
> This is Lord Petyr Baelish,
> of the king's small council.

Panel Two:
Littlefinger has taken Sansa's red rose in his fingertips.

> LITTLEFINGER:
> Your mother was my queen of beauty once.

> LITTLEFINGER:
> You have her hair.

Panel Three:
A wide panel. On the left, Sansa is with Septa Mordane and Jeyne Poole, still holding the red rose. In the middle of the composition is the rising moon on the horizon. On the right, Littlefinger is walking away.

Panel Four:
Another large image. We're looking at a large noble feast at the riverside. Six huge aurochs are on spits over a fire pit. There are tables spread out, with dozens of knights and noblemen walking around. The king and Cersei are at the table of honor. A juggler is tossing a cascade of burning clubs.

CAP: By then the moon was well up, so the king decreed that the last three matches would be fought on the next morning before the melee. The commons began their long walk home, and the court moved to the riverside to begin the feast.

Eventually, Daniel and I realized that starting with the tourney and ending with Tyrion would not work, as all the knights who aid Catelyn in Tyrion's abduction were on their way to the very tourney that would follow it, and all the hand-waving to try and explain this just did not feel convincing. So, in my edit, I returned everything to its original order and recommended replacing the full page of Tyrion's abduction, which had ended the issue, with a full-page image of the tourney, since that was our biggest spectacle. And Varys's last line in the Eddard chapter did make for a very effective ending for the issue overall.

PAGE SIX

NOTE TO THE COLORIST: All captions from page 6 to 17 should be in the color set to indicate Sansa. Please check previous issues and match.

Panel One:

A massive panel. On the left of the panel, we see one of the stands where the royalty sits, at enough of an angle that we can see a few of the faces. The focus of the spectators is Sansa, who is sitting with Jeyne Poole and Septa Mordane in among the high lords and ladies watching the games. Sansa is wearing a green gown that complements her (auburn) hair. Jeyne has her hand to her mouth in distress, but Sansa is looking on calmly.

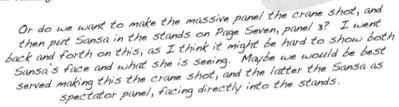

Or do we want to make the massive panel the crane shot, and then put Sansa in the stands on Page Seven, panel 3? I went back and forth on this, as I think it might be hard to show both Sansa's face and what she is seeing. Maybe we would be best served making this the crane shot, and the latter the Sansa as spectator panel, facing directly into the stands.

Stretching out from the stands, we can see a vast stretch of the tourney grounds laid out before us. The whole thing is gigantic. At least a dozen jousts are happening in different lanes, with a wide stretch of cleared ground for the melee, and tons of tents and pavilions. Banners are snapping in the breeze. The crowds are huge and, all around, commoners are kept apart from the royalty.
Part of what George thought the HBO series did poorly was to give the scale of the Tourney of the Hand. So let's pull out the stops here.

CAP: Sansa had attended the Hand's tourney with Septa Mordane and Jeyne Poole, and it had been better than the songs.
CAP: They watched the heroes of a **hundred** songs ride forth, each more fabulous than the last.

I added the emphasis to downplay the repeat of "songs."

NOTE ON PAGES 7–9: The close-ups are the highlights from the full tourney. We're seeing action shots, but in every one of them, we should also see action going on around them, giving the sense that this is all very much a part of some larger action.

Panel One:
Jaime Lannister on a huge white warhorse. He's wearing golden armor and the white cloak of the Kingsguard and has a fresh lance in his hand. He is beautiful, but he isn't playing to the crowd.

CAP: The Kingslayer rode brilliantly. He overthrew Ser Andar Royce and Marcher Lord Bryce Caron as easily as if he were riding at rings, then took a hard-fought match from Barristan Selmy.

Panel Two:
Sandor Clegane—the Hound—jousting against Renly Barratheon. Renly is wearing a helmet with golden horns and a full suit of armor, and the Hound is knocking him off the horse.

CAP: Ser Renly fell to the Hound with such violence he seemed to fly off his horse. His head hit the ground with an audible crack that made the crowd gasp, but it was only one golden antler on his helm snapping off.

Panel Three:
An image of the whole tourney grounds as seen from above, like a crane shot. The details are less important than the impression of size and extent. Unlike in the previous panel, where we see the spectacle that Sansa sees, here we get the whole scope of the thing—all the jousting lanes, as described before, and the melee area, and the tents, and the crowds.

Again, maybe this should be Sansa, Jeyne, and the Septa seen head-on... OR we could do Page Six as: Sansa as spectator, Jaime, Renly/Hound, then Page Seven as the full-page crane shot. Weigh in, guys. Now I like this LAST idea the best!

CAP: Later, a hedge knight in a checkered cloak disgraced himself by killing Beric Dondarrion's horse and was declared forfeit. Lord Beric put his saddle to a new mount and was promptly knocked off it by the warrior priest Thoros of Mount.
CAP: Ser Aron Santagar and Lothor Brume tilted thrice without result. Ser Aron fell afterward to Lord Jason Mallister, and Brune to Yohn Royce's younger son Robar.

PAGE EIGHT

Panel One:
We're close in on Ser Hugh of the Vale, recently knighted former squire of Jon Arryn. He's lying on his back, looking up at us. He's wearing full plate and a helmet (both of them bright and new) that lets us see his face. His cloak is sky blue, with a border of crescent moons where the blood hasn't soaked it. Where his gorget should protect his neck, there's an open space with a splintered length of lance in it. We're very close in on him, so we can watch him bloody and dying.

I cut the Hugh acking bit, because it looked a little silly. I think the image can carry its own weight.

CAP: The most terrifying moment of the day came during Ser Gregor Clegane's second joust, when the point of his lance rode up and struck a young knight from the Vale under the gorget.
CAP: Sansa had never seen a man die. She ought to have been crying, but the tears would not come.

Panel Two:
We're pulled back a little. Ser Hugh is on the ground of a jousting field, choking to death on his own blood. We can't see the full tourney, but we see two or three people standing over Ser Hugh. One of them is Gregor Clegane—the Mountain That Rides—whom we've introduced before. He is also wearing full armor and has the remains of a shattered lance in his massive hand. Ser Hugh's lance is on the green, abandoned. One of the horses in full armor and barding is in the background.

No, we haven't. This is his first appearance, so let's add in the description.

CAP: It would have been different if it had been Jory or Ser Rodrik or Father, she told herself. This young stranger from the Vale of Arryn was nothing to her.
CAP: The world would forget his name now. There would be no songs sung for him.

Panel Three:
This is a panel of Ser Loras Tyrell, the Knight of Flowers. We're seeing him out of the context of the tourney, almost like a portrait. He should be looking directly at the viewer, somewhat coyly. His plate mail is elaborately enameled with flowers. His hair is long, brown, and flowing. He's beautiful, with an almost anime-like androgyny.

Ramp up the description a bit for the colorist. Hair, eyes, what the armor looks like...

CAP: In the end it came to four: The Hound and his monstrous brother Gregor, the Kingslayer...
CAP: ...And Loras Tyrell, the Knight of Flowers.

PAGE NINE

Panel One:
An evening scene. The moon is rising on the horizon, and the sky is twilight purple. Sansa is sitting where we saw her before, and over her shoulder we can see the jousting lane. Ser Loras Tyrell is on his horse, his hand lifted to the crowd in victory. His horse is covered in a blanket of red and white roses. Another knight, wearing bronze armor inscribed with eldrich runes (Robar Royce) is on the ground, with his squire coming to help him sit up.

CAP: After each victory, Ser Loras would remove his helm, ride slowly around the fence, and finally pluck a white rose and throw it to some fair maiden in the crowd.
CAP: His last match of the day was against the younger Ser Royce, but Sansa's eyes were only for Ser Loras.

Panel Two:
Ser Loras on his horse stopped before Sansa. He's looking down at her gently. She's staring up at him, awestruck. It's still visibly an evening shot.

CAP: When his white mare stopped in front of her, she thought her heart would burst.

                    LORAS:
            Sweet lady, no victory is half
                so beautiful as you.

Panel Three:
Close on Sansa, looking down at her cupped hands. She's holding a red rose from Ser Loras.

CAP: To the other maidens, he had given white roses.   *Ser Loras*
CAP: She inhaled its sweet fragrance and sat clutching it long after ~~he~~ had ridden off.

Panel Four:
Sansa, in the foreground, looking worshipfully out over the fields. Littlefinger is behind her, looking at her.

                    LITTLEFINGER:
            You must be one of her daughters.
                You have the Tully look.

Panel Five:
Close on Sansa, turning to look over her shoulder. She's looking confused.

                    SANSA:
                I'm Sansa Stark.
            I have not had the honor, my lord.

PAGE TEN

Panel One:
Septa Mordane talking to Sansa. Littlefinger is behind them. He's smiling, but the expression doesn't reach his eyes.

> MORDANE:
> Sweet child!
> This is Lord Petyr Baelish,
> of the king's small council.

Panel Two:
Littlefinger has taken Sansa's red rose in his fingertips.

Or should we just have him touching her hair, which is much creepier and more intimate?

> LITTLEFINGER:
> Your mother was my queen of beauty once.

> LITTLEFINGER:
> You have her hair.

Panel Three:
A wide panel. On the left, Sansa is with Septa Mordane and Jeyne Poole, still holding the red rose. In the middle of the composition is the rising moon on the horizon. On the right, Littlefinger is walking away.

Panel Four:
Another large image. We're looking at a large noble feast at the riverside. Six huge aurochs are on spits over a fire pit. There are tables spread out, with dozens of knights and noblemen walking around. The king and Cersei are at the table of honor. A juggler is tossing a cascade of burning clubs.

CAP: By then the moon was well up, so the king decreed that the last three matches would be fought on the next morning before the melee. The commons began their long walk home, and the court moved to the riverside to begin the feast.

As you can see, we went back and forth on which image should be on the splash page, but eventually decided to do that as the big, establishing crane shot, then start focusing in more narrowly.

Daniel adds:

Once the reordering was done, and we lost the initial transition with Ser Hugh, the tourney really came into its own. We got the room for the full-page image of the whole place, which in retrospect, I think we really needed. We didn't get to focus as much on the death of Ser Hugh, but even that I don't really regret.

Like the man said, no songs will be sung for him.

And the final script took the form that you now see on the page.

Now the script goes to Tommy, who begins to work his magic. As Tommy describes it:

When drawing this book, I noticed I had to focus on getting better at expressions since the majority of what I've done so far has been conversation. So when an action scene comes along, I get superpumped and can't wait to dive into it.

As always, the only budget I have to worry about is time. I ended up taking two full months on this issue, if that tells you anything. Most pages' average people count is 10–15 per page. This scene was...well...start counting the people on the page six splash and I bet you get tired of counting and give up before you finish!

Drawing knights on horses is fun. It really is that simple. While every page had a labor-intensive crowd as backgrounds, it also had some killer action. One of the other challenges is not going too far with the laws of physics and having the scene feel like an issue of Ironman.

The splash was the most difficult, trying to show scope and action. I didn't want it to feel like a Where's Waldo poster. I also was testing out some drawing concepts based on a critique from one of my drawing buddies. The splash at the beginning of issue two had people spilling in, and while the page was nice, it didn't lead the eye. This time around I tried to gather people in groups, starting at the top and following counterclockwise and leading back to the top. I wanted it to feel like this was just a part of the tourney.

One last thing I'm proud of is page nine, panel two. I initially drew a terrible face on Sansa. As an artist, you can't see what a drawing is while you do it, so if you feel it's off, don't ignore that voice. A trick I do is to give a page a look before bed. If I feel a part is wrong, I erase it with no remorse and fix it the next day with fresh eyes. Now I think that redo is one of the best faces I've drawn. For what it's worth, I rarely use reference.

Here are the initial layouts, and what Tommy has to say about that stage in the process:

I can't state enough how all the thinking is done at the layout stage. I am trying to tell a story as clearly as possible. That is juxtaposed to the depth written into this EPIC story. Daniel, bless his heart, gets to do the first round of condensing. I can only imagine the fight in his brain about what to leave in and what to take out.

When I get the script, the hard part is done. He has clarified my course. I've gotten better at not being afraid to reduce the story even more. The neat thing about comic art is you have the whole page to view, so while a story is linear, unlike movies, you can see what just happened and what is going to happen. This allows me to avoid drawing a million people in each panel and I can zoom in on the people without losing the setting or tone. While Daniel and Anne have the final say, it took a few issues before I was able to make decisions that in my opinion help the book. As the issues have moved along, I am investing myself in the process in a way that only I can. The readers should feel themselves drawn in deeper as we progress, not just because of the story but also as the result of our execution as a creative team.

Daniel and I had only one comment on the layout, which was:

6: We're wondering about the complete top-down view of the tourney. We worry that by putting the angle directly above, it makes the whole look almost like a modern sports arena. So we'd recommend keeping the point of view high in the air but changing the angle at which we're looking at it to something more like forty-five to thirty degrees and making the whole thing look like a really big, violent Ren Faire. And when doing the lists, remember that the horses need to gallop out both ends. As shown, a few of them would be running into poles at the end of their run.

And that is it. We are SO excited to see your tourney in detail!

Tommy's initial pencils looked like this:

And Tommy says:

Because the layouts are where the thinking is done, the final pencils, on pages like this, are all about keeping the pencil moving. The willpower to draw crowds is draining. If you keep the pencil moving, it'll take care of itself if you've solved problems in the layout stage.

Save for comments of "Gorgeous, gorgeous, gorgeous!" our only issues were to remove the checkers from Beric Dondarrion's cloak in panel 7.3 and move the rose from Jeyne to Sansa in panel 10.3.

Here are the two corrected panels:

With the two panels corrected, Jason Ullmeyer and Joe Rybandt at Dynamite now take the reins, coordinating the talented Ivan Nunes and Marshall Dillon in colors and letters respectively. Here is how Jason describes the process:

> Once we get the final pencils, our Editor, Joe Rybandt, comes in and reviews the final pages, lets me know if they are good to format for the colorist and letterer, and advises Marshall and Dillon to look for the pages.
>
> Once the line art is approved, I get the hi-res files and, in cases like Tommy's art, where we are going straight from pencils to colors, I darken and clean up the line work if/when necessary. Luckily for me, Tommy's pages are pretty clean, so other than darkening up the lines a bit and brightening up some of the whites where necessary, there is not a ton to do. Then, once that is done, I make sure that all pages are sized/proportioned properly so that Ivan and Marshall are working on proper, print-size files. This way, when we get in the final letters and colors, there will be little to no alignment corrections to be made when assembling the final book. This helps us cut out some time on the back end of production.

Mostly, Marshall and Ivan make our jobs very easy, and there is usually little correcting on the letters-and-colors end. The initial letter files come in on the pencils—which I find makes it easier to concentrate on the new additions without the distraction of color. (It also means that Ivan is still busy working his magic.)

SANSA HAD ATTENDED THE HAND'S TOURNEY WITH SEPTA MORDANE AND JEYNE POOLE, AND IT HAD BEEN BETTER THAN THE SONGS.

THEY WATCHED THE HEROES OF A **HUNDRED** SONGS RIDE FORTH, EACH MORE FABULOUS THAN THE LAST.

THE KINGSLAYER RODE BRILLIANTLY. HE OVERTHREW SER ANDAR ROYCE AND MARCHER LORD BRYCE CARON AS EASILY AS IF HE WERE RIDING AT RINGS, THEN TOOK A HARD-FOUGHT MATCH FROM BARRISTAN SELMY.

SER RENLY FELL TO THE HOUND WITH SUCH VIOLENCE HE SEEMED TO FLY OFF HIS HORSE. HIS HEAD HIT THE GROUND WITH AN AUDIBLE CRACK THAT MADE THE CROWD GASP, BUT IT WAS ONLY ONE GOLDEN ANTLER ON HIS HELM SNAPPING OFF.

LATER, A HEDGE KNIGHT IN A CHEQUERED CLOAK DISGRACED HIMSELF BY KILLING BERIC DONDARRION'S HORSE AND WAS DECLARED FORFEIT. LORD BERIC PUT HIS SADDLE TO A NEW MOUNT AND WAS PROMPTLY KNOCKED OFF IT BY THE WARRIOR PRIEST THOROS OF MOUNT.

SER ARON SANTAGAR AND LOTHOR BRUME TILTED THRICE WITHOUT RESULT. SER ARON FELL AFTERWARD TO LORD JASON MALLISTER, AND BRUNE TO YOHN ROYCE'S YOUNGER SON ROBAR.

THE MOST TERRIFYING MOMENT OF THE DAY CAME DURING SER GREGOR CLEGANE'S SECOND JOUST WHEN THE POINT OF HIS LANCE RODE UP AND STRUCK A YOUNG KNIGHT FROM THE VALE UNDER THE GORGET.

SANSA HAD NEVER SEEN A MAN DIE. SHE OUGHT TO HAVE BEEN CRYING, BUT THE TEARS WOULD NOT COME.

IT WOULD HAVE BEEN DIFFERENT IF IT HAD BEEN JORY OR SER RODRIK OR FATHER, SHE TOLD HERSELF. THIS YOUNG STRANGER FROM THE VALE OF ARRYN WAS NOTHING TO HER.

THE WORLD WOULD FORGET HIS NAME NOW. THERE WOULD BE NO SONGS SUNG FOR HIM.

IN THE END IT CAME TO FOUR: THE HOUND AND HIS MONSTROUS BROTHER GREGOR, THE KINGSLAYER...

...AND LORAS TYRELL, THE KNIGHT OF FLOWERS.

Mostly, this is the stage that we are catching typos and mistakes that Daniel and I both missed while going exhaustively over the script. Astute readers may already have spotted the autocorrect error on page 7 of the final script that both Daniel and I were incapable of seeing until the letters first appeared. It is easy, when you know how something is supposed to read, to blip over how it is actually spelled. But sometimes, in a different context, it can leap out. So here is where we realized that Thoros had become from Mount instead of Myr, and where that got corrected.

The only other change we made to the letters was to separate Petyr's two speech bubbles in panel 10.2, to add a bit of a creepy pause. Often in the script, Daniel and I will break one character's lines into two or more segments—partially to cut down the number of letters in any one individual bubble or caption, but also to indicate a pause or change in direction. We sometimes specify in the script where the longest pauses should be by instructing that the split speeches should be in separate bubbles, but mostly we trust Marshall to walk that delicate line between establishing the drama and not blocking the art, which he does so well.

And then sometimes—as in this case—something that we did not see necessarily needing a long pause suddenly seemed to when art and letters both came together. And just by moving those two speeches apart, we were able to add more emotional weight to that panel and to that moment.

As for colors, Ivan's excellent work usually needs little to no correction. For the tourney scene, we had no comments at all, and here is what it looked like when it came in.

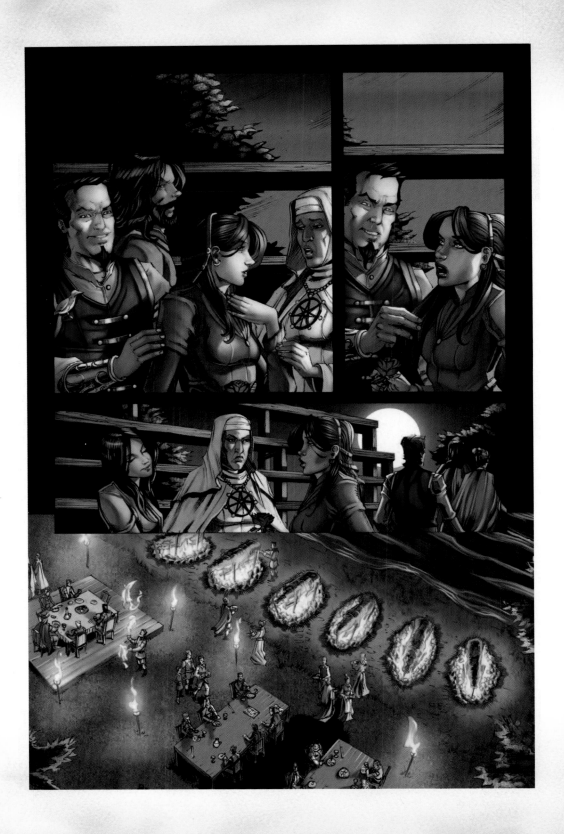

Our only color comment on this issue ended up being a heraldic issue in the Tyrion scene. Initally, the Bracken heraldry was showing a red stallion on grey-brown rather than brown, and the Frey heraldry colors were reversed, with a silver-grey castle on a blue background rather than the reverse. So we fixed that all up.

Jason Ullmeyer adds:

Once we have approved, corrected color files, we delete the older versions to ensure that no mistakes are made when assembling the final book. Then, once all colors and letters are approved, we assemble the book, double-check that all letters and colors align perfectly, give the book a final once-over, and send them on to the printer.

A little less than four weeks later, the final book appears in stores, then eventually gets folded into our compilation hardcovers.

The only thing left, then, is to give you a special advance preview of issue 13! Here is Mike Miller's cover and the lead-in to Tyrion's trial by combat....
Enjoy!

In A *Game of Thrones: The Graphic Novel, Volume 3,* we will take a look at the extensive gallery of character sketches Tommy has done to establish and distinguish the huge cast of characters that populates Westeros. I have a notebook containing hundreds of character sketches Tommy has created—especially since each issue introduces an average of about five new players whose appearances George must approve before they appear in the pages of the graphic novel. So I will share that with you next time.

In the meantime, we hope you have enjoyed Volume 2 of the graphic novels as much as we have enjoyed—and continue to enjoy—creating them!

—Anne Lesley Groell
　　Executive Editor
　　Random House, Inc.

We hope you've enjoyed this look inside the process and that you continue to have as much fun on this visual journey as we have!

GEORGE R. R. MARTIN is the #1 *New York Times* bestselling author of many novels, including the acclaimed series A Song of Ice and Fire—*A Game of Thrones, A Clash of Kings, A Storm of Swords, A Feast for Crows*, and *A Dance with Dragons*. As a writer-producer, he has worked on *The Twilight Zone, Beauty and the Beast*, and various feature films and pilots that were never made. He lives with the lovely Parris in Santa Fe, New Mexico.

DANIEL ABRAHAM is the author of the critically acclaimed fantasy novels *The Long Price Quartet* and *The Dagger and The Coin*. He's been nominated for the Hugo, Nebula, and World Fantasy awards, and has won the International Horror Guild award. He also writes as M. L. N. Hanover and (with Ty Franck) as James S. A. Corey.

TOMMY PATTERSON'S illustrator credits include *Farscape* for Boom! Studios, the movie adaptation *The Warriors* for Dynamite Entertainment, and *Tales from Wonderland: The White Night, The Red Rose*, and *Stingers* for Zenescope Entertainment.

THE BASKETMAKER'S ART

THE BASKETMAKER'S ART

CONTEMPORARY BASKETS AND THEIR MAKERS

EDITED BY ROB PULLEYN

Lark Books

Asheville, North Carolina

Library of Congress Catalog Card Number: 86-82336

ISBN 0-937274-63-1

Printed in Hong Kong.

Book acknowledgements are frequently tests of one's ability to say "no." So many people are involved in the creation of a book that it becomes nearly impossible to stop listing them.

This book, however, was primarily the creation of three people who each employed considerable creative talent and brought to the project a cooperative spirit and enthusiasm that made the entire endeavor a joyful experience.

Nancy Orban was the spirit and hand behind the book. She collected, organized and prepared the materials, worked with the artists and solved the problems.

Ron Zisman studied the artists' work, read their statements, asked questions and designed the book.

Thom Boswell gathered together the elements and prepared them for printing, always looking for the untied loose end.

A special thanks, as usual, to Val Ward who took the pressure, watched the schedule and pointed out the errors.

A book such as this also needs someone whose role it is to actively contemplate and to offer explanations. Both Shereen LaPlantz and Lillian Elliott are to be thanked for their willingness to have their observations committed to print.

Rob Pulleyn

CONTENTS

Introduction

2

Contemporary Baskets and Their Makers

Biographies

162

Yesterday's Baskets: Tradition and Form

Shereen LaPlantz

The art basket is a relatively new character in the ongoing tale of basketry, and it has a rich, diverse background to build upon. The craft of basketry has been part of most cultures throughout time. There doesn't appear to be a leader, or a single group that discovered basketry and then taught all the others. Instead, basketry seems to have developed simultaneously throughout the world, at the same and different times. This happened because the techniques are simple enough and obvious enough for anyone to stumble onto . . . and basketry filled a need.

Basketry is a living tradition—it is still being influenced, still changing with each basketmaker. So often we think of a tradition as something that is now preserved in a museum. But traditions need not be so strict that there is no room for change, or that they can't be influenced by a single individual. Today's artistic climate encourages all types of basketry. Many artists are reviving basketry traditions and learning from old artisans, old notes, or museum collections. Although they are trying to accurately duplicate work from the past, sometimes even calling themselves "replicators," their reproductions are achieved through an artist's hands and brain. The creative process is selective—artists, even replicators, make baskets in those styles that appeal to them, using materials and shapes they like. When they teach the tradition, they teach what appeals to them and their students comprise the next generation of artists, making baskets that interest *them*. The selective process continues and the tradition evolves.

To understand the current trends in basketry, let's go back a few hundred years and sketch a brief history. The North American Indian tribes had well-developed basketry traditions before the white settlers arrived, and many of their styles and techniques are still used today. The Pueblo Indians of the Southwest use both coiling and twining techniques. For coiling, their materials range from the rather wide bundles of grasses preferred by the Hopi, to the stiffer, thinner rod coils in the Apache baskets. Yucca leaves and devil's-claw are used for designs worked over core bundles of grasses or shoots. For twining, stiffer materials are used: willow shoots or the more readily available commercial reed. The Southwest is known for intricate motifs on its baskets; human, animal, and kachina designs are used, as well as geometrics.

There was also a high density of Indian tribes in California. The mild climate and availability of food allowed these tribes to develop as smaller groups; these groups in turn created a wide variety of basketry styles. Tribes of southern California work mostly in coiling, using grasses, rush stems, and yucca leaves. Designs are intricate, often geometric, with some human and animal forms. The Pima also use devil's-claw for design work. Some tribes use feathers to decorate the outermost edges of wide bottle forms. Coiling and twining both were used by tribes of central California, as far north as the wine country and Pomo. Their utilitarian seed beaters and fish traps are worked in open twining, and made with stiff shoots of materials such as willow and hazel. Finer twining is used in baskets meant for storage, carrying, and grinding. The patterns are geometric, and quite intricate. The coiling also tends to be finer, over a rod core. These patterns, too, are geometric, but are often decorated with beads, shell pieces, or feathers. The Pomo are also known for their miniature coiled baskets.

The tribes of northern California use a double-strand twining technique that makes a pattern on the outside of the basket, while the interior is left plain. Bear grass, maidenhair fern, and hazel are among the materials used. The patterns are geometric and intricate. The Hoopa Indians of that area make particularly fine basket hats that are simple bowl forms worn upside down on the head.

Tribes of the Pacific Northwest make twined and coiled baskets. Twining seems to be predominant. The Nez Perce make soft, flat bags out of corn husks and yarn. The patterning is geometric, worked as a wide band which covers most of the bag. The Washo make round, soft bags called "Sally bags," which are often decorated with highly stylized human and animal forms. The Tlinget make rattle-top baskets that are especially interesting. Enclosed in the basket's lid is a finial, with seed hulls sealed inside of it. The lid rattles softly when moved.

Top left: Tlinget basket, cedar root, false embroidery, separate compartment in lidded "rattle top" with seeds inside, twined.
Top right and bottom: Nootka/Makah, cedar root, grass, twined.

2

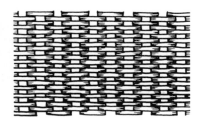

Plain weave. Over 1, under 1, pattern.

Twill. Over 2, under 1, pattern.

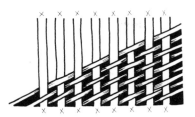

Randing. An over 1, under 1, weave with a new weaver tucked behind each upright or stake.

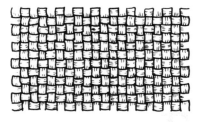

Plaiting. Over 1, under 1, pattern. The difference between plaiting and weaving is the tension. In weaving the stakes are spaced far apart and the weavers closely packed. In plaiting there is even tension between the stakes and weavers, leaving small square openings between the weave.

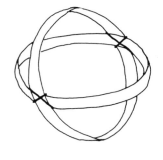

Rib construction. Two hoops are lashed together.

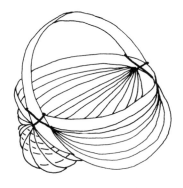

Ribs are added to make a frame. The frame is then woven over.

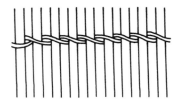

Twining. Twining uses two elements or weavers. They twist around each other as they weave over 1, under 1.

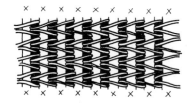

Twining one row with the twist underhand and the next row with the twist overhand.

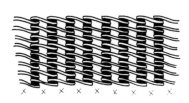

Reverse direction or arrow twining.

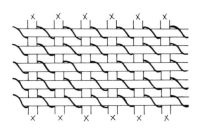

T twining. A rod is placed in front of the stakes and wrapped or lashed in place.

Figure 8 stitch.

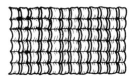

Coiling. The figure 8 stitch always covers the row below it, resulting in a double layer of stitches on each row.

Wrap stitch coiling.

Knotting. The finished look is the same as with figure 8 coiling.

Traditional Basketry Techniques

Right: Wedding cake basket by Mary Adams, sweet grass, split ash, curlicues.

Left: All Eastern Woodlands baskets. (Center:) Akwasasne Mohawk, black ash with sweet grass, curlicues. (Clockwise, from top right:) Mohawk, ash. Akwasasne Mohawk, strawberry baskets by Mary Leaf, ash, sweet grass, curlicues. Thimble basket, sweet grass, black ash. Akwasasne Mohawk, porcupine basket by Mary Adams, ash, sweet grass, curlicues. Akwasasne Mohawk, lidded basket by Mary Adams' mother, ash splint, curlicues.

Other tribes of the area make finely twined baskets out of cedar root and salt marsh grass. The patterns are delicate, and either geometric or figurative, often depicting people in boats, or whales and fish. The gathering baskets from this area have scalloped, open tops. To enclose the contents securely, a leaf was laid over the open top and laced in place. Coiled baskets from this region are sewn with stiff materials worked over a wide core, and are round or rectangular in shape. The coils are covered with a pleated, flat strip, an imbrication, which carries a geometric design.

The Northern Plains tribes are known for their twined utilitarian baskets, carrying or burden baskets, baby cradles, and seed beaters. These were made from stiff material, shoots and willow, using open and closed twining techniques. The Mandan made woven backpacks with thick U-shaped reinforcing rods at the corners. The patterns on

them were geometric.

The area from the Great Lakes to New England comprises the Eastern Woodlands, a region with its own specific styles of basketry. Tribes in this area make baskets of ash splints and sweet grass braids woven together. The pattern work is either in the form of added curlicues of ash, or, in some cases, block-printed geometric designs. The curlicue work is unique to the Eastern Woodland tribes and the eastern Cherokee. The Seneca in this area make masks and mats of braided cornhusks.

The southeastern Indians use cane and palmetto, and work mainly in plaiting. These tribes used designs ranging from simple geometrics to extremely complex geometrics. Some designs were highlighted by the use of three colors; others were executed in a single color, subduing the pattern. The Choctaw made unusually shaped baskets, referred

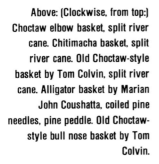

Above: (Clockwise, from top:) Choctaw elbow basket, split river cane. Chitimacha basket, split river cane. Old Choctaw-style basket by Tom Colvin, split river cane. Alligator basket by Marian John Coushatta, coiled pine needles, pine peddle. Old Choctaw-style bull nose basket by Tom Colvin.

Above: New England-style swing handle basket by John E. McGuire, black ash, hickory, double bottom construction.
Left: Shaker fancy basket by John E. McGuire, black ash, hickory.
Photos: Henry E. Peach.

to as "elbow," "bull nose," and "heart." Pine needle coiling also is done in this region by the Coushatta, who often shape their baskets to represent animals, such as alligators and turkeys.

A legend of Louisiana's Chitimacha tribe tells of a valiant warrior who slew a giant serpent that was threatening the tribe. As the serpent writhed in pain before its death, it gouged in the earth a great trough, now the Bayou Teche, where the tribe lives. The warrior's brave deed is still reflected in the distinctive serpentine designs on the double-woven lidded baskets made by the Chitimacha.

Many European immigrants brought their knowledge of basketry techniques and their preferred styles to the United States with them. Frequently they also brought seeds or starts of the basketry materials. As the newcomers settled, their basketry changed, adapting to local needs and materials, the new styles reflecting the influences of their new environment.

The British brought their woven splint baskets, such as market baskets, and willow baskets. The splint baskets have hand-carved handles and are quite sturdy. The Indians probably taught these settlers how to use black and white ash as basketry material.

The Shakers modeled their baskets after the woven ash splint baskets made by the Indians in the area. They worked closely with the Indians and learned from them. The Shakers were concerned with economy of materials, and became adept at working with thinner and thinner splints. Their baskets were made from fresh, green ash, which was formed over molds then left to dry and harden on the mold. This is not an uncommon practice; the Shakers just did it especially well.

5

The Appalachian ribbed baskets seem to be a combination of the Scottish, Irish, German, English, and Indian techniques. There is a controversy over who taught whom, but the resulting basket is a hybrid, different enough to suggest many influences. These baskets are white oak, woven over a rib skeleton. The length and positioning of the ribs dictate the shape of the basket, creating "fanny," "gizzard," "hen," "pie," and "key" baskets, as well as others.

White oak rod baskets also are made in the Appalachian region. The oak is drawn through dies to create thin rods for weaving. Willow work is done here too, as is rye straw coiling.

The Amana Colony, in Iowa, was a cooperative religious community settled by Germans. Each village in the colony had its own basketmaker and willow patch. Apparently there wasn't much interaction between people of Amana with Indians or with other settlers in the area, so their basketry changed little until the 1930's. Some of the basketmakers continued to work at the craft, and have taught their techniques to young basketmakers.

It should be mentioned that willow was—and is—one of the principal materials used for basketry. Willow starts were brought to North America with many of the British and European immigrants, and wil-

low patches can still be found dotting the countryside around older towns. There was even a movement at the end of the 1800's to teach the Indians to use willow for their basketry, but little came of it.

A great deal of willow work is being done today. In the small northern California town where I live, there are willow patches that were started by the Portuguese fishermen who settled here, and wild willow grows by the river. There are still some descendants of the Portuguese who make willow baskets, and there is a Catholic priest here who learned basketry in his native Rumania and still enjoys the craft. Contemporary basketmakers collect wild willow to make art baskets, and there is some willow furniture made here. If you look closely, you'll find that many towns have a similar variety of interest in willow basketry.

The art of basketry was not brought to this country by Europeans alone. The sea islands off South Carolina and Georgia were the first stop in America for many Africans, brought here as slaves. Some stayed on the sea islands, and put their knowledge of basketry to use. These Gullah baskets are made by coiling sea grass. The shapes can be complex, with tiers or undulations. The patterns are kept to simple stripes, and highlight, rather than interfere with, the forms.

Right: All Appalachian rib-style white oak baskets. (Clockwise, from top:) "Fanny" or "gizzard" basket with woven rim handle and spine, attributed to Mildred Youngblood. Market basket. Pie basket with woven rim and interior spine, attributed to a cousin of Mildred Youngblood.

Opposite, top: (Clockwise, from top:) Round tray, peeled and unpeeled willow by Joanna Schanz. Round Eastern basket, peeled and unpeeled willow with removable rim spikes on the outside by Joanna Schanz. Unpeeled wild willow basket by Joanna Schanz.
Opposite, bottom: Gullah baskets, coiled sea grass.

This short history of basketry has left out many traditions, glossed over others, and presented only highlights. It is meant only to suggest the richness and diversity of traditions from which contemporary basketmakers learn, and upon which they build. This history is not finished. Immigration continues. New traditions continue to come in, to adapt, and in turn to influence existing traditions. For example, the Southeast Asians and Ethiopians have brought with them their own basketry traditions. They are trying to learn about the basketry materials indigenous to the U.S., and adapt their techniques accordingly. Their work can now be seen in exhibits, and they teach at art centers and weaving supply stores.

The exciting basketry being done today, some of which you will see in this book, has been built on the foundations of past traditions. As artists, we have been influenced by what we see, but this doesn't mean we copy. It means we learn, adding new bits of information to our basketry vocabulary. With enough information we can develop our own styles and make exciting art baskets.

Untitled by Katsuhiro Fujimura.
Cut, affixed, piled cardboard,
7'6"x5'3"x8'6", 7'6"x4'7"x7'9",
1984. Photo courtesy of 12th
Biennale de la Tapisserie de
Lausanne.

Today's Baskets:
The Development of a Contemporary Aesthetic

Lillian Elliott

In the beginning, that is, in 7000 B.C., there were baskets, but when I began my career in art in 1960 there were none on the art scene. There were, of course, accomplished native American baskets, but the idea that a basket could be "art" had not yet occurred to anyone besides Ed Rossbach. He exhibited his non-traditional baskets along with other non-loomed textiles in 1968. When his book *Baskets As Textile Art* was published in 1973, the textile community was surprised by it. He did what is popularly called in California "consciousness raising."

Weavers had been searching for larger and larger materials so as to make architecturally scaled textiles, loom woven or not. There was a rather frantic feeling in the air, unnamed but exciting, of a new beginning. Weavers had just begun to think they might be creating art when they made their textiles. That first book of Rossbach's quietly and gently stepped back from the frantic action, and reported in an intimate, personal style on what was going on by observing ethnic baskets. His book was not only an appreciation of past art, but a studied commentary on the current scene. The finite size of baskets was reassuring when the "sculptural" textiles were constantly growing. Not only that, the book showed a broad range of three-dimensional forms constructed using the familiar textile techniques.

In 1976, when Rossbach's *The New Basketry* was published, the audience was there. As a matter of fact, weavers had become so interested in the subject that many of them were *in* the book. From the period when no one made baskets as personal expression to the time when such baskets could barely be presented in a single book had been just a few years. Amazing! It was not only that Rossbach wrote well, it was also that the field of textiles was at an exciting stage. Baskets had entered the textile scene. So had Art.

Jean Lurcat instituted the International Biennial of Tapestry in Lausanne, Switzerland, in 1962 in an effort to develop a greater appreciation of tapestry as an art form. Over the years the Biennial has become a rather prestigious international exhibit, underlining the importance of textiles. One way to trace the development of the art basket is to follow the progression of entries in the Biennial, as they changed from two-dimensional textiles to three-dimensional sculpture. At this point basketry was completely within the fiber world. The Lausanne Biennial was the most celebrated international fiber exhibition. The history of fiber *was* the history of the experimental basketry movement.

The 12th Biennial, in 1985, was titled Textiles as Sculpture. A number of entries related to basketry, but only three could be clearly defined as baskets, and even they used non-traditional materials and methods. One of those entries, Katsuhiro Fujimura's "Untitled" was in fact two enormous baskets constructed of sections of corrugated cardboard, assembled in a most untraditional manner. Anyone who saw these monumental baskets can testify to the potential power inherent in the basket form.

Perhaps the most lasting influence of the Lausanne Biennial has been its effect on the scale of work in textiles, since, until 1985, it required that work be a minimum of five square meters to enter. With this size restriction the Biennial was reinforcing a feeling for size that architects were beginning to look for in textiles. They had begun to seek out large-scale wall hangings to add warmth to the cold, clean architecture of public buildings. New yarns were needed for the larger wall hangings. Traditional fibers, such as wool, linen, and cotton, were available in yarn that was appropriate for the traditional purposes, such as clothing, blankets, and rugs. But when the total size of the work became considerably larger, the scale of the traditional fiber yarns was wrong. Even their feel was not appropriate in the finished work. Wool and cotton, spun so carefully and uniformly to feel good to the wearer, seemed too soft in these "public" textiles. The soft materials were particularly ill-suited to baskets, so the textile artists' search for firmer materials helped the new basketmakers when they began their work. It's worth noting that just when textile artists were hunting hard materials, sculptor Claes Oldenburg was beginning to work in earnest on his soft sculpture, and Christo was wrapping in polyethylene and canvas the fountain at Spoleto and Little Bay, Australia.

Raw fibers filled the need for firmer, larger "yarns." Sisal and jute, rope of all kinds, made an appearance in fiber shows. One exhibit, Deliberate Entanglements, at the UCLA Galleries in 1971, was full of sisal—I know because I'm allergic to it! Textile shows ceased to require that entrants submit three yards of cloth. Large hangings began to protrude from the walls and to be suspended from ceilings. Before long they also were standing in the middle of the room, supported by armatures of steel underneath.

Although the Lausanne Biennial received the most publicity, there were other important international exhibits. Two notable exhibitions of non-functional textiles took place in 1969: Perspectief in Textiel at the Stedelijk Museum in Amsterdam, and Wall Hangings, at the Museum of Modern Art in New York. Perhaps one more backward look is called for: Magdalena Abakanowicz of Poland won a gold medal in 1965 at the Biennial in San Paulo as an *artist* rather than as a fiber artist. And so, hope was instilled in the heart of every ambitious

Below: "The Pont Neuf Wrapped, Paris, 1975-85" by Christo. 444,000 square feet of woven polyamide fabric, 36,300 feet of rope, copyright Christo/C.V.J. Corp. Photo: Copyright Wolfgang Volz. Opposite: "Patti LaBelle" by Ed Rossbach. Newspaper, 16"x10", 1986.

textile student in this country.

As fiber artists began to "let it all hang out," new artists were beginning to make baskets. At the turn of this century, in his book *Aboriginal American Basketry*, Otis Tufton Mason describes how basketmakers always avail themselves of the materials in their immediate landscape. Some of the basketmakers, in looking for new forms, also looked for new materials. Pointing the way once again was Ed Rossbach, who used newspaper, a readily available fiber on today's landscape, and plastic, a mysterious, durable, and beautiful raw material. His ideas so influenced his students that they were unaware they hadn't always found plastic beautiful.

There are also artists who feel that the harvesting of their own raw materials is crucial to their art. This attitude toward natural materials may reflect a desire to return to simpler, happier times, or may be related to the recent environmental movement, a wish to connect with nature and all living things. Basically, it is the response of individual artists to various natural materials. It is an attitude similar to the response of a traditional craftsman to his material described in this touching passage from the book *Bamboo* (John Weatherhill, Inc.), which tells of a Japanese shakuhachi (flute) maker.

This maker, Kozo Kitahara, learned the trade from his father and continues it in his old studio. He is one of the fifteen who are left in Japan. He has been making shakuhachi for twenty-five years, even though he is still young. He brings to his work a sense of humbleness and dedication which is immediately communicated. Loving the bamboo itself, he cannot even bring himself to eat its shoots. In the winter, he goes up alone into the mountains, where he selects each one of the stems he will work. To stand thus in silence in the snow of a bamboo grove is his only holiday. The bamboo dries for three months on the roof of his house and then seasons in the dark for three years before its condition is ideal. During this season-

ing process the stems are trimmed into rough lengths, leaving a portion of flaring root at the base of each, which will become the bell end of the finished instrument.

When large sculptural forms began to appear, the attitude about looms changed. No longer was the loom seen as a tool which made the production of cloth faster and simpler. The loom suddenly was considered old-fashioned, a relic of the past to be disposed of quickly, so that textiles could be considered up to date and in step with the 20th century. Weaving seemed dreary and time consuming, while the newly celebrated off-hand glass blowing looked so effortless, almost magical. It was the time of the "happening" and the light show. Conceptual art and on-site installations began to appear.

The loom can most easily produce cloth of a rectangular shape. There also is a limited range of yarn sizes which it can handle comfortably. The new kind of work required expansion in many directions, shape among them. Explorations of off-loom processes began. It didn't seem to matter that twining, plaiting, netting, bobbin lace, and the myriad of techniques done without benefit of a loom were in fact more time consuming—they didn't *look* that way. Basketry was a direct outgrowth of this attitude. Interest in off-loom techniques for textile construction soon faded as artists moved on to felt and handmade paper. Only in basketry did that interest continue with vigor. The unrestricted shapes, materials, and techniques were focused by the notion of the vessel form. Here were individual three-dimensional forms which made use of the off-loom techniques and which had no need for elaborate hanging devices or armatures.

It is surprising how few substantial books have been written about baskets. There have always been how-to books, but rarely have there been books that deal seriously with the other aspects of basketry. Otis Tufton Mason's *Aboriginal American Basketry*, published in 1902, is an exception, though the book's publication was as much due to an interest in native Americans as to an interest in their baskets. Another exception is Bignia Kuoni's very fine book, *Cesteria tradicional iberica*, written in Spanish and published in 1981. It is an attempt to record the basketry traditions of Spain and Portugal while the traditions are still alive. I find myself wondering how much time will have to pass before the people who now make strictly utilitarian baskets will begin to make the art baskets which are now being made in this country. Will any of us, the contemporary American basketmakers, ever again make those useful baskets related to crops and harvesting?

I once had a Chinese-American student who was interested in doing research on traditional Chinese baskets (and there *is* a long tradition). She could read Chinese, so she was able to search the East Asian Library as well as the local university library. No luck. She finally located one book on the subject, written by a prominent Western anthropologist. There were a few photographs, and a short text which said in effect that there is nothing written on Chinese baskets. Chinese scholars apparently did not find baskets a worthy subject for books, no matter how important they are in ordinary life, or perhaps *because* they are omnipresent in ordinary life.

One really important influence on contemporary basketmaking has been the idea of the package. That concept covers a wide range, from packaging and presentation in advertising to the great wrapped sculptures of Christo. The exhibition, The Package, shown in 1959 at the Museum of Modern Art in New York, anticipated the interest in the vessel form. Books followed: *How to Wrap Five Eggs*, *How to Wrap Five More Eggs*, and *Tsutsumi*. The 1975 exhibition, The Art of the Japanese Package, was the first exhibit of Japanese packaging to appear outside Japan. It captivated a wide audience, including the textile world. Though the artists who currently are making baskets don't consider baskets packaging or advertising, we all present our thoughts on inside and outside space through our own finite personal bundles.

The question of labeling is always with us. A strange phenomenon has occurred in a number of exhibits in the last few years, most notably in a well-advertised exhibit, Intimate Architecture, at MIT. Clothing is described as architecture, textiles are described as painting, baskets as sculpture. There seems to be an attempt to make textiles or baskets more important. I don't think it works. The only people who pay attention to it are the textile people, or the basketmakers, and they already know that textiles and baskets are important.

In the spring of 1986 Lawrence Dawson, Senior Museum Anthropologist at the Lowie Museum of Anthropology, University of California, Berkeley, gave a lecture on the migration of basketry techniques and styles in prehistoric times from northern Asia to the north coast of the United States. One of the particularly interesting aspects of Dawson's lecture was a description of the types of prehistoric baskets which traveled. Twined baskets, such as fishing and berry baskets, had a tradition which remained unchanged in the new land as long as the specific need for which they were made continued to exist. Coiling, however, was a somewhat later technique, perhaps invented in North America. It spread in areas where twining traditions had long been entrenched, and it wasn't bound by the time-honored rules and regulations which governed twining. As Dawson stated:

> In areas where coiling spread in later times it was always received by the local people as a foreign idea, and therefore unencumbered by rules, whereas their traditional basketry was governed by nearly inflexible rules of manufacture and decoration. When coiling became known from neighbors or by trade it was welcomed as a novel technique in which one could freely extemporize and innovate.

In southern California designs were sometimes translated intact from twined baskets to coiled baskets with one important change: two stitches were used in coiling for every stitch in twining. If, however, the design was not taken directly from twining, there were no restrictions at all on the design. Therefore, coiling was much freer than the other techniques. Today's situation seems similar: ignorance of the traditional restrictions on design and technique leaves the contemporary basketmaker free to experiment with everything—materials, technique, color, scale.

It's curious to note that while basketmakers are now looking for firm, self-supporting materials, jewelers have begun to explore the use of textile techniques with wire. Generally, they have been attempting to construct precious and miniature basketry structures as body ornament. Potters, too, have taken to the basket form. The baskets of clay or metal make us aware of the essence of baskets. They emphasize the pliability, surface treatments, and complex structural variations possible in the more usual basketry materials. Seeing the "hard" baskets makes us see the softer ones, just as Oldenburg's soft typewriter makes us see all typewriters in a new way.

It is not the materials used which make contemporary baskets valu-

Scaffolding constructed from
bamboo. Photo: Dana Levy from
Bamboo (John Weatherhill, Inc.:
Tokyo, Japan).

departments at universities, or with textile exhibits at galleries and museums. There does not appear to be much of a parallel development outside this continent. Baskets obviously are being made elsewhere, but as part of an older tradition, not as part of this same movement. In 1981 an exhibit called Vannerie Traditionnelle d'Afrique et d'Asie et Nouvelle Vannerie was held in Lausanne. All the basketmakers represented in the small contemporary section were from the United States. The same was true of the exhibit Fibres Art 85, at the Musee des Arts Decoratifs in Paris.

There are signs of the "new" basket forms being made elsewhere. The catalog of a 1982 traveling exhibit in Japan, Art and/or Craft: USA and Japan, showed a few. In the summer of 1985 I met a student in Sweden who was interested in investigating non-traditional basket forms. She was encouraged by friends and teachers to come to the U.S. to study, since the audience here would be more receptive to her work. In England, judging from the magazine *Crafts*, there is also the beginning of interest in new basketry.

The interest in new basketry in this country seems to have originated mainly on the coasts, primarily in California. Perhaps this is because the coasts are most affected by trade and new ideas. It might be the adventurous spirit associated with the West, or it might simply be the influence of Ed Rossbach on his students.

In the art of the 20th century, there has been a break with tradition. New materials, new methods, new ideas are embraced. The new baskets, too, break with traditional basketry. If there is any relation to older work it is that the new baskets seem related to bridges and scaffolding in non-industrialized countries, rather than to the baskets of those countries.

Baskets. Even the word seems humble, self-effacing, and traditional. Some people have asked why I refer to my works as baskets when they aren't the traditional useful baskets. No one expects a concert of contemporary music to sound like Bach, or a contemporary painting to look like a Leonardo. Strange, then, that many expect contemporary baskets to be just like baskets of another century.

This attitude is not unique to the public, either. Critics and writers also suffer from a time lag. I once heard a thoughtful talk on basketry presented at the opening of an exhibition of contemporary baskets. The speaker called for an appreciation of contemporary baskets by describing the beauty of natural materials and traditional techniques as they continue in the new work. As admirable as the traditional baskets may be, following in that tradition is not the aim of most artists making baskets today. Not one piece in the above mentioned show was a traditionally constructed basket made of natural materials.

The new movement in basketry seems to have practitioners all over this country. There is a wonderful diversity among us, perhaps because of the solitary nature of the work. Most of the new basketmakers have had art training, but are self-taught in the skills of basketmaking. Just a few years ago there was no category in fiber shows for basketmakers. Now they are publishing books about us!

able. This quote from *Bamboo* describes traditional art, yet it also seems to refer to contemporary baskets.

> The crafts of bamboo are very pure. This is because bamboo itself as a raw material has almost no financial value. Thus the estimation set on those articles made by masters of the traditional crafts derives entirely from the skill and art of their hands. It is the craftsmanship itself which is prized. . . . This is more true since bamboo is not by its nature a material which can be expected to last a long time.

One of the interesting features of the contemporary basketry movement is that it is almost exclusively a North American phenomenon. Its growth so far has been associated with art schools and textile

 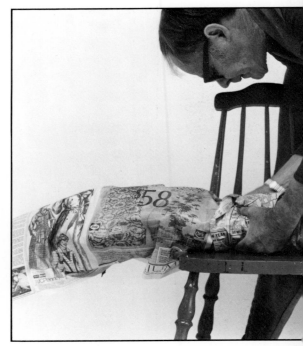

14

ED ROSSBACH

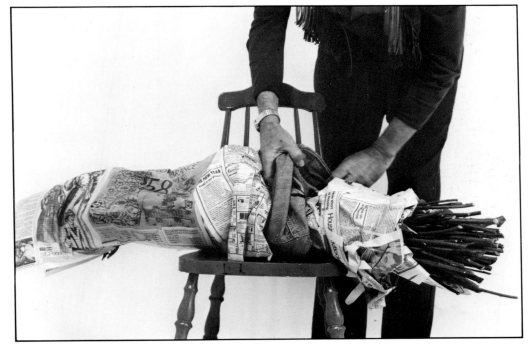

O ur house is full of baskets. They're stored everywhere in the place. They're in clothes closets, kitchen cupboards, cedar chests, and, mostly, in cardboard cartons. I am obsessed with them; they are oppressive, and yet my wife and I keep buying more, and I keep making more. They seldom depart the premises once they get inside, and if they do,

Below: Untitled,
newspaper and plastic,
10"x10", 1976.
Top right: Untitled, willow
with raw hide, 8"x10",
1968.
Bottom right: Untitled, corn
husk, 9"x12", 1969.

for an exhibition, they almost always return to find their places have been taken by other baskets. But they must be taken in. They are like the sentimental characters in a poem by Robert Frost.

I hate to part with baskets that I've spent time with. Someone—someone I don't even know—sent me an everyday basket (everyday, at least, to the people where it was made) and said that when I had lived with it for a while I should pass it along to someone else. I can't do it.

My own baskets—those that I have made—are kept in places where I need not look at them for any length of time. They are more satisfying when I see them only occasionally. I've heard that some people feel that way about their children. Once in awhile I'll take one of my baskets out of storage and look at it for about ten minutes. Usually that's enough. Sometimes the basket seems better than I remember it, sometimes it seems worse. I know within ten minutes. Memory, for me, is not a reliable tool of judging baskets; I have to confront the basket itself. Some-times there are baskets I have

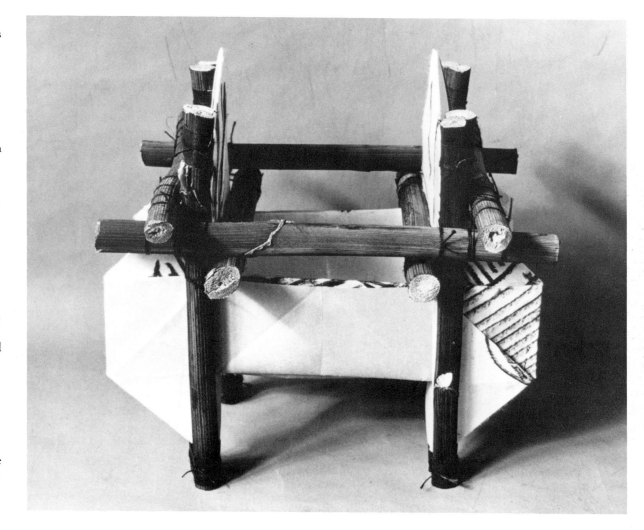

Opposite: "Lettuce Basket," lacquered newspaper, 10"x9", 1982.
Clockwise, from top right: Untitled, raffia, 10"x12", 1973. Untitled, ash splint with lacquer, 8"x8", 1985. "Mickey Mouse," sea grass, synthetic raffia imbrication, 10"x11", 1975. "Butterfly," silk screen, 14"x14", 1973.

made that I have completely forgotten. I realize that if they hadn't turned up by chance, I never would have thought of them again. They are like so many incidents in travel, never consciously thought of again.

The strange thing is that I'm not anxious to sell my baskets, or to give them away. If I really like a basket that I have made, even though I almost never look at it, I want to keep it. If I don't like it, I don't want to sell it, either. If I do sell a good basket I say, "That was a good basket. It has gone out of my life forever. I won't even be able to remember it. I probably shouldn't have sold it."

I guess what I am trying to say in this roundabout fashion is that baskets mean a great deal to me. I have a couple of Indian baskets from the Puget Sound area that keep speaking to me. They're around all the time. I never get tired of them. I also have some little plaited things from Indonesia made for tourists. I never get tired of them either. All these baskets seem so remarkable, I'm amazed that everyone isn't astonished—struck dumb—by their wonder.

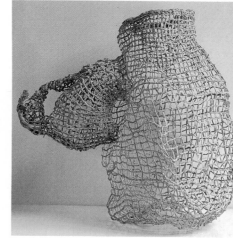

19

JAN BUCKMAN

When stories and legends are lost, culture dies. The baskets I make are legends growing out of the collective memories of my heritage. They speak of a culture, of honor, of commitment, of magic, of spirituality. They reflect a time when art and religion were fused with the demands of daily survival, a time when the sacredness of every object and every action was crucial to the blessings of that survival. My baskets express my effort to emulate these qualities in my own life.

My work comes from the contrast between what is, and what was; what is, and what can be. The following are some of the contrasts I attend to:

. . . watched a healthy young man on a riding lawn mower, plugged into a Walkman, mowing a postage-stamp-sized lawn. . . .

It took Datsolalee, a Washo Indian, one full year to complete a basket.

. . . stood in awe in a gourmet food store at the array of electric potato peelers, microwave merry-go-rounds complete with

Untitled, waxed linen, 3¾"x5½", 1985. Photo: Peter Lee.

painted horses, food processors that do everything but swallow the food for you . . .

My 75-year-old neighbor handed me a small maple branch that blew down in yesterday's wind, telling me it will boil a good cup of "coffee" on the cookstove.

. . . walked by the Hollywood Video store and stopped to watch a man and boy filling a grocery cart with movies for the weekend. I wonder what you might see in their eyes Sunday night. (And now I hear you can have videos delivered to your door) . . .

There was a man who dug himself a pond over a period of many years. As part of his daily walk, he carried away two buckets of dirt.

. . . there are elevators that express their desire that you have a nice day in the two dimensional voice of a computer. At least you don't have to come up with a Christmas gift to give this operator . . .

We should, I have read, ask forgiveness of the trees and plants as we harvest them. Spent a day cutting ironwood saplings to weave a trellis. Asked for a lot of blessings

Left: Untitled work in progress, waxed linen. Photo: Peter Lee.
Above left: Detail of untitled, waxed linen, 4½"x5⅜", 1982. Photo: Jim Christoffersen.
Photo of Jan Buckman by Thomas Paul.

Top: Untitled, waxed linen,
4⅜"x3", 1982.
Bottom: Untitled, waxed
linen, 5⅜"x4½", 1982.
Photos: Jerry Mathiason.

and felt rather foolish talking to the trees, yet as I sit in their shade, making a basket, I sense a presence.

. . . for people who have trouble keeping track of their lives and possessions, there is a keychain that reveals its location by playing Beethoven's Ninth Symphony . . .

My neighbor announced to me that it was one year ago today that my son helped tear down her old corn crib. She was freezing peas. As she reuses each plastic lid, she adds another piece of masking tape, recording not only the date, but some significant event or vision of the day. The sun rose red that morning.

. . . heard of a man who spent a full year inside the Water Tower in Chicago and never had to leave the building. He was fed, entertained, supplied, cooled, heated, all within the confines of two city blocks. The only sunshine came through tinted windows. Wonder what finally made him open the door to the outside . . .

An Indian woman, after drying berries, tanning deer hides, digging roots, gathering firewood, sits down

to work on a basket to be used for storing nuts. She feels the need to make this one precious, and twines the image of blackbirds as they fly through the now-leafless trees.

I am often asked how I can spend 70 hours making a basket. My answer is: How can people rush through their days in order to spend 70 hours of their time watching television? I try to make baskets an integral part of my life, in context and in harmony with the things I do every day, not something separate which demands a solitary space, behind the closed doors of an austere white-walled studio. Weeding the garden, splitting wood, and baking bread are not tasks that I must hurry to complete in order to return to a basket. The daily duties of my life feed my basketmaking and basketmaking nourishes my living. I am learning to honor time; not with a sense of urgency or pressure but through a respect for details.

I take pleasure in the process of making a basket, of persisting until it becomes precious. Beyond this, the baskets must speak for themselves. The imag-

ery on the pieces is not more or less than my curiosity and response to my environment. One could see horizons, or the silhouette of a scrub oak which has survived countless storms perched on a rocky cliff, or the trail and wanderings of seen and unseen beings on the forest floor.

Left: Untitled, waxed linen, 4¼"x7", 1985.
Below: Untitled, waxed linen, 4½"x3¾", 1984.
Photos: Peter Lee.

Left: Untitled, waxed linen, 5⅞"x4½", 1983. Photo: Jim Christoffersen.
Below left: Untitled, waxed linen, 4½"x3¾", 1984. Photo: Peter Lee.
Below right: Untitled, waxed linen, 4"x5", 1985. Photo: Peter Lee.
Opposite: Untitled, waxed linen, 4⅜"x4", 1986. Photo: Peter Lee.

25

JANE SAUER

"Polychrome III," waxed linen, paint, 4½"x10", 1985. Photo: Matthew Taylor.
Photo of Jane Sauer by Red Elf.

I began as a painter but was always more in love with the tactile qualities of the paint and canvas than with the formal concerns of painting. During my college years, there were very few schools offering anything but "fine arts." I actually believed that my desire to be more intimately involved in the artistic process was a negative regression to "women's handiwork," and that the only *real* artists were painters and sculptors. My work with textiles began in 1975 after I matured in my own thinking, thanks to the education given me by my children and the women's movement. I also was temporarily unable to paint at that time because of a back injury. After trying loom weaving, rug hooking, direct dyeing on fabric, and needlepoint, I found basketry and since then have not thought of doing anything else. I must admit that I am obsessive about my art form.

Because I have a husband, children, various animals, and the problems and responsibilities that surround us all, I am constantly involved in an inner struggle, or maybe I should say *war*, wanting to do my art work and wanting to take care of the rest of my life. I try to work at least eight hours each weekday and some over the weekend. When I am close to a deadline I do nothing and think nothing but art for several days or maybe even weeks. My family is very supportive, and I am not

26

"Nest/Nested/Nestle,"
waxed linen, silk, feathers,
4"x6", 1980.
Inset: "Ceremonial Basket,"
waxed linen, silk,
3½"x5½", 1979. Photos:
Huntley Barad.

expected to make dinner, clean house, or wash clothes. I think women my age (48) find it very difficult not to perform domestic duties naturally, because we were taught that this is our burden. I have had to train myself to share household responsibility with the rest of the family, so some of my struggle is within myself.

My studio is a converted unattached garage. I like being able to work where I can still have contact with my children. My eleven-year-old has a shelf for her own art supplies in my studio. My other children, who enjoyed doing art work when they were younger, check in occasionally. My husband frequently works in my studio also. The space seems at times to be very private and personal, and at times it is the center of our family life.

My art work is the container for the tensions of my life. I firmly believe in the power of a well-designed, beautifully executed basket, but I want to go beyond that. I am always saying something in my work, searching for forms that are archetypal. I find the struggle to go beyond is accompanied by fear of failing,

Opposite: "Earth Vessels," waxed linen, paint, 3"x6" and 5"x8", 1982. Photo: Huntley Barad.
Below left: "No Exit," waxed linen, paint, silk, 2"x18", 1985. Photo: Matthew Taylor.
Center: "Breaking Old Patterns," raw linen, waxed linen, 3¼"x10¼", 1984. Photo: Matthew Taylor.
Right: "Color Factor," raw linen, waxed linen, 3½"x10", 1984. Photo: Matthew Taylor.

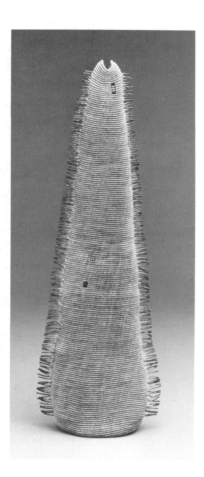

anxiety over wasting time and materials to create a failure, and concern that I am not pushing the boundaries far enough and fast enough. But at the same time there is the feeling of exhilaration during that magic time when things come together, when one idea is rushing toward another, when I am in new territory and have some control. I find agony and joy in searching to capture a sense of the unexpected, to reveal the magic of form, to bring order to chaos, to bring together the mind and the spirit. I hope to make forms that have a universal emotional impact, that are generic, yet are allegorical to my own life. My shapes have become progressively simpler, and I have become more involved with their messages. I have been using symbols to express circumstances, tensions, contradictions, and the interplay of relationships in my life.

My newest forms are directly related to a recent trip to Japan. The Eastern experience further deepened my resolve to "simplify and intensify." The Japanese have a wonderful way of isolating and simplifying an object so as to make it become more

important. Japanese art revealed visual language, a high regard for craftsmanship, and the stretch for perfection, coupled with mysterious, serene, meditative qualities.

Because I am making both baskets and sculpture I am concerned with structure. The process of knotting, the technique I am now using exclusively, is a slow procedure of building a structure row upon row. I have chosen to develop the possibilities of this one process instead of exploring a number of different

methods of building with threads. I enjoy the rhythm of the process, and there still are many new techniques to try. Knotting creates a grid and therefore almost begs to become a pattern.

Man's constant desire to make stripes and patterns is

Left, from left: "Entrance with No Exit," waxed linen, paint, 6½"x18", "Soft Daggers," waxed linen, paint, 4½"x10", "Subliminal Messages," waxed linen, paint, 2"x19", 1985-86. Photo: Red Elf.

Below: "Adolescent Fascination," waxed linen, dyed silk, 3¼"x10", 1984. Photo: Matthew Taylor. Opposite: "Celebration," waxed linen, paint, 4½"x7½", 1985. Photo: Matthew Taylor.

fascinating. Besides making baskets, every culture has made stripes and patterns. I share this same attraction and desire to decorate obsessively, and frequently I need to let this desire dominate my art. Patterns and stripes, of course, require the added dimension of color. I love the challenge of discovering what I can make color do. How can I make you *feel* by the color and shape I use? How can I make one color into several colors by varying the quantity and by changing the color's surroundings? I have begun to experiment with painting on threads so I can have all the colors I want, not just those available commercially. I feel as if I have come full circle; I am using the painting skills that I began with.

I am still awed by the fact that, without using any tools or technical equipment, but by just moving my fingers, I can make this object called a basket. I can make a sculptural form with patterns, stripes, grids, shape, symbols, gentle stain of the fibers, vibrant or pungent colors, and I have only just begun my exploration of the endless possibilities.

KARI LØNNING

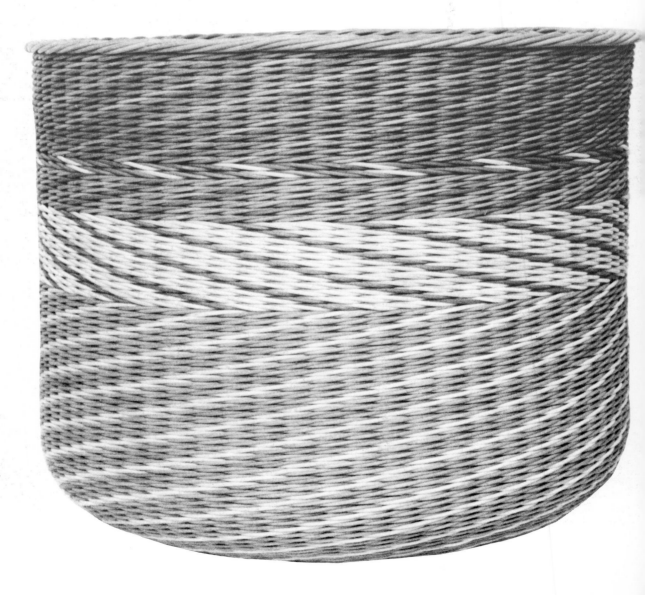

My interest in basketry began as an intrigue with construction. For my first basket, I used a copper wire tea strainer as my model. The woven wire held its shape and created a pattern of its own. Soon after, I took a basketry workshop. In the beginning, I made baskets for the fun of it—the process was so time consuming, I couldn't imagine doing it for a living. But within a few weeks, I was doing little else.

I always work in dyed or natural rattan. It is completely controllable, and has little character of its own. It allows me to weave the "thrown" forms which I was incapable of mastering as a potter. By using fiber-reactive dyes instead of glazes, I can achieve an infinite spectrum which I can count on reproducing. Since most of the work I do is by special order, this is a very important factor. At present, I am working with a muted palette of pale pink, lavender, gray, blue and green. With my production pieces, I know what each basket will look like before I begin it. In the other work, I hold a vague idea as to general form or color but, so far, no piece was completed as it was first conceived.

Opposite: "Bigger Diagonals," dyed rattan, 16"x19", 1984.
Left: "Shaped Container," dyed and natural rattan, 11"x13", 1976.
Above: "Cusine's Flat Hen Basket," dyed and natural rattan, 17"x11½", 1979.
Photo of Kari Lønning by Karen H. Fisher.

Many people still feel they have to put something into a basket to justify its existence. Although my production work is constructed to withstand use, its primary purpose is not its function. The essence of baskets, as containers, is still very important to me, but my reason for making baskets rather than working in another medium is that baskets are the most appropriate vehicle through which I can experiment with color, design and form. The double-walled pieces are constructed so that they *already* contain something—space. For me, they are an affirmation of what baskets can be.

My favorite place to work is on my front porch. It faces south, so even when it's cold, I can be in the sun, protected from the north winds. The porch is surrounded by flowering trees, bushes, and perennials, which bloom as soon as the ground thaws and last late into fall. Colors, scents and textures abound. When it does get too cold to be outside, I move into a studio off my kitchen. There, too, I am surrounded by flowering plants and bulbs and greenery. Though I have a room piled high with assorted dye lots of

reed, baskets in progress or finished, office materials and slides and photographs, I can't have clutter around me while I work. Even having the wrong music on while I work is distracting.

I am now working with more decorators, and have been trying to balance the colors I want with the colors they want. A subtle change toward more muted and gray tones seems to have satisfied us all. As the baskets got larger, I put more energy into using contemporary colors. The production pieces moved away from the craft market into the home furnishings market. Here it seems people can see the baskets first as design statements and then decide whether they are interested in them as baskets.

It's hard to make the shift from a production frame of mind to a more abstract, allow-it-to-come way of thinking. I don't know what my work will be like in the future; there are still many possibilities I want to explore. I may not even be working in basketry, but I will be making or designing something, and I will be addressing the fascination I have with space and color.

Photo of basket grouping at left courtesy Meredith Gallery, Baltimore, Maryland.

I am a basketmaker. I work
with metallic fiber, and
glean industrial metals from
alleys behind machine shops.
This started several years ago,
when one day I passed a machine
shop and saw long, thin strips of
iridescent blue and purple steel
turnings in the trash bin. Then I
discovered the recycling center
where I used to hunt for just the
right tin cans to snip and paint.
Later I found that the trash bins
at the farm implement store contained the most wonderful trac-
tor gears. I would pile them up
and paint them with transparent
paint.

When I started to work in
fibers I always tried to integrate
metal turnings with the natural
wool and linens as I used the
processes of weaving, felting, and
basketry. I had always thought of
the turnings as fiber, and found I
could use them in much the
same way as any other fiber.
First I made a basket of mostly
natural fiber with just a little
metal. Then the baskets became
mostly metal, sometimes with a
bit of natural material like date
fruit stalks, banana bark, or
mink.

I use industrial metals like

KAREN S. TURNIDGE

copper, stainless steel, brass, aluminum, tin, and other metals. I go to local machine shops and gather curling metal chips as they come rolling off the lathes. I search through trash barrels to find the particular turning I need. Lately I have been working with a machinist who has helped me develop turnings to use for specific purposes, because I could no longer find enough of the same kind of shavings in the trash. I also work with woven wire, and wire that I find.

I try to work with as many basketry techniques as possible and add to them metal techniques that enhance my basketry. I have found the unique characteristics of metal to be a challenge and a help in making baskets. Through experimentation I discovered metal could be used like vines or twigs in traditional basketry techniques like twining, plaiting, or coiling. The turnings are so flexible I can knit, crochet, or "felt" with them. I have knit both with wire and turnings, and have used crochet or looping as well. Felting, as I think of it, refers to the fiber technique in which wool fibers are meshed together using

heated water and pressure. The same technique can be used with turnings of a special shape.

Some of my plaited baskets and pulled warp baskets have the see-through qualities of woven metal. Metal can also be squishy and soft, as in my plaited tin baskets and some crocheted baskets. I like the contrast of soft and hard, and the visually similar characteristics of the linear wire, woven wire and spikey gleam of mink.

Opposite: "Fire Series: Kindle," copper, wire, 7"x17", 1985. Above: "Bristle," copper, wire, computer-aided design, 13"x8", 1984. Photos: Suzanne Coles-Ketcham. Photo of Karen Turnidge by Darrell Turnidge.

I treat the metal fiber in various ways before making it into baskets. Sometimes I pound on it with a hammer, or bend it, or rivet it, but mostly I enjoy baking it. I touch it with a torch, or use a kiln to color one strand at a time. If I'm really feeling brave, I bake the whole basket at once. I use heat coloring with stainless steel to bring out golds, purples, and blues (my cool palette). When heat is applied to copper, reds, oranges, pinks, and silvers are produced. Aluminum and titanium can be anodized for color, a process that I am now starting to use. By using impressionistic color and constructions that indicate action, I try to convey liveliness, energy, and celebration.

In 1983 I was introduced to the technique of pulled warp basketry and I have used computer graphics with a homemade program to design baskets to be made with this technique. Basically, a view of one edge of the basket is drawn, then the whole basket appears on the screen, then a pattern can be printed for weaving—all in a matter of minutes. I have also done some experiments with a computer

photographic process to make graphics for use with loom-woven baskets.

When planning a basket I think mostly of the material I have on hand, because that determines the size, color, and technique I will use. My drawings and sketches for my work are usually many variations of form, and rarely have I completely followed my drawing as I work. I have been inspired by the ancient and contemporary clay pottery that I have studied, and modeling with clay is a way to sketch ideas for baskets. Shapes can be formed with it very quickly. One of my small baskets was made in much the same way as a clay pot. I just pinched the copper to shape and added knitted copper as structural support on the outside.

In my studies of metal techniques I have gained endless inspiration from metalsmiths and jewelers and the richness of their ideas, their sculptural design, and their fine craftsmanship.

I am a basketmaker because I love the basket's symbolism. It reminds me of man's first action of gathering food and of the first temporary baskets. I like to push the symbol into the present by thinking of contemporary themes, like nutrition. The double helix construction has inspired some of my baskets. In my mind it symbolizes new life, as in the DNA molecule. The cell wall also gives inspiration for baskets. Various energy sources inspire my work: the sun, wave action, swirling water, and fire. As I work I think of solar energy and the power of energy in water. I think of light or of a hearth fire. The feminine symbolism of basketry for me as a mother is involved with the nurturing and protecting of life. The process of going round and round as one makes a coiled basket makes me think of the recycling of industrial metals. As a basketmaker I enjoy bringing the elements of metal, fiber, and names together to convey the symbolism of ideas that interest me.

My last name is Turnidge (Turn-edge), which originated in England and referred to a lathe worker. I never really thought about it much when I was developing my ideas, but the name is appropriate.

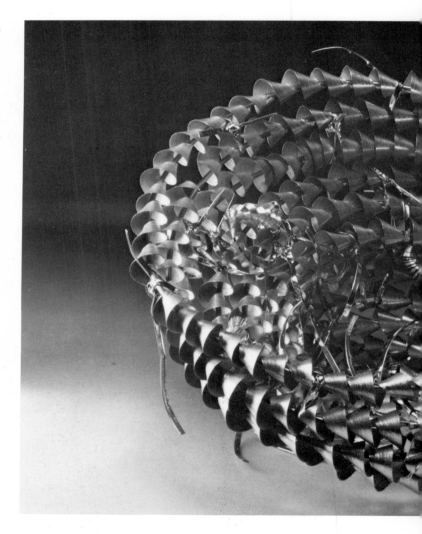

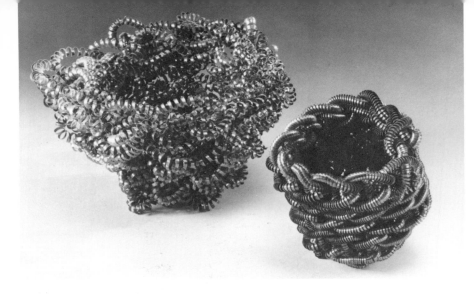

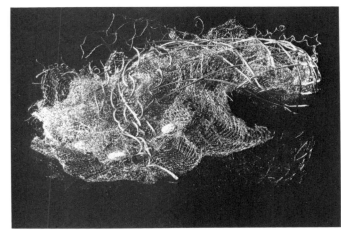

Clockwise, from above left: "Crochet Olé," stainless steel, 5"x8", 1984. "Crochet Cup," stainless steel, 3"x4", 1984. "Catching Agate Tides," copper, 16"x9", 1985. "Jittery (All Tied up in Knots)," brass, stainless steel, 13"x5", 1985. "Party Pizza, No Anchovies," copper, 14"x4", 1984. "Aluminating," aluminum, 12"x9", 1985. Photos: Suzanne Coles-Ketcham.

PATTI LECHMAN

I can't remember when I wasn't involved with fiber—it's always been a part of my life. I sewed by hand as a child, dressing dolls. By the time I was twelve, I was making a lot of my own clothes. My mother, sister and I developed the skill of copying garments when my sister and I were in college and wanted more "label" clothes than the family budget allowed. My mother taught me to knit, and we turned out intricate cable-patterned sweaters with tiny crocheted egg-shaped buttons.

An interest in art came early, too. My father loved to sketch and draw, and there were always books around, books on drawing the figure, lettering, drawing trees. When I was seven, he gave me an oil paint set, a beautiful wooden box filled with shiny little tubes of paint, small bottles of turpentine and linseed oil, brushes, and a palette. I began painting immediately, and "sold" my first painting to a school friend in exchange for a real coconut she had brought back from a family trip to Florida.

My interest in textiles and art continued through high school. In college I worked toward a degree in home economics. I took art courses as electives, but had little time to do any exploring. After graduation, I spent three years working on a Master of Fine Arts degree in ceramics, and kept trying to do clay and fiber pieces that worked together. They never really succeeded. A jewelry instructor with whom I did independent study said, "Patti, I don't know

Right: Work in progress. Photo of Patti Lechman by Bert Sharpe.

why you have to do those clay and fiber things. The fiber alone is so much better!" She was right and I knew it.

After finishing my M.F.A., I moved to Memphis, Tennessee, and taught art appreciation and ceramics at the community college. Having my hands in the clay all day at school made me crave working in fibers again. A happy accident pushed me back into fiber work and basketry. I decided to go to Penland for a summer session, but sent my registration too late to get into the clay course I wanted. My second choice was a class in fiber, and I worked 12 to 16 hours a day on baskets during the two weeks I spent there. The work was ideal for me because it was portable, clean, and I could work on the pieces even when I was physically tired from teaching. For several years I didn't have a studio space at home where I could do clay. Fiber was much easier to deal with.

Two summers later I went to a workshop that dealt with color and design on small scale knotted work. I wanted to use color more in my work, and from the first I loved working on a very

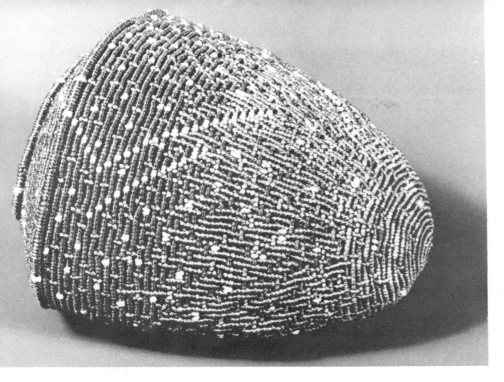

small scale. Everything about the new approach began to work for me. I experimented with various materials before settling on the lightly sized nylon that I use. It has a crispness that I couldn't find in linen, cotton, or silk though I liked the idea of those fibers. I take some comfort now, however, in knowing that the flag on the moon is made of nylon and should last forever. (After having problems with moths in wool, I appreciate that quality.) I just couldn't make the natural fibers do what I wanted them to do. I dye some of my colors because the yarn I use isn't available in a wide spectrum. When the pieces are completed, I treat them with a color preservative and thread sealer, so I'm satisfied about colorfastness now.

Most of my pieces begin with styrofoam forms that I rasp or carve out of large blocks. I sit in the studio making a powdery mess as I file away at the styrofoam, creating several forms at a time. The pottery influence is apparent at this stage as I translate the vessel form of clay to the vessel container form in fiber.

I do most of my work (the

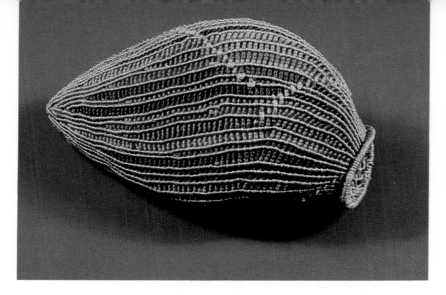

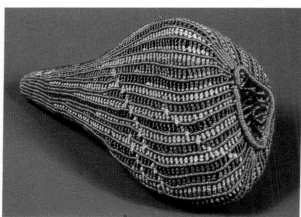

Clockwise, from left:
"Poseidon's Pouch," nylon,
4"x3", 1984. "Animated
Landscape," nylon,
4"x2½", 1986. "Fiesta,"
nylon, 4"x2½", 1985.

actual knot tying) at night while
sitting with my knees pulled up,
holding the work in place in
front of me, close to eye level.
The process is almost meditative
for me; I untie myself as I tie
knots in the pieces. I make
baskets because I love them. I
love the process and the product.

It all seems to be a logical
culmination of years of experiences with fiber, with vessels,
and with a quiet, contemplative
way of working privately, a contrast to the draining physical
work of the teaching which is my
livelihood.

In writing artist's statements
during the last few years I've

46

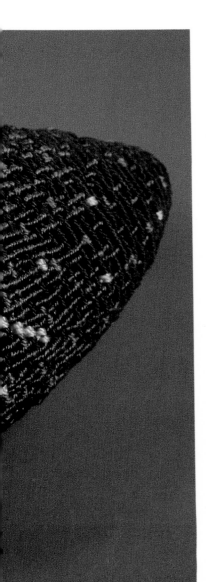

"City Lights," nylon.
2½"x4", 1984.

tried to be somewhat analytical about the pieces. What do they say? Are they really art? Does the fact that they are "pretty" make them trivial? We can't always know why something we do or see touches us, or speaks to us. Maybe all we know is that it does, or doesn't.

I want to do something visually delightful—a creation of grace and elegance, like a beautiful meal or a warm letter treasured by the recipient— something that makes a moment in time more beautiful, makes the spirit soar, maybe briefly, but joyfully. I also want to make art which doesn't reject the viewers, the participants, but envelops them with a warmth and doesn't require a level of understanding that only a mind educated in the aesthetics can respond to. I think I'm that way as a person. Maybe it has to do with "good manners." A friend once told me that I had "lovely" manners but that I was quite mysterious. I think the work has those qualities.

I do these baskets because I want to, because I love them, not to give other people pleasure but because they give *me* pleasure.

47

The traditional vessel is a functional form dealing with enclosure and access. While my vessels allude to function, they cannot be entered. I create structures that represent my thoughts and feelings about man's physical relationship to earth. As I look back to the late 1960's and early 1970's, I begin to understand how my work has evolved, and I can see a clear connection between the forms and the earth-rooted themes that have now emerged in my vessels. I now recognize that my work always concerned containment, protection and entrapment, and a particular curiosity about the passage from the known to the unknown.

In the late 1960's, I was making sculptural wall hangings and small stuffed fiber sculptures. I was searching for a technique that would enable me to use the fibers in a unique way; I was knitting, crocheting, embroidering, stuffing, beading, appliqueing and machine stitching, using these materials alone and sometimes combined. I leaned toward the three-dimensional and purely sculptural object. I experimented

Top: "Barren Vessel," fiber, rubber, acrylic, 8½"x10", 1985. Bottom: "Spirals," fiber, acrylic, straw, 6¼"x10½", 1985. Photos: Bob Hanson. Photo of Norma Minkowitz by Steve Minkowitz.

with fabric, using multilayer techniques such as reverse applique, which I taught and lectured on for a number of years.

The early 1970's were experimental years; I searched for ways of combining surface, structure, technique and content in a personal way. From that time to 1984 my work was divided between wearable and non-functional art. The female figure was common subject matter for both my wearables and sculptures, as were the box, egg, and ovoid shape.

From 1976 to 1978 I worked on a project which was a departure for me. I created a chair called "The Landscape of my Mind." I perceive it to be a combination of the functional art object and the purely non-functional object. The chair had a raised, linear quality inherent in knitting and crochet. It related the human form to the landscape. The body, when seated, would become one with the chair; the chair, which functioned as a container, was also a sculpture.

I did several non-functional boxes and a series of shoe forms that I feel were a bridge between

NORMA MINKOWITZ

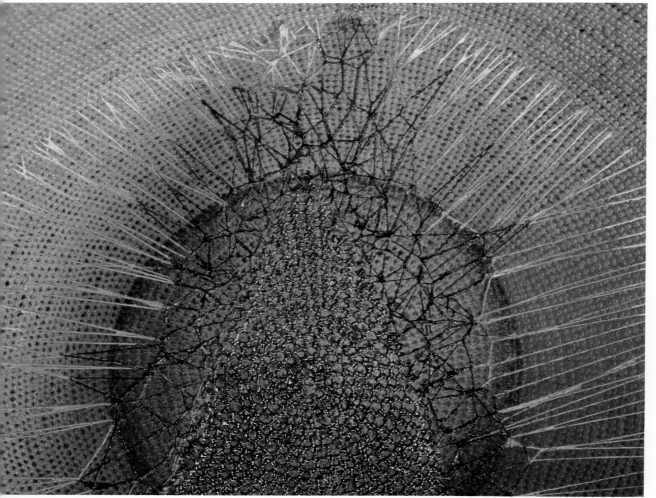

the wearables and containers. Shoes are also containers, but these shoes could not be worn; they were light, airy and incapable of holding any weight.

The new vessels have evolved and presently are my clearest form of expression. The structures of these "containers" also have gone through a transformation. At the beginning the vessels were more open, wide rims with narrow bottoms, expanding the idea of sky and air above and the bowels of the earth below. While some still retain that shape, they have become more closed, often cylindrical traps with no way in or out.

After years of stuffing, embroidering, crocheting, and knitting with solid closed forms, I have attained a method that best expresses my artistic goals in a way I feel is unique. My work is now open and delicate, but structural. This skeletal weightless quality is meant to express our fragile and vulnerable existence. My work continues to deal with containment, but in a new and exciting way. These vessels represent earth, they surround an inner space that alludes to function. The top is life, sky,

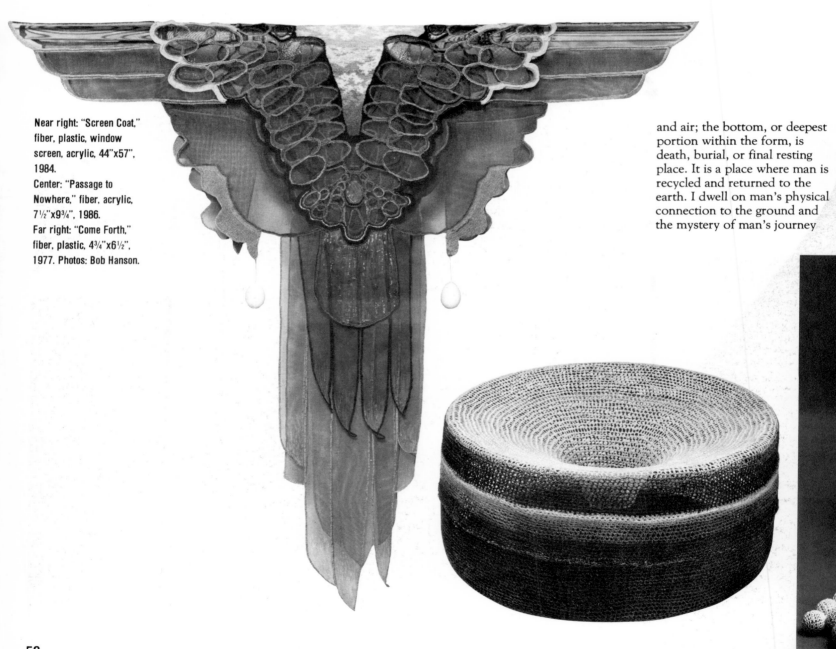

Near right: "Screen Coat," fiber, plastic, window screen, acrylic, 44"x57", 1984.
Center: "Passage to Nowhere," fiber, acrylic, 7½"x9¾", 1986.
Far right: "Come Forth," fiber, plastic, 4¾"x6½", 1977. Photos: Bob Hanson.

and air; the bottom, or deepest portion within the form, is death, burial, or final resting place. It is a place where man is recycled and returned to the earth. I dwell on man's physical connection to the ground and the mystery of man's journey from life to the unknown. My vessels address the lack of control one has over his destiny; a too-long life needing to end, or the premature end of one just beginning. This is the uncertainty and vulnerability I hope my work will communicate.

My artistic goal is to go beyond the structure and to impose meaning to my work. The choice of process has become part of the content. The quality of openness in combination with painting and drawing, and the utilization of earth-related found objects as well as man-made parts are just some of the personal additions.

Many of the new vessels are just of fiber, some are painted and drawn up with colored pencils, and some have different colors of threads creating patterns. I enjoy this play of materials, as it invites the viewer to dwell upon the illusion and perhaps not see any difference between paint and thread. For creating art, crochet is my primary technique; the flexible and organic qualities of fiber combined with the use of a single tool provide an intimate personal relationship with both the material and the object. I need to be responsible for, and have total control over, every aspect of the process and object in order to communicate my concepts, questions, and fantasies of man's physical and spiritual existence.

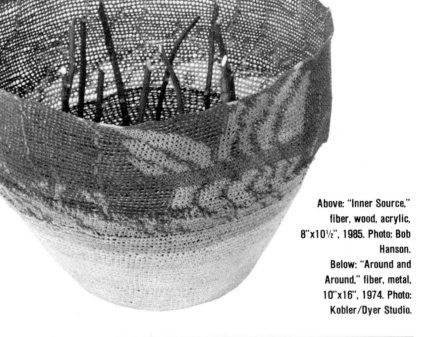

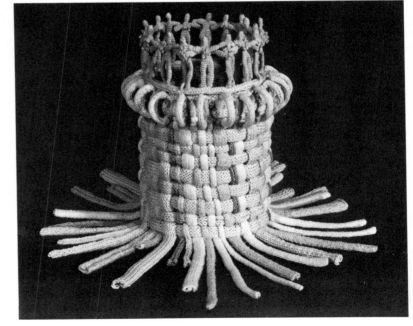

Above: "Inner Source," fiber, wood, acrylic, 8"x10½", 1985. Photo: Bob Hanson.
Below: "Around and Around," fiber, metal, 10"x16", 1974. Photo: Kobler/Dyer Studio.

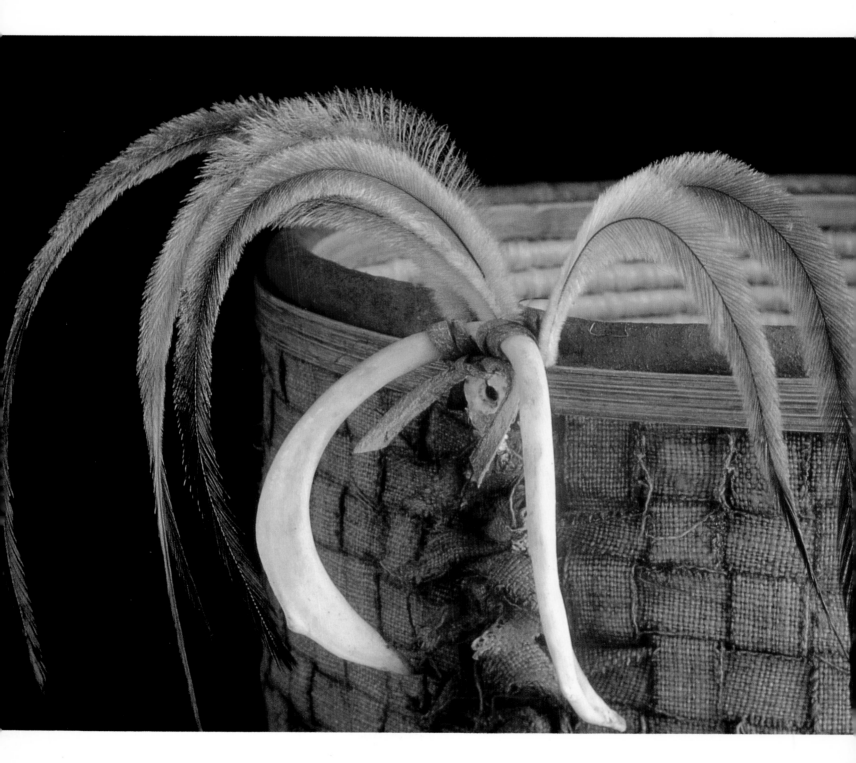

54

LISSA HUNTER

Early morning is a lovely time in Maine. The sky over the Back Cove near my house is pink or orange or lavender or sometimes all three. As light overcomes dark I am usually trying to put things in order in my studio, sipping coffee, planning for the day by making a list. I like this time. The work table before me is covered with correspondence, work in progress, drawings, ledger sheets, the tools of the trade.

The day's activities may include making paper, coiling, answering letters, packing work, drawing, photographing work, stitching, or any number of seemingly unrelated tasks. The list is always longer than the day accommodates and so the leftovers will go on the next day's list.

One of the pleasant things about being a basketmaker is the sheer beauty and friendliness of the materials. No metal shavings or anvils or motorized saws or turpentine odors or clay dust to live with. This is particularly important when your studio is an integral part of your home, as is mine. The living room has been given over to studio space.

An 8-by-8-foot section of wallboard is used for drawing and collage assembly. Work tables and shelves house jars of beads, spools of thread, paints, brushes, fabric yardage, books, feathers, leather. Spindles with raffia, splint, sweet grass and iris leaves hang over the west-facing window. All conspire to create an atmosphere of creative confusion and visual stimulation.

I like having visual juxtapositions and overlays around me all the time. Shelves and a bulletin board are covered with images

Opposite: Detail of "YooHoo Basket." Photo: Rob Karosis.
Photo of Lissa Hunter by Kirby Pilcher.

raffia, paper cord, handmade paper, leather, rayon braid, sealing wax, raffia braid, 9"x7", 1984. "Marble Basket," raffia, paper cord, handmade paper, leather, clay marbles, beads, copper wire, raffia braid, 9"x9", 1984. Photos: Rob Karosis. "Pocket Stilt Basket," raffia, paper cord, handmade paper, fabric, watercolor paper, sticks, beads, feathers, thread, 6"x15", 1984. Photo: Color Management Concepts.

Clockwise, from right: "Bone Latch Basket," raffia, paper cord, handmade paper, leather, linen fabric, bone, wood, raffia braid, beads, cane, 9"x7½", 1983. "Saddle Basket,"

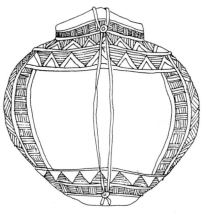

Below: "Sundown Basket,"
raffia, paper cord,
handmade paper, water-
color paper, fabric, leather,
thread, raffia braid,
feathers, beads, 7"x9",
1984. Photo: Craig Blouin.
Sketches by the artist of
work in progress, 1984.

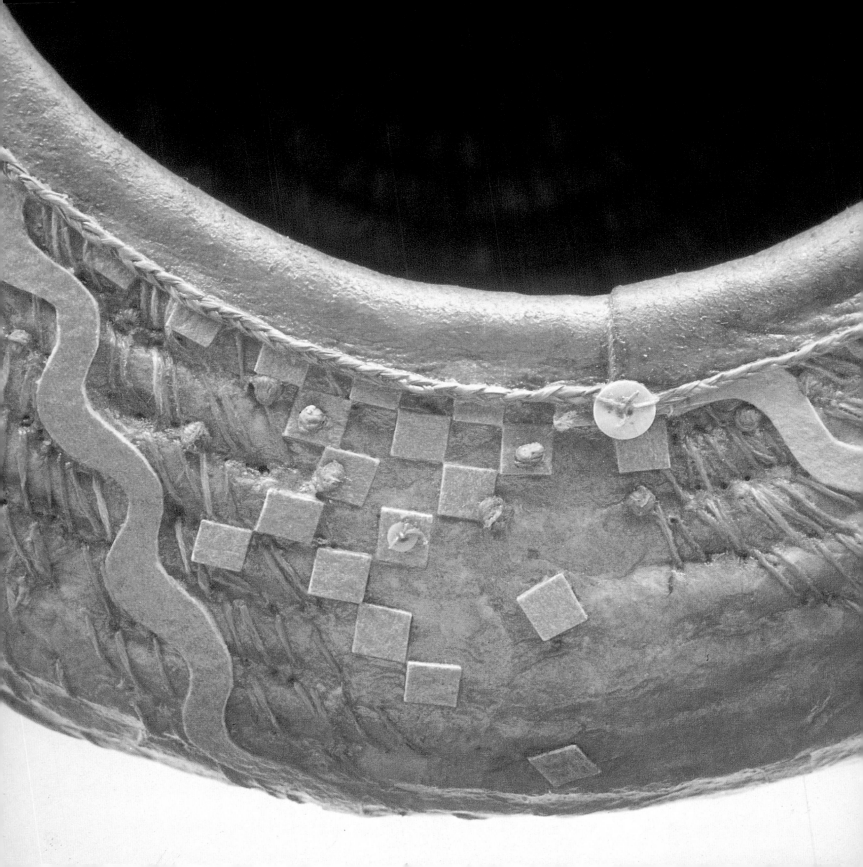

Opposite: Detail of "Halley's Basket," raffia, paper cord, handmade paper, watercolor paper, leather, cotton floss, beads, raffia braid, 9½"x5", 1986.

Below: "As Time Goes By Basket," raffia, paper cord, handmade paper, watercolor paper, beads, thread, 10½"x6¼", 1986. Photos: Stretch Tuemmler.

and objects which have caught my fancy for one reason or another. Occasionally I move them around to see what new combinations I can stir up. For instance, a postcard of a Japanese kimono, photographs of a Navajo blanket and Rousseau's "The Dream" present a provocative combination of histories and sensibilities that can lead to a new image. Sometimes the influence is subliminal and only later do I see the "parts" put together in a drawing. But all of that visual information finds a place somewhere, sometime.

I make both baskets and collages and so I often have quite a few pieces in progress at the same time. This allows the freedom to switch from the tedium of coiling to the more interesting and creative activity of drawing to the satisfying process of finishing a piece all in the same day.

Perhaps my choice of basketmaking as a medium of expression is the synthesis of a background in painting and in textiles. I think of my baskets as two-dimensional surfaces stretched around space. I am "painting" on the surface using the same design elements I

would use in a two-dimensional format, the "canvas" just happens to be curved around a basket.

But my choice of basketmaking goes beyond that. I love the materials, the manipulation of those materials. I love the idea of making a functional vessel within an ancient tradition but also extending that tradition. Who's to say that what we do today won't be thought of as traditional in fifty years?

There is something intimate and direct about a basket. The structure is right before you. No

tricks, no advanced technology. The forms of my baskets are coiled with raffia over paper cord. Coiling is a traditional technique, its formal possibilities more varied and subtle than some other construction techniques. The surface created, which is solid and closed and relatively even, is then covered with handmade paper and embellished with collaged paper, fabric, leather, stitching, beads, bones, whatever suits the desired image.

Drawing is the mechanism for putting together the mate-

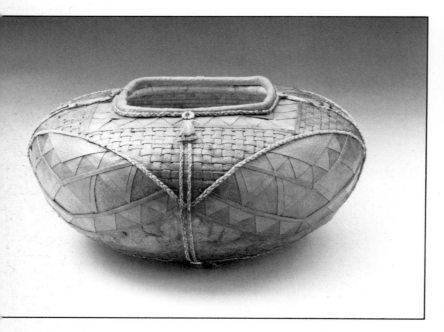

rials, techniques and images that I collect. Many times I don't know where an image is going until the drawing takes me there. It is a curious process that often surprises me and usually pleases me. Occasionally I will see a basket fully finished in my mind's eye and merely record it on paper, but more often the drawing process is integral to the creation of a basket or collage.

Sometimes I work directly on the surface of the basket with no drawing, but this is becoming increasingly rare.

I am not sure where the ideas, the images, come from. Sometimes I feel a little foolish in admitting this. In our age of rational thought, when intelli-gence is prized over intuition and every question is thought to have an answer, I must say that I truly

am not aware of all the influ-ences which produce the particu-lar combination of color, shape and form of my baskets. An admiration for American Indian and other tribal arts is apparent. But why am I drawn to them in the first place?

My father is a magician. My childhood was punctuated with amazing occurrences which seemed natural to me because

everything is new and surprising
to a child. I didn't realize that my
father was any different from
anyone's father. As I approached
twelve years of age, he was willing to teach me a few of his
tricks and I was more than eager
to learn them, so I became the
sorcerer's apprentice. But before
long I realized that being amazed
was far more exciting and satisfying than knowing how the trick
was done. Magic is magic only
when you can't comprehend the
illusion.

I feel that way about my
work. I don't really want to
know where it comes from. The
surprise of what comes out is far
more exciting and satisfying to
me than if I analyzed and
researched ethnic or historical
sources and could explain their
appearance in my baskets. Magic
is more fun.

Evening is a lovely time in
Maine. The low light contributes
to a gentle sunset signalling the
end of another day. I try to leave
the chaos on my desk in some
semblance of order for tomorrow, when I will make another
list and coil and draw and make
paper and feel fortunate that I
can do what I do.

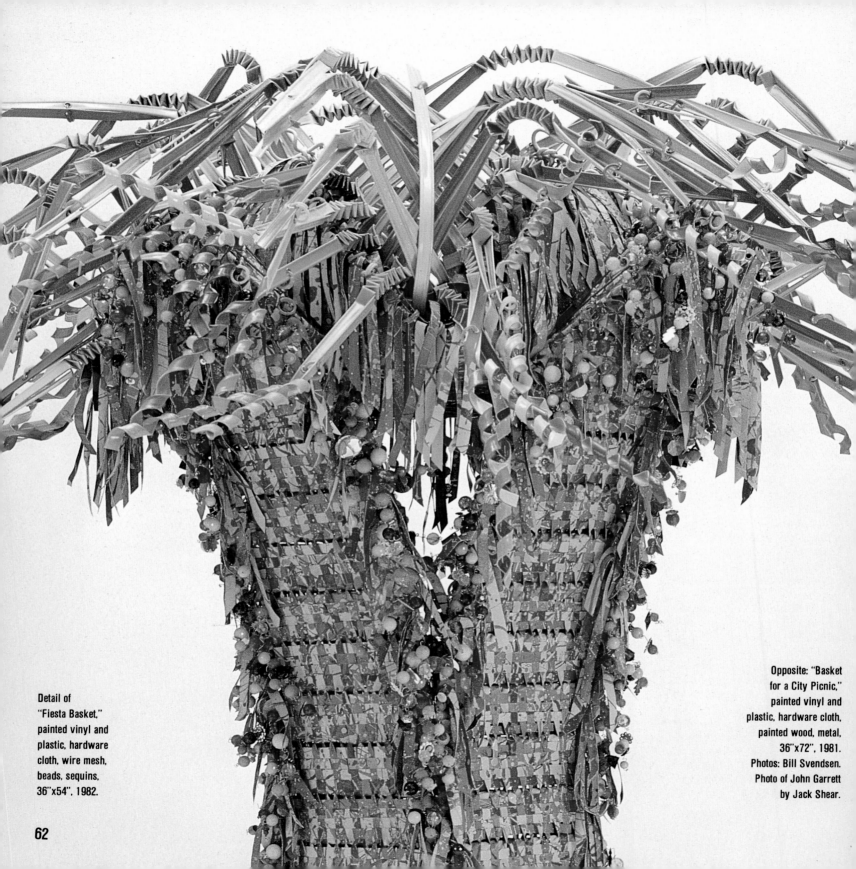

Detail of
"Fiesta Basket,"
painted vinyl and
plastic, hardware
cloth, wire mesh,
beads, sequins,
36"x54", 1982.

Opposite: "Basket
for a City Picnic,"
painted vinyl and
plastic, hardware cloth,
painted wood, metal,
36"x72", 1981.
Photos: Bill Svendsen.
Photo of John Garrett
by Jack Shear.

JOHN GARRETT

Although I have lived in the Los Angeles metropolitan area for many years, my formative years were spent in the semi-arid desert land of southern New Mexico. The desert environment of my youth influenced my early art work through the natural materials and subtle colors I used. As with textile constructions, the products of elements intermixed to create structure; my basket forms are a mesh of ideas and materials. Living in Los Angeles, I respond not only

to the lush, rich plant life superimposed here on a desert biome, but also to the glaring, plastic world of every commercial street. This is one of the reasons I have chosen to work primarily with synthetic materials, products of our technological era. In doing so I take anonymous materials and give them character.

I paint large sheets of fabric-backed vinyl with different colors in a variety of designs. The sheets are then cut into strips. I plait these strips into squares, and also use them to lash spray-painted plastic slats onto wire hardware cloth. The large plaited squares are attached to the lashed construction with plastic-coated wire. I form the basket shape by pulling the four corners of the square toward its

center, creating a three-dimensional form with an interior space. I add plastic beads and sequins to the wires. In my current work, I am manipulating rectangular forms into conical forms, and working with prefabricated metal conical armatures as well.

The geometric constructions make reference to screens, fences, lattices and the grid patterns of city streets and high-rise buildings. The designs painted on the vinyl suggest landscapes and atmospheric occurrences. Manipulating the ends of the plastic slats by braiding and with heat, and the addition of beads, sequins, and plastic cord, create curvilinear elements similar to flowers, tendrils and vines. The result is a synthesis of pattern and imagery between that which is geometric and manmade, and the wild, robust organic growth of nature. The synthesized organic materials which were neutralized in processing regain some of their original qualities. Much of my work is festive in quality. Bright colors and reflective surfaces bring to mind parties, fiestas, confetti, city lights, discos, and parades.

"Spider Basket," painted vinyl and plastic, hardware cloth, painted wood, metal, 36"x54", 1981. Photo: Bill Svendsen.

Below: "Wand Basket II," painted vinyl and plastic, hardware cloth, beads, sequins, plastic clothesline, 16"x48", 1985. Photo: Myron Moskwa.
Opposite top: "Time's Heart," bamboo, yarn, palm fiber, hardware cloth, 14"x16", 1983. Photo: Bill Svendsen.
Bottom: "Black Bouquet," painted vinyl and plastic, hardware cloth, beads, sequins, plastic clothesline, 16"x24", 1983. Photo: Bill Svendsen.

The addition of columns or bases constructed of the same materials as support structures for the baskets have further expanded the expressive possibilities of my work. The columns have become tree trunks; yet the square form relates strongly to architectural elements.

Also expanding the possibilities of the baskets is the dialogue between them and the concerns with materials and ideas found in my other work. After making some wire box construction which incorporated plastic toys, I created "Nesting Basket." It is made of everyday, domestic objects: hot pads, clothespins, plastic toys, and a feather duster. This is a basket which is about work, but is certainly not a work basket.

In exploring the expressive character of materials which are not generally used in connection with fiber work, I made a series of wall pieces with steel and aluminum. I used sheets of these metals and cut them into strips. The strips were then lashed to a welded wire grid. I employed these materials and a similar method of construction to creat "Vulcan Purse," a fantasy basket.

I live and work in a one-bedroom duplex on a hill in the Silverlake area of Los Angeles. From the windows of my apartment I can see Sunset Boulevard snaking its way to downtown Los Angeles. I drive several times each week to UCLA, where I teach, through many neighborhoods of varied social strata and ethnic composition. Although I am immersed in this city, I often think of my early life in New Mexico. Indeed, memories of those times often surface in my work. Where lies the relationship between the spines of the cactus and plastic sequins? I love the great foreboding calm of the desert, and I love the electrically charged urban atmosphere where I now live. I see my baskets, and all my work, as an attempt to make whole the disparate experiences and feelings of my life.

FERNE JACOBS

I have this image: A woman calls me to follow her. I do and we go very far, into another world. This is a world where people are standing still. No one is moving. She tells me to sit down, and I do. She shows me my work. I see pieces of mine and I see some that are unfinished. She tells me to pick up the unfinished work. I do. She then tells me to work, and I do. The people begin to move. It is as if the air begins to circulate again, and life returns to this place.

I would say that this place is a room in my soul. And it is interesting to me that when I work I often have the image that I am making a place for a breath or a stream of air.

Opposite: "Roots," thread and collage, 6"x21½", 1984-85.
Right: "Flame," waxed linen, 5"x48", 1982-83.
Photos: Janice Felgar.
Photo of Ferne Jacobs by Iggy Samuels.

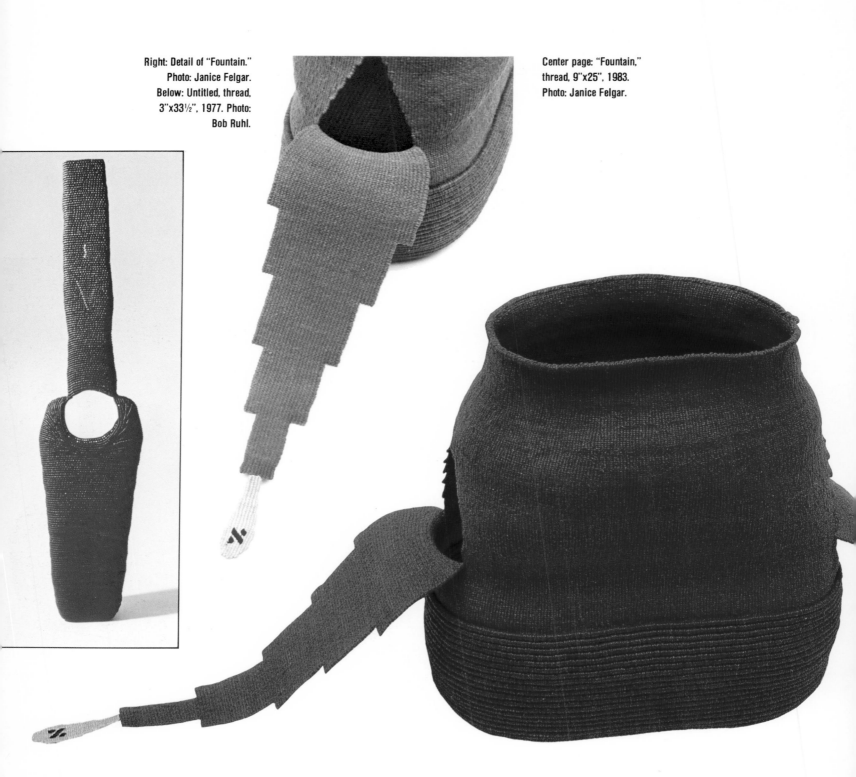

Right: Detail of "Fountain."
Photo: Janice Felgar.
Below: Untitled, thread,
3"x33½", 1977. Photo:
Bob Ruhl.

Center page: "Fountain,"
thread, 9"x25", 1983.
Photo: Janice Felgar.

Left: Detail of "Petals."
Below left: "Petals,"
thread, 5¾"x8½", 1984-85.
Below right: "Music,"
thread, 6¾"x9¼", 1984-85.
Photos: Janice Felgar.

RINA PELEG

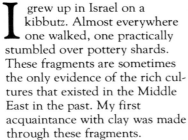

"Interweave," ceramic,
4'5"x9', 1985.

I grew up in Israel on a kibbutz. Almost everywhere one walked, one practically stumbled over pottery shards. These fragments are sometimes the only evidence of the rich cultures that existed in the Middle East in the past. My first acquaintance with clay was made through these fragments.

I actually began working with clay after high school, when I discovered the raw material and became aware that almost any shape could be made from the formless mass. At first, I worked with the potter's wheel, making functional pieces to be used at home. I gradually discovered the many possibilities inherent in the material, and began to make explorations in new directions.

Working with clay became a way for me to make contact with the world outside the kibbutz, and, ultimately, outside Israel.

While attending university, I worked on hand-coiled clay structures that involved a play of unconnected coils within the structures. This experience taught me that a coil could be used the way rope or string is used in weaving and plaiting, and I began to "weave" with these

clay coils. Since the woven coils were basically the same kind of coils I had used for traditional pinched pots, the transition was simple and involved only a slight change in technique. My weaving experience made this a natural transition for me.

The clay baskets I am doing

now are made up of plaited coils which are not pinched. The final shape is strong, yet light in feeling. The clay basket "breathes" as a natural basket would.

The connection between basketry and working in clay is, all in all, a very natural one. Both clay and the straw used in

73

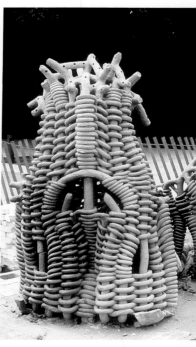

Left: Rina Peleg working on "Farrago."
Below: "Farrago," low-fire stoneware, 4'x5', 1985.
Opposite: "Yellow Gold Art II," clay, 18"x31", 1981.

making baskets are materials man has used for ages.

The commonly held theory is that the first clay container came about when a hole was dug in the (clay) ground and a fire made in it. After the fire was put out, it was discovered that the sides of the hole had been, in fact, fired. The hole had become a container for anything, including water. The next step, obviously, was to "remove" the hole from the earth by cutting around it, leaving enough clay to make the new shape self-supporting and portable.

Eventually, it was discovered that the best way to make a pot was not to remove a hole, but to create the structure from clay coils pinched together. The idea of using coils presumably came from baskets. An imprint of a basket made of coils was found in the clay floor of a home in Jericho from 7000 BC, a time when ceramics was in its infancy (there are no pots dating from that period).

The idea of making clay baskets is also connected to the ancient custom of imprinting mats or other woven material into the surface of clay pots

before they were fired. Baskets were also used as molds which were later burned with the clay. Coils resembling rope are often found as decorative elements on early pots. Large pots in ancient Egypt often had no handles and were carried in large baskets, much the way some Mexicans and Africans carry their wares to the market today.

A final example of the affinity between clay and basketry is the fact that clay and straw were probably the first materials used to make huts. The inside of these primitive structures was covered with slip and later burnished. The outside was covered with woven straw mixed with clay which stopped the rain from coming in.

Weaving in clay has become a natural activity for me. The difference between my pieces and the traditional, functional works made of clay or straw is that the pieces I make are quite impractical. At first glance, my work seems to relate to functional objects with which we are all familiar. However, the works were made without any function in mind.

I have been concerned for several years with the so-called right of the ceramist to create non-functional works. My own reason for making such works was, and is, a strong attraction to the basic, classical shapes. This clearly came about from a deep emotional need, not practical intent.

I began to understand some of the sources from which I drew my ideas upon learning about Jung's theory of the collective unconscious. Jung maintains that there is a collective unconscious of the whole human race which manifests itself in each individual through that person's dreams, or, as in my case, through creativity. The basic shapes of ceramic utensils which were used ages ago, as well as the technique of making them, form part of the collective unconscious.

My need to make the shapes I do can be explained in part, if such explanation is necessary, by Jung's theory, my own acquaintance with the history of arts and crafts, and a strong personal need.

Left: Woven clay structure,
porcelain, 25"x38", 1983.
Right: "Yellow Gold Art I," clay,
18"x31", 1981.

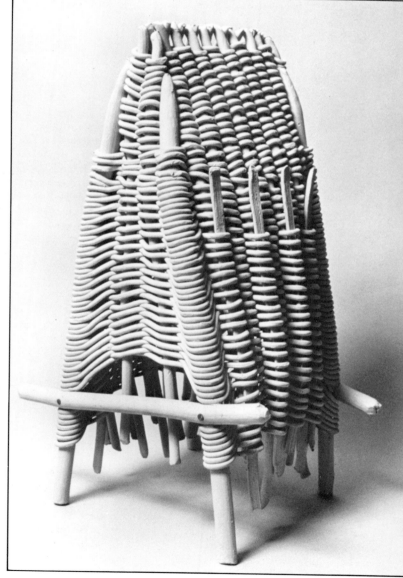

KARYL SISSON

My work is an exploration of the physical and metaphorical possibilities that result when materials, structure and form interact. Over the years, I have rummaged through basements, garage sales and junk stores, salvaging cloth, sewing notions and related elements from domestic life. These have provided the inspiration for my sculptures. Presently, my focus is the transformation of familiar objects through the building of form.

The zippers, clothespins and twill tape serve as building materials, while any number of basketry and needlework techniques provide the methods of construction. The basic structures are developed by interlocking the materials; no glue, nails or internal supports are used. Because the structures are flexible, they can be manipulated in various ways to create different forms which are linked to my interest in ancient and indigenous architecture, organic growth, and patterns in nature.

After traveling to the Yuca-

Photos of Karyl Sisson and her work by Myron Moskwa.

78

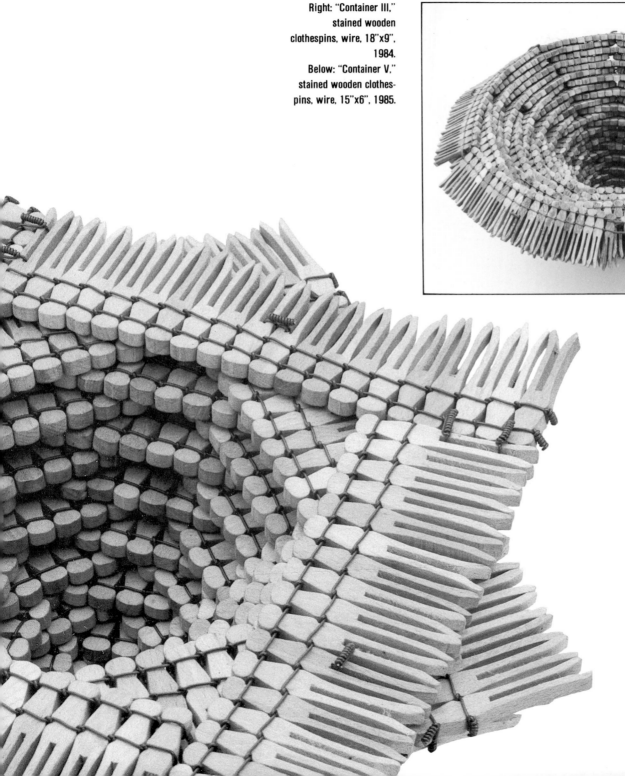

Right: "Container III,"
stained wooden
clothespins, wire, 18"x9",
1984.
Below: "Container V,"
stained wooden clothes-
pins, wire, 15"x6", 1985.

tan and experiencing Chichen Itza, I began making small pyramids constructed solely of clothespins and wire. By inverting the pyramid, I discovered a container form which exposed more of the pin and revealed an interior surface and space quite different from the exterior surface and shape. I was intrigued, and began working with my particular palette of materials to create other container forms. These have become another vehicle for me to explore personal and formal issues. What I find important is the aesthetic and suggestive quality of the form, rather than whether the form is defined as or serves as a functional object.

Clothespins and zippers are simple inventions that have endured in our complex society. My sculptures seek to mirror the beauty and simplicity of these individual elements.

Right: Side view of "Container III." Top: "Container II," wooden clothespins, wire, 19"x8", 1984. Bottom: "Vessel IV," dyed zippers, stained wooden clothespins, 18"x10", 1985.

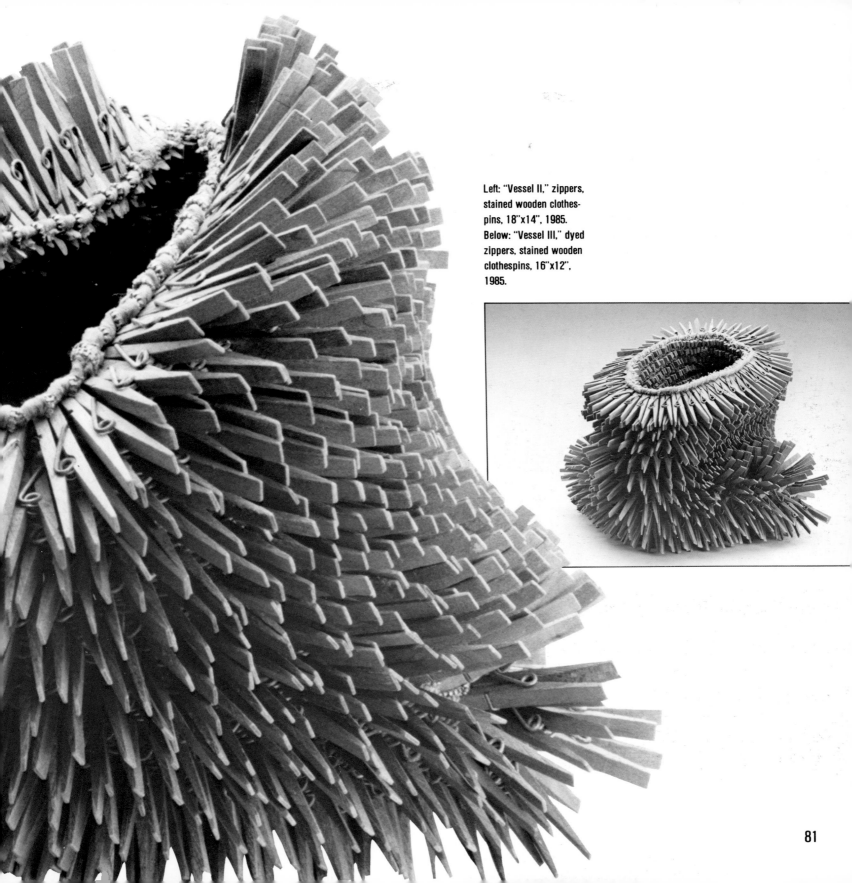

Left: "Vessel II," zippers, stained wooden clothes-pins, 18"x14", 1985.
Below: "Vessel III," dyed zippers, stained wooden clothespins, 16"x12", 1985.

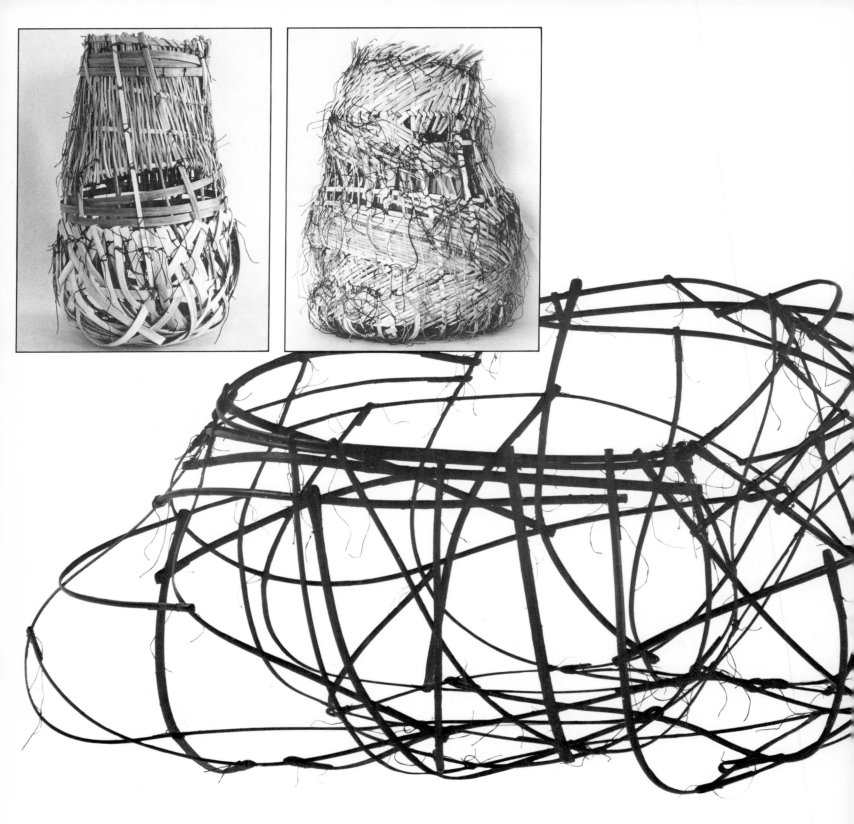

LILLIAN ELLIOTT

I'm not exactly sure when I began making baskets. I had been weaving flat tapestries for years. It was as though my sense of volume had been developing in secret while I worked on the flat surface. I was amazed to discover that I loved working in the round. I was elated at being able to work on large forms. Making baskets was like taking the best of ceramics, drawing, and textiles; it was direct, expansive, and it could even be meditative. One of the nice features was that I could work over the whole surface of an object simultaneously, and not from top to bottom as I do in weaving. Unlike tapestries, baskets didn't take years to finish.

Opposite top left: "Hayfield," rattan, linen, acrylic, 10"x19", 1981.
Opposite top right: "Tag Ends," reed, bamboo, linen, 15"x25", 1979.
Center: "Drawn Form," rattan, acrylics, linen, 48"x26", 1984.
Photo: Scott McCue.
Photo of Lillian Elliott by Roy Elliott.

Opposite left: "Yi Dynasty," rattan, linen, acrylics, 17"x25", 1985. Opposite right: "Shrine," reed, linen, paper, coconut mid-rib, 42"x42", 1981. Below: "Feather Basket," plastic tape, wire, 15"x12", 1979.

I began by buying strong cord that had a lot of spring. Without untying the package of cord I tied a number of strands together. Then, I cut here and there, and tied down again. I continued this procedure until I could see a form emerging from this mass, or mess, in front of me. Next, I bought a package of glossy, brilliantly dyed raffia. I twined the outer layer together, and began to hollow out the inside, as I might have done with clay, until what was left was a small, rather precious raffia bundle with a hollow core, a kind of simple basket. By this time I was hooked.

Then I did a series of large baskets, binding and tying, regarding the lines I made as drawings, simultaneously describing the inner structure and the outer shape. In the beginning I wanted the material to speak for itself. I was reluctant to direct it too much. Later, I came to enjoy making a strong statement, and the question of imposing my will just didn't seem to be an issue any longer.

I decided to work on a three-dimensional piece that would try to make a strong statement. I

mixed a number of materials
together, and painted the piece
black when I finished. It was so
strong that at one point in the
construction it actually frightened me. I called it "Goya,"
because it has some of the intensity of the "Horrors of War"
series. This is one of the few
baskets of that time that continues to interest me.

I think of my baskets as describing volume, as if I were
drawing on the surface of very
specific concrete forms. I think

of the form inside as pushing out
the outside walls. The elements
and shapes of my baskets are like
skeletal structures, and often I
paint them black to reinforce
that impression. I see them as
calligraphic shapes or gestural
drawings. They aren't drawings,
they are baskets, but they move
in space describing where one
perceived shape ends and
another begins.

As for technique, I mostly
use twining because it holds
together dissimilar elements,

because it makes such a strong
structure, and because I twine
without being conscious of the
technique. Sometimes, I also use
knotting or binding. I don't
much care what techniques are
used. I'm concerned with the
form and impact, that is, the visual impact of the work. I feel the
same way regardless of the
medium I work in.

My work has included col-
laborating on baskets with Pat
Hickman. I construct the form,
and Pat covers the surface. At
every stage we consult about the
results. Those baskets have a
very different character from the
ones I do alone. The most
obvious difference is that there is
always a surface on the collaborative work. The baskets I do
alone are generally linear, often
black, and usually rather
complicated.

I'm interested in combining
manmade and natural materials,
although I've not done much of
it so far. I think that I've been
more adventurous with form
than I have been up till now with
materials. I'm interested in using
more color. I once saw a traveling exhibit of Korean folk art
and could hardly wait to paint

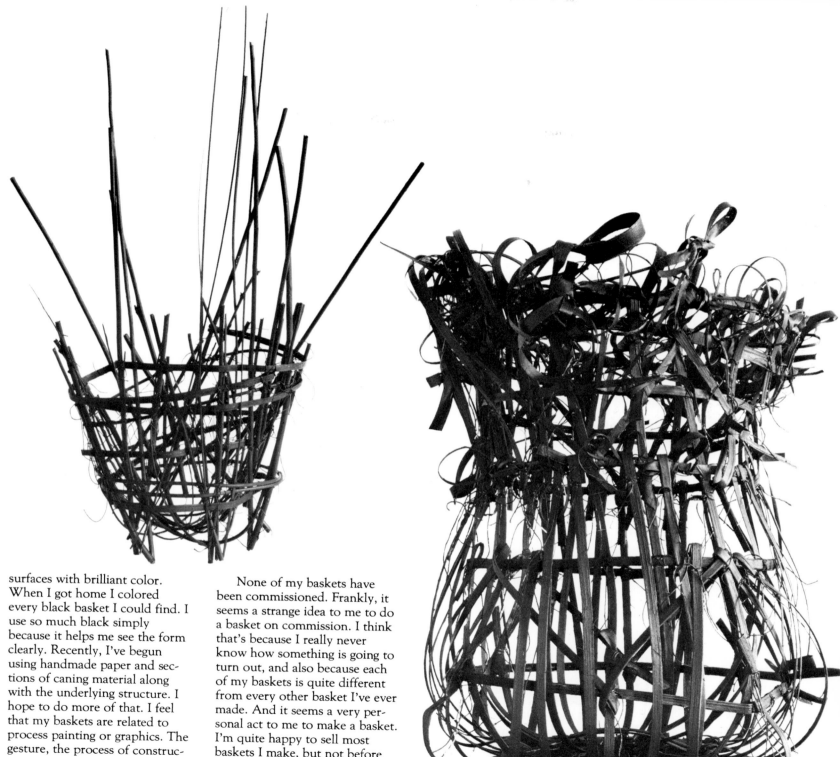

surfaces with brilliant color. When I got home I colored every black basket I could find. I use so much black simply because it helps me see the form clearly. Recently, I've begun using handmade paper and sections of caning material along with the underlying structure. I hope to do more of that. I feel that my baskets are related to process painting or graphics. The gesture, the process of construction is an important aspect of the final work.

None of my baskets have been commissioned. Frankly, it seems a strange idea to me to do a basket on commission. I think that's because I really never know how something is going to turn out, and also because each of my baskets is quite different from every other basket I've ever made. And it seems a very personal act to me to make a basket. I'm quite happy to sell most baskets I make, but not before I've made them. I think it would cramp my style.

MARY MERKEL-HESS

Until a few years ago I worked only as a metal-smith. I began making baskets in response to some things I was doing in metal. It was a case of a sudden inspiration causing me to leap into another medium. At the time, I was completing my graduate work in metal, and was folding metal and trying to recreate natural forms in a spontaneous way. I realized that I was imagining things that could be done more effectively with paper. It was an idea I couldn't resist, and in a few restless weeks of experimenting I developed the technique that I used to make my first baskets. It was a papier-mache technique utilizing a variety of glues and many layers of very thin paper applied over a mold. There were paper cord inclusions that strengthened the baskets and created a pattern on the interior surface. The first baskets were ribbed, and had ruffled leaf-like edges. They were both a contrast to and a further exploration of the ideas I had been working on in metal. The freedom the paper gave me was heady and the response to the paper vessels was positive. I had

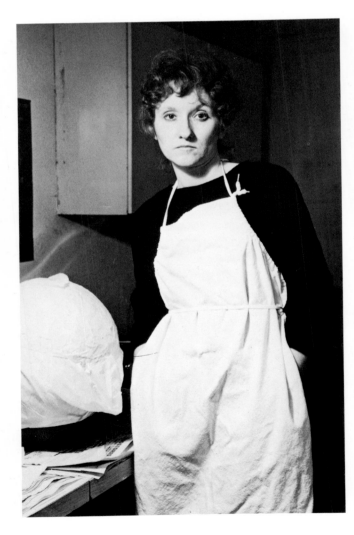

many ideas to explore. I continued making the baskets, and what had begun as a short-term project became a major involvement.

These first baskets were inspired by natural phenomena, the grasses, flowers, and life along the roadsides in my native Iowa. I grew up in rural, weedy places and then left to live in a city for ten years. When I returned to the country I was overwhelmed by the grassy memories of my childhood. I was interested in sticks and deeply folded plant leaves, and the paper was perfect for recreating these forms. The colors I used were natural and the paper cord inclusions looked like natural fiber. As I became more interested in basketry itself, there were references to traditional baskets, mainly woven patterns embedded in the paper. I have great appreciation for basketry, but little knowledge of traditional techniques.

During the next year I made many baskets at the same time that I continued to work in metal, always making vessels. Container forms have interested me from the very beginning of

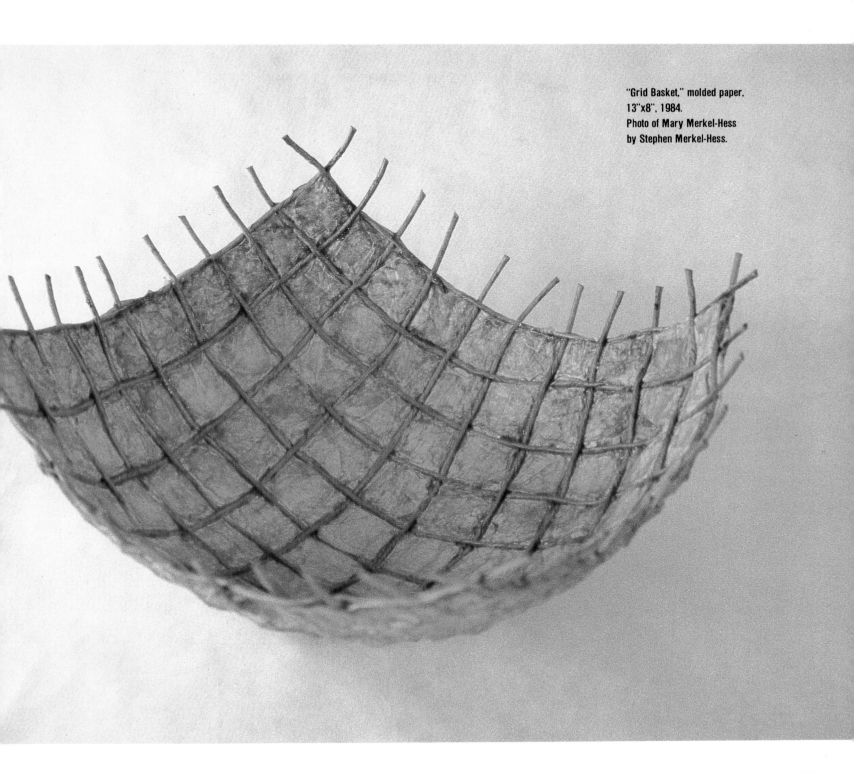

"Grid Basket," molded paper,
13"x8", 1984.
Photo of Mary Merkel-Hess
by Stephen Merkel-Hess.

my art career. It was that interest that led me to crafts, to metal-smithing and eventually to baskets. I like the fact that all sorts of containers, made of all sorts of materials, are part of our daily lives. They are so elemental. I always imagine that some kind of vessel was the very first tool of our species. In some prehistoric dawn an ancestor picked up a shell to carry water or used a hollow gourd to gather food. For me, vessels seem inextricably linked to what it means to be human. I think simpler societies felt this more keenly than we do. At least their ritual utensils for religious ceremonies have more importance than similar objects do in our society. But then, we think of many kinds of containers as art objects, which is also a way of conferring importance.

At the same time that I was making baskets with references to plant life, I was doing drawings of flowers, seed pods, and grasses as a means to visual understanding and inspiration. I kept a notebook and covered quite a bit of other paper with drawing and watercolor painting. It occurred to me that I was

Above: "Shadows of Flowers (3)," paper, wood, 6"x7", 1985.
Opposite top left: "Intoxicated with Shadows of Flowers," paper, wood, 10"x18", 1985.
Opposite top right: "Waving Grasses Basket," molded paper, 7"x14", 1983.
Opposite bottom: "Lotus," molded paper, 20"x8", 1983.

making drawings on paper and then making baskets of paper. Perhaps there could be a more direct connection. I began to use my drawings in the actual basketmaking. I rolled up some of my drawings. I was delighted with the delicate rolls, and the clues along their edges of the drawings inside. At first, I inserted the drawings through the center of the molded baskets. This led not only to more drawing but also more insertions—sticks, painted dowels, and wrapped wands. When I inserted something through the center of the basket I was challenging its function directly. Previously, the interior of a vessel had seemed inviolable to me—the space inside was the essence of its container-ness. But so often, when my baskets were displayed, they were hung with the interior surface outward. Even baskets that were quite deep were hung in this way and I began to feel that this space was mine to use. The interior of the basket became like a canvas to me—an area set apart for artistic composition. I experimented by piercing the baskets with rolls of paper or sticks, always trying to

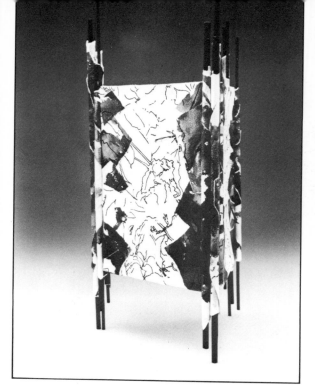

play the straightness and pattern of those elements off the curvilinear surface of the basket.

While I was first piercing baskets, I was also making another, quite different attempt to use my drawings directly in container-making. For these pieces I constructed a box of paper board and then covered it with fragments of a drawing. I chose a box shape because I wanted something archetypal, but not in any way earthy. The boxes have simple, geometric shapes that are entirely covered with drawing—on the interior as well as the exterior. The drawing-covered box is a repository for my visual feelings about a subject, an icon. When I have more drawings than I need to cover a box, I roll them up and attach them to the sides. I think of these paper vessels not only as icons, but also as three-dimensional drawings, much abstracted in their final form. The way they look may make them seem like a departure from my other baskets, but they are not. The boxes are another way of dealing with the same visual phenomena that inspired my first baskets.

Left: "Basket Figure,"
raffia, rattan, hibiscus, dye,
15½"h., 1983. Photo:
Andrew Gillis.
Below: "Basket V," rattan,
pigment, 8"h., 1982. Photo:
Hillel Burger.

JOANNE SEGAL BRANDFORD

I am not a basketmaker. My "baskets" are not really baskets. I think they are *images* of baskets.

For many years I have used netting technique to investigate light and space. With the "baskets" my goal was to bring an increased sense of volume to the nets, and to have them stand up on their own, independent, defying gravity.

I enjoy the directness of knotless and knotted netting. To make a net I simply begin, with minimal equipment and with no theoretical encumbrance. I bring idea to material as easily as pencil to paper.

The intimate nature of the process is also important to me. A net builds relatively slowly, one loop at a time, as I bring the active end of a working strand through a previously constructed loop. This encourages very close, moment-by-moment involvement. The creation of each new mesh demands active participation, with mental awareness and physical control. While working in the present, I must, at the same time, review and test what has gone before. I appreciate this direct and complex flow between

mind, eye, hand, material, front and back, past and present.

Making nets is like writing or musical composition. Words and notes, like the loops of a net, may appear to interconnect in simple and linear fashion, yet complexity and beauty are always possible.

These net "baskets" are the products of a three-way (material/method/me) conversation. Occasionally I have gone too far, stubbornly refusing to acknowledge material/method limits. At these times I have retreated, to repair, support, and revise. The trick is to know when to stop.

Photo of Joanne Segal Brandford by Andrew Gillis.

93

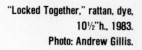

"Locked Together," rattan, dye,
10½"h., 1983.
Photo: Andrew Gillis.

"Burden," rattan, dye, paint,
19"h., 1985.
Photo: Andrew Gillis.

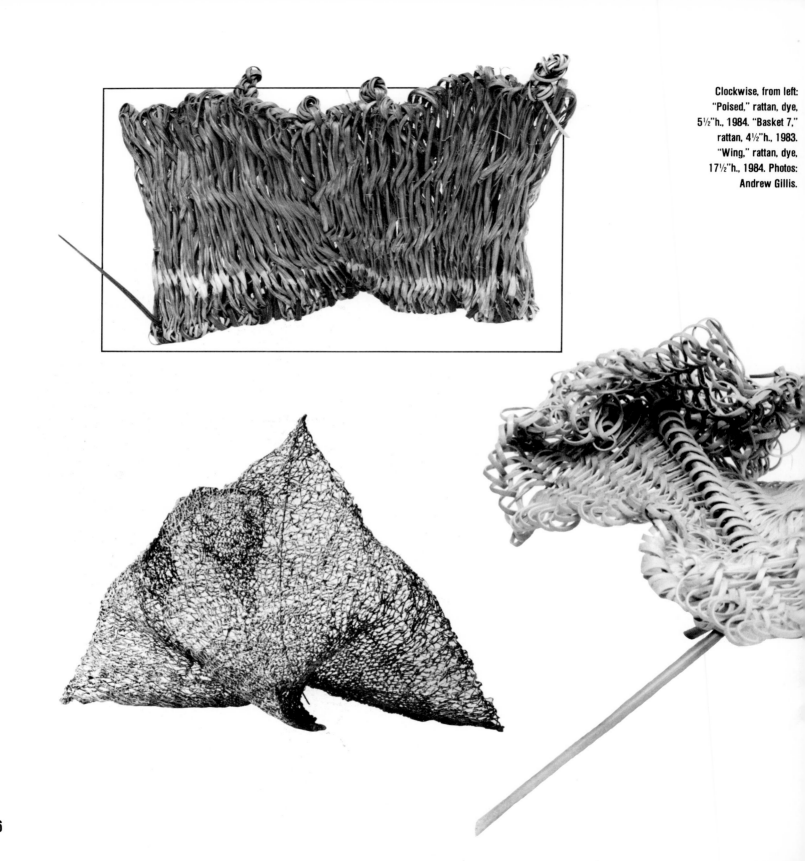

Clockwise, from left:
"Poised," rattan, dye,
5½"h., 1984. "Basket 7,"
rattan, 4½"h., 1983.
"Wing," rattan, dye,
17½"h., 1984. Photos:
Andrew Gillis.

"Brown Net," raffia, hibiscus, rattan, dyes, 7"h., 1983. Photo: Andrew Gillis.

MICHAEL DAVIS

In college I studied ceramics and painting, then I worked as a graphic artist and illustrator. I was dissatisfied with the creative progress in my painting and pottery, and knew that my emergence as an artist depended upon finding an avenue of expression that better suited my abilities.

My life as a fiber artist began quite by chance. I had taken a new job with a company whose offices were on the second floor of a gracefully restored Victorian house. On the first floor was a retail fiber shop, through which I had to pass to get to my office. The shop was a feast for the senses, filled with looms, hand-spun yarns, fleece, spinning wheels of all shapes and sizes, reeds, grasses, and materials from the world over, in a myriad of colors. This was my initial exposure to the fiber arts, and made me realize I had found my niche as an artist.

My formal instruction in basketry was through that shop, a class in the Appalachian style of weaving. I spent two years perfecting Appalachian forms before experimenting with color. I was amazed that with subtle structural changes these utilitarian baskets could become sculpture-like forms. I also realized that one could spend a lifetime on a single Appalachian form, creating endless interpretations.

To further develop my craft, and because of limitations of Appalachian forms, I began working with wicker, using the twining technique. I achieved fluidity by using very small ribs and weavers, and created sculpted architectural forms through control of the tension between the ribs and weavers.

Generally, I work on several baskets simultaneously, and spend eight to ten hours a day in my studio. This way of working keeps my interest at a maximum, provides stimulation, and strengthens the work in progress.

Sometimes I make preliminary sketches, especially for commissioned pieces, when sketches are used for the initial

Left: "Mahogany," dyed reed, Plexiglas, 11"x13", 1985. Photo: Jettie Griffin.
Below: "Namibia," dyed reed, acrylic paint, 19"x32", 1985. Photo: Daryl Bunn.
Photo of Michael Davis by Jettie Griffin.

Left: "Texture 11," unspun
canton rope, 12"x12",
1984.
Right: "Wooly Bully," dyed
reed, unspun hemp,
Plexiglas, 14"x15", 1985.
Photos: Daryl Bunn.

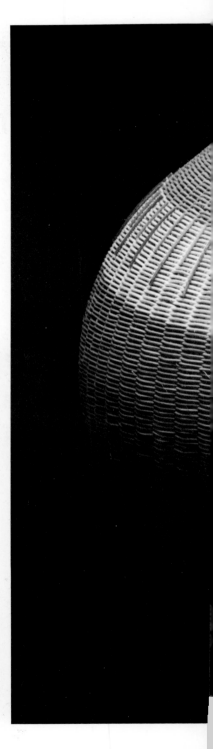

presentations. At other times I begin with a mental picture of what I want to achieve, and then let the controlled and uncontrolled weaving processes take over. Some of my most striking baskets have resulted from this approach. Availability of a material sometimes forces me to rethink a design. I may substitute new color or texture—or both—in order to maintain the integrity of the piece.

In my recent works I have been using geometric and abstract patterning imposed upon sculptural shapes, emphasizing color and achieving a multi-dimensional quality. I manipulate color through dye mixing, applied acrylic and

"Form Series 11," dyed reed, Plexiglas, glue, 14"x17", 1985. Photo: Daryl Bunn.

enamel paints, and a pointillism technique I've adapted for basketry. My shapes are deliberate and architectural.

My background in pottery has had a strong impact on my work, as has my interest in the work of tribal artisans. At one time I worked in a museum, and developed a profound interest in primitive tribal art. I was fascinated with the ceremonial masks and symbolic fiberworks. I spent every spare moment studying the exhibits of African and Indian work, especially the basketry.

I use intricate methods to add textural interest to a piece. The process is painstaking and time-consuming. For example, on one of my recent baskets I've applied over 2000 pieces of painted, clipped pine needles.

Sophistication of design, skill of technique, and refinement of execution are my aims in the exploration of basketry as an art form. In my work, frustration and exhilaration merge as I find myself reconciling opposing forces. This duality—this dichotomy—is resolved through the harmonious interplay of opposites: the old versus the new, restraint versus freedom, the natural versus the manmade, simplicity versus complexity. These conflicts and their resolution form the basis of my work.

There is a responsibility that comes with increased exposure and recognition. I continually study the field of fiber. I maintain a file on fiber artists who have shown in the last six to eight years. I keep an up-to-date notebook with names of galleries that advertise nationally, particularly those that have shown fiber recently, to provide me with leads as to possible markets.

I believe that, as basketmakers, we have a commitment to galleries, collectors, museums, and the public, to produce more complex and original works. I also feel that basketry should be promoted to a greater degree, as are other art forms. The idea of fiber as a fine art form is now being increasingly accepted by the contemporary arts community. There's an air of change and a growing awareness of the aesthetic possibilities of fiber arts. However, fiber commands insufficient interest and understanding from the art establishment, and consequently, from the public.

Left: "Pluto," dyed reed, enamel paint, Plexiglas, 17"x23", 1986. Right: "Spiritual Ritualism," dyed reed, Plexiglas, acrylic paint, 17"x24", 1986. Photos: Daryl Bunn.

"Indian Summer," dyed
reed, pine needles,
Plexiglas, 16"x24", 1985.
Photo: Daryl Bunn.

MARIAN HAIGH-NEAL

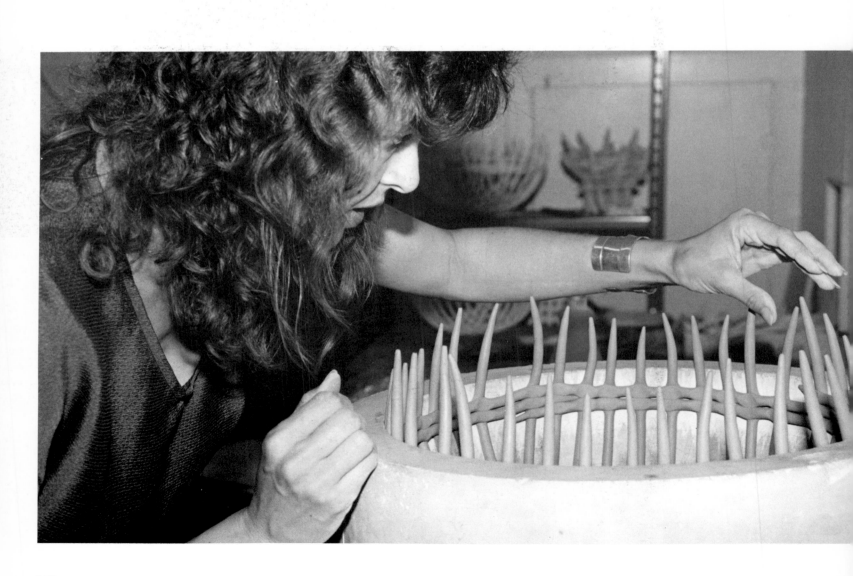

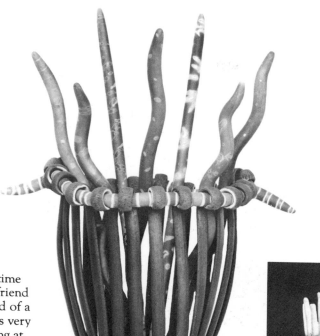

Left: "Guardian Basket," glazed stoneware, 18"x24", 1984. Photo: Phyllis Frede.
Below: "White Split Basket," glazed stoneware clay, 14"x8", 1980. Photo: Toni Altieri.
Photo of Marian Haigh-Neal by Russell Balch.

By 1979, I had been working in clay full time for several years. A friend gave me a large plaster mold of a hemisphere shape and I was very intrigued. I had been looking at some Italian majolica basketlike flower pots. So I decided to make some coils and experiment with the new round form. I loved the pure, sparse shape that resulted, and I was hooked on basketry.

The idea for a new basket starts slowly in my mind as a shape, color, or feeling. Occasionally I may have enough of an idea to actually sketch something on paper. More often I simply work with some shapes of clay, or coils, or sticks, and place them in ways I've never seen in order to strengthen that feeling I'm looking for. I have never used baskets as a direct influence, only as a departure for my forms. My influences are bones, landscapes, bare trees, coral, vines, thorns, ceremonial objects and rituals.

I have always been fascinated by ancient civilizations and their costumes, ceremonial objects, and everyday implements. There has, for me, always been a sense of what rich cultures lay in the past.

This richness I also find through nature in the twisting honeysuckle vines and cut-aways in hills that reveal different strata of rock, clay and colored earth. The landscape of the Southwest, where I have lived and explored for many years, is a very strong influence in my work. I love the vast open spaces, rolling hills, windswept trees, rivers, canyons, mountains and occasional bleached bones lying on the ground. This sometimes sparse landscape full of hidden strength is something I try to bring to my work.

My "Seagrass Basket" developed over a period of two years. One spring I spent a week on a tiny island on a coral reef and snorkled in the ocean for the first time. Even now I find it difficult to say how I was affected by the color and movement of the coral and fish suddenly before my eyes. The sound of the water lapping at my ears and my magnified breathing heightened the sense of a totally new world. When I came home all I had was a memory of those things. Eventually, I developed a pastel glaze I thought was suitable and added it to a basket form with wavy tines. This basket helps me recall my beautiful experience under water in another world.

I have always worked somewhat sporadically in my clay studio, partly because I have had to hold down part-time jobs from time to time and also because it seems to be a normal work cycle for me. Sometimes I spend several weeks or months thinking I'm not making any progress in my work. I have come to learn that most of the time I am processing information inside. This inner processing feels uncomfortable if it lasts a very long time. I continue other work I can do without much emotional investment.

Finally the new idea starts coming together, and I may be on a creative run for several months with ideas spinning out in many directions. Then after a period of time I scrap the new pieces which have not worked

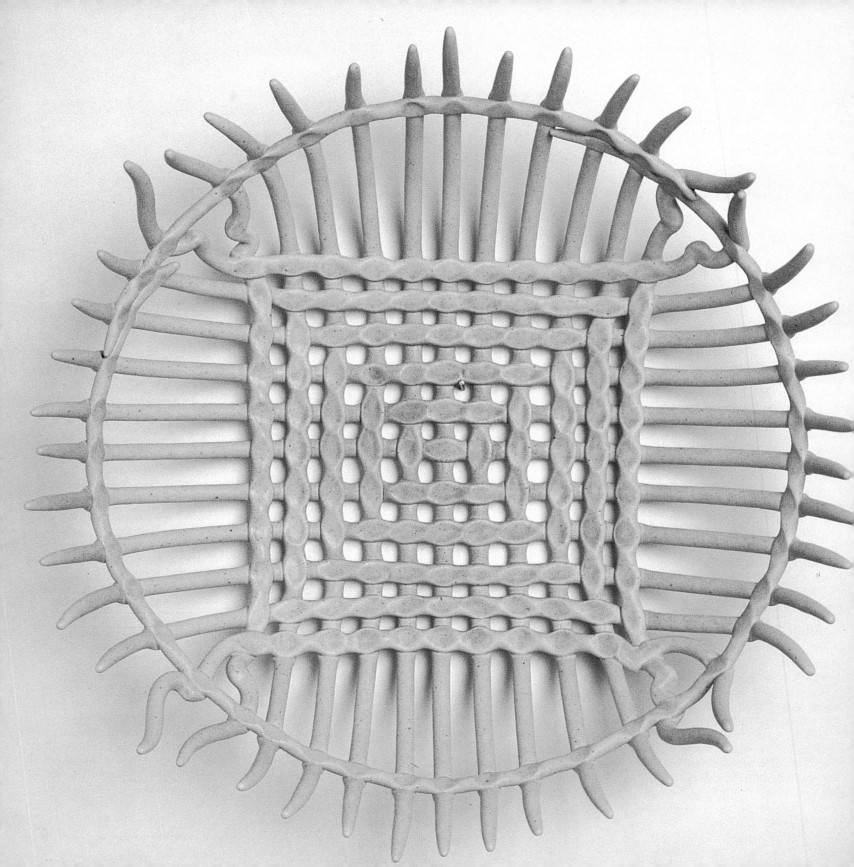

technically or have not fulfilled
the idea I had in mind. Most of
my rejects end up in the shard
pile at the base of the "wailing
wall" outside my studio door.
Occasionally I keep a piece that
didn't work, because it may lead
me to a new series. It might have
just a germ of an idea I need to
look at for a while, perhaps for
several years.

In 1984 I began looking for a
shape different from the round
form. I had always been interested in curvy, snakelike forms,
and in a fit of boredom made a
number of curved fat coils with
smoothly tapered ends. Then I
sat in my studio looking at all
my new snakelike coils, wondering why I had made them and,
more to the point, what I was
going to *do* with them. So I just
played around, and I painted
stripes and designs on some of
the coils and refined them. Then
I had a friend throw some large
bowl forms on his wheel for me
to use as new molds. I experimented and learned that with a
few tricks I could suspend basketlike forms from previously fired
clay structures. This has allowed
me to build larger, stronger
pieces in steps.

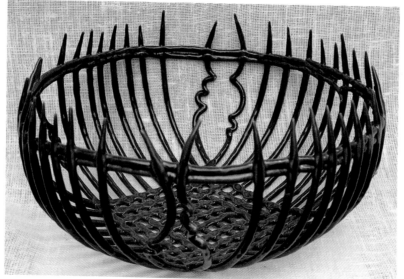

My black and gray colored
baskets are fired in an electric
kiln in a saggar with sawdust.
During the firing, the sawdust
heats to burning, but only
smolders and smokes because
the saggar chamber has a lid that
has cut off most of the oxygen. It
is the smoking sawdust, or
reduction atmosphere, which
gives the clay its characteristic
color range. The baskets that I
build in stages and finish in sawdust are fired from three to
seven times each.

Baskets throughout history
and even present day have acted
as helpers to mankind. It is usual
for a basket to be constructed

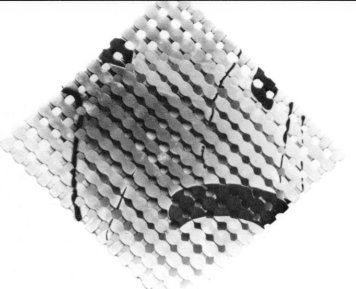

with the themes of beauty and function in mind. I am interested in beauty and other ideas as well. It is common for snake imagery, sharp tines, and thorns to surface in my work. It is fun for me to make a lovely, pastel colored "thorn basket" that contains an element of danger suggested by the sharp thorns. I am reminded of picking succulent blackberries to eat but having to watch carefully for thorns and snakes. It is not to say I find snakes and thorns bad things; quite the contrary: I celebrate them. These additional nuances in my work are the reminders of the capriciousness of life we see about us every day.

"Thorn Basket," with its reference to vines and thorns, and "Bone Basket" are my most recently completed baskets. I am still excited about building a form in pieces and watching it change through all the firings.

My newest work once again seems to be processing within. In the meantime I am working on some small landscape-like sculptures that hold small basins in their surfaces. I am looking at wood formations and listening to my dreams at night.

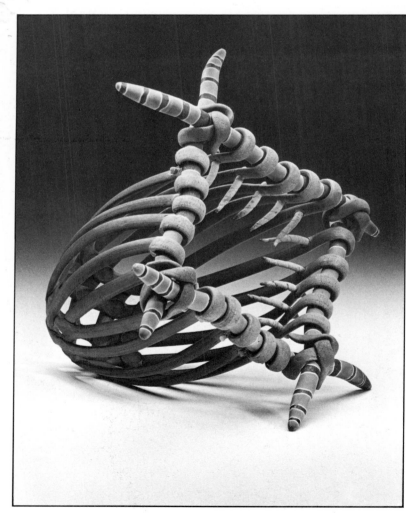

DOUGLAS E. FUCHS

My work reflects a long preoccupation with structure, form and control of materials. Beginning in 1978 I was taken with a cylindrical shape, quite totemic and phallic in nature. Originally wall hung, these forms began to dictate a life of their own. Legs developed, and my "spirit totems" were born. These smaller attempts led to "Forest Group" in 1980, and ultimately to "Sky Towers," an eighteen-foot form woven at Artpark in August, 1980. This imagery coalesced in "Floating Forest," a monumental fiber environment that contained seven woven totem forms ranging in height from six to thirteen feet, and a variety of other fiber constructions, all made of indigenous

Left: "Floating Forest Totem
Group" and "Survivor," reeds,
raffia, sea grass, pods, palm stalks,
aloe stalks. 8-9'h., 1981.
Photo of Doug Fuchs by
Catherine Snedecker.

111

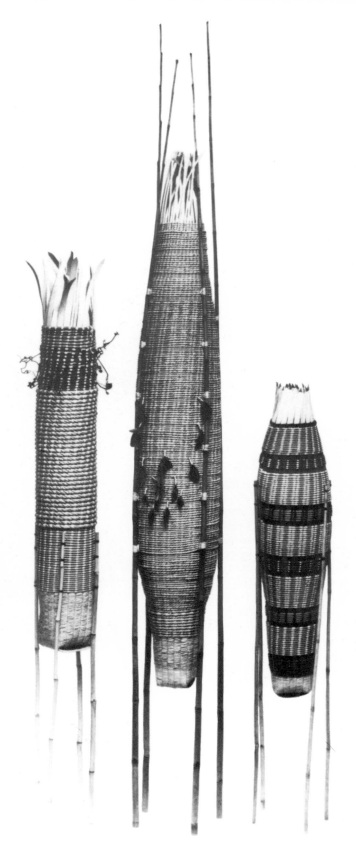

Australian materials. This work was commissioned by the Australian Crafts Council on a fellowship that involved living with the Aborigines, researching their baskets and teaching contemporary basketry all over Australia. An exciting and demanding year!

On my way home I traveled in Indonesia and Southeast Asia, notably Bali and Thailand. I was thunderstruck with color and festive vitality wherever I went. After returning home I began to work with dyed, painted, and synthetic materials, in a new series: "Chedis," forms inspired by Thai temples. This work evolved to "Quiet Cone" an eighteen-foot installation for a show at the Bronx Museum (and currently in the collection of the Museum of Contemporary Craft). I continued to make cone forms until a serious illness disabled me.

I see the body of my work as an expanding search for a spiritual sensibility in the fiber to which I am very drawn. I feel that it works as a body of symbolic form and emotion. It helped lead me into the world of international sensibility and a greater openness to life.

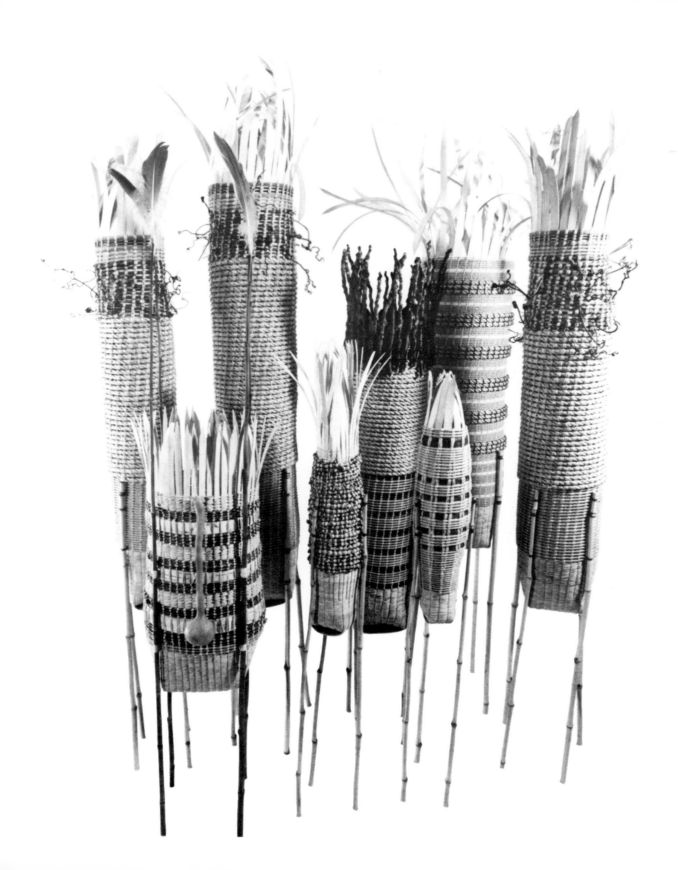

Opposite left: "Spirit
Baskets," raffia, reed, bone,
beads, feathers, 5'h., 1979.
Opposite right, from left:
"Tall Totem with
Tendrils," 4½'h. "Parrot
Totem," 7'h. "Painted
Totem 1," 4'h., reed, raffia,
sea grass, bamboo,
feathers, 1980-81.
Left: "Forest Totems," flat
reed, rolled paper, raffia,
grapevine, manila rope,
birch bark, twigs, bamboo,
3-6'h., 1981.

Above: "Sky Poles," plastic
stripping, PVC-coated rayon,
copper wire, 8'h., 1984.
Right: "Rainbow Forest," plastic
stripping, PVC-coated rayon,
copper wire, pin heads,
14"h., 1984.
Opposite: "Chedis/Cones," reed,
sea grass, plastic-coated wire,
dyes, paints, 3-4'h., 1983.

Opposite: "Floating Forest,"
reed, raffia, sea grass,
sisal rope, leather
stripping, tarred jute, pods,
palm stalks, aloe stalks,
pine wood, bamboo, paper
bark, 1981.
Near right: "Sleeping
Creature in Dream Vessel"
from "Floating Forest,"
paper bark, palm fiber,
palm branches, tarred jute,
pine, willow.
Far right: Detail from
"Floating Forest."

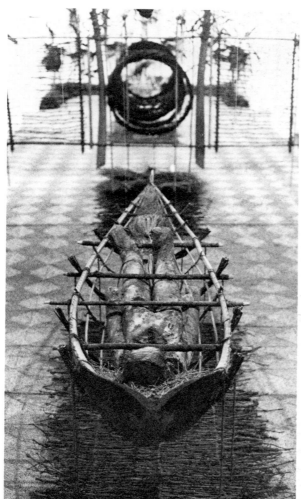

JAN YATSKO

"The Return of Halley's Comet," wool, painted reed, cotton thread, ribbon, plastic cylinder, 7"x15", 1982. Photos of Jan Yatsko and her work by Scott Kriner.

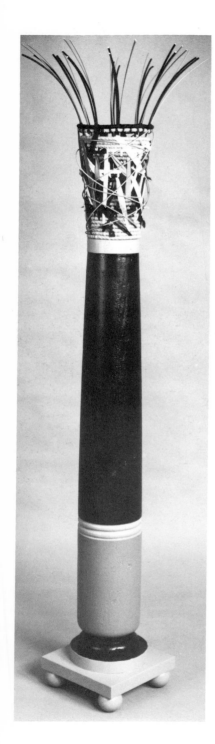

W hoever said "curiosity killed the cat" was just searching for a catchy phrase. That person didn't know cats and didn't know human nature!

There are several similarities between my creative experiences and a cat's behavior. Growing up as an only child, I learned to become comfortable with solitude and to think independently. Most of my childhood was spent finding within myself activities for a single participant. This continuous introspection enabled me to develop my own resources and to stimulate my own curious nature. My childhood experiences taught me that solitude, with good classical or jazz music in the background, gives the proper atmosphere in which to develop creative ideas and to carry them out.

Even as the cat sleeps, he is aware of the activity around him. My creative ideas are formulated through my own curiosity and my desire to interact with other creative individuals. The aura of mystery connected to the interior of a container fascinated me. The thought of what might be in the container, and the act of

finding out, inspired me to create my own irresistible containers. I made my first baskets in 1976, and my primary purpose was to produce well-constructed pieces. As my craftsmanship improved, I started to experiment with color and to stretch the limits of the coiling technique. The inclination to build upon previous experiences led me to expand my range of materials with each new basket. First, barks and vines were added to the wool basket structure; then with my "bird's nest" basket series, I began to select any material which would convey my idea. This process paralleled my choice of materials with a bird's random selection for its nest.

The bird's nest basket series is a continuing tribute to birds and their incredible capability to build nests that sometimes defy gravity, are well crafted, and are aesthetically good. Birds are truly the first basketmakers. The nest series led to a curiosity about bird behavior, and half a dozen bird feeders now fill my modest city backyard. Watching the birds is year-round entertainment for myself, my husband

Above: "Baobab Basket #2," tulip poplar bark strips, wool, 10½"x21½", 1980.
Left: "A New Direction," wool, linen, silk, white birch bark, plastic rods, 7½"x13½", 1981.
Opposite page, left: "Egyptian Nights," painted porch post, ripstop nylon, painted reed, 8"x4'8", 1985.
Opposite right: "For the Folks Back Home," bamboo, wool, leather, cotton thread, buttons, 9½"x11½", 1983.

121

Below: "Manhattan Loft,"
tulip poplar bark, bamboo,
wool, linen, buttons,
9½"x21", 1982.
Right: "Tree House," tulip
poplar bark, bamboo,
embroidery thread, plastic
screening, 18"x28", 1983.

Left: "Stacked in My Favor," wool, bamboo, tulip poplar bark, painted reed, 9½"x13¼", 1982.
Right: "Hidden Fantasies," wool, embroidery thread, painted reed, plastic screening, buttons, 7½"x11½", 1984.

and our three cats.

The nest series happened to coincide with a time of intense career and artistic reflection. After eight years of creating detailed pieces of sculpture, I was experiencing a decline in motivation. To counteract this, I began to work in brilliant colors, tried to find humor in my work and my life, and continued to create, but through different avenues. I felt it wasn't important which medium or technique I used as much as it was important to continue to create. My humor manifested itself in the bird's nest series. I thought of common bird nesting sites and interpreted them in a lighthearted way. A piece titled "Manhattan Loft" depicted a nest in the tall buildings of Manhattan, but it also referred to the artist's lofts in SoHo. Another basket entitled "A Nest for Big Bird" was five feet tall, and was constructed on site during a five-day craft show. When the basket was completed, I topped it with six colorful origami birds, each two feet long.

One of the avenues I pursued was banner commissions. I was attracted to the festive colors and to the airy qualities of the fabric I used. I continue to find satisfaction in my banner work, but it has also brought me back to baskets. Recently I have begun to coil with the ripstop fabric and to create banner environments around baskets. The banner environments were a natural creative progression. The result is that I can view one of my baskets as a part of a whole creative picture and not as a separate entity.

It is rather amusing that many of my interests begin with the letter "b": Baskets, banners, ballet, bicycling . . . chocolate, cats. Well . . . "c" does follow "b"! Besides, we have now come full circle to "cats" and their "curiosity" and a little curiosity never hurt anyone!

BRYANT HOLSENBECK

I am an urban basketmaker with my roots in nature. I study the way nature creates: the way tree roots grow and how they intertwine, or how pine cones are shaped and how they sow their seeds.

Yet the materials I use for my work come from people as well as from nature. I figure that what we produce industrially is just an extension of the natural formation process. So I take materials that I find or buy—yellow plastic strapping tape which wraps newspapers, old bottle caps and buttons, as well as honeysuckle, seaweed and rattan—and I weave objects out of them.

When I start to make something, I feel that I am in conversation with the materials. My vocabulary is my specific weaving or painting technique. When I teach beginning basketry, I always tell students not to worry about what their first baskets will look like. The job with the first one is to learn the specific technique and to get used to the feel of the materials. The finished piece and the process of making it are what I think art is all about—taking materials and putting them together to make a

whole. Birds do it all the time, and they don't even have fingers.

I began making baskets while living on Nantucket Island, a place with a rich basketry heritage. I was struck by the fact that almost all baskets are made from natural materials, though it is hard to tell because of the evenness with which the fibers are processed, bleached and dyed.

Well, I thought, I can make baskets which are useful, but which still look like the natural materials they were made from. When baskets were first made, people were fighting to survive, trying to separate themselves from the harshness of their surroundings. Now, we are all finding a need to put nature back into our lives, to protect it and keep it with us. So I began making baskets, which were functional pieces with the natural textures left intact whenever possible.

I am making my living now making new shapes—integrating classic techniques and designs with modern and traditional methods. The first time I made a basket which differed from the standard of what I thought a basket ought to be, I was afraid

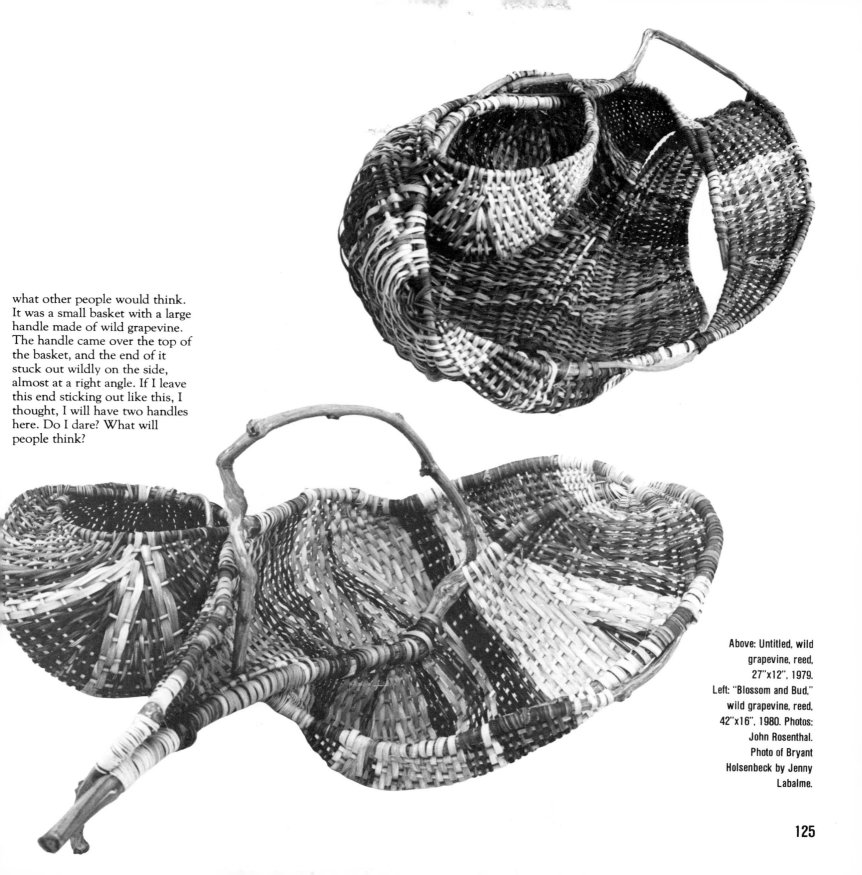

what other people would think. It was a small basket with a large handle made of wild grapevine. The handle came over the top of the basket, and the end of it stuck out wildly on the side, almost at a right angle. If I leave this end sticking out like this, I thought, I will have two handles here. Do I dare? What will people think?

Above: Untitled, wild grapevine, reed, 27"x12", 1979. Left: "Blossom and Bud," wild grapevine, reed, 42"x16", 1980. Photos: John Rosenthal. Photo of Bryant Holsenbeck by Jenny Labalme.

Because I was living on Nantucket, I had the opportunity to show my work with the Artist's Association there. I entered the basket in their summer show. To my great relief and delight, people were quite excited by it, and had a lot of fun with the two handles. It was a lesson in my own self-estimation and the opinions of others which I have never forgotten.

Over the years I have learned that my work is about change, about the evolution of form and ideas within the culture in which I live. Traditional techniques and materials are important evolutionary tools. I am learning that the choice for me is to take the risk, try as many new ideas as I can when they occur to me. And in the end, it has been good for business.

When I show my work, the two questions people most often ask me are number questions. How long, and how much? How long did it take me to do it? And how much does it cost?

These are practical questions. I have two answers I have worked up over the years concerning the how-long question. One of them is to tell them my

126

Left: "Lighted Windows," reed, found objects, mattress springs, paint, 40"x40", 1985. Photo: Richard Faughn.
Right: "Sea Clock," reed, wood, bamboo, buttons, Xerox, found objects, 48" diameter, 1982. Photo: Richard Faughn.
Below: "Clutz," reed, found objects, ribbon, paint, 6"x4", 1982. Photo: John McAllister.

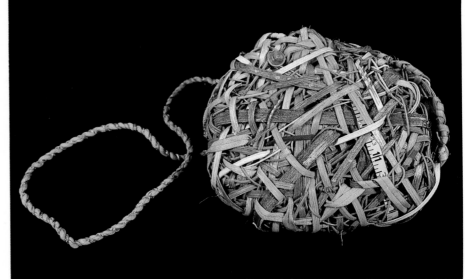

Below: Untitled basket, reed, found objects, paint, 14"x10", 1983. Opposite top: "Venus Opera Hat," reed ribbon, seed pods, electronic parts, beads, 8"x14", 1982. Opposite bottom: "Hatz," reed ribbon, seed pods, found objects, 7"x14", 1982. Photos: Richard Faughn.

age, whatever it is, because my work has grown as I have grown. What I do now, I could not have done five years ago. It is experience layered upon experience which makes people ask this question. They sense that a great deal of learning has happened here.

The other answer is one an old tobacco farmer once gave me with a twinkle in his eye, when I asked him how long he worked each day. "Honey," he said, "if I knew I wouldn't do it." I know in a larger sense—say how much time each week I spend—how long my work takes to complete. But I drop time expectations while at work; that is the only way I can be truly involved.

I like to tell my students that baskets are about time. More than anything else, that's the value of them. They are not made from precious materials like gold of silver. There is no dramatic chemical transformation as in ceramics or metallurgy. There is simply time. There is the time spent gathering and processing the materials— whether you are stripping honeysuckle bark or unraveling telephone wire. And there is the time spent weaving—thread upon thread, twist and turn, twist and turn, until finally, the basket is complete. Finished. Ready to hold whatever you may choose. A finished basket is a mark of time spent. Within that is its value.

Sometimes people are surprised at how much my work costs, especially if they compare the prices of my baskets to the price of beautiful rattan weavings made, for example, in the Philippines, where people still get much less for their labor. We live in a world where we no longer need to make things by hand. We don't need baskets. After all, we have paper bags, plastic bags and cardboard boxes. Hand labor that was once necessary has been supplanted by mass production. Yet I feel that the creation and use of handmade things is of primary importance to our well-being. By understanding how something is made, we gain self-esteem, and a broader understanding of our place in the world. Expressing my ideas in what I make, and selling the finished work, is one way that I can communicate my understanding.

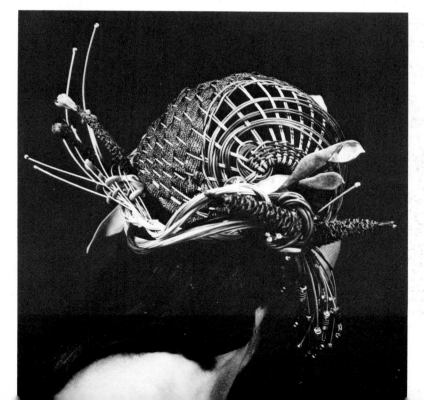

I find that what I have to say continues to expand. When I teach a class, my students always challenge me. I am forever seeing them do something that I didn't think could be done, or hadn't thought of before as a solution to a particular problem. I like to tell beginning students that they have the advantage of wide-open expectations. It is my job as an artist and a teacher to keep these expectations open, and to help them over the rough spots with techniques and encouragement. And I must constantly remind myself of the things I tell my students.

I have been weaving objects for almost a decade now. I see ahead of me risks that need taking—I need a new and larger studio, for instance, and need to expand my present work into still larger sculptures. If I think about how hard it will be I will remain where I am, with what I am already familiar with. But if I remember how important my own stepping ahead has been in the past, then the way is clearer. As an urban basketmaker, it seems that my job is to take the fabric of my past experience and keep weaving new possibilities.

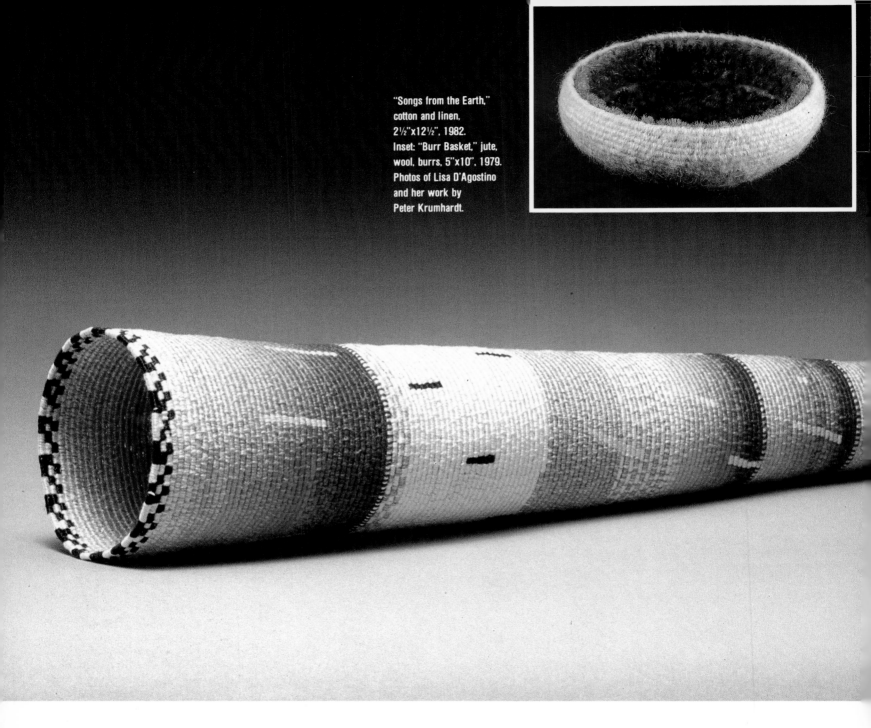

"Songs from the Earth,"
cotton and linen,
2½"x12½", 1982.
Inset: "Burr Basket," jute,
wool, burrs, 5"x10", 1979.
Photos of Lisa D'Agostino
and her work by
Peter Krumhardt.

LISA D'AGOSTINO

Initially, I chose the coiling technique because of its convenience. I had the desire to do art work, but no money to buy materials. Coiling was attractive because it did not require a special studio or equipment. I could sit in my bedroom with a needle, scissors, yarns, and a card table to set it all on. Some of the yarns came from the local department store. Others were leftovers from fiber courses I took in college. The yarns were not all the best quality, but that was not a main concern at the time. My intention was to work and experiment with shapes and colors. The results were varied, but I learned what I liked and didn't like and had new ideas as I continued to work. I liked the tactile, rigid quality of the coiling. To focus on this quality, I kept the forms of the baskets simple. The form I fell in love with is the cone. It seems to echo the shape of towers and monuments that have been erected by many cultures for thousands of years. Mayan, Aztec and Egyptian pyramids have always fascinated me with their history, their purpose and their mystery. I began to use cotton embroidery

Left: "Cygnus," cotton, jute,
silk, 3"x5", 1980.
Right: "Basket with Grid,"
jute, silk, cotton, wood,
2½"x4½", 1980.

floss as wrapping material. It is available in hundreds of colors. I kept the baskets neutral in color.

About seven years ago I moved to Iowa. For most of this time I lived on several acres in the country. The view from the house was open sky and land, and in some ways they have become part of my work. It was a very calm place, and enabled me to work with the tedious technique of coiling with the embroidery floss. I think it also had a somewhat spiritual effect on me. Since living there, I have had a craving to visit vast, open places like those in the West and Southwest.

Soon after I moved to Iowa a friend reintroduced me to astronomy. I say "reintroduced" because my father explained the stars to me years ago, and sparked an interest that has lasted. I was hooked on astronomy for several years. I bought a star chart, telescope, astronomy books and magazines, joined an amateur astronomy club, and went to a few "star parties." I spent a lot of time reading about astronomy, and spent hours outside at night learning the constellations and the locations of var-

ious stars. I was amazed that I could see the rings of Saturn and the four moons of Jupiter right from my backyard. I loved it, and its influence appeared immediately in the baskets. The places and events in the baskets have astronomical themes, as do all the titles. Some of the inspiration has come from magazine articles, the "Cosmos" television series, a billboard along the

133

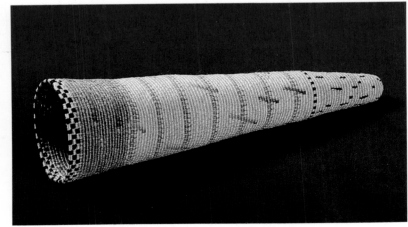

Right: "Return of the Comet," jute, cotton, 2½"x8½", 1981.
Top left: "Meteors Over Rt. 3," cotton, linen, 4"x18", 1984.
Top right: "Heroes and Falling Stars: He left the vivid air signed with his name," cotton, linen, 3"x12", 1985.

highway in Arizona, and current events.

While astronomy was working its influence, color slowly appeared. At first it was only little bits of complementary color with neutrals of black, gray and white. It didn't seem possible to work with many different colors and all the strands of cotton without going crazy. I started with a little color and slowly worked up to more complicated patterns, with more color in each new piece. I graded the colors from light to dark, usually on a neutral background. Eventually, I began to use graded color in the foreground and background as well.

With so many color gradations, the embroidery floss has to be organized in some way. I use three strands of floss. They might be three different colors, or two strands might be the same color and the third strand a shade or two lighter or darker. I arrange rows of these combinations from light to dark on heavy paper. Then I tape them down and number them. The color gradations are subtle, so it is important to number everything. When one color combination

"Welcome to Meteor City,"
linen, cotton, 2½"x5½"
each, 1984.

runs out it is easy to determine what is needed.

The process of coiling, with so many strands of color coming and going, is very slow. Working once around a five-inch circumference can take an hour, and completing one inch of coiling can take eight hours. It takes a lot of time but time is always a problem anyway. I work full time, and the only time for art work is in the evenings or on weekends. Most of the artists I know seem to have problems similar to this. Luckily, most of my summers have been free and they provide the only large blocks of time in which to work.

Accomplishing art work while holding a full-time job has made life hectic at times, as well as frustrating. When I am at my job, my thoughts continually turn to the baskets and what I could be doing, or what I want to do next. While I'm working on a basket, thoughts of the job sneak in and I wish I didn't have to go to work tomorrow. It's a double life! Overall, it has worked out pretty well. The important thing is to *make* the time to work, and to use that time in the best possible way.

RACHEL NASH LAW

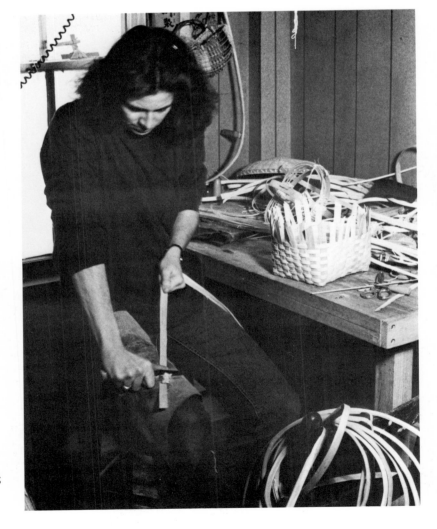

In 1985 my husband and I moved back to West Virginia after an absence of seven years. We both felt a need for the mountains, a more familiar environment and a place to settle down. The move has made such a difference! I have become a morning worker, excited about the quiet and the long stretch of time in the day that seems to come with an early rising. I feel that the serenity of these mountain surroundings and the richness of their basketry materials will eventually influence my work.

For the past four years I have been interviewing traditional basketmakers in conjunction with a writing project. Seeing so many different baskets and the details of their construction has shown me a whole new variety of technical features. This study has begun to manifest itself in my own baskets. The biggest influence on my work comes from three historical types of white oak baskets: the rib basket, made with a framework and ribs; the split basket, mainly made with flat weaving materials; and the rod basket, made with a round white oak rod woven in a variety of wickerwork forms. These baskets, along with the original purpose of baskets—practical containers for storage or transportation—are the inspiration for my work at this time.

I have worked with many materials, including bark, vines, and shoots, but I prefer white oak as my basketry material. Working with white oak in basketry is like working green wood, and requires a variety of tools and technical skills. White oak is a very elastic and pliable wood, and lends itself to many different basket forms. White oak is strong and durable, but lightweight and easy to maintain. The work itself is clean work, surrounded by fresh smells. The shavings are easy to sweep up and are used either as fire starter or in the garden as mulch.

A suitable white oak tree is not easy to find. It must be a young tree and free of blemishes or disease. The trunk must be straight, six to eight inches in diameter. Sometimes it takes the better part of the day to find good timber, but the ease in preparation and the beauty of the wood make the search for the right tree worthwhile. My

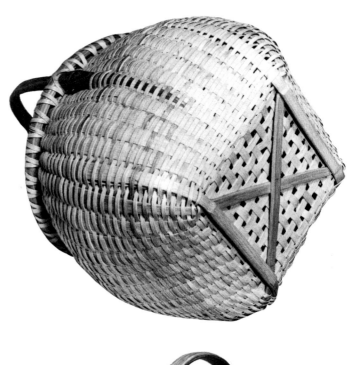

Below: Stacked hoop
basket #2, white oak rib,
8"x14", 1985.
Right: Footed split basket,
white oak, 11¾"x15¾",
1985.

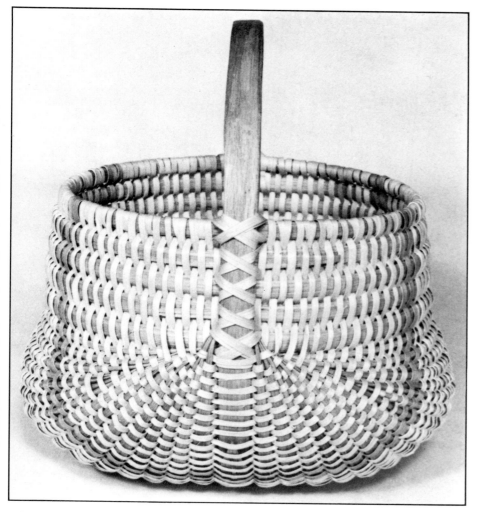

137

Left: Hamper, white oak, 24"x16", 1984.
Above left: Round lidded basket, white oak, 9¼"x19", 1985.
Above right: Basket skeleton, white oak, hickory bark, doweling, 12"x19", 1984.

husband helps with the search and with the physical labor of cutting and hauling the log out of the woods. He sometimes does the initial splitting of the log, dividing it into quarters, but I do the rest in order to get the exact size and shape I want.

Preparation of material and the actual basketmaking cannot be separated. Usually I have an idea of what I would like to make, but try to leave the form open for change, in case a better idea should surface, or the materials seem better suited to something else.

There is something new to learn on each piece of work. Each tree handles a bit differently, and some make better basket material. Much of the work depends on the tree's pliability and how it meets the needs of a particular basket. In addition to the wood itself, construction techniques often become strong design elements and act differently with different basket forms. Varying and combining techniques in different types of baskets leads to the creation of new forms and new ideas.

For the past six years most of

my work has been on a small production line, and I've made an occasional exhibition basket as needed. That work method now seems to be completely reversed, with more time spent on experimental and exhibition pieces than for production items. I feel that I am finally to the place where I can go ahead and work through a series with a single construction feature or a basket form as the idea base. The apprenticeship is over, but the commitment to the craft remains, and the search for ideas is definitely still in progress.

138

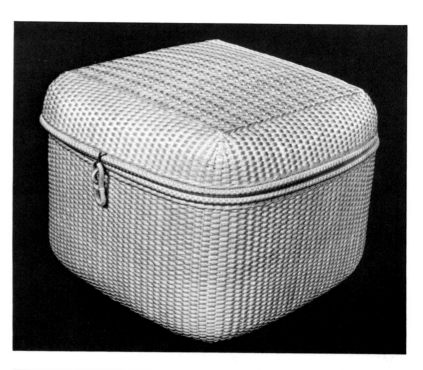

Right: Lidded willow basket, 10"x9½", 1979.
Below right: Single-V gathering basket, white oak, 17"x14", 1985.
Center page: Round rod and split basket, white oak, 17"x14", 1985.

SHEREEN LAPLANTZ

My baskets are an interaction of form and texture. They're an extension of the things I love, functioning one at a time, then in conjunction.

The above statement is the result of a lengthy discovery process. I have a thorough art background, as a weaver. I began studying art in junior high school and continued through graduate school. I learned that I wanted to be *good* as an artist. Being bad would have been okay, too; at least that could have been done with flair. What I feared was mediocrity—I knew I had to develop a solid technical background, an understanding of what I liked, why, and how to achieve it. I was patient, and clocked a great deal of time in the studio. Fortunately, I finally got tired of waiting and began to do only what I wanted to do—baskets.

Three elements have surfaced in my work and have become recurring themes: texture, architecture (I continually explore shapes), and layering. Each element has developed in each of my recent styles. And I do like a technical challenge. When a technique becomes too easy for me, I branch off into another style.

My first baskets were simply an exploration of textures. I used a form with straight sides, a square base, and a round top. I embellished it with surface curls in every way I could imagine. I made hundreds of these, and developed an intimacy with basketry. It was working on these curled baskets that hooked me, that finally lured me away from weaving and into basketry. It was also during this time that I developed my understanding of what a basket is, what sizes and proportions I like, and how I want to work with materials. The textural surface curls were no longer enough, and I became concerned with shapes.

I continued to play with textures, and tried thatching—like thatched roofs. The baskets, as extensions of me, became ladies, and I dressed them. Sometimes it felt like I was playing with dolls, but more often it related to the fashion history which I enjoy reading. I made hundreds of the thatched ladies, and learned that I don't like texture for texture's sake alone. No matter how much I like it, that one design element can't overwhelm the object. This style let me understand how I feel about texture, how much of it I want on a piece, how closely it should hug the object, and that it needs to work in conjunction with my other themes. I like those ladies, even now.

After the ladies, I no longer worked with just one design element at a time. Textures continue to be a primary concern in my work. And they are becoming more elaborate. They've become another layer of basket, or a network covering the surface, or even other baskets worked on the surface of the base basket. In my dreams I honeycomb the surfaces of my baskets with textures.

The second recurring element in my basketry is architecture. It actually started as shaping. I became bored with simple, basic forms and began to try more involved shapes. At first

Opposite: Untitled pyramid, flat paper, fiber splint, reed, waxed linen, 12"x9", 1981. Photo of Shereen LaPlantz by David LaPlantz.

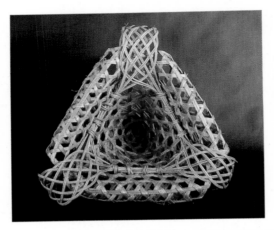

Left: Untitled double layer basket, flat and round reed, 15"x12", 1985.
Below: Untitled pyramid, flat paper, fiber splint, lauhala, waxed linen, 10"x10", 1981.
Opposite: Untitled double layer basket, flat and round reed, splint, 28"x23", 1983.

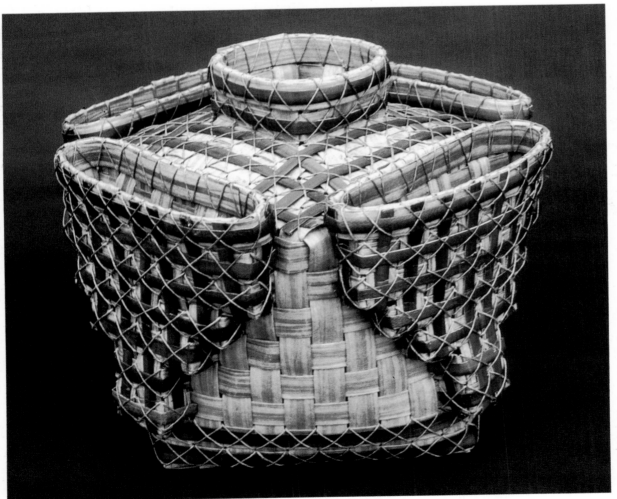

these were influenced by other baskets, primarily from Southeast Asia and the Philippines.

Then I began to work with pyramids. These were far more complex shapes, and didn't even *feel* like baskets. They were a direct response to architecture. These baskets looked like skylines, like contemporary buildings, or like a hillside covered with small houses. Through these baskets I was finally able to combine texture with a complex shape.

Shaping hasn't been a primary concern since I made the pyramids. Now it's simply part of the way I view a new technique. It's also an element that *must* be in my basketry. Like a "given" in math, it's automatically there to build on. When I find a new material, when I'm depressed and just want to make baskets, or when I feel dry, I play with shapes. Shapes are fun. They're also spontaneous. I might be working on one idea, and halfway through another presents itself, already finished. Shaping demands that I keep a fresh eye. Expecting to see only one thing eliminates all the discoveries along the way. And it's

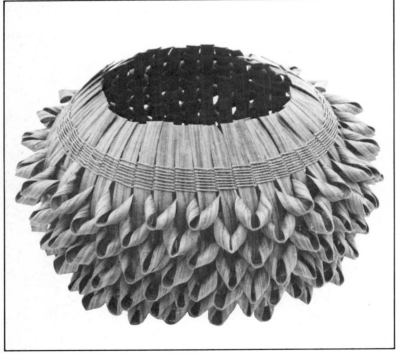

those discoveries that create the forms for my basketry.

The final element in my basketry is layering. This is almost texture, and its result is surface texture, but it's based on clothing. While dressing the thatched ladies I did some layering, with overskirts and aprons and head scarves. The thought of layering one thing over another kept nagging at me. I had tried some openwork baskets, but they were too open, a bit formless, and far too fragile. The obvious solution was to put the lace basket over a solid form, another basket.

This form is still fascinating to me. It has evolved through several styles. First I made openwork round reed baskets over some of my favorite solid forms. These were baskets I made for me: they didn't photograph well, so I couldn't get them in print. Next I tried openwork with openwork, which allowed me to have two baskets intersect, to actually weave in and out of each other. They didn't even have to be the same shape. I learned to think of two layers, two separate baskets as a single unit, a single skin that could be applied to another solid form. Since at least one of the openwork layers was always hexagonal plaiting, I also learned to think of weaving in three directions at once. Then there was the big breakthrough, when I learned that three-directional plaiting can be shaped the same as regular plaiting. What I knew about shaping would transfer. I could continue to play with shapes, but play with them in the current technique.

Finally I learned that mad weave, another three-directional weave, is a cousin of hexagonal plaiting. One is the open form, the other is solid. They mesh together. They cooperate so that hexagonal plaiting can flow over and tuck into the surface of mad weave. I've only begun to play with mad weave, and don't yet know what it has to offer me. But it sounds like a technique that's made to suit all my interests, as it fascinates me as much as each of the styles before it did.

With each new style I go through a specific process. I find that a technical problem has been fascinating me for a long time. First it sort of nags, then it arouses curiosity, causing me to dabble and be frustrated, finally

Below: Untitled double
layer basket, flat and
round reed, splint,
11"x8", 1983.
Opposite top: Untitled
thatched lady, reed, raffia,
palm bark, 12"x7", 1979.
Opposite bottom: Untitled,
paper, fiber, splint, waxed
linen, 14"x7", 1978.

it demands full attention. I learn the new technique through simple basic forms. When I was a beginning basketmaker I would then embellish the simple forms with texture. Now I make, in the new weave, some of the more complex shapes I enjoy, then add the textures. Working on surface texture helps me understand a technique. Through adding embellishment I learn what's happening with the technique; how it's shaped, what the rules are and which can be broken, and what numbers of elements work best for the things I try. This is when I learn to like the technique. Once I understand the technique, I start to work on shaping. When I can control the shape, and have developed the complexity I want, then I either compound or layer. To "compound" is to build a basket on top or on the side of a basket. It's at this point that the baskets come alive, develop personality, and become magical.

I make my baskets because I like them. It takes tons of perseverance and discipline to get them to a point where you might like them too. And it's worth every minute.

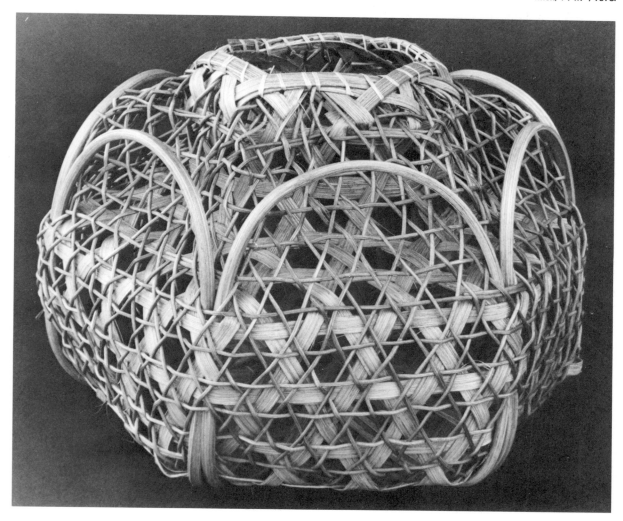

NANCY MOORE BESS

Ten years ago, I was concerned about mastering the traditional basketry techniques in the purest forms and in using those techniques to create containers. Everything was quite clear and linear. I taught a great deal and wrote a number of articles. But research into the cross-cultural aspects of basketry led me to realize basketry techniques, and related fiber techniques, had much broader application than I had realized.

I started reading books like *Shelter*, *The Houses of Mankind* and *Craftsmen of Necessity* and was fascinated to find wicker-work and plaiting used on the

Opposite: "Temporary
Basket," natural raffia,
9½"x6", 1984. Photo:
Doug Long.
Photo of Nancy Moore
Bess by Bob Hanson.

sides of homes, bamboo used for scaffolding, wicker techniques in the wattle fencing of England. I began collecting fans and brooms that incorporated basketry techniques, and I looked to the Pacific Basin area for further inspiration. A study of this region's techniques led me to the use of ropes and cordage: coir (coconut fiber), nawa palm, manila, and sisal. Ropes led me to dyeing, not because of a commitment to natural dyes, but because of a desire to control color and because coir and manila take color so well. Financial considerations led me to raffia, which I now consider my perfect material.

What began as a single focal point—basketry—became instead the core from which many other interests branch. Work developed, not singularly, but rather in related series. Increasingly, my work reflects my ongoing interest in fencing, screens, dwellings, fiber armor and thatching.

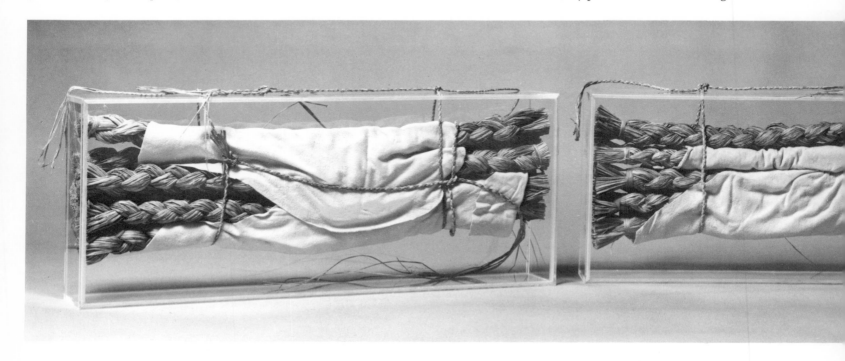

Opposite top: From "With a Nod to Japan" series, natural raffia, 4½"x4½", 1984. Opposite bottom: "Chamois Bundle," hand-dyed raffia, natural chamois, 22"x9", 1984-85. Photos: Bob Hanson. Top, this page: "Pink Palm Bundle," hand-dyed raffia, palm leaves, 22"x9", 1984. Bottom, this page: "Volcanic Ash," paper, 3"x4½", 1977. Photo: Bob Hanson.

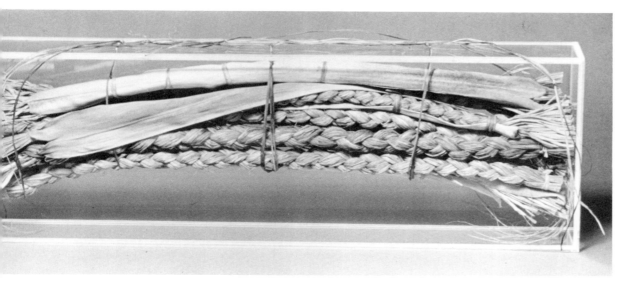

My current research style is a combination of "good ol' librarian" (that is what I did with my B.A.), clipping and recording, plus what I affectionately call "mental cataloging." It took years of looking at fiber armor before "Private Armor" and subsequent pieces evolved. What a sense of satisfaction when it came together!

I have no interest in trying to predict what will evolve in my work, but hope, instead, that I can remain flexible—something that doesn't come easily for me. I will finish the wattle fencing behind my herb garden and my perennial bed in Connecticut. I will continue to learn about bamboo until I can grow it myself and prepare it for basketry.

While doing these wonderful things, I will also iron my pillowcases (because I can't stand them wrinkled), hang my laundry out on the line whenever possible, have lots of animals around, read mysteries, buy old table linens at yard sales, cook the things my family loves while teaching them to love cumin as much as I do, be involved in peace projects, and share my energy with those I care for. Why? Because one doesn't work in isolation. Basketry is *part* of my life . . . not my life.

No more time to write. Back to the studio. I want to re-dye some reed for a new project and must meet Ivan's bus at three. It is raining, the cat, Grape Nuts, must now get off my lap, and I must say ENOUGH.

Left: "Post Modern Taos," hand-dyed rattan reed, corn husks, raffia, 17"x11", 1985-86. Photo: Bob Hanson.

Right: Untitled pair, dyed coir rope, hand-painted manila rope, 28" diameter. Photo: Doug Long.

Below: From "With a Nod to Japan" series, natural raffia, chamois, 5"x5", 1984-85. Photo: Doug Long.

JOHN MCQUEEN

The baskets I make are branches of trees rearranged and no longer real the way a tree is real. Yet, for the moment they are themselves and they carry their intent. Though they will straighten and release their uncertainty, they are now corners and condense what is around them. They are an organization bent on isolation; a rationalization for changes in nature wrapped around a hole. It is this hold, this insisting, this fortification held against momentum that these sticks guard against. It is an outside needing its inside. An obsession strong enough to change them from what they are into what they do. These are branches of trees arranged and unreal and for the moment they are having themselves.

Untitled basket, ash, grapevine, paint, 11"x25". 1985. Photo: Brian Oglesbee, courtesy of Bellas Artes Gallery. Photo of John McQueen by Jessie Shefrin.

Opposite: Untitled baskets, bark, wood, 18"x20", 1984. Photo: Brian Oglesbee, courtesy of Nina Freudenheim Gallery.
Top right: Untitled basket, malaluka bark, Spanish moss, 10"x11", 1977.
Bottom right: Untitled basket, basswood, pine bark, 15"x17", 1986. Photo: Brian Oglesbee, courtesy of Nina Freudenheim Gallery.
Below: Untitled basket, raspberry canes, red osier, 9"x32", 1985. Photo: Brian Oglesbee, courtesy of Bellas Artes Gallery.

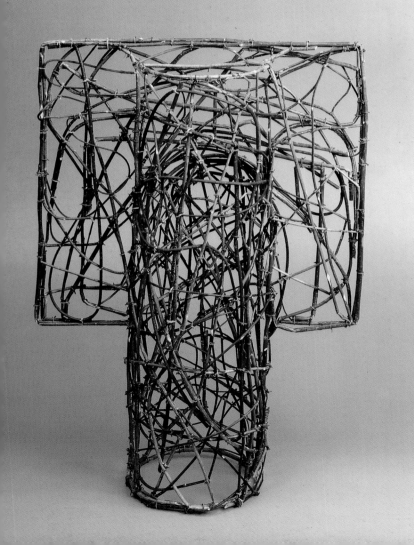

Left: Untitled basket, cedar bark, grapevine, 9"x29", 1986. Photo: Brian Oglesbee, courtesy of Bellas Artes Gallery. Below: Untitled basket, ash splints, red string, 9"x18", 1983. Photo: Brian Oglesbee.

Top left: Untitled basket,
white pine bark, 8"x18",
1977. Photo: Lester Mertz.
Below right: Untitled
basket, day lily stalks,
willow bark, 13"x15",
1980. Photo: Stephen
Myers.
Below left: Untitled basket,
birch bark, 13"x14", 1978.

DONA LOOK

My work with white birch bark began with an interest in the folded bark containers made by the Woodland Indians. These single-fiber baskets were constructed with minimal shaping and alteration of materials. I started weaving strips of handmade paper into small containers, and strips of birch bark into sheets to be used like fabric. Eventually,

Clockwise, from above:
Untitled, white birch bark,
waxed silk, 5½"x8½",
1985. Photo: Sanderson
Photography. Untitled,
white birch bark, waxed
linen, silk, 8"x9", 1984.
Photo: Ralph Gabriner.
Untitled, white birch bark,
waxed silk, 7½"x8¾",
1985. Photo: Sanderson
Photography. "Fish

Strainer," white birch
bark, waxed linen, 20"x7".
1983. Photo: Ralph
Gabriner.
Photo of Dona Look by
Ken Loeber.

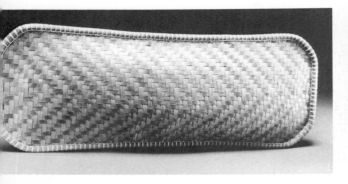

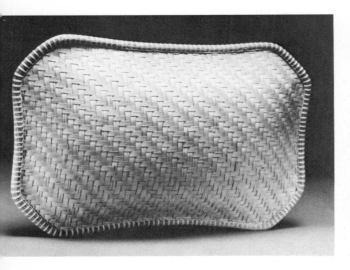

Top: "Derr Pillow," white birch
bark, waxed linen, balsam needles,
grass, 20¼"x3", 1983.
Bottom: "Tré Pillow," white birch
bark, waxed linen, balsam needles,
grass, 16¼"x3", 1983.
Center: Untitled three, white birch
bark, waxed linen, silk. Photos:
Sanderson Photography.

birch bark took the place of the handmade paper. Although I continue to experiment with other materials native to northern Wisconsin (cedar bark, black ash, willow), my present concern is exploring the use of white birch both in sheet and strip form.

Since the trees have individual characteristics, each piece of bark is unique. Often the process of gathering, sorting, and preparing the bark dictates appropriate use of the materials. A common element to emerge in this work is the juxtaposition of a consistent woven pattern with the irregular natural lines on the sheets of bark.

I consciously limit the techniques and materials I employ so that I can concentrate on form. I think of these pieces as vessels rather than baskets since practical function is of little interest to me. I minimize decorative elements, but the use of silk thread allows me to add subtle lines of color. I find the decorative qualities are inherent in the materials. While the resulting vessels are measured and monochromatic, I feel the essence of my work is in the forms and simplicity of construction.

NANCY MOORE BESS, New York, NY

Nancy Moore Bess earned a B.A. from the University of California-Davis, and studied at the State University of New York at Stony Brook, Columbia University, and the Fashion Institute of Technology.

Exhibitions of Bess' work include Works in Miniature at Elements, Greenwich, Connecticut; Basketry, Gallery North, Setauket, New York; and traveling exhibits, The New Basket: A Vessel for the Future, and Texture, Form and Style in the Marketplace.

Her work has appeared in *Contemporary Basketry, House and Garden, Fiberarts* magazine, *Craft Australia*, and she wrote the book *Step-by-Step Basketry* (Golden Press).

Bess is a member of the New York Textile Study Group, an association of nearly 50 artists who share a common interest in the fiber arts.

JOANNE SEGAL BRANDFORD, 200 Brookfield Road, Ithaca, NY 14850

Joanne Segal Brandford has exhibited widely, with solo shows in New York, California, and New Jersey. Group exhibits since 1975 include Fiberworks, the Tyler School of Art group invitational exhibition; In the Spirit of the Inca, Berkeley, California; IV Textile Triennale, Lodz, Poland; The New Basket: A Vessel for the Future, Brainerd Art Gallery, Potsdam, New York; Basketry Now, Lawton Gallery, Green Bay, Wisconsin; and After Her Own Image: Woman's Work 1985, Winston-Salem, North Carolina.

In 1986, Brandford had two-person shows at 15 Steps Gallery in Ithaca, New York, and at Swain School of Design in New Bedford, Massachusetts.

Ms. magazine, *The Village Voice, Fiberarts* magazine, *Craft Horizons, The Boston Globe*, and *AIA Journal* are among the publications which have featured her work, which also has appeared in several books, including *Textile Art* (Rizzoli), *The New Basketry* (Van Nostrand Reinhold), and *Weaving Off-Loom* (Regnery).

Brandford earned a B.A. in decorative art and an M.A. in design from the University of California at Berkeley. She has been a member of Amos Eno Gallery, New York, since 1977.

JAN BUCKMAN

Jan Buckman was a 1985 fellowship recipient from Arts Midwest and the National Endowment for the Arts. She studied basketry and rugs on the Navajo and Hopi Reservations in Arizona.

Group shows in which her work has appeared include Baskets and Quilts in Philadelphia; New Wisconsin Fiber, Lausanne, Switzerland; Fine Focus in St. Louis; ArtQuest '85, a traveling exhibit; and Basketry Today, organized by the University of Wisconsin.

LISA D'AGOSTINO

Lisa D'Agostino earned a B.F.A. from Bowling Green State University, and she works both in fiber and in metal. Invitational exhibits in which her work was shown include Fiber Miniatures, The Hand and The Spirit Gallery, Scottsdale, Arizona; Fine Focus, Craft Alliance Gallery, St. Louis; Northwest Southwest Influence, Sarah Squeri Gallery, Cincinnati; The Basketry Link, Mendocino (California) Art Center, and Fiber National '84, Dunkirk, New York. Her work has appeared in several books, including *Jewelry—Contemporary Design and Technique* (Davis) and *Jewelry—Basic Techniques and Design* (Chilton), and in a number of magazine articles as well.

MICHAEL DAVIS, 3520 Herschel Street, Jacksonville, FL 32205

Michael Davis studied painting, ceramics, graphic design, sculpture, printmaking, spinning, dyeing, and weaving while earning a B.A. from the University of North Florida.

His first solo show was in 1980, Jacksonville, Florida, and he has exhibited since then in group shows throughout the United States. His work has appeared in several publications, including *Fiberarts* magazine and *Shuttle Spindle & Dyepot*.

Davis has taught at the Brookfield (Connecticut) Craft Center, the Jacksonville (Florida) Art Museum, and at The Florida Tropical Weavers Guild State Convention.

LILLIAN ELLIOTT, 1775 San Lorenzo, Berkeley, CA 94707

Upon receiving an M.F.A. from Cranbrook Academy of Art, Lillian Elliott became a fabric designer for the Ford Motor Company. She has lectured and instructed since 1959, most recently at San Francisco Community College and at the Conference of Northern California Handweavers. Elliott has juried the Craftsmen's Association of British Columbia Exhibition, the San Francisco Art Festival, San Mateo County Arts Council Textile Exhibit, and the Marin Society of Artists 57th Annual Show.

Her work has appeared in many invitational shows, including Contemporary American Basket-Makers, Purdue University; Woven Structures, London; and Vannerie, Museum of Decorative Arts, Lausanne, Switzerland. She has had more than a dozen solo shows, and has exhibited at the Twelfth Tapestry Biennial in Lausanne. Since 1981 she has collaborated with textile artist Pat Hickman; their works have shown throughout the United States and in Europe and Japan.

Articles about Elliott and her work have been published in *Handweaver's Bulletin, The Goodfellow Review of Crafts, Fiberarts* magazine, and *American Craft*. Her work has appeared in *Ms., Craft Horizons, The San Francisco Examiner, Artweek*, and in more than 20 books. She has written for *American Craft, Arts and Activities*, and *Minnesota Weaver Quarterly*.

In 1985 Elliott received a travel grant from the Swedish Women's Educational Association International. That same year she was designated a California Living Treasure.

DOUGLAS ERIC FUCHS

Doug Fuchs earned degrees from Catholic University and Columbia University, and for ten years was a Christian Brother. He taught weaving and basketry in school and museum art programs.

Fuchs' work has shown at galleries throughout the country, including The Elements Gallery, Greenwich, Connecticut; The Artisans Gallery, New York; the Following Sea Gallery, Hawaii. His work has been featured in such publications as *Fiberarts* magazine and *Shuttle Spindle & Dyepot*.

A major installation, "Floating Forest," commissioned by the Crafts Council of Australia, was mounted in Adelaide, Melbourne, and Sydney to high critical acclaim. Fuchs also received a fellowship from the National Endowment for the Arts.

Doug Fuchs died as this book was being prepared.

JOHN GARRETT, 3212 Larissa Drive, Los Angeles, CA 90026

John Garrett has taught at several California colleges and at the Kansas City Art Institute. He currently is on the faculty at the University of California at Los Angeles. His exhibitions include Master Fiber Works, Kansas City, Missouri; Works in Miniature, New York; L.A.V.A. '81, Los Angeles; Basketry: Tradition in New Form, Boston; Fiber Miniatures, Scottsdale, Arizona; the Vehta Biennale, Brussels; The Art of Basketry, Louisville, Kentucky; California Basketry: Past and Present, San Diego, California; Fiber R/Evolution, Milwaukee, Wisconsin; Woven Works, University of Wisconsin; and Basketry: Tradition in New Form, Boston and New York.

Garrett has completed collaborative work with Neda Al-Hilali, Marilyn Anderson, and Mary Ann Glantz. He has traveled throughout Europe and to Ghana and Afghanistan. In 1983, he won a National Endowment for the Arts fellowship, and in 1984 completed a 16-by-40-foot mural on Santa Monica Boulevard in Hollywood.

MARIAN HAIGH-NEAL, 2600 Bridle Path, Austin, TX 78703

Marian Haigh-Neal studied at Kansas State University, then received her B.F.A. from Arkansas State University. She lectured to graduate students and faculty at The Royal College of Art, London, and in 1980 won a grant from the Texas Commission on the Arts to serve in the Arts in Schools program. Haigh-Neal's work has appeared in exhibits throughout the country, and articles about her work have been published by *Fiberarts* magazine, *American Craft*, and *Metropolis*.

BRYANT HOLSENBECK, Durham, NC

After earning an M.Ed. from the University of North Carolina-Chapel Hill in 1972, Bryant Holsenbeck worked in the citrus groves of Kibbutz Evron in Israel. She then attended workshops at Penland School, Arrowmont School of Crafts, and Alfred University.

Her work has appeared in Contemporary Basketry, Gatlinburg, Tennessee; Crafts Invitational at the Southeastern Center for Contemporary Art; National Invitational, The Art of Adornment, American Crafts in Iceland, National Basketry Invitational, Louisville, Kentucky; U.S.A.—Portrait of the South, Rome, Italy; Fiber Sculpture at the Green Hill Center for North Carolina Art; National Basketry Invitational, Brookfield (Connecticut) Craft Center; and Basketry Invitational, Kohler Arts Center, Sheboygan, Wisconsin.

Holsenbeck taught at workshops at the Greenville County (South Carolina) Museum of Art, Penland School, and the Haystack Mountain School of Crafts, and was a visiting artist in the Virginia school system.

In 1985 she received an Emerging Artist grant from the Durham (North Carolina) Arts Council. In 1986, Holsenbeck mounted a solo show at The Private Collection gallery, Cincinnati.

LISSA HUNTER, 89 Mackworth Street, Portland, ME 04103

Lissa Hunter's works are in several corporate collections. She has had solo and two-person exhibits from Maine to New Mexico, and has been included in group shows such as Fiber Forms 1979, Nashua, New Hampshire; Miniatures, Cannon Beach, Oregon; Within and Beyond the Basket, Berkeley, California; and Winter Artists, Taos, New Mexico. Hunter's work has shown at Clay and Fiber Gallery in Taos, New Mexico, Netsky Gallery in Coconut Grove, Florida, Gallery on the Green in Lexington, Massachusetts, and the Elaine Horwitch Gallery in Scottsdale, Arizona.

Hunter holds a B.A. in fine arts and an M.F.A. in textile arts from Indiana University.

FERNE JACOBS

Ferne Jacobs studied painting and weaving, and earned an M.F.A. from Claremont Graduate School. Exhibitions of her fiber work, drawings, and basketry include First World Crafts Exhibition, Toronto; First International Exhibition of Miniature Textiles, London; The Basket-maker's Art, New York; The Contemporary Basketmaker, Purdue University; Beyond Tradition: 25th Anniversary Exhibition of the American Craft Museum, New York; American Basket Forms, Brookfield (Connecticut) Craft Center; Poetry of the Physical, American Craft Museum; Fiber R/Evolution, Milwaukee; Fibre Structures, The Denver Art Museum; and Fiberworks, an international invitational fibers exhibition, the Cleveland Museum of Art.

Her work has been featured in a number of magazines, including Artweek, American Craft, Craft Horizons, and Fiberarts; and in books, such as A Modern Approach to Basketry, In Praise of Hands: Contemporary Crafts of the World (New York Graphic Society), Beyond Weaving (Watson-Guptill), and The Art Fabric: Mainstream (Van Nostrand Reinhold).

Jacobs has lectured throughout the United States, most recently at Rhode Island School of Design, Tyler School of Art, Philadelphia School of Art, and the University of Wisconsin. She has received two grants from the National Endowment for the Arts. Her work is included in a number of private and public collections.

SHEREEN LAPLANTZ, 899 Bayside Cutoff, Bayside, CA 95524

Shereen LaPlantz began her career as a weaver. She holds a B.A. from California State University at Los Angeles, and studied design at Cranbrook Academy of Art, Bloomfield Hills, Michigan.

She has written and published two books, The Mad Weave Book and Plaited Basketry: The Woven Form. She publishes The News Basket, and her work has appeared in The New York Times, Fiberarts magazine, Craft Australia, and The Washington Post. LaPlantz has exhibited widely in group and solo shows, including those at the Pacific Basin School of Textile Arts; Brookfield (Connecticut) Craft Center; the 1982 Convergence exhibit in Seattle; the History Museum in Gdansk, Poland; Mendocino Arts Center, California; and D W Gallery, Dallas.

She studied Maori fiber techniques in New Zealand, Aboriginal basketry in Australia, and participated in a workshop with the Akwesasne basketmakers on the Mohawk reservation in Hogansburg, New York.

RACHEL NASH LAW, P.O. Box 245, Beverly, WV 26253

A graduate of the National School of Basketmaking in West Germany, Rachel Nash Law also studied at the International School of Interior Design in Washington, D.C. and earned a B.S. from Virginia Polytechnic Institute and State University in Blacksburg.

Law's baskets have been seen at the Renwick Gallery, Washington; Arrowmont School of Arts and Crafts, Contemporary Art Gallery, Louisville, Kentucky; Hightower Art Association, and in the 1984 television movie, "The Dollmaker."

She is a member of American Craft Council, Ohio Designer Craftsmen, and the Southern Highland Handicraft Guild. Recipients of a 1985 Appalachian Studies Fellowship, Law and colleague Cynthia Taylor are researching white oak basketry and writing a book on their findings.

PATRICIA ANN LECHMAN, Department of Fine Arts, Shelby State Community College, P.O. Box 40568, Memphis, TN 38174

After receiving a B.S. from the University of Georgia, Patti Lechman earned an M.S. in design from Indiana University and an M.F.A. in ceramics from Michigan State University. She studied fiber at Penland School of Crafts and at Miami University.

Her works have been featured in Southern Accents,

American Craft, and are included in The Fiberarts Design Book II. Lechman received a first place award from The Basketry Link, a show organized by the publishers of The News Basket, and merit awards in the Tennessee Artist-Craftsmen's Biennial Exhibition and The Path of the Handweaver show in Memphis, Tennessee. She received the American Craft Council award at the Spotlight '83 show in Winston-Salem, North Carolina, and has exhibited in Fiber R/Evolution.

KARI LØNNING, 36 Mulberry Street, Ridgefield, CT 06877

Kari Lønning earned a B.F.A. degree in ceramics from Syracuse University, and studied tapestry weaving in Sweden. She worked with cloth and wood before she turned exclusively to basketry.

Her work has shown at The Hand and Spirit Gallery, Scottsdale, Arizona; Detroit Gallery of Contemporary Art; Brookfield (Connecticut) Craft Center; The Kohler Arts Center, Sheboygan, Wisconsin; The Elements Gallery, Greenwich, Connecticut; and at the Pacific Basin School of Textile Arts.

In the 1985 Designed and Made for Use Competition at the American Craft Museum she received a design award. She was awarded a grant to study basketry by the Connecticut Commission on the Arts.

DONA LOOK, P.O. Box 204, Algoma, WI 54201

Currently a partner in the Loeber/Look Studio, Dona Look is a self-taught basketmaker. Her work appeared in these shows: Art for the Table, an American Craft Council benefit in New York; Washington Craft Show, Smithsonian Institution; Basketry Today, an invitational traveling exhibit; Crafts/National, Buffalo; Fiber Individualists, Wustum Museum of Fine Arts in Racine, Wisconsin; Designed and Made for Use, American Craft Museum, New York.

She won first prize in the 1984 Philadelphia Craft Show at the Philadelphia Museum of Art. Look was featured in the portfolio section of American Craft magazine.

JOHN MCQUEEN, RD 1, Box 119A, Alfred Station, NY 14803

After earning an M.F.A. from Tyler School of Art, Temple University, John McQueen received a National Endowment for the Arts grant. In 1980, he was a Japan-United States Friendship Commission Exchange fellow.

His group exhibits include Fiberwork of the Americas and Japan, Kyoto, Japan; The Contemporary Basketmaker, Purdue University; The New Basket—A Vessel for the Future, Potsdam, New York; The Art Fabric: Mainstream Exhibition, San Francisco; and the 1986 Inaugural Exhibition at the American Craft Museum in New York. He has had solo shows at galleries across the country, including the Hadler/Rodriguez Galleries, New York; Helen Drutt Gallery, Philadelphia; and Fiberworks Center for the Textile Arts, Berkeley, California. A collaborative installation with Jessie Shefrin was mounted at the Kohler Art Center in 1986.

McQueen has lectured and taught at schools and museums throughout North America, the Textile Museum, the Cleveland Museum of Art, Nova Scotia College of Art and Design, and the Art Institute of Chicago.

MARY MERKEL-HESS, 1110 Cottonwood Ave., Iowa City, IA 52240

Mary Merkel-Hess earned degrees from Marquette University, the University of Wisconsin-Milwaukee, and the University of Iowa.

She has exhibited throughout the United States, and won awards in Paper/Fiber VII, Johnson County (Iowa) Arts Center; The Basketry Link, Mendocino (California) Art Center; Fiber 83/Off the Loom, Rock Island, Illinois; and Banc Iowa Invitational Art Fair, Cedar Rapids, Iowa.

Merkel-Hess had solo exhibits in 1985 at Middle Tennessee State University and the Iowa Artisans Gallery in Iowa City.

NORMA MINKOWITZ, 25 Broad View Road, Westport, CT 06880

A Cooper Union Art School graduate, Norma Minkowitz has been a professional fiber artist since 1972. She has lectured and taught throughout the eastern United States.

Her work has been shown in numerous exhibitions, including The Great American Foot, Clothing To Be Seen, and Baroque '74, all at the Museum of Contemporary Crafts, New York; Fibre Structures, Pittsburgh; The Flexible Medium: Art Textiles from the Museum Collection, the Renwick Gallery, Washington; Second International Competition of Miniature Fibre Work, British Craft Centre, London; and Fiber R/Evolution, Milwaukee. She has had a number of solo exhibitions at Julie: Artisans' Gallery, New York.

Her work has appeared in periodicals such as *American Craft*, *Hartford Magazine*, and *Fiberarts Magazine*, and in a number of books: *Textile Collector's Guide* (Monarch), *The Container Book* (Crown), *Design Principles and Problems* (Holt, Rinehart & Winston), and *Contemporary Crafts of the Americas* (Regnery).

Minkowitz has received awards from New Britain Museum of American Art, Wesleyan University, Convergence '76, Penn State University, and Pittsburgh Center for the Arts. She was first place winner in fiber in ArtQuest '86, at Parsons School of Design in New York, and she placed first in Objects and Images In-Of-Under-About Surface at the Monterey Peninsula (California) Museum of Art.

RINA PELEG, 640 Broadway, Apt. 8E, New York, NY 10012

Rina Peleg's work is represented in the collections of the Greenville County (South Carolina) Museum of Art, the Ceramic Museum, Tel Aviv, Alfred University, and in other institutions. She has been featured in *Kunst & Handwerk*, *Ceramics Monthly*, *American Ceramics*, and *Fiberarts* magazines.

Peleg has exhibited in group shows, including Westwood Clay National, Los Angeles; Basketwork, Renwick Gallery Anniversary, Washington; and the 40th International Ceramic Art exhibition, Faenza, Italy. She has had solo shows at the Ceramics Museum in Tel Aviv, the Branch Gallery, Washington, Theo Portnoy Gallery and Heller Gallery in New York, and at the Everson Museum of Art in Syracuse.

Peleg studied at Bezalel Academy of Arts and Crafts in Jerusalem, and earned her M.F.A. from Alfred University. She taught ceramics at Haifa University and at the Teacher Training College for the Kibbutz Movement in Oranim.

ED ROSSBACH, 2715 Belrose Avenue, Berkeley, CA 94705

Ed Rossbach received a B.A. in painting and design from the University of Washington, holds an M.F.A. in ceramics and weaving from Cranbrook Academy of Art, and an M.A. in art education from Columbia University. His work is in the collections of the Museum of Modern Art, New York; Stedelijk Museum, Amsterdam; American Crafts Museum, New York; Renwick Gallery, Washington; Trondheim (Norway) Museum; The Brooklyn Museum. He has exhibited at Nouvelle Vannerie, Musee des Arts Decoratifs, Lausanne, Switzerland; the Triennale, Milan; the Brussels World's Fair; Museum of Contemporary Crafts, New York; Oakland (California) Art Gallery; Fiberworks Gallery, Berkeley, California; Fiber Constructions, The Textile Museum, Washington.

He has written three books on basketry: *The Nature of Basketry* (Schiffer Pub., Ltd.), *The New Basketry* (Van Nostrand Reinhold), and *Baskets as Textile Art* (Van Nostrand Reinhold). His articles have appeared in *American Craft*, *Craft Horizons*, *Handweaver*, and *Fiberarts* magazine.

Rossbach is Professor Emeritus of design at the University of California at Berkeley, and an Honorary Fellow of the American Craft Council.

JANE SAUER, 6332 Wydown Boulevard, St. Louis, MO 63105

Jane Sauer earned a B.F.A. from Washington University, and has been a full-time studio artist since 1979. She was artist-in-residence at New City School in St. Louis, and conducted workshops at Arrowmont School of Crafts, The Cleveland Museum of Art, Haystack Mountain School of Crafts, and at the Craft Alliance Art Center.

Sauer has had solo and two-person shows at The Hand and Spirit Gallery, Scottsdale, Arizona, The Works Gallery, Philadelphia, Gayle Wilson Gallery, Long Island, and B.Z. Wagman Gallery, St. Louis. Additional exhibitions include The New Basket: A Vessel for the Future, a traveling exhibit; Other Baskets, Craft Alliance Gallery, St. Louis, and the Worcester (Massachusetts) Art Center; 4th International Exhibition of Miniature Textiles, British Crafts Centre, London.

The *St. Louis Dispatch*, *American Craft*, *Fiberarts Magazine*, *The Crafts Report*, and *The New York Times* are a few publications which have covered Sauer's work. She received the Best of Show award at the Maryland Crafts Council Juried National Biennial in 1986, and received a Visual Artist's grant from the National Endowment for the Arts.

KARYL SISSON, 1750 North Beverly Glen, Los Angeles, CA 90077

Karyl Sisson earned an M.F.A. from the University of California at Los Angeles, and a B.S. (magna cum laude) from New York University. She was a merit award winner in ArtQuest '86. Her work has appeared in Artists Craftsmen '82, Braithwaite Fine Arts Gallery, Cedar City, Utah; Fiber Structure National II, III, and IV, Downey (California) Museum of Art; Within and Beyond the Basket, Pacific Basin School of Textile Arts, Berkeley, California, and Fiber R/Evolution, Milwaukee, Wisconsin.

KAREN TURNIDGE, 1270 Denise Drive, Kent, OH 44240

Karen Turnidge studied at the University of Oregon and Kent State University. She attended numerous workshops on basketry techniques, metalworking, and jewelry making.

Exhibitions of her work include New Outlook/Fibers, Kent State University; Best of '85 Ohio Designer Craftsmen, Contemporary Metals USA, Downey Museum of Art, Los Angeles; Basketry Today, an invitational traveling exhibition, and Crafts National, State University of New York, Buffalo. She received an award for jewelry design from Fashion Group Inc. in 1985.

JAN YATSKO, 313 E. Frederick Street, Lancaster, PA 17602

Jan Yatsko studied papermaking and basketmaking at Peter's Valley Craft School, Layton, New Jersey, and Mannings Handweaving School, East Berlin, Pennsylvania.

Yatsko's work has been featured in *The New York Times*, *Fiberarts* magazine, *The Goodfellow Catalog of Wonderful Things* (Berkeley), and *The Fiberarts Design Book II* (Lark Books).

Her exhibits include Basketry Form & Function, Los Angeles; Contemporary American Fibers, Reykjavik, Iceland; Basketry Invitational, Wilmette, Illinois; and the Winter Show, The Society of Arts and Crafts, Boston.